T0270924

Turbulence, Coherent Structures, Dynamical Systems and Symmetry

Turbulence pervades our world, from weather patterns to the air entering our lungs. This book describes methods that reveal its structures and dynamics. Building on the existence of coherent structures – recurrent patterns – in turbulent flows, it describes mathematical methods that reduce the governing (Navier–Stokes) equations to simpler forms that can be understood more easily.

This Second Edition contains a new chapter on the balanced proper orthogonal decomposition: a method derived from control theory that is especially useful for flows equipped with sensors and actuators. It also reviews relevant work carried out since 1995.

The book is ideal for engineering, physical science, and mathematics researchers working in fluid dynamics and other areas in which coherent patterns emerge.

PHILIP HOLMES is Eugene Higgins Professor of Mechanical and Aerospace Engineering and Professor of Applied and Computational Mathematics, Princeton University. He works on nonlinear dynamics and differential equations.

JOHN L. LUMLEY is Professor Emeritus in the Department of Mechanical and Aerospace Engineering, Cornell University. He has authored or co-authored over two hundred scientific papers and several books.

GAHL BERKOOZ leads the area of Information Management for Ford Motor Company, covering all aspects of Business Information Standards and Integration.

CLARENCE W. ROWLEY is an Associate Professor of Mechanical and Aerospace Engineering at Princeton University. His research interests lie at the intersection of dynamical systems, control theory, and fluid mechanics.

Established in 1952, this series has maintained a reputation for the publication of outstanding monographs covering such areas as wave propagation, fluid dynamics, theoretical geophysics, combustion, and the mechanics of solids. The books are written for a wide audience and balance mathematical analysis with physical interpretation and experimental data where appropriate.

RECENT TITLES IN THIS SERIES

Turbulence, Coherent Structures, Dynamical Systems and Symmetry

SECOND EDITION

PHILIP HOLMES
Princeton University

JOHN L. LUMLEY
Cornell University

GAHL BERKOOZ
Information Technology Division, Ford Motor Company

CLARENCE W. ROWLEY
Princeton University

CAMBRIDGE
UNIVERSITY PRESS

Shaftesbury Road, Cambridge CB2 8EA, United Kingdom

One Liberty Plaza, 20th Floor, New York, NY 10006, USA

477 Williamstown Road, Port Melbourne, VIC 3207, Australia

314–321, 3rd Floor, Plot 3, Splendor Forum, Jasola District Centre, New Delhi – 110025, India

103 Penang Road, #05–06/07, Visioncrest Commercial, Singapore 238467

Cambridge University Press is part of Cambridge University Press & Assessment,
a department of the University of Cambridge.

We share the University's mission to contribute to society through the pursuit of
education, learning and research at the highest international levels of excellence.

www.cambridge.org
Information on this title: www.cambridge.org/9781107008250

First published 1996
First paperback edition 1998
Second edition published 2012

A catalogue record for this publication is available from the British Library

Library of Congress Cataloging-in-Publication data
Turbulence, coherent structures, dynamical systems and symmetry / Philip
Holmes . . . [et al.]. – 2nd ed.
p. cm. – (Cambridge monographs on mechanics)
Rev. ed. of : Turbulence, coherent structures, dynamical systems, and
symmetry / Philip Holmes, John L. Lumley, and Gal Berkooz.
ISBN 978-1-107-00825-0 (hardback)
1. Turbulence. 2. Differentiable dynamical systems. I. Holmes, Philip, 1945–
II. Holmes, Philip, 1945– Turbulence, coherent structures, dynamical systems, and symmetry.
QA913.H65 2012
532´.0527–dc23
2011041743

ISBN   978-1-107-00825-0   Hardback

Contents

Preface to the first edition

On physical grounds there is no doubt that the Navier–Stokes equations provide an excellent model for fluid flow as long as shock waves are relatively thick (in terms of mean free paths), and in such conditions of temperature and pressure that we can regard the fluid as a continuum. The incompressible version is restricted, of course, to lower speeds and more moderate temperatures and pressures. There are some mathematical difficulties – indeed, we still lack a satisfactory existence-uniqueness theory in three dimensions – but these do not appear to compromise the equations' validity. Why then is the "problem of turbulence" so difficult? We can, of course, solve these nonlinear partial differential equations numerically for given boundary and initial conditions, to generate apparently unique turbulent solutions, but this is the only useful sense in which they *are* soluble, save for certain non-turbulent flows having strong symmetries and other simplifications. Unfortunately, numerical solutions do not bring much understanding.

However, three fairly recent developments offer some hope for improved understanding: (1) the discovery, by experimental fluid mechanicians, of coherent structures in certain fully developed turbulent flows; (2) the suggestion that strange attractors and other ideas from finite-dimensional dynamical systems theory might play a rôle in the analysis of the governing equations; and (3) the introduction of the statistical technique of Karhunen–Loève or proper orthogonal decomposition. This book introduces these developments and describes how the three threads can be drawn together to weave low-dimensional models that address the rôle of coherent structures in turbulence generation.

We have uppermost in our minds an audience of engineers and applied scientists wishing to learn about some new methods and ways in which they might contribute to an understanding of turbulent flows. Additionally, applied mathematicians and dynamical systems theorists might learn a little fluid mechanics here, and find in it a suitable playground for their expertise.

The fact that we are writing for a mixed audience will probably make parts of this book irritating to almost all our readers. We have tried to strike a reasonable balance, but experts in turbulence and dynamical systems may find our treatments of their respective fields superficial.

Our approach will be somewhat schizophrenic. On the one hand we hope to suggest a broad strategy for modeling turbulent flows (and, more generally, other spatio-temporally complex systems) by extracting coherent structures and deriving, from the governing

Navier–Stokes equations, relatively small sets of ordinary differential equations that describe their dynamical interactions. We freely admit that there is much speculation and there are few firm results in this, although a number of (partial) successes have been achieved. In collecting our thoughts and those of others, we hope to stimulate research which might ultimately put some of these ideas on a firmer footing. This is the "vision" side of the book. In contrast, and since we need these methods to analyze our low-dimensional models, we provide a brief introduction, with many simple examples, to relevant and well-established ideas from dynamical systems theory. This is the "technical manual" side of the book. We occasionally switch from vision to technical mode, or vice versa, with scant warning. We guarantee that we have a mode to annoy every reader.

Our (tedious) working of simple examples may make for impatience on the part of those hurrying to get to the main attraction, or attractor. In our defense we remark that full appreciation of an "application" as complex as turbulence must rest on a firm understanding of simpler cases. Equally, our use of the symbolic and abstract notation of dynamical systems theory may be a stumbling block for some. We encourage them to stagger on to the examples. (A glossary of technical terms and notations is provided at the end of Chapter 1.)

But while we may irritate, we hope not to confuse. The term "low-dimensional model" is already problematic. Our models ideally contain enough "modes" to permit reasonable spatial as well as temporal behavior of the larger scales in the flow: those dominant in the sense of average turbulent kinetic energy. Our models do not contain, nor shall we be concerned with, the inertial or dissipative ranges. We have in mind sets of ordinary differential equations containing perhaps 10–100 dependent variables: substantially larger than that of, say, Lorenz. This is drastically low in comparison to the number of modes (or nodes) necessary even in a large eddy simulation, let alone a direct numerical simulation, but it is high in the context of dynamical systems, in which we have relatively complete understanding only of systems of dimension ≤ 2! We approach the analysis of such "low but high"-dimensional models by building on yet lower dimensional models, for which more complete analyses are possible. In this, one of our prominent illustrative examples is provided by the heteroclinic attractor: a strongly nonlinear type of solution that occurs robustly in systems possessing certain symmetries. Heteroclinic attractors lead to "bursting" behavior in which systems exhibit relatively long quasisteady phases involving few modes, interrupted by violent events in which other groups of modes become active. The reader should not interpret our emphasis to mean that we think turbulence *is* a heteroclinic attractor, although such attractors do appear to represent some key features of the burst/sweep cycle in the boundary layer. Rather, the study of these attractors provides a nice example of the power of qualitative methods applied to equations which are, in dynamical systems terms, of high dimension (≥ 4).

A second area of potential confusion is in our use of linear spaces and linear analysis for the description and study of nonlinear objects. This is, of course, a quite normal tactic. Linear theory is well developed, relatively complete, and powerful. There is no contradiction in defining a nonlinear differential equation on a linear state space, or in representing a spatio-temporal field $u(x, t)$ as a linear combination of basis functions or "modes" $\varphi_j(x)$ multiplied by suitable time-dependent coefficients $a_j(t)$. The Fourier representation is a prime example. Of course, if $u(x, t)$ is a solution of a nonlinear partial differential equation, linear superposition will fail in the sense that the sum of two solutions will not generally produce a third, but we can still represent individual solutions via such series. Moreover, in spite of all the recent advances in nonlinear analysis, the tools of linear operator theory, including Fourier analysis and linearization of partial and ordinary differential equations, are still crucial in the study of nonlinear systems. The proper orthogonal or Karhunen–Loève decomposition, one of our major tools, also relies on linear theory and produces representations of functions and fields in linear spaces, within which we may then construct strongly nonlinear dynamical systems whose attractors quite happily display their nonlinear character.

The book falls into four parts. In the first – Turbulence – we introduce our general strategy and recall some key ideas from fluid mechanics and "classical" turbulence theory, which establish basic properties of some canonical turbulent flows. We describe coherent structures from the viewpoint of an experimental observer and follow this with a description of the Karhunen–Loève decomposition, with sufficient mathematical detail that the reader can appreciate the advantages and limitations of the low-dimensional, optimal representations of turbulent flows in terms of empirical eigenfunctions that it affords. We conclude this part by describing how the Navier–Stokes equations can be projected onto subspaces spanned by a few empirical eigenfunctions to yield a low-dimensional model, and outlining some of the additional "modeling" that must be done to account for modes and effects neglected in such radical truncations.

The second part – Dynamical systems – contains a review of some aspects of dynamical systems theory that are directly useful in the analysis of low-dimensional models. We discuss local and global bifurcations and important ideas such as structural stability and strange attractors. Symmetries play a central rôle in our ideas, and we devote a chapter to showing how they influence dynamical and bifurcation behaviors. We then gather our methods for a dry run: a study of the Kuramoto–Sivashinsky partial differential equation which, while much simpler than the Navier–Stokes equations, displays some of the same features and allows us to illustrate our techniques. The final chapter introduces some basic ideas from the theory of stochastic differential equations: ideas that we need in dealing with systems subject to (small) random disturbances. This part is a fairly relentless essay in the technical manual mode.

The third part – The boundary layer – returns to the Navier–Stokes equations and a specific class of models of the turbulent boundary layer. This is the problem to which our approach was first applied and it is probably still the most widely studied from this viewpoint. It is certainly the one that we understand best, shortcomings and all. We offer this description and critical commentary partly in the hope that it may help others avoid our mistakes.

In the final part – Other applications and related work – we briefly review a number of applications of this and similar strategies to other turbulent open flow problems. We do not consider applications in other areas, such as pattern-forming chemical reactions, flame dynamics, etc., although many such are now appearing. We close by speculating on the kind of understanding of turbulence that this approach is likely to yield, and on how some recent developments, such as inertial manifolds, are related to it.

We promise not to mention the word "fractal" in this book.

Preface to the second edition

Much work has been done on low-dimensional models of turbulence and fluid systems in the 16 years since the first edition of this book appeared. In preparing the second edition, we have not attempted a comprehensive review: indeed, we doubt that this is possible, or even desirable. Rather, we have added one chapter and several sections and subsections on some new developments that are most closely related to material in our first edition. We have also made minor corrections and clarifications throughout, and added comments in several places, as well as correcting a number of errors that readers have pointed out. Here, to orient the reader, we outline the major changes.

Clancy Rowley (the new member of our team) has contributed a chapter on balanced truncation, a technique from linear control theory that chooses bases that optimally align inputs and outputs. Over the past ten years this has led to the method of balanced proper orthogonal decomposition (BPOD), which is especially useful for systems equipped with sensors and actuators. Since low-dimensional models provide a computational means for studying control of turbulence, we feel that BPOD has considerable potential. This new chapter (5) now closes the first part of the book (readers familiar with the first edition must therefore remember to add 1 to correctly identify the following eight chapters). The only other entirely new sections are 7.5, a discussion of traveling modes in translation-invariant systems, 12.6, a review of work on coherent structures in internal combustion engines, and 12.7, which gathers a miscellany of recent results.

New materials also appear in Chapter 3, where we modestly generalize the derivation of the POD in Section 3.1, adding subsections on specific function spaces, and in Section 3.4, where the relationship between the method of snapshots and the classical singular value decomposition is described, where we introduce an inner product for compressible flows, and where we comment on using a fixed set of empirical eigenfunctions to represent data over a range of parameter values (e.g. Reynolds numbers). In Chapter 4 we now provide more details on Galerkin projection (Section 4.1), give an example of a PDE with time-dependent boundary conditions and explain how quadratic nonlinearities, such as those in the Navier–Stokes equations, permit analytical determination of coefficients in the projected ODEs (Section 4.2). We also describe the important notion of shift modes in Section 4.4. Section 8.4 now ends with remarks on spatially-localized models of the Kuramoto–Sivashinsky equation, Section 10.7.2 notes a model that uses shift modes to couple a time-varying "mean" flow and secondary modes, and in Section 12.4

we summarize low-dimensional models of unsteady wakes behind cylinders. We have also revised Section 13.4 to reflect the fact that the results on spatially-localized models with pressure rather than velocity boundary conditions described there are incomplete and do not completely resolve well-posedness of the Navier–Stokes equations with mixed velocity and pressure boundary conditions. Finally, the index has been substantially expanded and improved, and we have added over 80 references.

Acknowledgements

Many people have contributed their ideas, support, criticism, and time to help make this book possible. We wish first to thank our former and present students, postdoctoral fellows, and colleagues who, over the past ten years, worked directly on the project that led to the first edition of this book: Dieter Armbruster, Nadine Aubry, Peter Blossey, SueAnn Campbell, Hal Carlson, Brianno Coller, Juan Elezgaray, John Gibson, Ziggy Herzog, Berengère Podvin, Andrew Poje, Emily Stone, and Edriss Titi. As anyone who does them knows: teaching, learning, and research are inextricably joined, and without these students' and colleagues' demands that we explain what we mean, we would not have made the first halting steps upon which they could then improve. Many of the results and ideas in this book originated in their work.

Among our immediate colleagues, John Guckenheimer and Sidney Leibovich have been particularly helpful. Steve Pope helped with some of the probabilistic ideas in Chapters 3 and 13. At a greater distance, Keith Moffatt and Larry Sirovich have been useful critics, forcing us to examine our assumptions more closely. Ciprian Foias, Mark Glauser, and Dietmar Rempfer shared their expertise, and explained their insights and results to us. The opportunity to give lectures and short courses on this work has also clarified our understanding and, we believe, improved our presentation. PH would like to thank Klaus Kirchgässner, Jean-Claude Saut, John Brindley, Colin Sparrow, and Silvina Ponce Dawson and Gabriel Mindlin for arranging courses at Universität Stuttgart, Université de Paris-Sud, the University of Leeds, the Newton Institute, Cambridge, and the Fourth Latin American Workshop on Nonlinear Phenomena, San Carlos de Bariloche, Argentina, respectively. JLL would like to thank Yousuff Hussaini and Jean-Paul Bonnet for arranging courses at NASA Langley Research Center and The International Center for the Mechanical Sciences, Udine, respectively. CWR would like to thank Tim Colonius, the late Jerry Marsden, and Richard Murray for introducing him to this field.

The manuscript was typed in LaTeX, much of it by ourselves, but with the able assistance of Gail Cotanch and Phebe Tarassov. Alison Woolatt of CUP also helped us with LaTeX and CUP formats. Harry Dankowicz computed some of the figures in Chapter 8. Teresa Howley patiently redrew and improved all the figures. Jonathan Mattingly and Ralf Wittenberg read the manuscript and suggested numerous corrections and improvements. Jo Clegg's patient copyediting kept us on the (fairly) straight and narrow. Our thanks go to all of them.

Numerous Governmental Agencies have supported and encouraged this work: the Office of Naval Research has been with us from the beginning, joined subsequently by the Air Force Office of Scientific Research, the National Science Foundation, the Department of Energy, and NATO. GB was partially supported by ONR contract N00014-94-C-0024, NASA contract NAS1-20408, and AFOSR contract F49620-95-C-0027. A fellowship from the John Simon Guggenheim Memorial Foundation in 1993–94 gave PH the leisure to begin this project and to harass his co-authors for their contributions. We thank these organizations for their support.

In preparing the second edition, and especially Chapter 5 and Section 12.6, we have benefited from the work of Sunil Ahuja, Mark Fogelman, Miloš Ilak, and Zhanhua Ma, and the copy-editing of Richard Smith.

Finally, we are grateful to Simon Capelin of the Cambridge University Press, who encouraged us to complete this book almost from its beginning, and who invited us to prepare a second edition.

Philip Holmes, John L. Lumley, Gahl Berkooz, and Clancy W. Rowley

Princeton, NJ, Ithaca, NY, and Ann Arbor, MI

PART ONE

Turbulence

1

Introduction

1.1 Turbulence

Turbulence is the last great unsolved problem of classical physics.[1] Although temporarily abandoned by much of the community in favor of particle physics, the current popularity of chaos and dynamical systems theory (as well as funding problems in particle physics) is now drawing the physicists back. During the interim and up to the present, turbulence has been avidly pursued by engineers.

Turbulence has enormous intellectual fascination for physicists, engineers, and mathematicians alike. This scientific appeal stems in part from its inherent difficulty – most of the approaches that can be used on other problems in fluid mechanics are useless in turbulence. Turbulence is usually approached as a stochastic problem, yet the simplifications that can be used in statistical mechanics are not applicable – turbulence is characterized by strong dependency in space and in time, so that not much can be modeled usefully as a simple Markov process, for example. The nonlinearity of turbulence is essential – linearization destroys the problem. Many problems in fluid mechanics can be approached by supposing that the flow is irrotational – that is, that the vorticity is zero everywhere. In turbulence, the presence of vorticity is essential to the dynamics. In fact, the nonlinearity, rotationality, and the dimensionality interact dynamically to feed the turbulence – hence, to suppose that a realization of the flow is two-dimensional also destroys the problem. There is more, but this is enough to make it clear that one faces the turbulence problem stripped of the usual arsenal of techniques, reduced to hand-to-hand combat. One is forced to find unexpected chinks in its armor almost by necromancy, and to fabricate new approaches from whole cloth. This is its fascination.

At the same time, turbulence is of the greatest practical importance. The turbulent transport of heat, mass, and momentum is usually some three orders of magnitude greater than molecular transport. Turbulence is responsible for the vast majority of human energy

---

[1] Remarks of this sort have been variously attributed to Sommerfeld, Einstein, and Feynman, although no one seems to know precise references, and searches of some likely sources have been unproductive. Of course, the allegation is a matter of fact, not much in need of support by a quotation from a distinguished author. However, it would be interesting to know when the matter was first recognized. In this connection, similar sentiments were expressed by Horace Lamb in his *Hydrodynamics*, beginning in the second edition in 1895, and continuing through the sixth (and last) edition in 1932. We are indebted for this reference to Julian Hunt, citing its use by George Batchelor in his book *The Life and Legacy of G. I. Taylor*, Cambridge University Press, 1996.

consumption, in automobile and aircraft fuel, pipeline pumping charges, and so forth. It is responsible for the wind chill factor. In the atmosphere and ocean it is responsible for the transport of gases and nutrients and for the uniformization of temperature that make life on earth possible. For example, oxygen and carbon dioxide are not produced in the same places – oxygen comes largely from the equatorial rain forests and carbon dioxide is manufactured in industrial and urban centers such as New York City. Some mechanism is necessary to bring the carbon dioxide to Brazil, and the oxygen to the Big Apple. Radiation from the sun heats the surface of the earth; something is necessary to transfer the heat quickly and uniformly to the atmosphere where we can benefit from it. Without turbulence our speedy demise would be a race between frying our feet and freezing our heads, gasping in an atmosphere with too much or too little oxygen and/or carbon dioxide.

These practical aspects are, of course, responsible for most of the funding for turbulence research. It is absolutely essential as a design tool to be able to predict accurately the forces on and heat transfer from aircraft and automobiles. For regulatory purposes it is essential to be able to predict the results of siting of power plants and incinerators under various synoptic conditions. Manufacturers cry out for the ability to predict fluctuations in dopant distribution in the billets of silicon from which chips are formed. The military is concerned about the information loss in battlefield communication links induced by index of refraction fluctuations due to thermal turbulence. The list is endless.

From five centuries of observation and experiment, in many ways a reasonable physical understanding of turbulence has emerged. It is no longer a complete mystery. We can cite many simple physical arguments that shed light on common situations. When it comes to accurate predictions, however, we are in trouble. Aircraft manufacturers, for example, want accuracy corresponding roughly to the effect of adding one passenger to a Boeing 747. Automobile manufacturers want accuracy corresponding to the effect of adding one outside rear-view mirror. Regulatory agencies want assurances of comparable accuracy before going to court. Although our ability to calculate is improving constantly, we are not yet close to this level of accuracy.

Direct numerical simulation is not a realistic possibility in most cases of practical importance. In the foreseeable future, the cost of such simulation will remain far beyond our means, and will be limited to very low Reynolds numbers and simple geometries. In any event, simulation by itself does not bring understanding.

In a given practical problem, there may be many things that one wishes to know. The most common goals of computation are the mean forces and/or the mean heat transfer at various locations in the flow. These involve knowledge of second order quantities, the mean fluxes of momentum and heat. That is, the mean flux of j-momentum through a surface with a normal in the i-direction is $-\rho \langle u_i u_j \rangle$, where u_i is the fluctuating turbulent velocity, $\langle \cdot \rangle$ denotes an average, and ρ is the mean density. The flux of heat into an i-surface is $-\rho c_p \langle u_i \theta \rangle$, where c_p is the specific heat at constant pressure, and θ is the fluctuation in temperature. Both involve mean values of products of no more than two fluctuating quantities. Computation of index of refraction fluctuations in the atmosphere involves knowledge of the probability densities of fluctuating quantities, but an assumption about the form of the densities, plus knowledge of the variances, is usually enough. Hence, again, second order quantities are sufficient. A similar statement can be made about the dopant fluctuations in the silicon billet. There are more complex questions, however, that require more complex information. For example, suppose we wish to simulate the fluctuating pressure field on a panel, due to the presence of a turbulent boundary layer over the surface, perhaps to predict the spurious noise field generated on a sonar dome. This requires much more sophisticated modeling of the field.

It was in an effort to answer such deeper questions, that depend on a knowledge of the structure of the flow, that we embarked on the work described here. As we shall see below, many turbulent flows are characterized by considerable structure, and in particular by characteristic recurrent forms that are collectively called coherent structures. These are energetically dominant in many flows. We feel that, for flows in which these structures are dominant, it should be possible to build a relatively realistic, low-dimensional model of the flow by keeping only the dominant coherent structures, and simulating the effect of the smaller, less energetic, apparently incoherent part of the flow in some way. In this book we describe our tentative steps in this direction.

1.2 Low-dimensional models

Perhaps the first attempts to bring a dynamical systems perspective to turbulence studies were those of Landau (e.g. [204]) and Hopf [163]. They suggested that the continuous Fourier spectrum of temporal frequencies typical of turbulence might be produced via bifurcations occurring as the Reynolds number is increased (which Hopf, betraying his backgound, called μ rather than R_e). They envisaged a sequence in which at first periodic and then quasiperiodic attractors with increasing numbers of independent frequencies were created. In the language of modern dynamical systems theory, we would say that the resulting fluid flow corresponds to a *phase flow* on an n-dimensional torus in the state (or phase) space of the dynamical system. Hopf even constructed a model problem which exhibited just such a bifurcation sequence: what we might call a "route to chaos," except that we now realize that quasiperiodic flows are not strongly chaotic, since these solutions do not depend sensitively on initial conditions. Perhaps more significantly, Hopf also proposed that "to the flows observed in the long run after the influence of the initial conditions has died down there correspond certain solutions of the Navier–Stokes equations. These solutions constitute a certain manifold $\mathcal{M}(\mu)$ in phase space invariant under the phase flow. Presumably owing to viscosity $\mathcal{M}(\mu)$ has a finite number $\mathcal{N}(\mu)$ of dimensions." ([163], p. 305.) Hopf envisaged a finite-dimensional attractor.

Some twenty years after Hopf's paper, Ruelle and Takens [322] built on this suggestion. They observed that the quasiperiodic flows proposed by Landau and Hopf are not *structurally stable* and so would be expected to appear only in unusual circumstances. Drawing on the qualitative theory of (finite-dimensional) dynamical systems, which Anosov, Smale, Arnold, and others had extensively developed in the meantime, they gave an example of a structurally stable "strange" attractor that can appear after two or three quasiperiodic bifurcations, and so can live on a torus of only four dimensions (subsequently this was reduced to three: Newhouse *et al.* [258]). In connection with one of our themes, a footnote in their introduction is also noteworthy: "If a viscous fluid is observed in an experimental setup which has a certain symmetry, it is important to take into account the invariance of [the dynamical system] under the corresponding symmetry group." ([322], p. 168.) Ruelle gives an interesting account of the genesis of and tribulations encountered by their paper in [321].

In none of this work was a clear connection made between a particular fluid flow modeled by the Navier–Stokes equations with specific boundary conditions, and the "abstract" dynamical systems which exhibited quasiperiodic or strange attractors. However, unknown to Ruelle, Takens, and virtually all other mathematicians and physicists in 1971, Lorenz [217] had provided an example almost ten years before. A meteorologist and a former student of the dynamical systems pioneer, George Birkhoff, Lorenz was interested in the problem of weather prediction. He took a drastic truncation to three Fourier modes of the coupled Navier–Stokes and heat equations for Boussinesq convection in a two-dimensional layer (a Rayleigh–Bénard problem) and investigated them numerically and analytically. He found strong evidence for a strange attractor, unfortunately far beyond the Rayleigh number range in which his truncation was reasonable. Nonetheless, after its general discovery in the early 1970s, due largely to the mathematician Jim Yorke, Lorenz's paper has had an enormous influence. In Chapter 6 we give a sketch of what is now called the Lorenz attractor.

The events which first began to persuade fluid dynamicists that low-dimensional models and strange attractors might have some practical interest for them were probably the experiments of Gollub, Swinney, and their colleagues in the mid 1970s (see Swinney and Gollub [361]). Working with small, closed fluid systems, and especially with the Taylor–Couette flow between counter-rotating cylinders and thermal convection in small boxes, they found striking experimental evidence of sequences of bifurcations leading to "low-dimensional" chaos as the Taylor and Rayleigh numbers respectively were raised modestly above the initial onset of linear instability. Power spectra displaying jumps from two or three frequency quasiperiodic motions to broad band chaos were measured. Subsequently it became possible to link some of these results tightly with bifurcation analyses of the governing equations, particularly in the Taylor–Couette problem (see Golubitsky and Stewart [135], Golubitsky and Langford [133], Golubitsky *et al.* ([136], case study 6), Iooss *et al.* [75, 95, 173, 174], and Laure and Demay [206], for example). There is an enormous literature on this system: in his 1994 review, Tagg [363] estimates nearly 2000 papers while citing some 350 himself. Chossat and Iooss have published a book on the mathematical aspects of the problem [76].

In some cases, previously unknown classes of solution were predicted which were subsequently observed experimentally (e.g. Andereck *et al.* [6], Tagg *et al.* [364]). Again, the symmetries of the experimental apparatus were crucial in this. It is probably fair to say that the tools and viewpoint of dynamical systems theory are now acknowledged to have a useful rôle to play in the study of such closed fluid systems, in which relatively few spatial modes are active. These methods, including invariant manifold techniques, bifurcation theory, and the unfolding of degenerate singularities, have joined more classical asymptotic and perturbation methods for the study of hydrodynamic stability and "weakly" nonlinear or pre-turbulent interactions.

In this book we want to take a further tentative step. We propose that low-dimensional dynamical systems can also provide models for, and hence contribute to the understanding of, certain fully developed, open turbulent flows. As we remarked in the Preface, our "low" is not so low in dynamical systems terms: we are thinking of sets of 10–100 ordinary differential equations (ODEs). But in the fluid mechanical context this is *very* low and we clearly cannot expect to reproduce fine-scale spatial details of the flow. Consequently, to be sure of capturing the key behaviors, we will have to pay particular attention to the manner in which the fluid velocity field in physical space is represented in the phase space of the dynamical system. We shall focus on flows with predominant coherent structures, and use the proper orthogonal or Karhunen–Loève decomposition (POD) to extract, from experimental or simulated ensembles of data, those "modes" or empirical eigenfunctions that carry the greatest kinetic energy on average. This procedure will provide us with the basis for a sequence of subspaces, of increasing dimension, onto which the Navier–Stokes (or other) equations can be projected by Galerkin's method to yield sets of ODEs. In this procedure we represent the fluid velocity field by a superposition of the empirical spatial modes multiplied by (as yet unknown) time-dependent coefficients. Substituting this representation into the governing equations and taking the inner product with each basis function in turn yields a set of nonlinear ODEs for the modal coefficients. These entirely deterministic dynamical systems will be the foundations for our low-dimensional models.

Our main goal is not to reproduce accurately the results of a direct numerical simulation with fewer, more efficient modes. The fact that such empirical basis functions are adapted to a particular flow geometry and Reynolds number, and are only available at the end of extensive data collection and computation, probably makes them a poor choice for efficient simulations in any case. Rather we are interested in *understanding* the fundamental mechanisms of turbulence generation in "simple" flows such as shear layers, jets, wakes, and boundary layers. In this quest for understanding we often want to reduce the dimension of our models to a minimum. Thus, even with the optimal bases of the POD, our truncations are typically so severe that a bare projection is unsatisfactory and some sort of additional modeling is needed to account for neglected modes and/or spatial locations. Such modeling might include relatively simple "eddy viscosity" energy transfer of the sort proposed by Heisenberg and Smagorinsky (see Batchelor [32] or Tennekes and Lumley [368]) as well as models to account for slow variations of the mean shear which drives the turbulent fluctuations in flows such as boundary and shear layers, due to the turbulence itself. Ideally, and in greater generality, we envisage the introduction of a probabilistic element to our deterministic ODEs to reproduce the conditional probability measures that describe the activity of the neglected modes as a function of the state of those modes included in the model. Very little appears to be known about this issue, but we have encountered and partially resolved a crude version of it in our treatment of the outer part of the boundary layer in models of the wall region.

After determination of a "good" subspace, projection of the governing equations, and modeling to account for neglected modes, we have a set of ODEs, for an understanding of which we can appeal to the methods of dynamical systems theory along with other, more widely known mathematical tools. If done properly, the projection and modeling preserve the underlying symmetries of the fluid flow and of the governing equations and boundary conditions. Such symmetries may include spanwise translations and reflections for a shear layer or a boundary layer on an infinite flat plate, and rotations and reflections for a circular jet or wake. Thus the ODEs will exhibit a corresponding symmetry: in the language of dynamical systems theory, they will be *equivariant under some group* Γ, and we have to take this into account in studying the bifurcations and other dynamical behavior of the system. Behavior that is structurally unstable and hence rare in general may be stable and relatively prevalent for such Γ-equivariant systems. We have already mentioned heteroclinic attractors in the Preface, and the reader will find several more examples later in the book.

The result of our dynamical systems analyses of the low-dimensional models is a (partial) understanding of the structure of solutions in phase space and in particular of attracting sets and how they change through bifurcations as external and modeling parameters are varied. The final tasks are to map those results back into physical space, reconstructing the space–time velocity field of the fluid flow from the empirical basis functions and their time-dependent coefficients, to compare the resulting instantaneous and averaged quantities with experiments, and to translate the understanding achieved in state space into insights about the fluid flow itself.

This is the general strategy we propose: find good basis functions for the turbulent flow in question, model to account for neglected effects, project the governing partial differential equations onto a low-dimensional subspace spanned by the most energetic modes, analyze the resulting low-dimensional model, and finally return to the physical domain to interpret that analysis. As we see in Chapter 2, not all turbulent fluid flows are energetically dominated by coherent structures, and so the approach we describe here is far from offering a complete solution to "the problem of turbulence." We believe, nonetheless, that it provides one more approach and a set of new tools, or even weapons, for the unequal combat referred to in the introduction to this chapter.

1.3 The contents of this book

As noted in the Preface, the book has four parts. The first two, which constitute well over half the book, are fairly general in nature. We introduce key ideas from fluid mechanics, turbulence theory, and dimension reduction methods in the first five chapters, and from dynamical systems theory in the following four. Turbulence experts can probably skip pieces of Part One, and dynamicists can certainly skip most of Part Two, but in both places readers may find new viewpoints recommended and unfamiliar connections drawn. We hope that these parts of the book will be of fairly lasting and general interest. The remaining two parts are more specific and more speculative, for in them we focus on our own work on the turbulent boundary layer and on other attempts to derive low-dimensional models for turbulent and transition flows. We offer our own work mainly in the spirit of an extended example, since it allows us to discuss and illustrate difficulties and limitations as well as successes of the approach. We are far from claiming a complete understanding of boundary layer turbulence via our models, but we hope that the reader who accompanies us to the end of Chapter 13 will agree that some new things have been said.

In the remaining four chapters of this first part, we give some background on turbulence, describe coherent structures from an experimental viewpoint and summarize some of the major findings relevant to shear dominated flows. In Chapter 2 we sketch some experimental methods by which coherent structures in developed turbulent flows may be found and characterized, and describe their relation to instabilities of simpler laminar and transition flows. We also review the "classical" approach to turbulent flows, via the averaged Navier–Stokes equations and careful order-of-magnitude and scaling estimates. We discuss in some detail the cases of turbulent mixing and boundary layers (the main illustrative application of our approach is to the latter). We close with a brief preview of how coherent structures might appear as attractors in simple dynamical systems.

Chapter 3 is devoted to the proper orthogonal decomposition. We provide the basic mathematical results, with enough elements of their proofs to illustrate both the scope and limitations of representing turbulent fields by finite- (low-) dimensional projections. We pay particular attention to the influence both of symmetries, in the physical flow and in the particular data ensembles used, and of the ensemble averaging on which the method is founded, on the sets of basis functions that it produces. We also describe the relation of these empirical modes to certain other statistically based techniques for prediction and analysis of turbulence, such as stochastic estimation.

In Chapter 4 we discuss the Galerkin method, and show how the Navier–Stokes equations, or in general any evolution equation, may be projected onto a finite-dimensional subspace, and in particular onto a subspace spanned by empirical modes, to produce a finite set of ODEs. We also discuss various "modeling" issues such as those mentioned in Section 1.2.

Chapter 5 introduces recent work on the balanced proper orthogonal decomposition, and on the balanced truncation method for linear systems on which it is based. This method is particularly useful for systems with control inputs based on observation of specific flow quantities, and we provide examples to illustrate its superiority to the proper orthogonal decomposition in such cases.

The second part of the book is a mini-treatise on dynamical systems theory. Since we are concerned only with low-dimensional models, we restrict ourselves to finite-dimensional ODEs and iterated maps. In Chapter 6 we sketch the main ideas and tools, including linearization, invariant manifolds, structural stability, the center manifold theorem, normal forms, and local and global bifurcation theory. We end the chapter with a discussion of attractors, the main example being the strange attractor of Lorenz. Throughout this and the remaining chapters in this second part, we illustrate the theory with many simple and *very* low-dimensional examples.

Chapter 7 deals with symmetries, bifurcations, and local and global dynamical behavior of equivariant ODEs, leading up to an important example derived from spatial translation and reflection invariance, which can be understood in the context of Fourier mode representations of traveling waves. This example, an $O(2)$-equivariant normal form for the interaction of wavenumbers in the ratio 1:2, is a four-dimensional ODE possessing heteroclinic attractors, which, while not strange, have an interesting structure which seems relevant in models of many fluid flows with symmetries. The chapter ends with a brief description of how the POD method can be extended to provide empirical modes that represent uniformly translating structures (traveling modes).

In Chapter 8 we exercise our new methods on a simple model problem: the one-space-dimensional Kuramoto–Sivashinsky partial differential equation. We find the $O(2)$-equivariant normal form of Chapter 7 buried in this system. In the final chapter of this part, Chapter 9, we consider the effects of stochastic and other symmetry-breaking perturbations on systems with heteroclinic cycles.

The third part of the book is devoted to a description of attempts by ourselves and our students and colleagues, to apply our strategy to the wall region of the turbulent boundary layer. Most of Chapter 10 contains discussions of the Galerkin projection and modeling issues, introduced in Chapter 4, in the specific case of the wall region. We describe the choice of specific subspaces and the resulting hierarchy of nested systems of increasing dimension that results as more modes are included. The chapter contains a description of the various symmetries that the low-dimensional model ODEs inherit from those of the boundary layer itself, and ends with extensive discussions of the validity of the models for the mean flow and losses to neglected modes.

In Chapter 11 we bring dynamical systems techniques to bear on the model ODEs and provide relatively complete analyses of the bifurcations and dynamical behavior of the boundary layer models. We illustrate and supplement our analyses with numerical simulations of models of various dimensions and, here and in Chapter 10, we offer a critical interpretation of the results, showing how the use of empirical basis functions can sometimes lead to paradoxical effects. The chapter includes reconstructions of the fluid velocity fields and interpretations of our findings in phase space, in terms of the turbulent flow itself.

The first of the two chapters of Part Four contains brief reviews of work by other groups in which the same general approach is taken. We do not pretend to give a complete survey of this rapidly developing field, but the examples of "laboratory" open flows that we have chosen, including jets, wakes, and transition in boundary layers, illustrate that our methods have wide applicability. Related ideas have been and continue to be used in the meteorological community (cf. [216]) – the work of Farrell and Ioannou is an interesting case in point [105–107] – and there are clearly applications to many other problems involving the dynamics of spatio-temporal patterns. In this second edition we have added references to some recent work, including new sections on time periodic flows in internal combustion engines and other applications (12.6 and 12.7).

In the closing Chapter 13 we speculate in broader terms on the place and uses of low-dimensional models among the many other approaches to turbulence. It seems clear that such models offer new *understanding* of turbulence generation involving coherent structures, and so contribute to the intellectual challenge alluded to in the second paragraph of the present chapter. Can they also be of help in answering technological questions such as those mentioned towards the end of Section 1.1? A particular interest of our own is in the use of such models in formulating strategies for the active control of turbulence and, in addition to the material in Chapter 5, we provide a brief description of our ideas at the end of Chapter 12. In Chapter 13 we also mention a number of other recent developments that are related to our story, including mathematical ideas such as inertial manifolds and other reduction methods which offer new approaches to the Navier–Stokes equations.

1.4 Notation and mathematical jargon

By now the reader knows that in this book we propose the application of ideas in the qualitative theory of dynamical systems to the description and analysis of turbulent flows. While qualitative theory had its beginnings in Poincaré's studies of problems in celestial mechanics about one hundred years ago [282], it was soon thereafter hijacked by pure mathematicians and only in the last ten to twenty years has it begun to see broad applications in the sciences and engineering. The explosion of interest in "chaos theory," encouraged by books such as Gleick's [128], has certainly sparked a general awareness that there are new ideas and methods out there, but we recognize that many of the basic concepts and technical issues may remain mysterious for potential users, including the intended readership of this book. Rather than try to skate over what may be unfamiliar mathematical material for some readers, we have tried to introduce it with simple examples drawn from the world of low-dimensional ordinary differential equations. By working such examples in some detail, we hope to leave our readers in a position to fill in missing steps in more complicated cases and to tackle new ones that may arise in their own research. But this is emphatically *not* a dynamical systems textbook: we do not state, much less prove, even the most basic theorems in the field, and those formal definitions that are included are given in passing, usually indicated by *italics*.

Even with an approach based on examples, so foreign to "pure" dynamical systems theorists, we cannot avoid introducing and using a modicum of mathematical jargon. Our defense of this is twofold: (1) we believe that, once learned, the symbolism of dynamical systems theory, largely drawn from topology, makes the precise description of key ideas such as invariant manifolds and attractors much simpler and more compact than is possible with the English language alone, and (2) we hope that this book might be the beginning of an exploration of the current research literature, in which case the symbolism will have to be mastered anyway. After each new excess of jargon in the text, we try to give a (rough) characterization in words, and we encourage readers who are repelled by abstract formulae to clench their teeth and read on to get to the examples and pictures.

Nonlinear analysis is built on linear analysis and, to avoid doubling the length of this book, we must assume some familiarity with the fundamental ideas of finite-dimensional linear vector spaces, spanning sets of vectors, bases, norms, inner products, linear subspaces, eigenvalue problems, and the like. Similarly, one of the major applications of this beautiful theory is to the solution of linear ordinary differential equations, and we assume a basic knowledge of that as well. Books such as Strang's *Linear Algebra and its Applications* [358] or Boyce and DiPrima's *Elementary Differential Equations and Boundary Value Problems* [56] provide the necessary background. More geometrically oriented introductions to nonlinear ordinary differential equations are Hirsch *et al.*'s *Differential Equations, Dynamical Systems and an Introduction to Chaos* [156], Arnold's *Ordinary Differential Equations* [15], and Glendinning's *Stability, Instability and Chaos* [129], all of which are written from a more mathematical viewpoint. The last of these is a good introduction to many of the modern concepts presented in Part Two of the present book.

Here, to prepare for the onslaught, we recall some of the mathematical notations we shall use. First there are the standard names for some commonly encountered spaces:

$\mathbb{R}^n$: *n*-dimensional real Euclidean space, the elements of which are vectors $\mathbf{x} = (x_1, x_2, \ldots, x_n)$, with each x_j a real number. The real line $\mathbb{R}^1$ is simply written $\mathbb{R}$.

$\mathbb{C}^n$: *n*-dimensional complex Euclidean space; as above, but each x_j is a complex number. $\mathbb{C}^1$ is written $\mathbb{C}$.

We normally denote vectorial quantities by boldface letters $\mathbf{x}$ and scalar quantities by italic letters x. Single bars $| \cdot |$ denote the Euclidean norm or absolute value of whatever is inside them; they also denote the modulus in the case of a complex number. Other norms are generally indicated by double bars: $\| \cdot \|$. We occasionally use the *supremum norm*, written sup $|\mathbf{x}|$, which indicates the least upper bound. If A is a subset of $\mathbb{R}$, the number M is an upper bound for A if $a \leq M$ for all a in A. When M is the smallest such number, it is the *least upper bound*. The *infimum* inf $|\mathbf{x}|$ or *greatest lower bound* is defined analogously.

The *inner product* (also *scalar* or *dot product*) of two elements $\mathbf{u}$, $\mathbf{v}$ in $\mathbb{R}^n$ or $\mathbb{C}^n$ is given by:

$$(\mathbf{u}, \mathbf{v}) = \sum_{i=1}^{n} u_i v_i^* = \mathbf{v}^* \mathbf{u},$$

where in the second expression $*$ denotes the complex conjugate, and in the third expression * denotes complex conjugate transpose.

A set V is *open* if and only if for every point $x \in V$ there is a neighborhood B_x of x with B_x contained in V. A set U is *closed* if and only if for each point y not in U there is a neighborhood B_y of y entirely disjoint from U. Alternatively, a set is closed if and only if it contains all its limit points. Examples are given directly below.

$[a, b]$: the closed interval of the real line $\mathbb{R}$, delimited by the points $a < b$: all points x satisfying $a \leq x \leq b$. A curved parenthesis denotes that the endpoint is not included, thus (a, b) denotes the open interval $(a < x < b)$ and $(a, b]$ the half open interval $a < x \leq b$. This notation extends to higher dimensions; thus $[0, 1] \times [0, 1]$ or $[0, 1]^2$ denotes the (closed) unit square in $\mathbb{R}^2$ with corners at $(0, 0)$, $(1, 0)$, $(1, 1)$, and $(0, 1)$; here $\times$ means the direct product.

L^2: the (Hilbert) space of square integrable real or complex-valued functions, an example of an infinite-dimensional inner product space. Often the domain of definition is indicated in parentheses: thus $L^2([0, 1])$ denotes the space of functions defined over the unit interval $0 \leq x \leq 1$. Square integrable means that the functions $f(x)$ belonging to $L^2([0, 1])$ satisfy

$$\|f\| = \left[\int_0^1 |f(x)|^2 dx \right]^{\frac{1}{2}} < \infty. \tag{1.1}$$

In general the integral is taken over the domain of definition, Ω. The boundary of the domain is customarily written as $\partial\Omega$. Note that L^2 is an inner product space, the inner product being defined by

$$(f, g) = \int_\Omega f(x) g^*(x) dx, \tag{1.2}$$

where * denotes the complex conjugate. Note that $(f, g) = (g, f)^*$ and that the L^2-norm $\|f\|$ of f can also be written

$$\|f\| = (f, f)^{\frac{1}{2}}. \tag{1.3}$$

For functions $f(\mathbf{x})$ of several variables one uses a multiple integral, and for vector-valued functions, such as the velocity field in a fluid flow $\mathbf{u}(\mathbf{x}, t) = [u_1(x_1, x_2, x_3, t), u_2(\ldots), u_3(\ldots)] \in L^2(\Omega)$, the inner product is defined by

$$(\mathbf{f}, \mathbf{g}) = \int_\Omega (f_1 g_1^* + f_2 g_2^* + f_3 g_3^*) d\mathbf{x}, \tag{1.4}$$

where Ω denotes the spatial domain occupied by the fluid (e.g., $\Omega = [0, 1]^3$). The space L^2 is a natural one in which to do fluid mechanics since, from the above, it is the space of flows having finite kinetic energy. In fact we simply have (for constant density ρ):

$$\text{kinetic energy} = \frac{1}{2}\rho \|\mathbf{u}\|^2. \tag{1.5}$$

Adjoint: if $A : V \to W$ is a linear mapping between two inner product spaces V and W, the *adjoint* of A is a mapping $A^* : W \to V$ such that $(Av, w)_W = (v, A^*w)_V$, for all $v \in V$ and $w \in W$, where $(\cdot, \cdot)_V$ and $(\cdot, \cdot)_W$ denote the respective inner products on V and W. For example, if $\mathbf{A}$ is a real $n \times m$ matrix, viewed as a mapping $\mathbf{A} : \mathbb{R}^m \to \mathbb{R}^n$, with the standard inner products on $\mathbb{R}^m$ and $\mathbb{R}^n$, then for any $\mathbf{v} \in \mathbb{R}^n$, $\mathbf{w} \in \mathbb{R}^m$, we have

$$(\mathbf{Av}, \mathbf{w}) = \mathbf{w}^T \mathbf{Av} = (\mathbf{A}^T\mathbf{w})^T \mathbf{v} = (\mathbf{v}, \mathbf{A}^T\mathbf{w}), \tag{1.6}$$

so the adjoint of $\mathbf{A}$ is $\mathbf{A}^T$, where the superscript T denotes transpose.

Next there are the relation symbols:

$\in$: is an *element* of, thus $x \in \mathbb{R}$, $a + ib \in \mathbb{C}$.

$\subset$: is a *proper subset* of; thus $[0, 1] \subset \mathbb{R}$. Proper means that the subset is strictly smaller than the set it belongs to.

$\subseteq$: is a *subset* of, thus $(-a, b) \subseteq \mathbb{R}$. Here the subset may be the whole thing: $(-\infty, \infty) = \mathbb{R}$.

The binary operation symbols used are:

$\cup$: *union* (of sets): $[-1, 0] \cup [0, 1] = [-1, 1]$.

$\cap$: *intersection* (of sets): $[-1, 0.5) \cap [0, 1] = [0, 0.5)$.

$\backslash$: the *complement* of; thus $A \backslash B$ denotes the complement of the set B in the set A, as in: $[0, 1] \backslash (\frac{1}{3}, \frac{2}{3}) = [0, \frac{1}{3}] \cup [\frac{2}{3}, 1]$.

$\{A|B\}$ denotes the set of objects A which satisfy the condition(s) specified by B; thus $\{(x_1, x_2) \in R^2 | x_2 > 0\}$ is the upper half plane, excluding the x_1-axis.

In Chapter 3 we discuss relations between the proper orthogonal decomposition and attractors of infinite-dimensional dynamical systems. This requires some technical notions, including that of a compact set. A set A is *compact* if every covering of A by open sets contains a finite subcover. For subsets of finite-dimensional vector spaces, this is equivalent to A being closed and bounded.

Finite-dimensional vector spaces are used throughout the book: they can usually be thought of as $\mathbb{R}^n$ or $\mathbb{C}^n$, by referring to a specific basis and coordinate system. For two elements $\mathbf{u}, \mathbf{v}$ of such a space we write the inner product $\mathbf{v}^*\mathbf{u} = (\mathbf{u}, \mathbf{v})$, employing the same notation as in L^2. The *outer* or *tensor product* is the $n \times n$ matrix written $\mathbf{u} \otimes \mathbf{v}$ or $\mathbf{u}\mathbf{v}^{\mathrm{T}}$.

span$\{\mathbf{v}_1, \ldots, \mathbf{v}_n\}$ denotes the linear (sub)space spanned by the vectors $\mathbf{v}_1, \ldots, \mathbf{v}_n$: all linear combinations of the form

$$\mathbf{v} = \sum_{i=1}^{n} a_i \mathbf{v}_i. \tag{1.7}$$

This notion applies also to spaces of functions. For example,

$$\text{span}\{1, e^{2\pi i x}, e^{4\pi i x}\} \subset L^2([0, 1]) \tag{1.8}$$

is the set of functions of period 1 that can be written as the sum of the mean and the first two Fourier modes.

The superscript $\perp$ denotes the orthogonal complement of a (proper) subspace of a vector space. Symbol $\oplus$ denotes the direct sum, so if V is a vector space and $W \subset V$ we have $V = W \oplus W^{\perp}$.

There are several pieces of more-or-less standard notation for common operations and functions:

A function or map $f : X \to Y$ between two spaces X and Y is said to be *Lipschitz* if it satisfies a bound of the form

$$\|f(x) - f(y)\|_Y \leq K\|x - y\|_X, \tag{1.9}$$

for all $x, y \in X$, where $\| \cdot \|_X$ and $\| \cdot \|_Y$ denote norms on X and Y respectively, and K is called the *Lipschitz constant*. Linear functions and functions with uniformly bounded first derivatives are clearly Lipschitz, but Lipschitz functions need not be differentiable; for example, $f(x) = |x|$ is Lipschitz, with Lipschitz constant 1.

$\langle f \rangle$ denotes an average of the quantity or function f. For turbulent fields, as described in Chapter 2, this is usually an ensemble average over a number of realizations f_j:

$$\langle f \rangle = \langle f \rangle(x, t) = \frac{1}{N} \sum_{j=1}^{N} f_j(x, t). \tag{1.10}$$

Sometimes we employ time or space averages:

$$\langle f \rangle(x) = \frac{1}{T} \int_0^T f(x, t)dt \quad \text{and} \quad \langle f \rangle(t) = \frac{1}{L} \int_0^L f(x, t)dx, \tag{1.11}$$

which we make clear at the appropriate places. In Chapter 10, for brevity, we write the time average as $\bar{f}$.

For linear systems, we also require the notion of an *operator norm*. A linear input–output system may be defined as

$$\left.\begin{array}{l} \dot{\mathbf{x}}(t) = \mathbf{A}\mathbf{x}(t) + \mathbf{B}\mathbf{u}(t) \\ \mathbf{y}(t) = \mathbf{C}\mathbf{x}(t) + \mathbf{D}\mathbf{u}(t) \end{array}\right\}, \tag{1.12}$$

where the vector $\mathbf{u}(t)$ is the *input*, $\mathbf{y}(t)$ is the *output*, and $\mathbf{x}(t)$ is the *state* vector. One may view such a system as a mapping $\mathbf{G}$ from the input $\mathbf{u} \in L^2([0, \infty))$ to the output $\mathbf{y} \in L^2([0, \infty))$. The *induced (operator) norm* (induced by the L^2 norm on $\mathbf{u}$ and $\mathbf{y}$) is called the *infinity norm* of $\mathbf{G}$, and is given by

$$\|\mathbf{G}\|_\infty \overset{\text{def}}{=} \max_{\mathbf{u}} \frac{\|\mathbf{G}\mathbf{u}\|_2}{\|\mathbf{u}\|_2} = \max_\omega \bar{\sigma}(\hat{\mathbf{G}}(i\omega)), \tag{1.13}$$

where $\hat{\mathbf{G}}(s) = \mathbf{C}(s\mathbf{I} - \mathbf{A})^{-1}\mathbf{B} + \mathbf{D}$ is the *transfer function*, and $\bar{\sigma}$ denotes the maximum singular value of a matrix. We also use the two-norm on systems, defined as

$$\|\mathbf{G}\|_2 \overset{\text{def}}{=} \int_{-\infty}^{\infty} \text{Tr}(\hat{\mathbf{G}}(i\omega)^T \hat{\mathbf{G}}(i\omega))\, d\omega, \tag{1.14}$$

where Tr denotes the trace of a matrix.

Probability measures are used in several places. A non-negative function $\mu : X \to \mathbb{R}$ defined on a space X is a (normalized) *probability density* if $\int_X \mu(x)dx = 1$. Often X will be the phase space of a dynamical system, in which case it may be finite-dimensional (e.g. $\mathbb{R}^n$) or infinite-dimensional (e.g. $L^2(\Omega)$). We use the shorthand $\int f(x)d\mu$ to denote $\int f(x)\mu(x)dx$: integration with respect to the density or measure μ.

A measure μ is *invariant* for an iterated mapping $g : X \to X$ if, for every set $A \subset X$,

$$\mu(g^{-1}(A)) = \mu(A). \tag{1.15}$$

See Section 6.5.

The abbreviation a.e. stands for "almost every" or "almost everywhere," in the sense of an appropriate measure; that is, the property in question holds for all except possibly a set of measure zero. If no specific measure is specified, Lebesgue measure (length, area, volume, etc.) is assumed, but in Chapter 3, a.e. frequently refers to the measure associated with ensemble averages.

$1_A(x)$ denotes the indicator function, equal to 1 if the variable x belongs to the set A and equal to zero otherwise.

$\delta(t)$ is the Dirac delta (generalized) function, satisfying $\delta(t) = 0$ for $|t| \neq 0$, $\int_{-\infty}^{\infty} \delta(t)dt = 1$ and $\int_{-\infty}^{\infty} f(t - s)\delta(s)ds = f(t)$ for any continuous function f.

The Kronecker delta is

$$\delta_{jk} = \begin{cases} 1 & \text{if } j = k \\ 0 & \text{if } j \neq k. \end{cases} \tag{1.16}$$

In writing the partial differential equations of continuum mechanics it is sometimes convenient to refer explicitly to the components of vectors with respect to a particular (fixed) basis. To do this we use the conventional tensor notation, with summation implied on repeated indices (Einstein notation). Thus the incompressible Navier–Stokes equations,

$$\mathbf{v}_t + \mathbf{v} \cdot \nabla \mathbf{v} = -\frac{1}{\rho}\nabla p + \nu\Delta\mathbf{v} + \frac{1}{\rho}\mathbf{f}, \qquad \nabla \cdot \mathbf{v} = 0, \tag{1.17}$$

may be written, in Cartesian coordinates with respect to the standard orthonormal basis $\{\mathbf{e}_i\}_{i=1}^3$, as

$$v_{i,t} + v_{i,j}v_j = -\frac{1}{\rho}p_{,i} + \nu v_{i,jj} + \frac{1}{\rho}f_i, \qquad v_{i,i} = 0. \tag{1.18}$$

Occasionally we use Einstein notation to indicate sums over indices in modal decompositions, but we generally indicate these by an explicit summation symbol.

2

Coherent structures

In this chapter we present a brief introduction to those parts of fluid mechanics and classical turbulence theory that are relevant to the description of coherent structures by low-dimensional models. We sketch the landscape in which our models are to be built. The coherent structures we shall be concerned with are, after all, physical phenomena that are observed in flowing fluids, and in order to appreciate fully what we are doing, it is necessary to know just what phenomena have been observed, how they have been observed, what interpretations have been suggested for the observations, and what the characteristics of the flows are, in which the structures have been observed. The chapter is written primarily for readers with a background in applied mathematics, but without much knowledge of fluid mechanics. Readers with a background in turbulence will find much of the material familiar, but even they may find a few things they have not seen before.

2.1 Introduction

Turbulent flows are described by the Navier–Stokes equations, which are (in Cartesian coordinates and in the absence of body forces and compressibility):

$$\tilde{u}_{i,t} + \tilde{u}_{i,j}\tilde{u}_j = -\frac{1}{\rho}\tilde{p}_{,i} + v\tilde{u}_{i,jj}, \tag{2.1}$$

$$\tilde{u}_{i,i} = 0, \tag{2.2}$$

supplemented by appropriate boundary conditions. Here we use $\tilde{u}_i$ and $\tilde{p}$ to denote the instantaneous fluid velocity and pressure fields. We use the notation $\tilde{u}_{i,j} = \partial\tilde{u}_i/\partial x_j$, etc. and repeated indices imply summation. The parameter v is the kinematic viscosity, and ρ is the density, assumed to be constant. We consider only the incompressible case in this book; in fact, it is difficult to make compressible turbulence. Since the turbulent fluctuating velocities are of the order of a few percent of the mean velocity, and the mean flow Mach number for most applications of technological interest is seldom more than 5, the fluctuating Mach number is usually comfortably less than unity. Turbulence is thus basically incompressible, although there are compressibility effects when turbulence passes through a shock wave, for example.

In the brief account that follows, we can only sketch some of the key themes in classical turbulence theory. For more background and information on this material, see [368].

We will write $\tilde{u}_i = U_i + u_i$, where $U_i = \langle \tilde{u}_i \rangle$ is the *mean velocity*, $\langle \cdot \rangle$ is a suitable averaging operation, and u_i is the *turbulent fluctuating velocity*. The *turbulent kinetic energy per unit mass* is given by

$$\langle q^2/2 \rangle = \frac{1}{2} \langle u_i u_i \rangle. \tag{2.3}$$

The *dissipation of energy per unit mass*, ϵ, is given by $2\nu \langle s_{ij} s_{ij} \rangle$ as we shall see a little later. Here s_{ij} is the *fluctuating strain rate* of the turbulent motion, $s_{ij} = \frac{1}{2}(u_{i,j} + u_{j,i})$. It is of the utmost importance to understand that, although the dissipation of energy is due to viscous stresses and local strain rate, both of which depend on velocity gradients, and so take place at the smallest scales in the turbulent flow, the amount of dissipation is, in fact, determined by the large, or energy-containing, scales in the flow. Very crudely put, energy moves from the mean flow into the large, or energy-containing scales, and from them into the next smaller eddies, and so on until it reaches the scales at which the dissipation takes place. The process is not quite as straightforward as this simplistic explanation pretends, but the rate at which energy is withdrawn from the mean flow and enters the turbulence, and at which it leaves the energy-containing scales to enter the next smaller scales, is determined by the large-scale dynamics. By the time energy reaches the smallest scales, where it can be dissipated to heat, the rate of dissipation can no longer be influenced; it is like a flow entering a spectral pipeline, one end of the pipe being the energy-containing scales, and the other the dissipative scales. Changing the viscosity, for example, has no influence on the dissipation of energy – it simply changes the scale at which dissipation takes place.

We should say a few words about the Reynolds number. There are a very great many Reynolds numbers in any flow. They all have the same form: some velocity multiplied by some length and divided by the kinematic viscosity. They are usually all related. As a velocity we can use the mean velocity U, the root mean square (r.m.s.) turbulent fluctuating velocity $u = \langle u_i u_i \rangle^{1/2}$, or (in boundary layers) the *friction velocity* u_τ which is defined by $\tau_w = \rho u_\tau^2$, where τ_w is the *shear stress* at the wall. In turn, $\tau_w = \mu \partial U_1 / \partial x_2 |_{x_2=0}$, where U_1 is the mean velocity in the streamwise direction, x_2 is the distance normal to the wall, and $\mu = \rho \nu$ is the dynamic viscosity. Note that in what follows we frequently use u as an estimate for the magnitude of any of the components of the turbulent fluctuating velocity. The length scale can be the distance from the inflow boundary (or from the leading edge of a plate, or the upstream end of a body), usually designated by L, or the thickness of a mixing layer or boundary layer, usually designated by δ or θ. Such a thickness may be defined in several ways – geometrically, as by the point at which the mean velocity has reached 99% of its free-stream value, or by an integral of the momentum deficit, for example. We may also use ℓ, the scale of the energy-containing eddies in the turbulence, usually defined as the *integral scale* (that is, as the integral of the autocorrelation function, a measure of the distance within which motions are correlated).

This dizzying collection of possibilities is not as confusing as may seem at first sight. We use the Reynolds number that has the greatest dynamical significance for the problem at hand. We can usually show by analysis that the various Reynolds numbers can be expressed one in terms of another. Often there is a simple numerical factor relating them (as for the Reynolds numbers based on the geometrical thickness, the displacement thickness, and the momentum thickness, when discussing boundary layers; for a given mean velocity profile these three thicknesses are proportional. The different definitions appear because, for an arbitrary profile, they emphasize different physical aspects.) When talking about turbulence, we usually use $R_\ell = u\ell/\nu$, although we might (in the case of a boundary layer) use $u_\tau\delta/\nu$, since u and u_τ are numerically very close, as are ℓ and δ, and u_τ and δ are substantially easier to measure.

Turbulence is a continuum phenomenon. Except in certain circumstances (such as interstellar gas clouds) the smallest turbulent length scales are very much larger than molecular scales. The smallest turbulent length scale is the *Kolmogorov microscale*

$$\eta = (\nu^3/\epsilon)^{1/4}, \tag{2.4}$$

where ϵ is the dissipation of turbulent kinetic energy per unit mass. If we take the ratio of this to the mean free path ξ, and use the fact (to be justified below) that $\eta/\ell = R_\ell^{-3/4}$, where $R_\ell = u\ell/\nu$ is the Reynolds number of the turbulence, u is the r.m.s. turbulent fluctuating velocity, and ℓ is the size of the energy-containing-eddies, then we have

$$\eta/\xi = R_\ell^{1/4}/m. \tag{2.5}$$

Here m is the turbulent fluctuating Mach number. In any flow that is turbulent, R_ℓ is large, while m is almost never larger than unity. Even in interstellar clouds of neutral hydrogen, the ratio η/ξ is still larger than unity (≈ 6) (see [368]). There is thus no question that the Navier–Stokes equations adequately describe turbulence under most circumstances, although this is a debate that arises anew every few years.

In fact, turbulence is an inertial phenomenon. That is, turbulence is statistically indistinguishable on energy-containing scales in gases, liquids, slurries, foams, and many non-Newtonian media. These media have markedly different fine structures, and their mechanisms for dissipation of energy are quite different. This observation suggests that turbulence is an essentially inviscid, inertial phenomenon, and is uninfluenced by the precise nature of the viscous mechanism. It is necessary to have a mechanism for dissipation, since energy is extracted from the mean motion by the turbulence, and ultimately dissipated to heat, but as we shall see later, changing the dissipation mechanism does not have any influence on the amount of energy that is dissipated, merely on the precise scale and manner of its dissipation.

In graduate courses in fluid mechanics it is customary to present the known exact solutions of the Navier–Stokes equations. There are 11 of which we know, and eight of these are trivial, which is to say, they can only be obtained because of some special circumstance that causes the nonlinear terms to vanish identically. Of exact solutions of the full equations with the nonlinearity we know only three, all of which are simplified by some symmetry or similarity. Since turbulent flows are in general essentially three-dimensional and rotational and unsteady (the opposite of the simplifications that are usually introduced in fluid mechanics to make the equations tractable) there is little possibility of finding exact solutions to turbulent problems. Of course, the Navier–Stokes equations can be solved by direct numerical simulation (DNS) for various turbulent flows. This is frequently done, and is a valuable source of data. The DNS approach does not, however, give much *dynamical* insight. Our attempt to construct low-dimensional models of the most energetic modes in turbulent flows, as described in this book, is a direct attack on this difficulty – a way of converting the problem into something at least partially solvable.

In the classical approach to turbulence, the problem is simplified in various other ways, although it is difficult to find ones that do not vitiate it. One of the successful ways (which we examine in greater detail later) is order-of-magnitude analysis. If we write equations for the mean velocity vector and the various moments of the fluctuating velocity vector, these equations are not, of course, closed, since the nonlinearity of the Navier–Stokes equations introduces in each equation the next order moment. Hence, they cannot be solved, even less than the Navier–Stokes equations, which are at least closed and, with suitable boundary conditions, well posed. However, if we make certain scaling assumptions in various regions in each physical problem, it is possible to determine the sizes of the various terms in the equations, and in many cases of interest, one finds that many of these terms can be discarded. Often there is a small parameter, which allows an orderly asymptotic analysis, so that a first order equation can be obtained, and then a second order correction, and so forth. Although still nothing can be solved, often the order can be reduced by one or two, and simple expressions can be found connecting the various elements, making clear what the physical mechanisms are.

This is still a good way to begin, even if we intend to proceed to the construction of a low-dimensional model. The latter is not immune to scaling arguments developed by order-of-magnitude analysis, and knowing the scaling before beginning often suggests possible simplifications, as well as appropriate normalizations.

On rigid, impermeable surfaces, $\tilde{u}_i = 0$, the so-called no-slip condition, is generally applied as a boundary condition. This is a reflection of the fact that molecules of the fluid are trapped among the molecules adsorbed on the surface, and by the time they leave again, they have forgotten all directional information. Every few years, someone claims to have found a surface treatment that will violate the no-slip condition, but close examination has always shown these claims to be mistaken or fraudulent. From the point of view of physical chemistry, there is no reason to expect that the no-slip condition would ever be violated at normal temperatures and pressures.

In fluid mechanics, we often discuss inviscid fluids, things that exist only in our imaginations. An inviscid fluid is not required to satisfy the no-slip condition, and in fact it cannot, since the order of the equations is reduced by one when the kinematic viscosity v is set equal to zero. The usual boundary condition for an inviscid fluid is the no-penetration condition, $\tilde{u}_i n_i = 0$, where n_i is the normal vector on the surface. We mention this here, because many real fluids have very low viscosity, and away from the boundary the inviscid solution provides a good approximation. Non-dimensionalization of the Navier–Stokes equations indicates that the viscous terms are small. However, the real fluid must meet the no-slip condition, and there is a thin layer near the surface in which the tangential component is brought to zero; this is the boundary layer. The viscous terms $v \tilde{u}_{i,jj}$ are made important again in the equations by the introduction of a new, much smaller, length scale normal to the wall, within the boundary layer.

In unbounded flows we can consider turbulence of finite energy, more-or-less confined to a region of space. As we see later, turbulence is separated from irrotational fluid by a thin interface (with thickness of order η), and outside this interface the irrotational fluid is disturbed, but only by larger scales. Analyses have been done of the asymptotic behavior of the variables as they approach infinity (e.g. [33]), and these can be used as boundary conditions. However, this situation does not match many experiments. Such turbulence, having no energy source, must decay. Experimentally, we are more interested in turbulent flows that have a source of energy, such as uniform inflow at one boundary. In the case of a wind tunnel this uniform inflow would be provided by a fan, and the flow would be conditioned by passage through various honeycombs and screens, to remove all traces of disturbances. A boundary layer can then be formed by forcing the stream to flow past a plate, or a wake can be formed by holding an obstacle in the flow. In directions normal to the mean flow, the disturbance introduced (wake or boundary layer) will go to zero at infinity (and its behavior can be predicted by asymptotic analysis); in the downstream direction, the flow will leave the region, taking the turbulence with it. In such a flow work is done on the fluid in flowing past the obstacle or the plate, and this work appears in the turbulence, and ultimately in heat. The intensity of the turbulence can be steady in time.

Turbulent flows contain motions with a broad range of scales. The scale at which most of the energy resides is called ℓ, the integral scale of the turbulence; that at which dissipation takes place, the smallest scale in the flow, is the Kolmogorov microscale $\eta = (\nu^3/\epsilon)^{1/4}$. By using the expression $\epsilon = u^3/\ell$ (which we shall discuss in Section 2.4 below), we find that $\ell/\eta = R_\ell^{3/4}$, as claimed earlier. The higher the Reynolds number, the broader this range of scales. These scales have, as we shall find below, greater or lesser structure depending on the inflow and boundary conditions and geometry. It is usually the energy-containing scales that exhibit the most evident structure, and these are called coherent structures.

2.2 Flows with coherent structures

Figure 2.1 is taken from the notebooks of Leonardo da Vinci. On the right, he has sketched the wakes formed on the surface of flowing water behind obstacles. The structure of these wakes is dominated by a pair of counter-rotating vortices. Both the wakes and the vortices are turbulent, so that the whole is unsteady and does not repeat in detail, yet the pair of counter-rotating vortices is always there in more-or-less the same place, and maintains roughly the same size.

Structures like this are called *coherent structures*, and many turbulent flows contain them. We will see below that the relative rôle they play in the flow is highly variable, ranging from dominant, as in the flow sketched by Leonardo, to nearly undetectable, depending on flow geometry and history.

To avoid confusion in later chapters, we must make clear a usage which is nearly universal in fluid mechanics. Most of the open flows that have been studied in detail, and which we will use as examples, were initially produced in wind tunnels: these are the jets, wakes, and shear layers. They are thought of as entering the philosophical domain through a left-hand, or inflow, boundary and leaving through a right-hand, or outflow, boundary, as they

Figure 2.1 Leonardo da Vinci: seated man and water studies. Royal Library, Windsor Castle; Leoni Volume (12579).

did in the wind tunnel. These flows are steady on the average, so that the overall structure is not changing, although instantaneously unsteady, so that details do change. The statistics of the flows vary rapidly from top to bottom, and slowly from left to right. The scale describing the changes from top to bottom is small relative to the scale describing the changes from left to right (the first is perhaps a few percent of the latter, and the ratio can be used as an expansion parameter – these are called almost-parallel flows). Such flows obey a simple similarity, so that, when expressed in terms of local scales of length and velocity, the structure is similar at every cross section. That much is clear.

We suggested that there might be a source of confusion. This might arise because, universally among fluid mechanicians, one also speaks as though *time* runs from left to right. That is, one thinks of the initial state of a material region as being its state when it entered the philosophical domain through the inflow boundary, and one follows its evolution as it crosses the domain from left to right, carried by the mean flow, seeing it expire as it leaves the field at the outflow boundary. In fact, in many computations, the real problem (described above) is replaced by a problem that is statistically homogeneous from left to right, but is evolving in time. The presumption is that the small statistical inhomogeneity from left to right in the real problem is dynamically irrelevant, and that the two problems are equivalent. For many years there were arguments in the fluid mechanical literature about whether this was so, but all test cases that were compared supported the contention that they were equivalent. Hence the universality of this way of thinking among people in fluid mechanics. There may yet, of course, be found ways in which they are not quite equivalent, and below we shall try to be meticulous in avoiding this way of speaking. However, it is nearly impossible to discuss the physics of these flows without taking into consideration the fact that material elements are convected by the mean motion from left to right, and evolve as they do so, and hence may be found in their youth near the inflow boundary, and in their old age near the outflow boundary. This is a Lagrangian point of view, following the history of the same material region, the same lump of matter, rather than sitting at a point in laboratory coordinates and letting matter be convected past.

In contrast, the dynamical systems approach which we develop in this book typically implies an Eulerian viewpoint, the state of the fluid at a given time being specified *everywhere* in the spatial domain of interest by a single point in a suitable phase space. As the dynamical system evolves, its solution describes a path or orbit in this phase space, each point of which corresponds to a new velocity field in the physical domain. An explicit relation between physical space and phase space is often provided by a modal decomposition. Using this, the *initial condition for the dynamical system* describing a fluid flow is given by the flow conditions throughout the domain and not only at the inflow boundary.

We need a word here about mean values. There are many possible means: time, space, ensemble, phase, and so forth. From a theoretical point of view the ensemble mean involves the fewest complications. Simply imagine performing the same experiment many times, under superficially identical conditions. Imagine, for example, a vertical wind tunnel, producing a horizontally homogeneous turbulent flow into which is released every few seconds a tiny particle, which is carried up by the flow, its position being photographed at closely spaced intervals [349]. Each recorded trajectory is a realization, and the collection is an ensemble. Something approaching 1000 realizations are necessary for second order statistics to be within 10%. Mathematically there are niceties which need not concern us here. Unless we specify otherwise, we will usually be thinking of ensemble means.

Time and space averages make sense only if the quantity being averaged is statistically stationary or homogeneous, respectively. That is, for a time average to make sense, it is essential that it be impossible to tell from the statistics what time it is. There must be no overall development of the statistics as time runs on. Similar statements can be made for a space average, in whatever direction the average is taken. A space average, of course, can be taken in one, two, or three dimensions, in which case the property being averaged must be homogeneous in one, two, or three dimensions. In practical situations, of course, nothing is ever completely stationary, or completely homogeneous, nor can averages be taken over infinite intervals. Then one is faced with a two-variable problem. Take a time average, for example. There are two time scales – a time necessary for the average to converge to a required accuracy, and a time characteristic of the non-stationarity of the process. One must be small relative to the other for the approximation to be satisfactory. This is a process which is very familiar in physics in, say, the definition of density in an inhomogeneous medium. Density is defined as an average over a volume; this volume must be large enough to contain enough molecules to give a sufficiently accurate value of the density, while being small enough not to average large-scale inhomogeneities of the field. The same process is at work when the eye interprets a half-tone photograph of a face in a newspaper. The eye must see enough benday dots in a given region to give a reliable local value of the gray tone, while not averaging out the features of the face. This sort of consideration has been with us for a long time – it is involved, for example,

Figure 2.2 Side view of a mixing layer between two streams with equal densities, velocity ratio $U_2/U_1 = 0.38$, and Reynolds number $(U_1 - U_2)L/\nu \approx 10^6$. From [49]; photo courtesy A. Roshko.

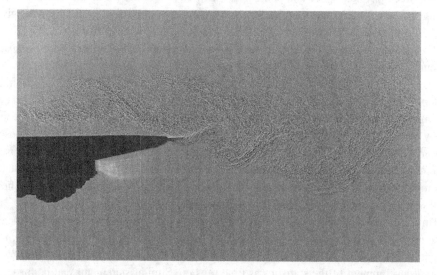

Figure 2.3 Supersonic mixing layer. NASA Langley Research Center; photo courtesy T. Gatski.

in the distinction between "weather" and "turbulence" in the atmosphere, and brings up eminently arguable questions about the existence of a gap in the spectrum of atmospheric motions justifying averaging over one scale but not over the other. To avoid all these sticky questions, we eschew space and time averages where possible.

Examining pictures of various turbulent flows, we discover that the proportion of organized and disorganized turbulence in each flow is different. For example, if we look at mixing layers that are allowed to develop from undisturbed conditions at the left-hand boundary (with only thin laminar boundary layers on the splitter plate; see Figure 2.2), we find that there is an energetically large organized component, which only relatively slowly becomes three-dimensional and disorganized as the material region moves to the right, although the nearly two-dimensional organized structures have from their entry on the left a stochastic component, so that their occurrence across the field is not precisely periodic, nor are their strengths equal. On the other hand, if we examine a mixing layer evolving from quite disturbed entry conditions (Figure 2.3), with a thick, turbulent boundary layer on

Figure 2.4 Jet. See text for description. Photo copyright Imperial College, Department of Aeronautics. Ref. PB3/8; courtesy P. Bradshaw.

the splitter plate, we find that the proportion of the organized component is considerably less – although the organized component is still visible, it is no longer dominant. Thus, in the same type of flow, we find that the entry conditions change the relative strength of the organized and disorganized components. (In Section 2.3 we describe how pictures such as Figures 2.2 and 2.3 are obtained.)

Considering a flow with a different geometry, such as a jet, we find a similar contrast. If the entry conditions at the orifice are undisturbed, as in Figure 2.4, with only thin laminar boundary layers on the inner surface of the nozzle, then there is found near the entry point a laminar instability which becomes three-dimensional and undergoes transition as it is carried by the mean flow from left to right, leaving in the turbulence near the outflow boundary the remnants of the instability structure. On the other hand, if the boundary layers on the inside of the nozzle are thick and turbulent at entry through the inflow boundary (Figure 2.5), no laminar instability is visible there and there are no visible organized remnants in the turbulent motion near the outflow boundary (although organized components can be found using statistical techniques, which we discuss more fully in later chapters). More than this, however, there is evidently a substantial difference between this flow and the mixing layer. Even in the undisturbed state, the organized component of this flow is a very great deal weaker than that in the mixing layer under the same circumstances. With relatively clean inflow conditions, it is barely discernible; when the inflow conditions are disturbed, the organized structure becomes nearly undetectable. Thus, we can conclude that different flows, even under similar conditions, exhibit different relative strengths of organized and disorganized components.

We must pause for a moment to discuss what we mean by instabilities. Many flows are unstable in certain parameter ranges; that is, small disturbances grow and ultimately change the flow. The initial growth of the small disturbances may be exponential, and

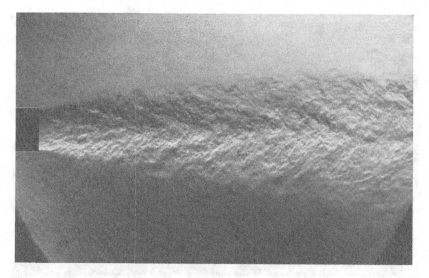

Figure 2.5 Jet. See text for description. Photo copyright Imperial College, Department of Aeronautics. Ref. PB3/7; courtesy P. Bradshaw.

the flow may be unstable to disturbances, no matter how small, if the parameter range is correct. In other cases, the flow may be stable to disturbances below some threshold, while disturbances above that threshold grow. Such behavior may be interpreted in terms of bifurcations of various sorts from steady or time-periodic solutions of the Navier–Stokes equations that occur as a parameter is varied. The growing disturbances usually approach another attractor in the phase space of the system, where they settle down, resulting in a new flow. In this state, they would hardly be referred to as instabilities, since they are no longer unstable. However, in fluid mechanics, even in this mature, energetically steady state, they are often loosely referred to as instabilities, since they grew as such on the initial flow, and caused it to evolve to the new state.

Many of the most studied instabilities of laminar flows are initially two-dimensional. For example, the Tollmien–Schlichting instability (see [97]) of the laminar boundary layer consists of disturbances in the form of two-dimensional waves with lines of constant phase parallel to the wall and perpendicular to the free stream. As this disturbance is carried downstream, it grows in amplitude; nonlinearities become important, and the waves become three-dimensional, the structures becoming more and more complex as the material is carried further and further downstream, until finally it is turbulent. The regions that are turbulent do not initially appear uniformly across the flow, but in patches, which spread until the entire flow is turbulent. Again, this must be seen as happening as a material region is carried from left to right; at any instant, at any given cross section, on a line across the flow, there will be intervals of turbulence and intervals of various stages of development of the laminar instability.

Another important case is the Kelvin–Helmholtz instability of a plane mixing or shear layer (cf. [203]; also see [250,311]). Again, this consists, near the inflow boundary, of two-dimensional waves with lines of constant phase parallel to the plane of the mixing layer and perpendicular to the free stream. As they are carried from left to right, the waves grow and roll up and nonlinearities become important. Three-dimensional structures called braids develop, lying in the streamwise direction, connecting the cross-stream rolled-up waves in the plane of the mixing layer (see Figure 2.12, below). Sufficiently far downstream this flow too becomes turbulent, but as we have seen above, it retains the structure of the laminar instability that appeared near the inflow boundary. In the jet with undisturbed inflow conditions, the Kelvin–Helmholtz instability arises on the shear layer formed as the jet leaves the orifice. Hence, this instability, in a slightly different cylindrical geometry, is also responsible for transition of the jet to turbulent flow.

In turbulence we are usually less interested in the initial phase of exponential growth of the instability than we are in the later development of nonlinearity and structure associated with the new attractor, since it is this long-term structure that gives texture to a turbulent flow.

So far it seems that the organized structures, to the extent that they are present in a turbulent flow, may be the remnants of some large-scale instability of the laminar flow near the inflow boundary. We shall show, however, that the turbulent flow itself is subject to large-scale instabilities, distinct from those to which the laminar flow is subject; these instabilities of the turbulent flow also grow, become nonlinear and settle down on an attractor, producing coherent structures that may be different from those produced by the instabilities of the laminar flow.

To explore this possibility, let us look first at the wake of a circular cylinder oriented across the stream. The precise cross section of the body has only a secondary influence on the form of the wake; one speaks of the wake of a *bluff body*, meaning any body about which the flow separates, because all such wakes are superficially similar in structure. The laminar flow separates as it passes the widest part of the cylinder (the precise location of the separation point depending on the state – laminar or turbulent – of the flow on the front of the cylinder), and this laminar shear layer is subject to a Kelvin–Helmholtz instability. The shear layers on the two sides of the cylinder interact, and organized vortex shedding results. The resulting flow is subject to smaller-scale instabilities, and becomes turbulent. If the wake is visualized close to the cylinder (Figure 2.6(a)), we see one kind of organized structure, which evidently is the remnant of the initial instability of the laminar flow around the cylinder, the vortex shedding. However, if the wake is visualized farther from the body (Figure 2.6(b) and (c)), we see somewhat different, though similar, organized structures.

We hypothesize that the organized structures present close to the cylinder have decayed as the material has been carried across the field, and new, distinct, though similar, structures have arisen in the material when it is far from the cylinder. To test this conjecture, we can examine the wake of another body, a flat plate also set normal to the stream, which is sufficiently porous that the laminar shear layers formed at the edges are not unstable, and there is no vortex shedding (Figure 2.7). Of course, the resulting flow is still unstable to small-scale modes, and near the plate we find a wake which has small-scale turbulence, but no organized structures. However, after the material has been carried some distance downstream, organized structures spontaneously appear in this wake, which had none near the plate. These structures strongly resemble those that appeared far from the cylinder in the former case, in which very different structures appeared close to the cylinder. It

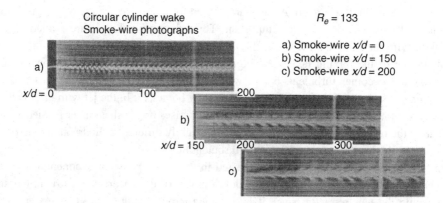

Figure 2.6 Circular cylinder wake at $R_e = 133$; photo courtesy of Roshko, Cimbala, and Nagib.

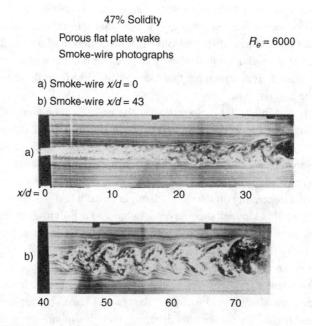

Figure 2.7 Wake behind a 47% solidity porous flat plate, $R_e = 6000$; photo courtesy Roshko, Cimbala, and Nagib.

seems reasonable to conclude that these organized structures, which appear in the wake far from the body in either case, represent a type of instability of the developed turbulent flow, drawing on the mean velocity profile to obtain energy, and giving up energy to the turbulent transport. Their precise form then, will be a function of the mean velocity profile, as well as of the distribution of the turbulent stresses.

The energy budget for such organized structures is complex, because they have reached a nonlinear energetic equilibrium. In addition, these organized structures usually involve velocities at right angles to the direction of the mean motion, and they therefore transport matter, mean and fluctuating momentum, and energy in that direction. This transport in turn modifies the mean velocity profile, as well as the distribution of turbulent stresses. If the organized structures are regarded as resulting from instabilities of a turbulent flow without organized structures, to investigate these instabilities some way must be found to predict the form of the hypothetical structure-free turbulent mean velocity profile and turbulent stress profile. We return to these ideas later when we discuss prediction of these organized structures.

Hence, in any given situation we may expect to find organized structures, the relative strengths of which are a function of the inflow conditions, of the type of flow, and the spatial location in the flow (distance downstream). These structures may be remnants of instabilities of the laminar inflow, or they may have developed from instabilities of the turbulent flow itself.

If we restrict our attention to narrow, almost-parallel, two-dimensional shear flows (jets, wakes, mixing layers), we find that the organized structures occur with more-or-less the same orientation to and distance from the flow centerline each time; that is, their orientation and position in the cross-stream, inhomogeneous direction is largely fixed by the boundaries of the flow. In the streamwise, or homogeneous direction, the location is more random; the existence of one structure seems to suppress the presence of another, but as soon as we are sufficiently distant from one structure, another one appears.

Organized structures are much more difficult to find in homogeneous flows, mostly because it is not clear where to look. There is nothing in the flow to pin them down to a particular location. Consider, for example, homogeneous shear flow. In direct numerical simulations Moin and his co-workers [242, 310] have found hairpin vortices throughout the flow. (The simulations of these homogeneous flows use periodic boundary conditions in one or more directions, but the period is large compared to the turbulent integral scale, and is felt to be irrelevant to the turbulent dynamics.) The orientation of these hairpin vortices is determined by the direction of the mean velocity shear, but their location in three dimensions is random. They are thought to arise from a type of nonlinear instability connected with the same instability that produces Langmuir cells in the ocean surface mixed layer (see [208] for a review), or that produces streamwise rolls in a turbulent boundary layer [209]. Both of these flows are inhomogeneous in the direction normal to the surface (the free surface in the case of the ocean, the wall in the case of the boundary layer), and hence the location of the organized structures is fixed by this inhomogeneity. The instability mechanism depends on transverse vorticity associated with the mean shear being deflected vertically (in the direction of the mean gradient), and then being transported in the streamwise direction at different rates at different distances from the surface, due to the gradient in the mean Lagrangian transport velocity, resulting in a stretching and intensification of the streamwise component of vorticity. This interaction is pictured in Figure 2.8. The fluctuations responsible for the vorticity deflections in the oceanic case are due to the random field of surface waves, while in the boundary layer they are due to the turbulence, but both produce a fluctuating Lagrangian transport. This nonlinear instability mechanism is not dependent on the presence of a surface (free or otherwise), and will work as well in the interior of the fluid. All that is required is a gradient in the mean Lagrangian transport velocity.

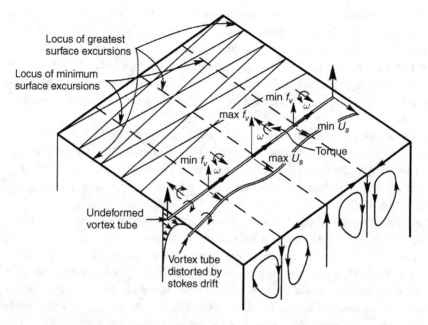

Figure 2.8 Sketch of interactions responsible for development of Langmuir circulation in the ocean surface mixed layer. From the work of Craik and Leibovich, cf. [208].

We may probably conclude that *any* turbulent flow will have a more-or-less organized component, the strength of which in a particular material region will be a complex function of the large-scale geometry of the flow, the age of the material region, and the inflow conditions for the material region, and which may, depending on the situation, be random in orientation and location in up to three dimensions.

The question whether it is necessary to take into account the presence of organized structures in a turbulent flow, when considering the dynamical behavior of the flow as a whole, has different answers in different situations. When we study a turbulent flow for practical purposes, we are often interested in little more than the Reynolds stress, $-\rho\langle u_i u_j\rangle$, since this is the mean momentum flux due to the turbulent fluctuations, and hence is the actual stress generated by the turbulent part of the flow. This is not a very sophisticated property of a turbulent flow; it is uninfluenced by subtle changes in the structure of the flow, and tells us little about the flow. Often, if the coherent structures in the flow scale in the same way as the disorganized motions, they can all be lumped together and the evident differences ignored. For example, in the turbulent mixing layer, if both the turbulence and the organized structures in a material region moving with the mean motion begin growing at the same place, say the inflow boundary, and have been growing together since that point as they have moved across the domain, then they will scale in the same way, and need not be considered separately. See [328], for example, where the growth, mean velocity profile, and profiles of Reynolds stress and component energies in a compressible turbulent mixing layer were successfully predicted using a second order turbulence model. The model completely ignored the presence of coherent structures, simply lumping their capacity to transport momentum across stream in with that of the smaller-scale turbulence.

However, if matter in turbulent motion, carried by a mean flow, produced under one set of circumstances and consequently having a given set of scales, is convected into a different set of conditions (a new strain rate field, for example) it is quite likely that the new conditions will give rise to a new instability of the turbulent profiles, giving rise in its turn to a new organized structure. Until the scales have had a chance to equilibrate, we will have a situation consisting of background disorganized turbulence with one set of scales, on which are growing organized structures with a different set of scales. The transport of momentum produced by this combination will be quite difficult to predict unless explicit account is taken of the organized structures. We must also consider that the matter in turbulent motion, before it was convected into the different set of conditions, may already have had an organized structure with which it was in equilibrium, and which will now compete with the new one in transporting momentum.

Historically, until approximately the time of the Second World War, work in turbulence emphasized a statistical approach, at the expense of structure. Liu [213] has documented the first appearance of the idea of coherent structures sometime around the outbreak of the war. It was probably first articulated by Liepmann [211], and was thoroughly exploited by Townsend [371], whose book shows how the presence of coherent structures in classical shear flows can be inferred from statistical evidence. Briefly, if a double structure is postulated, consisting of large-scale coherent structures of a particular form, and small-scale relatively-disorganized turbulence, then the form of the correlation functions can be predicted. By comparing this predicted form with measurements, suitable velocity and length scales can be chosen for the coherent structure models, and relative strengths assigned. This analysis was, of course, not foolproof, since not all components had been measured, and different structures can give rise to certain components of similar structure, but it was very instructive, and the conclusions drawn appear not to have been far from the truth.

For nearly thirty years the realization that turbulent flows contained coherent structures had relatively little impact on the field, workers with only rare exceptions continuing to use a statistical approach that ignored their presence. However, about 1970, at roughly the time of first appearance at meetings of the work of Brown and Roshko [60, 61], it became very popular to consider the rôle of coherent structures, and has remained so until the present time.

Initially Lumley and his co-workers took an observational point of view toward coherent structures, like naturalists observing a new insect. That is, they focused on the problem of unambiguous objective identification of the presence of a coherent structure in a turbulent flow, and ignored the dynamical behavior of the coherent structures; in doing so they ignored the questions of their origin and rôle. It was at this stage that the Karhunen–Loève, or proper orthogonal decomposition (discussed in the next chapter) was introduced as an objective way of identifying coherent structures in shear flows.

Not all workers in the field approached the identification question in the same way; Hussain, see for example Bridges *et al.* [58], has always felt that nothing should be identified as a coherent structure that does not possess certain dynamical properties, e.g. the presence of vorticity. Other workers, for example G. Brown, favored kinematic, but

somewhat subjective, detectors; that is, they selected a criterion for the presence of a coherent structure based on their conception of the dynamical behavior of the structure, and then
studied the statistics of the structures that met that criterion. Lumley [223] discusses the pitfalls of this approach, which runs the risk of detecting the presence of non-existent objects.
In [224] extensive discussion can be found that gives a measure of the diversity of opinion
as to the proper definition of coherent structure in the community at large.

2.3 Detection of coherent structures

Flow visualization has a bad reputation, like a disreputable uncle. On the one hand, of
course, it permits one to see things in flows that are otherwise invisible; on the other, it can
be misleading. Under the auspices of the National Committee for Fluid Mechanics Films,
Steve Kline made a film called *Flow Visualization* which is still distributed by Encyclope-
dia Britannica Films. This is probably the best introduction to flow visualization. We will
first discuss the various techniques, and later describe some of the pitfalls.

Schlieren and shadowgraph techniques are non-intrusive, and give excellent pictures in
flows with density variations, which is to say, compressible flows and flows involving mixing of fluids of differing densities. Figure 2.2 represents the latter situation, but with a subtle
twist: one stream being pure nitrogen and the other a mixture of helium and argon with net
density equal to that of nitrogen. It was visualized using shadowgraph techniques: the film
being placed directly behind the glass wall of the test section and the flow illuminated by a
spark of a few milliseconds duration (see [49, 61] – at high pressures the dissimilar gases
provided the necessary difference in refractive indices). Both Schlieren and shadowgraph
techniques involve refraction of light by density gradients. The pictures obtained, consequently, represent the first derivative of the field in the case of Schlieren, or the second
derivative in the case of shadowgraph, and hence strongly emphasize the small scales – the
spectrum is multiplied by the square or the fourth power of the wavenumber respectively.
The energy of the small-scale component in Figure 2.2 is consequently considerably less
than it appears, and the coherent component is relatively much stronger.

Smoke wires and hydrogen bubbles are similar techniques that are used in air and water
respectively. In the former, a wire placed across the direction of the mean flow is coated
with a volatile oil; the wire is then heated, and smoke is generated, which trails in fila-
ments downstream. Figures 2.6 and 2.7 were made using this technique. In the case of
hydrogen bubbles, a wire is placed across the flow in water, and a voltage applied, which
causes electrolysis of the water, generating fine hydrogen bubbles which trail downstream
from the wire. In both cases, a flat sheet with striations is created, which can be distorted
by the flow downstream, as in Figure 2.14, below. One of the difficulties with this type
of visualization is the possibility that the smoke or bubbles may be captured by a feature
of the flow that subsequently decays, leaving the tracer (smoke or bubbles) to mark the
site of a feature which no longer exists. For example, in the same experiments that produced Figures 2.6 and 2.7 a smoke wire was placed immediately behind a cylinder at low
Reynolds number (see Figure 2.9). A beautiful Kármán vortex street was obtained (part
(a)), which appeared to persist far downstream. However, by placing the wire across the

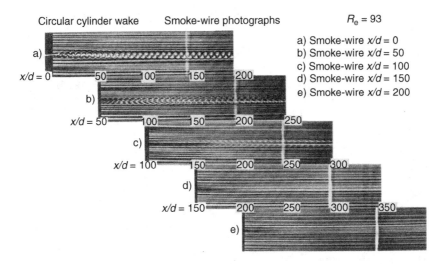

Figure 2.9 Circular cylinder wake at a Reynolds number of 93; photo courtesy of Roshko, Cimbala, and Nagib.

flow at increasing distances downstream ((b)–(e)), it was discovered that the vortices in the street had decayed at only moderate distances from the cylinder, and the smoke that had been trapped in the vortices remained to indicate their former positions, creating the appearance of lasting vortices.

In both the smoke wire and hydrogen bubble techniques, particles (bubbles) are introduced into the flow. We must ask whether they faithfully follow the flow, since otherwise this could be a source of error. In fact, this is a fairly easy calculation. Particle inertia can be estimated, and from this one can estimate a cut-off frequency below which the particle may be expected to follow the flow. The spectrum of turbulent fluctuations is known. In this way it is found that the smoke particles never present a problem, and a size limitation can be placed on the bubbles. The wire size and voltage can be adjusted to produce bubbles of the requisite smallness, see [222].

A related technique is the injection of dye. In water, colored milk is often used; in air, smoke generated in various ways. A general problem with this type of visualization is the possibility that the tracer may not be going where the operator thinks it is, or that its being there may not mean what the operator thinks it does, or that the shapes seen do not have the dynamical meaning the operator attaches to them; that is, the operator draws certain dynamical conclusions about the flow based on what he or she sees; this usually involves assumptions about why the tracer arrived where it did, and what it means. In a number of cases this has been found to be subtly misleading.

For example, Figure 2.10 shows a jet with a wide splitter plate extending completely across the diameter. It was assumed that such a splitter plate would completely decouple the two sides of the jet. Yet, the reader will note that the vortex rings are still in phase on the two sides. At this low Mach number of 0.1, they have been locked together by pressure fluctuations propagating upstream on one side of the splitter plate, across to the other side in the plenum, and downstream on the other. In a different situation, Figure 2.11

Figure 2.10 Jet with splitter plate. Photo copyright Imperial College, Department of Aeronautics. Ref. No. PB51/18; courtesy P. Bradshaw.

Figure 2.11 Visualization of surface stress near a wing root. Photo copyright Imperial College, Department of Aeronautics. Ref. No. PB2111/5; courtesy P. Bradshaw.

shows visualization of the surface stress around a wing root, by applying pigmented oil to the surface. Here the direction of the surface stress is quite different from the direction of the flow outside the boundary layer. This effect is much greater in the laminar boundary layer than it is in the turbulent boundary layer, which is more strongly mixed from top to bottom. In Figure 2.11, the boundary layer undergoes transition from laminar to turbulent flow at the midpoint, and the direction of the surface stress changes abruptly, although the direction of the flow outside the boundary layer does not.

Flows seeded with smoke, bubbles, dye, or other reflective particles can be selectively illuminated in various ways to probe particular features. Thin sheets of intense laser light, for example, reveal two-dimensional cross sections of three-dimensional structures.

All these techniques have been used to make coherent structures visible. But beyond showing that there are indeed large coherent fluid motions present, it is probably fair to say that not much has been learned directly about the dynamics of these structures from visualization.

In addition, there are various techniques for detecting the presence of coherent structures from the signals produced by various probes introduced into the flow, which essentially measure quantities at a single point. Hot wire and hot film probes are usually used. These make use of very small heated platinum wires or films, with dimensions usually considerably less than a millimeter. The heat removed from the film is dependent on the flow velocity of the liquid. The films or wires are smaller than the smallest disturbances in the flow (of the order of the Kolmogorov microscale), and the probes are oriented so that the flow disturbed by the probe is swept downstream, and does not influence the part of the flow where the measurement is being made. Hot wire and hot film probes can be arranged in what are called rakes of five or ten probes, giving simultaneous values of the velocity at five or ten closely spaced points. Now the operator has to decide what spatio-temporal properties a coherent structure might possess, arrange the probes in a suitable geometrical array, and treat the signals appropriately, so that the array will indicate when a coherent structure is passing by, according to the criterion adopted. Supplementary measurements can then be made relative to the time at which the coherent structure was identified. These supplementary measurements can be given various statistical treatments using the time of identification as a datum, to produce conditional averages, for example.

Laser–Doppler velocimeters, which measure fluid velocities in a non-intrusive manner by scattering coherent light from particles in the flow, can also be used in this way. A disadvantage is the difficulty of measuring simultaneously more than two signals, although in an incompressible flow with sufficient spatial information the third velocity component can be deduced from the other two by appeal to the divergence-free condition (2.2).

These techniques have been very useful, and have produced a great deal of information. The only difficulty with them is their dependence on the criterion invented by the operator. There is nothing objective about this criterion; it depends intimately on the dynamical picture that the operator has in mind of the coherent structure. Of course, one tries to pick something that is unique to the coherent structure, so that the detection system will not be triggered falsely. Since the purpose of the measurement is to discover the dynamical behavior of the coherent structures, however, this is a little difficult, for the criterion must be set at a time when the operator knows relatively little about the structure.

2.4 The mixing layer

The low-dimensional models developed in this book generally illuminate only some aspects of a particular turbulent flow: for example, the intermittent behavior of the coherent structures in the wall region of a turbulent boundary layer. Such models must be placed in a larger setting, representing the rest of the flow. For example, what is the general size and shape of a turbulent boundary layer? How does it scale? What are the characteristics of the wall region? This kind of information is useful in establishing the relatively local low-dimensional model, setting (for example, in the case of the boundary layer) boundary conditions at the wall, asymptotic behavior far from the wall, scaling and so forth. In this and the next section we develop this kind of information for specific flows, using traditional techniques of turbulence theory. Although we apply these techniques only to two flows, the way in which the approach can be applied more generally will be clear.

There are several turbulent flows that share some simple characteristics. These are the almost-parallel, boundary-free, flows: the wake, jet, and mixing layer (see [368]). We deal with the boundary layer in the next section. These are all *thin* flows, which take place in regions the transverse dimensions (say, ℓ) of which are small relative to their length (say, L), $\ell/L \ll 1$. We define ℓ and L more precisely using spatial derivatives in the two directions later. Wakes and jets, of course, may be either plane or axisymmetric. The mixing layer can be plane or curved in the direction normal to the mean motion; however, the radius of curvature is nearly always large compared to the thickness of the layer, so that the layer does not differ appreciably in structure from the plane layer, which is taken as the standard. We take the mixing layer as an example, in particular the plane mixing layer; the approach can be applied equally well to the other flows, and the findings will be related. We give a brief introduction here to some of the basic things that may be said about the mixing layer without reference to coherent structures.

Let us pause for a moment and consider the physics of the mixing layer. On the high-speed side it is like the edge of a wake, while on the low-speed side it is like the edge of a jet. In a two-dimensional wake the mean velocity normal to the centerline of the wake vanishes at infinity. In a two-dimensional jet, the mean velocity normal to the centerline of the jet does *not* vanish at infinity, and hence neither do the volume nor mass fluxes, which are all proportional in a constant density flow. In an axisymmetric jet, the velocity normal to the centerline vanishes at infinity, but at a rate proportional to $1/r$, where r is the distance from the centerline; hence, the *volume flux*, or *mass flux*, integrated on cylindrical surfaces concentric with the axis does *not* vanish at infinity. This non-vanishing volume or mass flux at infinity, referred to as entrainment from infinity, is unique to the jet and jet-like flows; the jet is entraining fluid from the environment, engulfing it, and dragging it along with the jet fluid. It is this phenomenon that makes jets useful in many practical devices, such as jet pumps used in deep wells.

All flows with free boundaries entrain surrounding fluid; even the wake has a net volume or mass flux toward the axis near the outer edges of the flow – outside fluid is being mixed into the flow, and it is in this way that the flow spreads. However, in the wake, unlike the jet, the mass flux drops to zero at infinity. There is no net mass flux from infinity into the wake. The mixing layer entrains fluid from infinity on the jet-like side, and does not entrain fluid from infinity on the wake-like side. This asymmetrical entrainment from infinity requires that the mean velocity profile (or rather, its gradient) be asymmetric also. Many of the features of the mixing layer that are thought to be associated with coherent structures (e.g. the asymmetry of entrainment from infinity at the two sides of the mean velocity profile) are in fact mandated by the general asymmetry of the situation, and must be produced by whatever physical mechanism is responsible for transport of momentum, whether organized or disorganized.

In trying to understand turbulent flows, numerical solution of the Navier–Stokes equations has only relatively recently become possible. In any event, direct numerical simulation does not shed much light on mechanisms. Before the advent of this approach and the one considered in the present book, however, workers in turbulence usually considered the averaged equations for the mean velocity and the moments of the fluctuations, which (of course) are not closed (since the equation at order n contains a term of order $n + 1$ arising from the quadratic nonlinearity in the convective derivative). Since these equations could not be solved, it was necessary to find other ways of extracting information from them, preferably by making the weakest possible assumptions. Usually, this involved careful order-of-magnitude analyses of particular flows, making use of convenient characteristics peculiar to these flows. Here, the fact that this group of flows is almost parallel is useful. This analysis allowed identification of the most important terms, and often led to better physical understanding of the dynamics; this was ordinarily followed by assumptions of self-similarity (again, made likely by the long, thin geometry of this limited class of flows), which permitted further delineation of the mechanisms. Proceeding in this way, it is possible to obtain a great amount of rough structural information about many flows without actually solving a single equation. This basic information about the flow is necessary to provide a philosophical structure before carrying out any more refined analysis.

Returning to the plane mixing layer, we first carry out a careful analysis of the orders of magnitude of the various terms in the equations of motion. The layer is formed by the mixing of two streams; before these streams begin to mix at the inflow boundary, we imagine them to be separated by a "splitter plate" of zero thickness. We take x_1 in the streamwise direction, beginning from the end of the splitter plate, x_2 across the stream normal to the splitter plate, and x_3 normal to x_1 and x_2 along the edge of the splitter plate (to form a right-handed coordinate system). The mean velocity profile is specified by the component $U_1(x_1, x_2)$ in the streamwise direction; the spanwise component $U_3 \equiv 0$, but we shall see that the normal component U_2 does not vanish, because the flow is not perfectly parallel. We suppose that the mean flow is steady. A mixing layer is defined by $U_1(x_1, +\infty) = U_s, U_1(x_1, -\infty) = 0$, where U_s is not a function of x_1. All mean quantities are invariant in x_3, the spanwise direction. Of the two streams entering the mixing layer at $x_1 = 0$, one has a uniform profile $U_1(0, x_2) = U_s, x_2 > 0$, and the other has $U_1(0, x_2) = 0, x_2 < 0$. They have been kept separate for $x_1 < 0$, but are allowed to mix for $x_1 > 0$.

Note that the fluid from the high-speed stream, carrying high streamwise momentum, is being mixed transversely into the stationary fluid, bringing its speed up from zero; at the same time, the fluid from the stationary side, having zero momentum, is being mixed into the higher-speed flow, slowing it down. We expect this mixing process to be monotone; that is, we expect that the fraction of zero-momentum fluid will decrease monotonically with increasing x_2 (since it has had further to travel), and the fraction of high-speed fluid will be monotone in the other direction (for the same reason). To a first order approximation, the instantaneous flow consists of a coarse mixture of regions of high-speed fluid and regions of zero-momentum fluid. Hence, we expect that $U_{1,2} \geq 0$ everywhere, and that $0 \leq U_1 \leq U_s$.

Instantaneous pictures of turbulent flows (see Figures 2.1–2.7) make it clear that the turbulence is confined to regions of relatively limited extent with fairly sharp boundaries. The layer separating the turbulent fluid from the irrotational fluid outside is very thin, of the order of the Kolmogorov microscale $\eta = (\nu^3/\epsilon)^{1/4}$, the so-called viscous super-layer [87]. As we follow material elements of the flow on their journey downstream, the turbulent state gradually propagates into the irrotational state, so that increasing quantities of fluid which were irrotational at the inflow boundary become turbulent; this comes about by physical extrusion of pseudopodia of turbulent material into the irrotational fluid, and vice versa. This process takes place on all scales along the boundary between the turbulent and irrotational fluid, from the largest down to viscous scales of order η. At the smallest scales, viscous propagation of vorticity fills in the gaps. In this way, the extent of the turbulent region gradually spreads. Since the rate at which the largest-scale pseudopodia are extruded is of order u, the root-mean-square turbulent fluctuating velocity, this is also the rate of spreading of the turbulent region.

We are concerned in this section only with averages. Ordinarily, we speak of the *inter-mittency* γ, the average proportion of time that a point a given distance from the x_1-axis spends in turbulent fluid. The location of the turbulent/irrotational boundary is, of course, a stochastic process, and the boundary makes large excursions. A boundary to the average flow could perhaps be defined as roughly the point where $\gamma = 0.5$; beyond this point γ drops rapidly, although in principle it may not reach zero until $x_2 = \infty$. Of course, γ is an experimental quantity, a property of the flow, but its asymptotic behavior as $x_2 \to \infty$ is beyond experimental resolution; all we know about it comes from various statistical models of turbulent front structure. Since it is the transverse turbulent trans-port of streamwise momentum that is responsible for non-uniformity of the mean velocity profile, the rapid changes in the mean velocity profile take place where the turbulence is found, on average. We expect generally that γ and the mean velocity profile will approach their limits exponentially fast at the edges of the flow. We arbitrarily define the edges of the flow by taking the point where $(U_s - U)/U_s = 0.01$ and $U/U_s = 0.01$ (we could have used equally well $\gamma = 0.01$), and say that, roughly speaking, the turbulence, and the major variation in the mean velocity, is essentially confined to this region. We refer in what follows to "outside the flow," meaning at values of x_2 well outside these edges.

We begin by writing the instantaneous velocity as the sum of the mean and fluctuating velocities (we imagine ensemble averages for simplicity):

$$\tilde{u}_i = U_i + u_i, \tag{2.6}$$

where $\langle \tilde{u}_i \rangle = U_i$, $\langle u_i \rangle = 0$, and $\langle \cdot \rangle$ indicates an ensemble mean. The equation for the mean velocity can be obtained by substituting this decomposition in the Navier–Stokes equations (2.1) and averaging:

$$U_{i,j} U_j + \langle u_i u_j \rangle_{,j} = -P_{,i}/\rho + \nu U_{i,jj}. \tag{2.7}$$

Now, not all the terms in Equation (2.7) are of equal importance. Let us take $\partial(\cdot)/\partial x_1 = \mathcal{O}((\cdot)/L)$, while $\partial(\cdot)/\partial x_2 = \mathcal{O}((\cdot)/\ell)$, where $\ell/L \ll 1$; that is, we suppose that the flow is narrow, or slowly developing. We take $U_1 = \mathcal{O}(U_s)$. Here we mean by $A = \mathcal{O}(B)$ that A is no larger than B throughout the flow, although it might be much smaller.

First, we examine the averaged equation of continuity

$$U_{1,1} + U_{2,2} = 0, \tag{2.8}$$

and from it deduce the magnitude $\mathcal{O}(V_s)$ of U_2 :

$$U_s/L = V_s/\ell \tag{2.9}$$

or

$$V_s = U_s\ell/L. \tag{2.10}$$

This simply says that, if the flow is slowly developing, the transverse velocities must be small. In fact, if the flow is parallel ($U_1 = U_1(x_2)$, $L = \infty$) they must vanish.

We also need a magnitude for the turbulent quantities: we take $\langle u_i u_j \rangle = \mathcal{O}(u^2)$. The size of u relative to that of U_s will be determined later.

We are interested ultimately in the equation for U_1, the 1-component of Equation (2.7). We begin by determining the magnitude for the mean pressure term in that equation. Our assumptions above do not give us direct information about the size of this term. However, since we presumably know the variation of the mean pressure outside the turbulent part of the flow (this would form part of the boundary conditions of the flow), we should be interested in the variation of the pressure across the stream. Hence, we examine the 2-component of Equation (2.7) for the transverse mean velocity U_2, which describes the transport of cross-stream momentum. Assigning orders of magnitude as discussed above, we find that

$$-P_{,2}/\rho = \mathcal{O}(U_s^2\ell/L^2, u^2/\ell, vU_s/L\ell) \tag{2.11}$$

or

$$-P/\rho = -P_\infty/\rho + \mathcal{O}(U_s^2\ell^2/L^2, u^2, vU_s/L), \tag{2.12}$$

where we have integrated the order-of-magnitude equation with respect to $x_2 = \mathcal{O}(\ell)$, and P_∞ is the mean pressure outside the mixing layer. The expression $A = \mathcal{O}(B, C, D)$ means that A is no larger than B, C, or D, whichever of them is largest. At this point it is not possible to determine which is the largest, but it will not matter. The first term indicates the cross-stream pressure differences that would be induced by streamline curvature, the second those induced by the variations in turbulent intensity, and the third, those due to viscous stress. If we wish a simple situation, it is now clear that we must suppose that P_∞ is not a function of x_1, since otherwise the variation of P_∞ could induce any behavior desired in the mixing layer, making it grow or shrink at any desired rate.

If the expression (2.12) is now differentiated with respect to x_1 and substituted in the equation for U_1, we find that, for each term in (2.12), there is a term in the U_1 equation that is larger. Thus, the term $P_{,1}/\rho$ can be neglected. This is true of all almost-parallel flows – the streamwise pressure gradient, which is zero outside the flow, is also negligible within the flow. We can put this another way, which gives more physical insight, by saying that at high Reynolds number (so that the viscous term may be neglected) the cross-stream variation of the streamwise pressure gradient in any flow is negligible so long as the streamline curvature is small; a substantial gradient in pressure across the stream would be required to hold the flow on a curved path.

Considering now the magnitudes of the remaining terms in the equation for U_1, we find that both the inertial terms are of order U_s^2/L, the larger of the two turbulent Reynolds stress terms is of order u^2/ℓ, while the smaller is of order u^2/L; the larger of the two viscous terms is of order $\nu U_s/\ell^2$, and the smaller of order $\nu U_s/L^2$. The magnitude of the smaller Reynolds stress term, relative to the larger, is ℓ/L, while the magnitude of the smaller viscous term, relative to the larger is ℓ^2/L^2. Since $\ell/L \ll 1$, we are justified in neglecting the smaller Reynolds stress term, as well as the smaller viscous term, relative to the larger of each. We will presume, subject to later verification, that we can raise the Reynolds number sufficiently (decrease the viscosity sufficiently) to make the viscous terms negligible. We are now left with the two inertial terms, both of the same size, and the remaining Reynolds stress term. Thus, the inertial terms and the Reynolds stress term must be of the same size:

$$U_s^2/L = u^2/\ell \tag{2.13}$$

or

$$u^2/U_s^2 = \ell/L. \tag{2.14}$$

We can immediately draw an interesting conclusion from this: if the mixing layer is to develop in a self-similar way, even approximately, so that, rescaled, the statistical properties of the layer will look much the same at every stage of development, then u/U_s must not be a function of x_1; for, if it were, then the strength of the turbulence relative to that of the mean flow would differ at different cross sections, and the flow could not be self-similar. If u/U_s is not a function of x_1, then ℓ/L cannot be either. If we presume that the only possible length scale in the x_1-direction is x_1 itself, so that $L = x_1$, we have determined that ℓ must be proportional to x_1, or that the layer must grow linearly.

Whether or not the layer will grow in a self-similar way is, of course, a matter of conjecture. Self-similar behavior in general of all the nearly parallel flows has been called into question by the results of Marasli et al. [235] for the wake, which indicated that the two-dimensional turbulent wake remembered its initial conditions for a very long time (evidently beyond the range of the experiment, in any event). At the moment (1996) neither direct numerical simulations nor experiments can achieve sufficiently large downstream distances to seriously probe the question of whether self-similarity for the wake would asymptotically set in. In the case of the mixing layer, we can reason that the mixing layer growing from undisturbed inflow conditions, that is, from a uniform profile without turbulent fluctuations (something that can be approximately achieved in the laboratory) would have no other length or velocity scales introduced, and hence should be self-similar, and indeed it is found to be so experimentally. In this respect the mixing layer, beginning as it were from a point source, is simpler than the wake, which begins from a finite profile which must necessarily introduce other parameters.

We have not yet justified our neglect of the larger viscous term $\nu U_{1,22} = \mathcal{O}(\nu U_s/\ell^2)$ in the 1-component of Equation (2.7). The ratio of this viscous term to the remaining Reynolds stress term $\langle u_1 u_2 \rangle_{,2}$ is $(L/\ell)^{1/2}/R_\ell$, where $R_\ell = u\ell/\nu$, the Reynolds number of the turbulence. For this viscous term to have the same size as the largest term already neglected in the 1-component of Equation (2.7) (the smaller Reynolds stress term $\langle u_1 u_1 \rangle_{,1}$, of order ℓ/L relative to the larger Reynolds stress term), we need $(L/\ell)^{1/2}/R_\ell = \ell/L$, or $R_\ell = (L/\ell)^{3/2}$. Experimentally, it is found that $\ell/L = 6 \times 10^{-2}$. Hence, $R_\ell > 68$ is required to neglect the viscous terms. Since R_ℓ grows linearly with x_1, this is soon achieved. This suggests that there will be a region near the origin of the flow (the lip of the splitter plate, which separates the two streams before they begin to mix) which will be viscously dominated, rather than properly turbulent, and that for a truly turbulent flow we must look only at points well downstream of the point $L = 278\nu/U_s$, obtained from use of the value $\ell/L = 6 \times 10^{-2}$ and the associated value $u/U_s = \sqrt{\ell/L} = 0.245$. In air ($\nu = 15 \times 10^{-6}$ m²/s), with $U_s = 10$ m/s, this distance is only of order 0.4 mm. Hence, at distances of the order of a centimeter, the layer will be uninfluenced by viscosity, and described by Equation (2.15) below, and we might expect self-similarity.

In any event, our reduced equation for the mean profile, the 1-component of (2.7), keeping only the largest terms, has become

$$U_{1,1}U_1 + U_{1,2}U_2 + \langle u_1 u_2 \rangle_{,2} = 0. \tag{2.15}$$

If this equation were multiplied by the mean density, it would be the equation for the mean budget of streamwise momentum, so all the terms represent a contribution to that budget from some source. The first term describes advection in the streamwise direction by the streamwise mean velocity of the streamwise mean momentum; the second term describes advection in the cross-stream direction by the cross-stream mean velocity of the stream-wise mean momentum; and the third term describes mean advection in the cross-stream direction of streamwise fluctuating momentum by cross-stream fluctuating velocity. Hence, Equation (2.15) states that advection of mean momentum in the streamwise and cross-stream directions by the mean velocity is balanced by advection of fluctuating streamwise momentum by the fluctuating cross-stream velocity. The advection of fluctuating stream-wise momentum by the fluctuating streamwise velocity, $\langle u_1 u_1 \rangle_{,1}$ is negligible compared to the advection of mean streamwise momentum by the mean streamwise velocity.

The reader will note that we have not mentioned coherent structures, and so far everything we have said is equally applicable to any fluctuations, coherent or incoherent.

By looking for integral invariants, we may obtain more information about this flow. Equation (2.15) may be rewritten in terms of $U_1 - U_s$, since U_s is not a function of x_1. We may also make use of the continuity Equation (2.8) to obtain

$$[U_1(U_1 - U_s)]_{,1} + [U_2(U_1 - U_s)]_{,2} + \langle u_1 u_2 \rangle_{,2} = 0. \tag{2.16}$$

This equation can now be integrated across stream, since U_1 vanishes as $x_2 \to -\infty$, and $U_1 - U_s$ vanishes as $x_2 \to +\infty$. We also have that $\langle u_1 u_2 \rangle \to 0$ as $|x_2| \to \infty$, since there is no turbulence outside the mixing layer (i.e. for $|x_2| \gg \ell$). We are left with

$$\frac{\partial}{\partial x_1} \int_{-\infty}^{+\infty} U_1(U_1 - U_s)\,dx_2 + U_s U_2 \mid_{-\infty} = 0, \tag{2.17}$$

since the second term vanishes at the upper limit, and U_1 vanishes at the lower limit. Now let us assume that the flow is self-preserving, or self-similar; this means that a velocity profile at any cross section x_1, when normalized by the local value of the length scale and the local value of the maximum velocity, will be the same. Hence, we have $U_1 = U_s f(\xi), 0 \le f \le 1$, where $\xi = x_2/\ell$, and the dependence on x_1 enters only through the dependence of ℓ on x_1. Then (2.17) becomes

$$U_s^2 \ell_{,1} \int_{-\infty}^{+\infty} f(f-1)d\xi + U_s U_2 \mid_{-\infty} = 0, \tag{2.18}$$

since U_s is not a function of x_1 for the mixing layer. The integral is not a function of x_1. Hence, we have determined that $U_2 \mid_{-\infty}$ must be non-zero, and in fact has the value (independent of x_1, since ℓ is linear in x_1):

$$U_2 \mid_{-\infty} = U_s \ell_{,1} \int_{-\infty}^{+\infty} f(1-f)d\xi. \tag{2.19}$$

We do not worry about existence of the integral, since f is presumed to go to its limits 0, 1 exponentially fast outside the turbulent part of the flow (i.e. for $\mid \xi \mid \gg 1$). Note that the integrand is strictly positive, since the mixing process is presumed monotone. This non-zero value of $U_2 \mid_{-\infty}$ is entrainment. The mixing layer flow, at least on this side, is dragging in fluid from infinity, causing a mean flow normal to the axis of the mixing layer. Note that our order-of-magnitude analysis has not been contradicted; $\ell/L = \ell_{,1}$, and the integral may be presumed to be of order unity. In fact a simple piecewise linear approximation,

$$f = \begin{cases} 0, & \xi \le -1/2, \\ \xi + 1/2, & -1/2 < \xi < +1/2, \\ 1, & \xi \ge +1/2, \end{cases}$$

gives a value of $+1/6$ for this integral.

We can explore the transverse velocity a little further if we write $U_2 = (U_s \ell/L)g(\xi)$, so that g, like f, will be of order unity in magnitude. Differentiating the expressions for U_1 and U_2, and substituting in the continuity Equation (2.8), we find that

$$g' = \xi f', \tag{2.20}$$

where $(\cdot)'$ denotes differentiation with respect to the variable ξ, on which f and g solely depend. Then we can write

$$g(\xi) = (U_2 \mid_{+\infty} /U_s)(L/\ell) - \int_{\xi}^{+\infty} \xi f' d\xi. \tag{2.21}$$

We have picked $+\infty$ as our upper limit for reasons which will become evident in a moment. Recalling that ℓ is linear in x_1, and $L = x_1$, we have from (2.19) that

$$g(-\infty) = \int_{-\infty}^{+\infty} f(1-f)d\xi. \tag{2.22}$$

Now, if f' is even, then $\xi f'$ is odd; hence g' is also odd, and g is even. Thus, $g(+\infty) = g(-\infty)$. But, while $g(-\infty) > 0$ corresponds to entrainment from infinity, $g(+\infty) > 0$ corresponds to "extrainment" at infinity, if such a word exists – to the mixing layer rejecting fluid to infinity on that side. This is never observed physically in any flow. In fact, the upper side of the mixing layer is like a wake: that is, there is a dominant uniform background velocity U_s, the departures from which satisfy $U_s - U_1 \ll U_s$. The lower side, on the other hand, is jet-like: that is, there is no background transport velocity, and the mixing layer velocity goes to zero. In wakes, there is no entrainment from infinity, positive or negative, while in jets there is entrainment from infinity on both sides. Entrainment is observed only on the jet-like side of mixing layers. Hence, it is reasonable to take $U_2 \mid_{+\infty} = 0$, or $g(+\infty) = 0$, so that, combining (2.21) and (2.22), we have

$$ -\int_{-\infty}^{+\infty} \xi f' d\xi = \int_{-\infty}^{+\infty} f(1-f) d\xi. \tag{2.23} $$

We conclude that f' cannot be even, but must have a small odd component; in fact, it must be skewed toward the negative ξ side; that is, the values of f' must be larger for $\xi < 0$ than for $\xi > 0$. Hence, the velocity U_1 approaches 0 on the jet-like side faster, producing larger gradients as it does so, than it approaches U_s on the wake-like side.

None of this, of course, addresses the fundamental question of *why* jet-like flows entrain and wake-like flows do not, an experimental observation which we have used in setting $g(+\infty) = 0$; in addition to this, we have assumed only self-similarity.

We can extract a bit more information from Equation (2.15). We have determined that the two advective terms scale in the same way, and are the same order of magnitude. However, this is only a global estimate, and either term can be smaller in some region of the cross section of the mixing layer. In particular, let us look at the edges of the mixing layer, $x_2 \gg l$, where the mean velocity U_1 is approaching its limits, and $\langle u_1 u_2 \rangle$ is going to zero. In practical terms, ξ need not be very large; experimentally we find that 2 is large enough. We know that on the high-speed, wake-like side of the mixing layer, well away from the core of the flow, $U_1 \approx U_s$, while U_2 is small; on the other hand, on the low-speed, jet-like side of the mixing layer, again well away from the core, $U_2 \approx U_2|_{-\infty}$, while U_1 is small. Hence, on the high-speed wake-like side we can write approximately, far from the axis

$$ U_s U_{1,1} + \langle u_1 u_2 \rangle_{,2} = 0. \tag{2.24} $$

Using the averaged continuity Equation (2.8) this can be written as

$$ -U_s U_{2,2} + \langle u_1 u_2 \rangle_{,2} = 0, \tag{2.25} $$

which may immediately be integrated to give

$$ -U_s U_2 + \langle u_1 u_2 \rangle = 0. \tag{2.26} $$

The constant of integration has been set equal to zero, since we know that $\langle u_1 u_2 \rangle$ and U_2 must both vanish on this side of the mixing layer.

On the low-speed, jet-like side of the mixing layer, far from the axis, we can write approximately

$$ U_2 \mid_{-\infty} U_{1,2} + \langle u_1 u_2 \rangle_{,2} = 0 \tag{2.27} $$

and, again, this may immediately be integrated to give

$$ U_2 \mid_{-\infty} U_1 + \langle u_1 u_2 \rangle = 0, \tag{2.28} $$

where again we have set the constant of integration equal to zero because we know that $\langle u_1 u_2 \rangle$ and U_1 both vanish on this side.

Using the self-similar expression for U_2, and (2.26) we can write on the high-speed, wake-like side, far from the axis

$$\langle u_1 u_2 \rangle = (U_s^2 \ell / L) g(\xi), \tag{2.29}$$

and from the corresponding expression (2.28) for U_1, we can write on the low-speed, jet-like side, far from the axis

$$\langle u_1 u_2 \rangle = -U_2 \mid_{-\infty} U_s f(\xi). \tag{2.30}$$

If we further use the expression (2.19) for $U_2 \mid_{-\infty}$ and set $A = \int_{-\infty}^{+\infty} f(1-f) d\xi$, (2.30) becomes

$$\langle u_1 u_2 \rangle = -(U_s^2 \ell / L) A f(\xi). \tag{2.31}$$

Note that our estimates of the scaling are borne out. Expressions similar to these can be written for all the almost-parallel, self-preserving flows (sometimes they are valid across the whole flow: e.g. in the wake) and are often used experimentally as a check on self-preservation – that is, experimental values of the two sides of (2.29) or (2.30) are compared with each other.

A great deal more can be learned about these flows by looking at the equations for the turbulent kinetic energy. We find some important fundamental truths about the physics of turbulent kinetic energy dissipation. Also, we can show that production and dissipation are in approximate balance everywhere, and that the advection and turbulent transport are also in approximate balance. At the edges of the flow, where turbulence is propagating into the surrounding irrotational fluid, the turbulence is transporting itself outward, and being blown back by the advection. This is important not because it is characteristic of mixing layers, but because it is characteristic of all turbulent flows with free boundaries.

Again, there is no hope for explicit solution, since the equations for second order quantities involve third order quantities. However, we can use the same tools, namely the order-of-magnitude analysis based on the knowledge that the flow is narrow, and the assumption of self-similarity or self-preservation. The equation for the turbulent kinetic energy is

$$\underbrace{\langle q^2/2 \rangle_{,i} U_i}_{(1)} + \underbrace{\langle (q^2/2 + p/\rho) u_j \rangle_{,j}}_{(2)} = -\underbrace{S_{ij} \langle u_i u_j \rangle}_{(3)} - \underbrace{2\nu \langle s_{ij} s_{ij} \rangle}_{(4)}$$

$$+ \underbrace{\nu \langle q^2/2 \rangle_{,jj}}_{(5)} + \underbrace{\nu \langle u_i u_j \rangle_{,ji}}_{(6)}, \tag{2.32}$$

where $S_{ij} = \frac{1}{2}(U_{i,j} + U_{j,i})$ is the mean strain rate, $s_{ij} = \frac{1}{2}(u_{i,j} + u_{j,i})$ is the fluctuating strain rate, and $q^2/2 = \frac{1}{2} u_i u_i$ is the local turbulent kinetic energy per unit mass. Reading from the left, the terms of Equation (2.32) are respectively: (1) the advection of the mean fluctuating kinetic energy by the mean motion; (2) the transport by the fluctuating turbulent velocity of the turbulent enthalpy (the whole equation can be considered an equation for the mean fluctuating enthalpy – instead of the kinetic energy – since the mean value of the

fluctuating pressure is zero); (3) the production of turbulent energy, the so-called deformation work: the work done by the mean strain rate against the stresses due to the presence
of the turbulence; and (4) the dissipation (conversion to heat) of turbulent kinetic energy
by working against the viscous stresses (actually the entropy production), referred to as ϵ
in Section 2.1. The last two terms (5) and (6) represent molecular transport of turbulent
kinetic energy.

We can now carry out an order-of-magnitude analysis similar to that which we applied
to the 1-component of Equation (2.7) for U_1. Without going into the details, we know from
experience with the equation for U_1 that cross-stream gradients will be the only important
ones. The last two terms will be of order vu^2/ℓ^2, while the production term will be of
order $U_s u^2/\ell$. The ratio of the two is $v/\ell U_s = (\ell/L)^{1/2}/R_\ell$. This is even smaller than
the viscous terms in the equation for U_1 (by a factor of ℓ/L), and hence, if the viscous
terms in the equation for U_1 are negligible, the last two terms in (2.32) are also. We shall
neglect them.

The orders of magnitude of the other terms, from left to right, are (1): $u^2 U_s/L$, (2):
u^3/ℓ, (3): $U_s u^2/\ell$, and that of (4), $2v\langle s_{ij}s_{ij}\rangle$, is unknown. We cannot estimate directly the
magnitude of the dissipation term (4) using our simple arguments, but we can determine
it indirectly. The production term (3) is clearly the largest among the first three, and if
the magnitudes of the other terms are taken relative to that one we have respectively (1):
ℓ/L, (2): $(\ell/L)^{1/2}$, (3): 1, and (4): unknown. It is thus clear that, to lowest order, we have
simply

$$- S_{ij}\langle u_i u_j\rangle - 2v\langle s_{ij}s_{ij}\rangle = 0. \tag{2.33}$$

Bear in mind that this approximation neglects terms of order $(6 \times 10^{-2})^{1/2} = 0.24$, and
hence is rather crude. We may simplify the production term, the various components of the
strain rate having the following magnitudes: $S_{11} = \mathcal{O}(U_s/L)$, $S_{22} = \mathcal{O}(U_s/L)$, S_{12} and
$S_{21} = \mathcal{O}(U_s/\ell + U_s\ell/L^2) = \mathcal{O}(U_s/\ell)$. Hence, the largest term by ratios of, respectively
ℓ/L and $(\ell/L)^2$, is $U_{1,2}\langle u_1 u_2\rangle$. Thus we may write to lowest order:

$$- U_{1,2}\langle u_1 u_2\rangle - 2v\langle s_{ij}s_{ij}\rangle = 0. \tag{2.34}$$

This balance between production and dissipation is approximately true in many turbulent
shear flows. It is basically a statement that the flow is not far from some sort of global
equilibrium.

Now we must pause for a moment to consider the dissipation from an experimental
point of view. Although the expression for the dissipation involves small-scale components
of the velocity field, in fact the value of the dissipation is not determined by the small
scales, but rather by the energy-containing, or large scales in the field. Experimentally, if
the value of the viscosity is changed, this has no influence on the value of the dissipation;
it simply changes the scale at which dissipation takes place. This surprising result can be
explained by considering the energy cascade, mentioned in Section 2.1. In crude physical
terms, turbulent energy is taken from the mean motion to the largest turbulent eddies,
passes from them to the eddies next in size, and so on from eddy to eddy, progressively
decreasing in size until it reaches the scale at which dissipation takes place. These are

ideas of Kolmogorov, see [117]. The matter is in fact a little more complicated, since the energy is not all passed to the next eddy, but rather most is passed to the next eddy, with the remainder being passed to all smaller eddies, the closest in size getting the lion's share. It is also complicated by the fact that, locally and over short times, the energy is sometimes passed in the opposite direction, from small eddies to larger ones. This is referred to as *backscatter*.

The fact remains that, averaged over long times or large regions, the direction of energy transfer is from large to small. None of this changes the fact that the rate at which energy is passed into this spectral pipeline is determined by the large-scale end. It is thus proportional to the amount of energy that is passed into the spectral pipeline, u^2, and the rate, which is proportional to the inverse time scale of the energy-containing-eddies, u/ℓ_t, where ℓ_t is the size of the most energetic turbulent eddies, the so-called energy-containing-scales. This is not necessarily quite the same as ℓ, which was defined here in terms of gradients of the mean velocity profile. Thus we see that

$$2\nu\langle s_{ij}s_{ij}\rangle = cu^3/\ell_t, \tag{2.35}$$

where c is a coefficient of order unity. It is of the utmost importance that the physical significance of this relation be grasped: the correct order-of-magnitude estimate for this term does not involve the viscosity and the small scales, because the rate is determined at the other end of the spectrum, even though the dissipation is ultimately carried out by the small eddies. In fact, Equation (2.35) is more than an order-of-magnitude estimate: the coefficient in (2.35) is observed experimentally to be within 10% of unity in all flows in approximate equilibrium. The relation (2.35) can be upset only by making an abrupt change in the rate at which energy is fed to the spectral pipeline; until the energy has arrived at the dissipative scales, which takes a time of the order of ℓ_t/u, there will be an imbalance in the equation.

We may now use (2.35) and the known scale of the production term in (2.34) to obtain

$$U_s u^2/\ell = u^3/\ell_t, \tag{2.36}$$

or

$$\ell_t/\ell = u/U_s = (\ell/L)^{1/2}, \tag{2.37}$$

which is about 0.24, as we noted above.

If we want a slightly better approximation to (2.32) than (2.34), we should probably include not only the transport term which is $\mathcal{O}(\ell/L)^{1/2}$, but also the advection term, which is $\mathcal{O}(\ell/L)$. This is on grounds of the physics, since both of these terms represent transport, one by the fluctuating velocity and the other by the mean velocity. If we keep these terms, we have to put back (in the expression for the production) the terms $U_{1,1}\langle u_1^2\rangle + U_{2,2}\langle u_2^2\rangle = U_{1,1}(\langle u_1^2\rangle - \langle u_2^2\rangle)$, (making use of continuity, Equation (2.8)), since these are of order ℓ/L relative to the leading term. Hence, to this order our equation becomes

$$\langle q^2/2\rangle_{,1}U_1 + \langle q^2/2\rangle_{,2}U_2 + \langle (q^2/2 + p/\rho)u_2\rangle_{,2}$$
$$= -U_{1,2}\langle u_1 u_2\rangle - U_{1,1}(\langle u_1^2\rangle - \langle u_2^2\rangle) - 2\nu\langle s_{ij}s_{ij}\rangle. \tag{2.38}$$

We expect that the balance between production and dissipation will now not be quite exact, and the imbalance will be of order $(\ell/L)^{1/2}$, which will be balanced by the advection and transport terms.

If we now consider what happens near the edges of the flow again, we can extract still more information. We make use of the forms (2.29) and (2.30) for the behavior of $\langle u_1 u_2 \rangle$ near the edges of the flow. On the high-speed, wake-like side, far from the center, the only important advection term is the first, which is essentially $U_s \langle q^2/2 \rangle_{,1}$. Let us write $\langle q^2/2 \rangle = (U_s^2 \ell/L)h(\xi)$, $\langle u_1^2 \rangle - \langle u_2^2 \rangle = (U_s^2 \ell/L)k(\xi)$, and $\langle (q^2/2 + p/\rho)u_2 \rangle = [U_s^3(\ell/L)^{3/2}]m(\xi)$. Then the terms in Equation (2.38), in the same order, may be written as

$$- (\ell/L)h'\xi + (\ell/L)^{1/2}m' = -f'g + (\ell/L)f'\xi k - D, \qquad (2.39)$$

where $D = (L/U_s^3)2\nu \langle s_{ij} s_{ij} \rangle$. From Equation (2.20) we have a relation between the asymptotic behavior of f and g far from the axis. The asymptotic behaviors of $h(\xi)$ and $k(\xi)$ will be similar to that of g on this side, since these are various components of the same tensor, and the correlation coefficients are observed to be of order unity. Hence, the first term behaves roughly as $\xi g'$, the third as gg'/ξ, and the fourth as gg'. It is clear that the two production terms can be neglected as we move far out on this side, since they are quadratic in these functions that are falling rapidly. There is no reason to believe that $m(\xi)$ has a behavior much different from that of $h(\xi)$ or $g(\xi)$.

The behavior of $D(\xi)$ is a little harder to predict. We have first to develop a physical picture of the mechanism at work here. Recall the more-or-less empirical statement that the dissipation $\epsilon = u^3/\ell_t$. This can be written as $\epsilon = u^2(u/\ell_t)$. That is, the rate at which energy (per unit mass) enters the spectral pipeline is proportional to the energy (per unit mass) present, and inversely proportional to the time scale of the turbulence, since it is the strain rate of the energy containing-eddies that is responsible for starting the energy moving across the spectrum. When we first wrote this expression, we were thinking of u and ℓ_t as global values, a single number for the entire cross section. However, we can also consider local values, associated with each point in the cross section. Now, the (inverse) time scale of the turbulence at a particular laboratory point, u/ℓ_t, will be determined by the history of the material points passing through that laboratory point; in particular, the wandering material will try to bring its (inverse) time scale into equilibrium with the (inverse) time scale of the regions through which it has passed, the (inverse) time scale being determined ultimately by the local value of the mean strain rate. Hence, we might expect the value u/ℓ_t to be determined by an integral back over the mean path through the laboratory point, with a spreading awareness and a fading memory, of the mean strain rate. As the laboratory point changes, the path changes, and hence the values of strain rate appearing in the integrand. We might write this as $[(S_{ij}S_{ij})^{1/2}]$, where we use the square brackets to symbolize this moving averaging process. A local value of u^2 would certainly be proportional to $\langle q^2/2 \rangle$. Hence, we expect a physically realistic approximation to $D(\xi)$ to be something like $[(S_{ij}S_{ij})^{1/2}]\langle q^2/2 \rangle$; if the flow is out of equilibrium, this might also be averaged back along the path, as $[[(S_{ij}S_{ij})^{1/2}]\langle q^2/2 \rangle]$. To first order, this would scale like $h(\xi)f'(\xi)$, and consequently would also be small far out on this side of the mixing layer.

We conclude from this reasoning that, in the outer reaches of the mixing layer on the high-speed, wake-like side, the production and dissipation are of the same order (and we know that they approximately balance each other where they are large, in the center of the layer), and both are negligible relative to the transport and advection, which consequently must balance each other in this region. In fact, it is observed in all the nearly parallel flows that D approximately balances production all across the flow, and falls as fast as production falls near the edge, leaving advection to balance transport far from the axis. Hence we have, roughly (on the high-speed wake-like side, far from the axis)

$$m'(\xi) = (\ell/L)^{1/2}\xi h'(\xi). \tag{2.40}$$

Physically, this says simply that the turbulent fluctuations are trying to carry the turbulent energy into the irrotational fluid beyond the instantaneous edge of the mixing layer, but are being blown back by the component of the axial velocity normal to the edge of the mixing layer. The balance between these two effects keeps the edge of the mixing layer in the same place.

On the low-speed, jet-like side of the mixing layer, it is the other advection term that is important. Otherwise, our considerations are similar, and the conclusion is again that this advection term is balanced by turbulent transport, but in this case the efforts of the turbulent fluctuations to carry the turbulent kinetic energy into the irrotational fluid outside the instantaneous edge of the mixing layer are countered by the entrainment wind normal to the axis of the mixing layer, blowing the energy back toward the axis.

The conclusions that we have drawn here are valid for most other turbulent flows that are not strongly out of equilibrium, certainly for the nearly parallel flows. In particular, dissipation is crudely like production in size and shape, with this exception: production drops to zero near maxima of the mean velocity profile, and at points of symmetry of the flow, while dissipation does not, because dissipation involves several averages over the history of material points arriving at the laboratory point, which smear out these details. Near the edges of a flow, both dissipation and production are small, and the only important terms are advection and transport – transport carrying turbulent kinetic energy out into the irrotational fluid, and advection blowing it back again. None of these conclusions is influenced in the least by whether the turbulent fluctuations are coherent or incoherent, so long as they scale the same way.

As far as coherent structures in the mixing layer are concerned, we have already seen some of these in Figure 2.2. The mixing layer is subject to Kelvin–Helmholtz instability at every stage in its development. That is, if there were no coherent structures, any lateral displacement of the plane of the mixing layer would cause them to develop, since a displacement toward the high-speed side is associated with a reduced pressure, which causes the displacement to increase, while a displacement toward the low-speed side is associated with an increased pressure, which causes the displacement to increase also. Since the layer grows in a geometrically similar way, what is true at one stage of growth is true at all stages of growth. Initially, these instability waves are two-dimensional. As they grow in magnitude, they become nonlinear, at first still retaining their two-dimensional form. The material that has been carried furthest into the high-speed stream is now carried downstream faster than the material just under it, and hence overtakes this slower material, causing the wave to roll over itself like an ocean wave approaching the beach. This process continues, and the wave rolls over and over, creating a structure like a jelly roll, with thin layers of the exterior fluid sandwiched between thin layers of the interior fluid. As the rolling process continues, the layers are stretched, and become thinner and thinner. Eventually, the layers are thin enough for molecular transport to smear them out. This process is neater and more picturesque in the initial laminar phase of development; after the layer has acquired small-scale turbulence, the rolling and stretching process is disturbed by the smaller-scale turbulence, and the process is quite unsteady and irregular. Nevertheless, it is this process that is partly responsible for mixing of the exterior and interior fluid, for gradual contamination of the exterior fluid by the interior fluid, and for a slow linear increase in width of the mixing layer. In addition, on the low-speed side, it is this process that is responsible for the entrainment wind. We shall refer to these structures as rollers.

Since the layer grows linearly on the average, the rollers must also grow linearly on the average. The progressive incorporation of external fluid makes them grow laterally. In addition, the number of rollers per unit length in the x_1-direction must steadily decrease; that is, on average, the length of x_1-axis per roller must increase linearly, or the number of rollers per length of x_1-axis must be proportional to x_1^{-1}. This is achieved by coalescing or pairing of rollers: adjacent rollers move toward each other and become wrapped up in each others' velocity fields (since they are all rotating in the same sense). In a short time they have become a single roller. There are indications that the fluctuating pressure field from the rollers downstream is responsible for initiating the pairing process; that is, at location x_1 there are n rollers passing per unit time; at a second location further downstream, say x_1', there are $n' = n/2$ rollers passing per unit time. This pressure signal is sensed at x_1, and since it is subharmonic, it results in pushing alternate pairs together.

So far we have not discussed the three-dimensional aspects of these structures. The two-dimensional configuration is unstable. All approximately two-dimensional shear flows that are perturbed by waves or turbulence are subject to an instability probably similar to that which produces Langmuir cells in the ocean surface mixed layer. That is, the disturbances perturb the lines of cross-stream vorticity (associated with the shear) toward the high-speed and low-speed sides. The Lagrangian streamwise transport also has a gradient across stream, and the perturbed line of vorticity is carried forward faster on the high-speed side than on the low-speed side. As the perturbed vortex line is tipped forward by this transport gradient, there is now a streamwise component of vorticity. The progressive tipping and stretching results in intensification of this streamwise vorticity. In the case of the mixing layer, the vortex lines get caught up in the rolling up of the rollers, and this produces additional stretching and intensification. The result is counter-rotating vortex pairs, more-or-less in the streamwise direction, running from the low-speed side of one roller to the high-speed side of the next. These are known as braids: see Figure 2.12. They also are responsible for mixing of external and internal fluid.

This state – rollers and braids – is the final one to which the initial Kelvin–Helmholtz instability evolves, the new attractor on which the mixing layer has settled down, and the stable form of the coherent structures which are found in the mixing layer. Of course, in a fully developed, high-Reynolds number mixing layer with small-scale, less-organized

Figure 2.12 The same mixing layer as in Figure 2.2, but viewed from the top – the plane
of the picture is the plane of the mixing layer. From [49]; photo courtesy A. Roshko.

turbulence, the roller-braid structure is perturbed, and is somewhat disorganized but, as we
have seen, still much more organized than in other flows.

2.5 The turbulent boundary layer

A boundary layer is the turbulent flow that is formed next to a boundary, at which the
velocity must vanish. A boundary layer has a beginning, where it is of zero thickness, and
grows in the downstream direction.

Let us begin with a few simple ideas. As for the shear layer, we use a coordinate system
with x_1 in the direction of the free stream, x_2 normal to the wall, and x_3 spanwise, i.e.
parallel to the wall and normal to the free stream, to form a right-handed coordinate system.
The flat plate, on which the boundary layer forms, is supposed to be of zero thickness, and
to lie in the half-plane $0 \leq x_1 < +\infty$, $-\infty < x_3 < +\infty$. We can still suppose that outside
the viscous region near the wall, there is a cross-stream (normal to the wall) length scale ℓ,
and a streamwise length scale L, and that $\ell/L \ll 1$. These scales are defined, as they were
before, via the derivatives, so that $\partial(\cdot)/\partial x_2 = \mathcal{O}((\cdot)/\ell)$, and so forth.

Boundary layers are more complicated than boundary-free flows because two scales
are involved. In the boundary-free flows such as the mixing layer of the previous section
we could introduce $\ell(x_1)$ as the single relevant scale. Recall that ℓ was defined from the
gradient of the mean velocity profile. In a boundary layer, because of the presence of a
surface at which the velocity must vanish, we have a layer near the wall in which viscosity
is important, and we have another length scale, ν/u_τ, where u_τ is defined from the shear
stress at the wall:

$$\tau_w = \rho u_\tau^2 = \mu U_{1,2} \mid_w . \tag{2.41}$$

Recall that $\mu = \rho \nu$, where μ is the dynamic viscosity, ρ the density, and ν the kinematic
viscosity. The term u_τ is called the friction velocity, and it is in general a function of x_1.
Hence, we have everywhere two length scales. The region where the viscous length scale
is important is relatively thin; when $x_2 \gg \nu/u_\tau$, viscous effects can be neglected.

The quantities ν and u_τ are the fundamental scaling quantities in the wall region, and quantities non-dimensionalized by these scaling quantities are called wall units, usually indicated by a superscript $^+$:

$$y^+ = \frac{x_2 u_\tau}{\nu}, \qquad U^+ = \frac{U_1}{u_\tau}. \qquad (2.42)$$

We explore these units in greater detail later in this section; we introduce them here because of their ubiquitous appearance in any discussion of boundary layers. They are useful because mean and fluctuating quantities measured in the wall region, expressed in wall units, are universal: their profiles do not change with position x_1 in a given boundary layer, and they are the same in different boundary layers, under different conditions. In fact, the wall region is *defined* as the region that is invariant in wall units. The exact extent of this region depends on the Reynolds number of the boundary layer (several definitions are possible; for the present, we can speak of the Reynolds number based on the free-stream velocity and the thickness, which we define more precisely below). The *outer* or *wake* region of the boundary layer is that part which is *not* universal when expressed in wall units. The wall region and the wake region may be joined by asymptotic matching techniques, and the matching point and extent of the overlap region depend on the Reynolds number. Even at the lowest Reynolds number, the wall region extends roughly to $y^+ \approx 50$, and as the Reynolds number rises its extent increases.

We must consider the scaling velocity carefully. In a turbulent boundary layer, the changes in mean velocity from the free stream are small until x_2 is near the wall. In the mixing layer we could write that $U_{1,2} = \mathcal{O}(U_s/\ell)$, where U_s is the velocity on the high-speed side. In the boundary layer, however, if we use the free-stream velocity $U_s = U_1(x_1, +\infty)$ to write the same thing, $U_{1,2} = \mathcal{O}(U_s/\ell)$ would very much over-estimate the mean velocity gradient. It is only near the wall, in the region in which viscosity is important, that large gradients are found. We get a much more accurate picture of the structure of the boundary layer if we take the velocity difference $U_s - U_1 = \mathcal{O}(u_\tau)$ in the turbulent part of the layer that is uninfluenced by viscosity. This is the part of the boundary layer we call the outer region; it should not be confused with the flow beyond the edge of the boundary layer in the free stream, which we sometimes refer to as "outside the boundary layer."

This somewhat surprising choice of scaling velocity can be justified in three ways. The first justification is this: we have recognized that $U_s - U_1$ is not of order U_s, but is much smaller. We could leave it undetermined, and designate it as $\mathcal{O}(\mathcal{U})$, say. Now, the inner region, the region near the wall, where viscosity is important, scales by u_τ. When the two regions are matched, we will find that $\mathcal{U} = u_\tau$. The second justification is this: in any velocity profile, mean velocity differences are of the order of the r.m.s. fluctuating velocity. Experimentally we find that the r.m.s. fluctuating velocity in the boundary layer is of the order of u_τ. The third justification is this: there are only two velocity scales in the boundary layer, U_s and u_τ. Since we have found empirically that scaling the velocity differences in the outer region with U_s does not work because U_s is too big, we have no other choice than to use u_τ.

Note that this means that we have two scales of velocity in the outer region: $U_1 = \mathcal{O}(U_s)$ where it is not differentiated, and $U_{1,2} = \mathcal{O}(u_\tau/\ell)$ where it is. The ratio $u_\tau/U_s \ll 1$

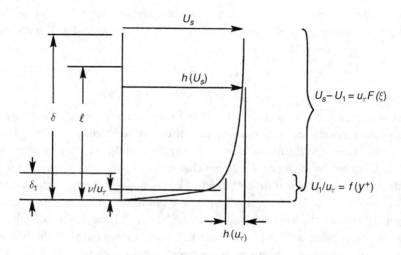

Figure 2.13 A turbulent boundary layer profile indicating the various velocity and length scales. As the Reynolds number increases, the thickness of the viscous region (near the wall) will become smaller relative to the whole. See text for full description of quantities.

(experimentally it is found to be about 1/30 for a broad range of Reynolds numbers, and slowly decreases as Reynolds number increases), see Figure 2.13.

We consider here only boundary layers on plane surfaces parallel to a uniform, parallel, steady unvarying external flow, for which $U_s = $ const. Note that, well outside the boundary layer, where the turbulence vanishes (just as it did outside the mixing layer), we can write

$$U_{1,1}U_1 + U_{1,2}U_2 = P_{,1}/\rho. \tag{2.43}$$

Suppose $U_{1,1} = 0$, so that the flow does not vary in the x_1-direction, and $U_{1,2} = 0$, so that it does not vary in the x_2-direction either; it is therefore a uniform, parallel flow, since $U_2 = $ const. from continuity (2.8). We suppose in addition that $U_2(x_1, +\infty) = 0$, so that the flow outside the boundary layer is a uniform flow parallel to the x_1-axis; then $P_{,1} = 0$. This is the simplest possible case: uniform flow without pressure gradient parallel to a flat surface, the so-called zero-pressure gradient flat plate boundary layer. Of course, boundary layers also form over curved surfaces, with pressure gradients in both the streamwise and cross-stream directions, and in free streams that are diverging or converging, in two or three dimensions, and may be unsteady. We do not do so here, but it is possible to show that much that we will say is broadly true of boundary layers in general, which is why the concept of the boundary layer is useful.

We may apply the scaling in terms of u_τ, ℓ, and L to the continuity equation in the outer region of the layer, giving $U_2 = \mathcal{O}(u_\tau \ell/L)$. First, consider the mean spanwise component of vorticity, $\Omega_3 = U_{2,1} - U_{1,2}$. Again we have that $U_{2,1} = \mathcal{O}(u_\tau \ell/L^2)$, while $U_{1,2} = \mathcal{O}(u_\tau/\ell)$. The ratio of the two is $\mathcal{O}(\ell/L)^2$, so the first term may be neglected. Hence, to a good approximation $\Omega_3 = -U_{1,2}$. Note that this approximation is also valid in the viscous region near the wall, since as we approach the wall $U_{1,2}$ goes to a constant $(= u_\tau^2/\nu)$, while $U_{2,1}$ goes to zero. Now consider the total vorticity in any cross section:

$$\int_0^{+\infty} \Omega_3 \, dx_2 \approx -\int_0^{+\infty} U_{1,2} \, dx_2 = -U_1(x_1, +\infty) = -U_s. \tag{2.44}$$

Since U_s is not changing, the total vorticity in any cross section is not changing. It must all have been created at the leading edge of the flat plate, where the boundary layer began. Just before the uniform free stream strikes the flat plate, the streamwise velocity is $U_1(x_1, x_2) = U_s$; immediately after striking the leading edge of the plate the streamwise velocity at the surface vanishes, but elsewhere is still uniform. Hence, the non-zero value of $U_{1,2}$ is confined to an infinitesimal layer next to the plate. This (i.e. the leading edge) is where the vorticity is created, and the growth of the boundary layer corresponds to the diffusion of the vorticity away from the surface. In the absence of a pressure gradient, no new vorticity is created, and none is destroyed.

Note that, in a region near the leading edge of the plate, the thickness of the boundary layer (suitably defined) will be less than, or of the order of, the viscous length ν/u_τ. In this region we may expect the boundary layer to be laminar and viscously dominated. As it grows, however, the thickness will eventually be large enough compared to ν/u_τ for the layer to become unstable and undergo transition to turbulence, after which the layer will be turbulent. This is observed to happen at approximately $U_s x_1/\nu = 5 \times 10^5$. In air $(\nu = 15 \times 10^{-6} \text{ m}^2/\text{s})$ with $U_s = 10$ m/s, we get $x_1 = 0.75$ m. This is too large to ignore; when we discuss self-similarity in the turbulent part of the boundary layer, we have to displace the origin to a virtual value to account for the initial laminar part of the layer. Henceforth, we shall redefine x_1 to have its origin at the apparent origin of the turbulent part of the layer.

Let us consider the location of the centroid of the vorticity:

$$\int_0^\delta x_2 \Omega_3 \, dx_2 = -\int_0^\delta x_2 U_{1,2} \, dx_2. \tag{2.45}$$

Now, because of convergence problems we cannot integrate to infinity, but rather must integrate to the outer edge of the boundary layer, which we will call $\delta(x_1)$, and which we will define by, say, $[U_s - U_1(\delta)]/U_s = 0.01$. Then we can write for the normalized centroid of vorticity

$$\frac{\int_0^\delta x_2 \Omega_3 \, dx_2}{\int_0^\delta \Omega_3 \, dx_2} \approx \int_0^\delta (1 - U_1/U_s) \, dx_2 = \delta_1. \tag{2.46}$$

This is known as the displacement thickness, and the designation as δ_1 is conventional. Imagine that the vorticity is all at a distance x_2^* from the wall, so that the mean velocity is zero for $x_2 \le x_2^*$, and equal to U_s for $x_2 > x_2^*$. Then, the displacement thickness $\delta_1 = x_2^*$. This piecewise constant velocity profile is like an inviscid profile over a wall at x_2^*, a wall displaced to that position; hence the name. Also, streamlines outside the boundary layer are displaced outward by an amount δ_1. Note that δ_1 is a great deal smaller than the geometric thickness, δ; using a crude power-law approximation for the turbulent mean velocity profile, we find that $\delta_1 = \delta/8$, as indicated in Figure 2.13.

Although we do not pursue more general boundary layers further in this book, to properly orient the reader we should place boundary layers (including our simple one) in their correct fluid-mechanical context. In formal boundary layer theory by matched asymptotic expansions (which can only be carried out for a laminar boundary layer, although the qualitative conclusions are equally valid for a turbulent boundary layer), the lowest order outer solution for the flow around an arbitrary body is the inviscid flow around the body. We ignore here the complications introduced by separation; we suppose that this is a streamlined body. The first order inner solution is the boundary layer driven by the pressure gradient due to the outer, inviscid flow. The second order outer solution is the inviscid flow around a body with the outer boundary displaced by the displacement thickness calculated from the first order inner boundary layer. The second order boundary layer is the boundary layer generated by the pressure gradient from the second order outer flow. This process can be continued indefinitely.

We can carry out an order-of-magnitude analysis of the mean momentum equations in the outer part of the flow exactly as we did for the mixing layer. Without going through the details, it is clear that only the derivatives normal to the wall are important, except in the advection term, and that we must retain the viscous term $\nu U_{1,22}$, since this will be important near the wall. The treatment of the pressure term is exactly the same as for the mixing layer, for the same reasons; unless the boundary layer is strongly curved, there is no reason to have a significant pressure gradient normal to the wall: the streamwise pressure gradient felt at the surface is the same as that felt in the irrotational part of the flow outside the boundary layer. We finally obtain

$$U_{1,1}U_1 + U_{1,2}U_2 + \langle u_1 u_2 \rangle_{,2} = \nu U_{1,22}. \tag{2.47}$$

In the outer part of the layer, of the first two terms, only the first is important, since U_1 is $\mathcal{O}(U_s)$, while $U_2 = \mathcal{O}(u_\tau \ell / L)$. As we approach the wall, however, the two terms assume equal importance. Equation (2.47) is valid throughout the boundary layer. If we confine ourselves to the part outside the viscously dominated region, we have

$$U_{1,1}U_1 + U_{1,2}U_2 + \langle u_1 u_2 \rangle_{,2} = 0. \tag{2.48}$$

We can obtain an integral of the motion by manipulating Equation (2.47). Subtracting U_s from U_1, which is permissible since U_s is constant by assumption, and using the continuity equation to bring the advection velocity inside the derivatives, we can then integrate the equation from the wall to infinity in x_2 to give:

$$\frac{\partial}{\partial x_1} \int_0^{+\infty} U_1(U_s - U_1)\, dx_2 = \nu U_{1,2} \mid_0 = u_\tau^2, \tag{2.49}$$

much as in the derivation of (2.17) above. Here we have used the fact that $U_1 = U_s$ at infinity, and that $U_1 = U_2 = 0$ at the wall; also that $\langle u_1 u_2 \rangle = 0$ at the wall and at infinity. To paraphrase Equation (2.49), momentum is not conserved here: it is constantly being removed by the wall.

Now, in the outer region of the boundary layer (away from the viscous region) where ℓ is the important scaling length, $x_2 \gg \nu/u_\tau$, we can assume a self-similar behavior. We take u_τ as the scaling velocity, as discussed above, because the gradients of U_1 in the turbulent part are rather small. As we remarked above, if we took an unknown scaling velocity $\mathcal{U}$, we would find in the matching (below) that $\mathcal{U} = u_\tau$, see Figure 2.13. We thus write

$$U_s - U_1 = u_\tau F(\xi), \tag{2.50}$$

where $\xi = x_2/\ell$, $\ell = \ell(x_1)$, and $u_\tau = u_\tau(x_1)$. Near the wall, in the viscous region, where $x_2 \ll l$, we can write

$$U_1/u_\tau = f(x_2 u_\tau/\nu) = f(y^+), \tag{2.51}$$

using the traditional designation (2.42) for wall units: $x_j^+ = x_j u_\tau/\nu$, with $x_1^+ = x^+$, $x_2^+ = y^+$, and $x_3^+ = z^+$. We consider a limiting process in which the local Reynolds number $u_\tau \ell/\nu = R_\ell \to +\infty$. If we assume that there is a region of overlap toward the outer edge of the inner layer (2.51), $y^+ \to +\infty$, and the inner edge of the outer layer (2.50), $\xi \to 0$, where both descriptions are valid, then in this region matching the derivatives $\partial/\partial x_2$ requires that

$$U_{1,2} = -u_\tau F'/\ell = u_\tau^2 f'/\nu. \tag{2.52}$$

Multiplying by x_2/u_τ, we obtain

$$-\xi F'(\xi) = y^+ f'(y^+) \tag{2.53}$$

in the overlap region. We consider an intermediate limit, so that $y^+ \propto R_\ell^\alpha$, where α is some power between zero and one. (Since $\xi = x_2/\ell = (x_2 u_\tau/\nu)(\nu/u_\tau \ell) = y^+/R_\ell \propto R_\ell^{\alpha-1}$, and we require that $\xi \to 0$, then $\alpha - 1 < 0$, or $0 < \alpha < 1$, as claimed.) Since α is arbitrary in this interval, if the two sides of (2.53) are not constant, they can be made to behave very differently by different choices of α. Hence, both sides of (2.53) must be constant, so that F and f are the same logarithmic function up to an additive constant. This produces the laws

$$(U_1 - U_s)/u_\tau = (1/\kappa) \ln \xi + b, \qquad U_1/u_\tau = (1/\kappa) \ln y^+ + a, \tag{2.54}$$

both valid only in the overlap region, of course. Here a and b are universal constants and κ is von Kármán's constant, also a universal constant. This overlap region, which is observed experimentally, is called the inertial sublayer. If one law in (2.54) is subtracted from the other we obtain

$$U_s/u_\tau = (1/\kappa) \ln R_\ell + a - b, \tag{2.55}$$

in which the Reynolds number $R_\ell = u_\tau \ell/\nu$ is the same as that we have used before. The term U_s/u_τ is a fundamental quantity. Of course it is the dimensionless ratio of the two scaling velocities in the boundary layer, and that makes it important; however, it is also related to the drag coefficient, and the drag produced by the boundary layer is the quantity of greatest practical interest. The drag coefficient is given by

$$C_D = \frac{\tau_w}{\frac{1}{2}\rho U_s^2} = \frac{\rho u_\tau^2}{\frac{1}{2}\rho U_s^2} = 2\left(\frac{u_\tau}{U_s}\right)^2. \tag{2.56}$$

Hence,

$$\frac{u_\tau}{U_s} = \sqrt{\frac{C_D}{2}}. \tag{2.57}$$

Equation (2.55) is known as the logarithmic friction law, since it gives a value of U_s for a given R_ℓ and u_τ.

In the viscous region near the wall, the mean velocity profile is given by

$$U_1/u_\tau = y^+. \tag{2.58}$$

This region is known as the viscous sublayer. The region defined by (2.58) (extended) and that defined by (2.54) (extended) intersect approximately at $y^+ = 12$. This region in which neither is valid is called the buffer layer. Note that, as $R_\ell \to +\infty$ the value of ξ representing the boundary between viscous and non-viscous regions approaches zero.

Let us return to Equation (2.49). We can split the integral into two regions, say $y^+ \leq 30$ and $y^+ > 30$, (the break point being chosen somewhere in the region of overlap between the inner and outer regions, called the inertial sublayer) and, using the scalings developed above, write

$$\frac{\partial}{\partial x_1}(U_s u_\tau \ell) \left[\int_{30/R_\ell}^{+\infty} F\, d\xi - (u_\tau/U_s) \int_{30/R_\ell}^{+\infty} F^2\, d\xi \right.$$
$$\left. + R_\ell^{-1} \int_0^{30} f\, dy^+ - (u_\tau/U_s)R_\ell^{-1} \int_0^{30} f^2\, dy^+ \right] = u_\tau^2. \tag{2.59}$$

Now, f is of order unity in the interval $[0, 30)$, and F is of order unity in the interval $[30/R_\ell, \delta)$, F is exponentially small for $x_2 \gg \delta$, $R_\ell \gg 1$, and $u_\tau/U_s \ll 1$ (as we have noted, the latter is about 1/30 over a broad range of Reynolds numbers). Hence, to a reasonable approximation, we have

$$\frac{d}{dx_1}\left(U_s u_\tau \ell \int_{30/R_\ell}^{+\infty} F\, d\xi \right) = u_\tau^2, \tag{2.60}$$

where the integral is a constant, say C. Now, we have assumed U_s to be independent of x_1. The ratio U_s/u_τ (Equation(2.55)) varies only slowly with Reynolds number R_ℓ; the range of values of R_ℓ observed in nature and technology is rather limited, so that for practical purposes a very crude, but acceptable, approximation is that $u_\tau/U_s = $ const. Then, since U_s is also constant for our simple zero pressure gradient layer, we find that $U_s u_\tau$ is not a function of x_1, and we may conclude

$$\frac{d\ell}{dx_1} = u_\tau/U_s C. \tag{2.61}$$

That is, the layer grows linearly. This is, of course, a bit too simplistic. If we want a better approximation, we must use the logarithmic drag law equation (2.55), that relates u_τ to U_s and Reynolds number. It turns out that the layer grows more like $x_1^{4/5}$ (to a slightly improved approximation), the difference arising from the fact the u_τ/U_s is not constant, but slowly decreases with Reynolds number. Equation (2.61) says nothing more than this: at the instantaneous edge of the boundary layer, the large eddies responsible for propagating the turbulent material into the irrotational fluid are moving with velocities of the order of u_τ (velocity differences in this region are of order u_τ, and the turbulent velocities therefore are of the order of u_τ). Hence, $d\ell/dt = U_s\, d\ell/dx_1 = u_\tau$, roughly, which simply says that the (inverse) time scale of the turbulence u_τ/ℓ is approximately the same as the (inverse) time scale of the mean flow U_s/x_1.

If we look at the region near the outer edge of the boundary layer, we have $U_1 \approx U_s$, while $U_{1,1}$, U_2, and $U_{1,2}$ are all going to zero. Hence, we have exactly the same situation as on the high-speed, wake-like side of the mixing layer, and we can write approximately

$$U_{1,1}U_s + \langle u_1 u_2 \rangle_{,2} = 0, \qquad (2.62)$$

or, via the continuity equation,

$$- U_{2,2}U_s + \langle u_1 u_2 \rangle_{,2} = 0, \qquad (2.63)$$

which may be immediately integrated to give

$$- U_s U_2 + \langle u_1 u_2 \rangle = 0, \qquad (2.64)$$

where again the constant has been set equal to zero, since both U_2 and $\langle u_1 u_2 \rangle$ vanish in the irrotational part of the flow.

We may use the continuity equation again to obtain an expression for U_2. Assuming self-similarity in the turbulent part of the layer, we can write

$$U_2 = u_\tau (\ell/L) G(\xi), \qquad (2.65)$$

which produces approximately (taking u_τ essentially constant – note that $L = x_1$ as before)

$$G' = +\xi F'(\ell' L/\ell), \qquad (2.66)$$

where the term in parentheses is nearly unity if ℓ is linear in x_1. Since $-\xi F' = 1/\kappa$ over a large part of the inviscid part of the layer (from (2.53) and (2.54)), we can write approximately

$$G = (\ell' L/\ell)(1/\kappa)(\xi - \delta/\ell), \qquad (2.67)$$

where we have used the fact that G must vanish at the edge of the layer, $x_2 = \delta$ and we are defining δ, the geometric thickness, by $[U_s - U_1(\delta)]/U_s = 0.01$ as above. From (2.64) and (2.65) we have

$$\langle u_1 u_2 \rangle = U_s u_\tau (\ell/L) G(\xi) \qquad (2.68)$$

in this outer part of the flow. This is almost identical to the mixing layer relation (2.29). With it, we can consider the balance of turbulent kinetic energy, which will be very much like that in the wake-like part of the mixing layer, and so obtain essentially the same conclusions; i.e. production and dissipation are very much alike in this part of the flow, and both are small at the edge of the flow; advection and transport are approximately in balance, and they represent the turbulent transport trying to carry the turbulent kinetic energy out into the irrotational fluid, and its being blown back by the component of U_s normal to the mean edge of the boundary layer (which is slightly tilted to the streamwise direction, as is that of the mixing layer).

Close to the wall, we have $U_1 = \mathcal{O}(u_\tau)$, $U_{1,2} = \mathcal{O}(u_\tau^2/v)$, $U_{1,1} = \mathcal{O}(u_\tau/L)$, $U_2 = \mathcal{O}(v/L)$ (scaling distances normal to the wall by v/u_τ). If we scale the various terms in (2.47), we find that the first two are of the same size, u_τ^2/L, while the second two are also the same size, u_τ^3/ν. The ratio of these is the Reynolds number $u_\tau L/\nu \gg 1$. Hence, to a good approximation, in this region we have

$$\langle u_1 u_2 \rangle_{,2} = \nu U_{1,22}, \tag{2.69}$$

which may be integrated immediately to give

$$\langle u_1 u_2 \rangle = \nu U_{1,2} - u_\tau^2. \tag{2.70}$$

The constant of integration is u_τ^2, from the definition of the friction velocity (2.41). In the viscous sublayer, where $U_1 = u_\tau^2 x_2/\nu$, we have $\langle u_1 u_2 \rangle \approx 0$. Further out, in the logarithmic layer, we have (2.54) which, combined with (2.69) gives the behavior of $\langle u_1 u_2 \rangle$ in this part:

$$\langle u_1 u_2 \rangle / u_\tau^2 = 1/\kappa y^+ - 1. \tag{2.71}$$

Yet further out, we have (2.67) and (2.68), which give approximately

$$\langle u_1 u_2 \rangle / u_\tau^2 = (\xi - \delta/\ell)/\kappa, \tag{2.72}$$

where we have made use of Equation (2.61) and the paragraph following it to write $u_\tau/U_s \approx \ell'$. Again, we must underline that all this is independent of whether the fluctuating motions in the boundary layer are coherent or incoherent. No matter what the motions are, they must obey these scaling arguments.

There is an enormous amount of data on the form of the coherent structures in the turbulent boundary layer, that has been accumulated over nearly 30 years. The data come both from laboratory observations and from direct numerical simulation. Much of this has been summarized by Kline and Robinson (see Robinson [307]). In an attempt to clarify these observations, Kline has set up what he calls a *ringi*, using the Japanese word for a process in which all interested parties contribute observations which are commented on by all interested parties, and so on *in perpetuum*. Unfortunately, the process does not show convincing signs of converging. There are good reasons for this: observations, whether real or numerical, are made in a highly disturbed environment. The controversy is not so much over what is seen, but over its significance: what is fundamental, and what is secondary? What is driving what? Which is chicken, and which egg? Which is random irrelevancy? (The process is probably complicated by the fact that many people have worked on the problem, and each wishes to stake out for himself an identifiable claim.)

There seems to be a desire (with which one can find a certain sympathy) to divorce the observations from considerations of dynamics, presumably with the thought that to do so is in some way more objective. We feel that this is a mistake. The boundary layer is, after all, constrained to obey the Navier–Stokes equations; knowing that, we are in possession of a sort of Kalman filter with which we can sort out the observations. That is, a collection of observations at various times must be consistent with the dynamical behavior of a fluid governed by the Navier–Stokes equations at those times, taking the earlier observations as part of various possible initial conditions, from which the later observations may be predicted. This formal process is used in weather prediction to replace missing data; it can also be used to interpret surface measurements in the boundary layer to give information about the state of the fluid-mechanical system in phase space. Here we wish merely to

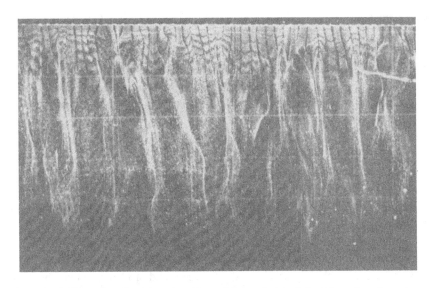

Figure 2.14 Flow visualized in a low Reynolds number turbulent boundary layer by a bubble-wire placed at $y^+ = 4.5$. From [195]; photo courtesy S. J. Kline.

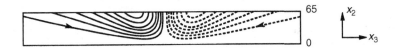

Figure 2.15 Eigenfunctions of the proper orthogonal decomposition in the wall region of a turbulent channel flow. From Moin [241].

apply it informally, in the sense that we accept only observations that are consistent with a dynamical model.

There is no basic disagreement about certain aspects of the form of the coherent structures that are observed in the turbulent boundary layer. Figure 2.14 shows a turbulent boundary layer. The flow is from top to bottom, and the reader is looking down on the layer; the surface is in the plane of the page. A bubble wire is placed across the top, at a height in the boundary layer of 4.5 in wall units; small spots of varnish, uniformly spaced, prevent the wire from making bubbles at a periodic array of points. The current is turned off and on periodically, to make blank lines. In undisturbed laminar flow the result resembles a sheet of graph paper with dark, bubble-free spanwise and streamwise lines. In the turbulent flow, it is evident that the bubbles are being swept into windrows by lateral velocities, and that in these windrows, the streamwise velocity is reduced. Among many other observations, probably the most reliable are those in which the proper orthogonal decomposition was used to identify recurrent coherent patterns in the data, as described in the next chapter (e.g. [241, 244]). Figure 2.15 shows the leading empirical eigenfunctions of this decomposition in the wall region; it is clear that the coherent structures that are sweeping the bubbles together into windrows are rolls. More complete analysis shows that these take the form of rolls approximately in the streamwise direction, not quite parallel to the wall, but with the downstream end slightly further from the wall. The rolls have diameters of a few tens in wall units, say no more than 50. Just what angle is made with the wall is a matter of controversy. These rolls sometimes occur alone, or with a companion of opposite rotation which might be very different in strength. The fact that the eigenfunctions of the proper orthogonal decomposition are obtained from a symmetric kernel (see Section 3.3, below) constrains them to be symmetric, giving rolls of equal strength, but the instantaneous fields typically do not obey this symmetry.

There is associated with these rolls a region of reduced speed in the x_1-direction, roughly as long as the rolls, and on the side of the roll having velocities in the x_2-direction away from the wall; we shall refer to this as the positive side. The region of reduced stream-wise velocities is about as narrow as a roll, and is known as a low-speed streak. There is some controversy in the literature about whether streaks occur alone, or always with a roll. The rolls are observed to evolve: they gradually strengthen, and in particular the updraft on the positive side strengthens, ejecting slow-moving fluid from the wall region. This part of the evolution is known as the ejection phase. Some sort of secondary instabil-ity is observed to take place, associated with ejection, and producing a burst of Reynolds stress, which is largely responsible for the maintenance of the turbulence in the boundary layer, but there is considerable controversy about the exact nature of this instability. Fol-lowing the Reynolds stress burst, the intense updraft on the positive side is observed to be replaced by a gentle downdraft, bringing higher-speed fluid down toward the wall from the outer part of the boundary layer. This is called the sweep phase. There is no general agreement on the etiology of these rolls.

We can make a certain amount of sense of these observations by introducing some dynamical considerations. Let us consider first the behavior of a streamwise invariant roll in a laminar shear flow, having a perturbation velocity $u_j(x_2, x_3; t)$, with no x_1-dependence. It is easiest to consider a periodic array of such rolls, rotating in alternate senses. It is straightforward to show that such rolls are globally stable, and must ultimately decay (see the discussion of Moffatt's paradox in Section 10.7.1). The equations for cross-stream velocities do not contain the mean velocity gradient, and hence have no source of energy. The streamwise velocity equation has an energy source consisting of the mean veloc-ity gradient multiplied by u_2. Thus, while u_2 and u_3 simply decay, u_1 at first grows, passes through a peak, and then decays. Physically, the velocity components in the x_2- and x_3-directions sweep slow-moving fluid up from the vicinity of the wall, ejecting it into the outer flow, forming a low-speed streak. These velocities lead the velocity in the x_1-direction, so that the last part to disappear, like the grin of the Cheshire cat, is the low-speed streak in the x_1-direction. We believe that this is the explanation for the persis-tent claims that the low-speed streaks are found without rolls. These conclusions are not dependent on the artificiality of the periodic structure; an isolated roll will go through the same life cycle, and will eject slower-moving fluid from the wall on the positive side, in just the same way.

The transient streamwise velocity profile that is formed in connection with the low-speed streak has an inflection point in the (x_1, x_2)-plane, and two such points in the (x_1, x_3)-plane as shown in Figure 2.16. Either the latter pair, or the former one, could be responsible for instability, since an inflectionary profile is inviscidly unstable [245]. Presumably the

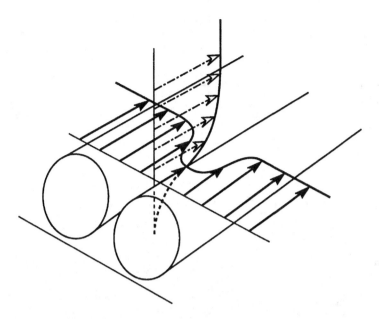

Figure 2.16 The wall-normal and spanwise instantaneous velocity profiles between two rolls, showing locations of the three inflection points.

growth rate of the instability associated with the inflection point in the (x_1, x_2)-plane is dependent on a number such as $U_{1,2}\nu/u_\tau^2$, where $U_{1,2}$ denotes the shear at the inflection point; the instability associated with the inflections in the (x_1, x_3)-plane is dependent on a similar quantity, but involving $|U_{1,3}|$ evaluated at the inflection point. As the intensity of the maximum x_1-velocity defect grows in the early part of the cycle, these numbers will grow, and presumably reach a critical value at which the secondary disturbance begins to grow. This growing disturbance presumably corresponds to the burst of Reynolds stress observed in the boundary layer following the ejection phase. Current indications [229] are that it is the inflections in the (x_1, x_3)-plane that become unstable first.

The bursting process is observed to be quite intermittent both in time and space. If T_B is the mean time between bursts, then we observe that roughly $T_B u_\tau^2/\nu = 3.6 \times 10^2$ [226] although there is a great scatter in the experimental data. This bursting time can be related to the wall shear stress, or drag coefficient. Using the definition of the drag coefficient in Equation (2.56) we can write $C_D = 7.2 \times 10^2 / (T_B U_s^2/\nu)$. Hence, for a fixed mean velocity, if T_B is doubled, then the drag coefficient is cut in half. We return to this when we discuss drag reduction in Section 10.6.

Figure 2.17 shows a realization from a direct numerical simulation. The vortices are made visible by showing the reduction in pressure in their cores (which occurs due to the rapid rotation). The Reynolds number is necessarily rather low, although it is not clear that this presents a problem. What is clear is the difficulty of making sense of structures in stochastic fields without the aid of sophisticated statistical techniques. Although the wall region, being governed by the Reynolds number $u_\tau x_2/\nu = y^+$, is basically a low Reynolds number flow, with relatively few degrees of freedom excited – indeed the first

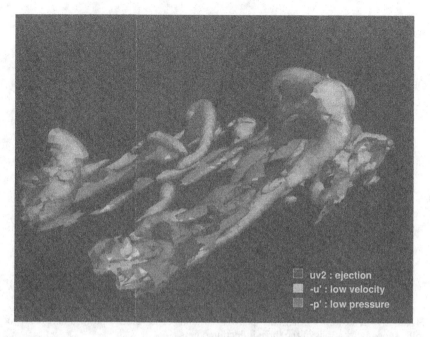

Figure 2.17 Direct numerical simulation of the wall region of a turbulent boundary layer; vortices visualized by pressure reduction in their cores. From Robinson [307].

eigenfunction of the proper orthogonal decomposition represents nearly 80% of the energy: see Section 4.4 below – the field is more cluttered and more difficult to interpret than that in the mixing layer, for example. However, the streamwise orientation and elongation of low pressure vortex cores is clear.

The source of these streamwise rolls is a matter of great controversy. The school of thought to which we belong believes that they result from the same sort of instability that was described in the mixing layer, that is, an instability of the sort that produces Langmuir cells in the ocean surface mixed layer. The mean velocity profile produces cross-stream vorticity, as we have seen: $\Omega_3 \approx -U_{1,2}$. These vortex lines are approximately convected with material lines (to an inviscid approximation). Turbulent fluctuations in the layer perturb these vortex lines. There is an x_2-gradient in the mean Lagrangian transport velocity in the x_1-direction. The Ω_3 lines that are perturbed upward are carried downstream faster than the parts of the Ω_3 lines that are perturbed downward. The parts of the perturbed vortex lines that join the upper and lower segments are now tilted, so that there is a component of vorticity in the streamwise direction. As the upper segments are carried downstream faster than the lower ones, the streamwise parts are stretched and intensified. These stretched and intensified streamwise parts become the rolls. The lines are not parallel to the surface, but are tipped; they become more nearly parallel, the longer the stretching process goes on. At any instant, there will be a population of such lines at various points in their life cycle, and hence, making various angles with the surface. The average of this population will give the observed angle. These vortices are sometimes referred to in the literature as "lambda," or "hairpin" vortices, because of their shape, although it is rare that the "head" – the upper segment, between the two rising legs – can be seen; since this has not been stretched, it is much weaker, see, for example [307].

Much of the above reasoning is supported in a paper [151] presenting the results of direct numerical simulation of a plane Couette flow: that is, a plane flow, parallel in the mean, driven by parallel boundaries moving at equal speeds in opposite directions (also see [382]). The flow near the wall is, of course, universal, and will be structurally the same in a boundary layer, a Couette flow, or a channel flow. The authors take a domain of small spanwise and streamwise extent, similar to the minimal flow unit of [175], and a low Reynolds number, so that the bursting process is more orderly than in a typical turbulent boundary layer, and examine the dynamics in terms of individual Fourier modes. As expected, they find that low-speed streaks are formed by the streamwise rolls, and that the inflectionary profile formed in the streamwise velocity by the movement of slow fluid away from the wall is unstable, leading to bursting. The most interesting part of their study concerns the regeneration of the streamwise vorticity after a burst. We shall translate the simplified scenario of vortex line perturbation and stretching described above into Eulerian terms, and imagine an initial value problem.

Suppose initially we have a perturbation (u_2, u_3) in the form of streamwise vortices and that there is no streamwise variation, so that $(\cdot)_{,1} = 0$. In this case the inviscid evolution equations for vorticity are as follows:

$$\left. \begin{aligned} \dot{\omega}_1 &= -\omega_{1,2} u_2 - \omega_{1,3} u_3 + u_{1,2}\omega_2 + u_{1,3}\omega_3, \\ \dot{\omega}_2 &= -\omega_{2,2} u_2 - \omega_{2,3} u_3 + u_{2,2}\omega_2 + u_{2,3}\omega_3, \\ \dot{\omega}_3 &= -\omega_{3,2} u_2 - \omega_{3,3} u_3 + u_{3,2}\omega_2 + u_{3,3}\omega_3. \end{aligned} \right\} \tag{2.73}$$

The perturbation constitutes streamwise vorticity ω_1, and it moves this vorticity around, contributing to $\dot{\omega}_1$ through $-\omega_{1,2} u_2 - \omega_{1,3} u_3$. Of these two terms, $-\omega_{1,2} u_2$ is the larger, since non-zero cross-stream derivatives are present only due to the perturbation, while all derivatives normal to the wall are of order one. However, let us consider creation of *new* streamwise vorticity ω_1^n (over and above the perturbation, and other than simple advection) by the inviscid evolution equation. For *new* vorticity at $t = 0$ we have $\omega_1^n = \omega_2 = 0$, $\omega_3 \neq 0$, $u_{1,3} = 0$, $\dot{\omega}_1^n = 0$. Hence, the equations are now:

$$\left. \begin{aligned} \dot{\omega}_1^n &= 0, \\ \dot{\omega}_2 &= u_{2,3}\omega_3, \\ \dot{\omega}_3 &= -\omega_{3,2} u_2 + u_{3,3}\omega_3. \end{aligned} \right\} \tag{2.74}$$

The initial lifting of the vortex lines to create the low-speed streaks is described by $\dot{\omega}_2 = u_{2,3}\omega_3$. Hence, vorticity normal to the wall is created ($\omega_2 = u_{1,3}$) which can be identified with the low-speed streaks. At the same time the transverse vorticity is advected, stretched, and shrunk by the transverse velocity field, producing fluctuations in the transverse vorticity: $\dot{\omega}_3 = -\omega_{3,2} u_2 + u_{3,3}\omega_3$. Now that there is vorticity ω_2 normal to the wall associated with non-zero values of $u_{1,3}$, streamwise vorticity can be created by tilting, turning, and stretching of this vorticity; this is described by $\dot{\omega}_1^n = u_{1,2}\omega_2 + u_{1,3}\omega_3$.

Of course, in the real process, all these things go on simultaneously. In addition, the flow is not streamwise invariant, so that there are also contributions from the streamwise modes.

As suggested in our earlier remarks, the authors of [151] also find that the primary contribution to the growth of new streamwise vorticity is due to the term $-\omega_{1,2}u_2$, which involves convection (redistribution) of existing streamwise vorticity, in agreement with our scenario; after that, $u_{1,3}\omega_3$ and $u_{1,2}\omega_2$ are the primary sources of the streamwise invariant terms, also in agreement with our suggested mechanism. The streamwise terms that contribute are $u_{1,1}\omega_1$ and $-\omega_{1,1}u_1$. We feel that the results in [151] and the present discussion provide strong support for the primacy of a Langmuir type instability as a source of the coherent structures.

There are other possibilities. When there is a burst of Reynolds stress associated with the secondary instability, this will strongly decelerate the flow locally, since it is a sudden, local increase in drag. This local deceleration resembles a phantom obstacle (Figure 2.18), which may produce a necklace vortex. Necklace vortices are produced around bridge piers and trees: the vorticity in the boundary layer is bent around the obstacle and stretched, producing a streamwise vortex pair with a downdraft behind the obstacle. Each of these legs, in collaboration with one leg of an adjacent necklace vortex, can form a new pair of rolls with an updraft between them. There are uncertain observations that rolls move laterally following a burst [195] and this scenario would explain that observation. It also happens to be in agreement with the low-dimensional model discussed in Chapters 10 and 11, which also predicts lateral motion.

There are more complex and elaborate instability models, e.g. that of Perry and co-workers, which also lead to lambda vortices. In addition, groups at Michigan State and at Lehigh believe they see motions approaching the wall from the outer part of the layer, and in the impact with the wall forming vorticity around the point of impact, the streamwise segments of which are stretched to form new streamwise rolls. These various ideas are summarized, with references, in [307].

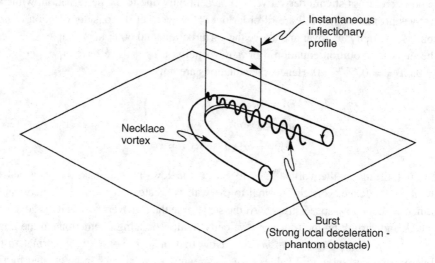

Figure 2.18 Necklace vortex produced by a phantom obstacle.

In evaluating these many possibilities, it seems that we should accept the simplest physical model that explains the largest number of observations. Unfortunately, it is not yet clear which this might be. We, of course, are prejudiced in favor of our own model (the Langmuir circulation instability, mentioned above), but all the models outlined here are consistent with the Navier–Stokes equations, in that they involve possible vortex motions and known interactions. In any event, we have streamwise rolls and bursting, no matter what their etiology. We can study the interaction of rolls regardless of their origin and this is what the low-dimensional models help us to do.

2.6 A preview of things to come

The question of the rôle and function of coherent structures, of why they are present in a turbulent flow, can, of course, have many answers. We have hinted above at some of these: they probably have grown from an instability in the past or present life of a material region; their location and orientation may be constrained by the geometry of the flow; turbulent material regions have long memories, and the coherent structures now present will in any event be strongly influenced by the past history of the material region.

These are not very sophisticated answers, however, from either the dynamical or mathematical points of view. In order to see how some of these rather imprecise notions might have clear mathematical analogs, it is helpful to restrict ourselves for a moment to functions of a single variable, time (as opposed to fields in space–time), and consider a variable produced by a low-dimensional dynamical system, one in particular that displays chaotic behavior. Gol'dshtik [131] edited a pioneering study of such systems (see also [63, 64]). Such systems are often characterized by the chaotic recurrence of repeatable patterns in their temporal evolution, which might be considered one-dimensional (temporal) versions of coherent structures. As in the shot effect expansion of Campbell (see, e.g. Rice [303]), the signal can be adequately represented by a summation of these characteristic patterns, occurring at statistically independent times. It was Gol'dshtik's contribution to recognize that these characteristic patterns might correspond to trips around a particular type of attractor in the phase space of the dynamical system that produced the signal.

In the simplest situation there is an unstable fixed point and a homoclinic orbit in phase space. The system's state hovers in the neighborhood of the fixed point for some time. This duration is determined by the amount of noise or other external perturbations acting on the system. Eventually, the noise is sufficient to kick the state away from the fixed point. It leaves the fixed point and takes a trip around the attractor, along the homoclinic orbit, returning to the fixed point. The noise, which influences the distribution of times at which the system leaves the fixed point, determines when the characteristic pattern appears. The trip around the homoclinic orbit produces the characteristic pattern itself. See Chapter 9 for more information.

Now, it is tempting to consider the possibility that coherent structures in turbulent space-time fields may also correspond to attractors in some sense. The bulk of dynamical systems theory was developed in the context of finite- and generally low-dimensional systems. While many of the ideas do extend to infinite-dimensional evolution equations such as the Navier–Stokes equations, it is very hard to obtain clear analytical or even numerical evidence. But there are some hints that may illuminate the space-time situation. Deissler [94] has carried out direct numerical simulations of two homogeneous turbulent flows with slightly different initial conditions, and has demonstrated that a norm for the difference between the two fields grows exponentially. According to at least one definition of a strange attractor (see Section 6.5), this is sufficient to demonstrate that a turbulent flow does correspond to a strange attractor in the phase space of the governing equations.

In addition, consider the low-dimensional models for the wall region of the turbulent boundary layer, which we lay out in more detail in Chapters 10 and 11. Admittedly, they are merely models, and not the turbulent boundary layer itself. However, we find that such systems have in their phase spaces circles of fixed points that correspond to arrays of quiescent streamwise rolls, which are the observed coherent structures in this flow. Relatively broad ranges of parameters exist for which typical solutions of the system make repeated trips along heteroclinic orbits connecting fixed points on this circle. These trips loosely correspond to bursting in the turbulent boundary layer. Notice that there is a subtle difference between this space–time situation and the one-dimensional situation of Gol'dshtik: there, the states in which the solution lingers near the fixed point correspond to the (featureless) durations between coherent structures, when essentially nothing happens; the homoclinic orbit corresponds to the temporal form of a coherent structure. In the space-time situation, the fixed point corresponds to a quasisteady flow field spatially dominated by coherent structures, and the heteroclinic orbit corresponds to the bursting described in the previous section. However, the solutions lingering near the fixed point still correspond to the rather quiescent time between abrupt events which occur at unpredictable times. It is more a question of which event is labeled a coherent structure. Doubtless, the bursts in the space–time situation also have their own characteristic structure, but this has not been measured exhaustively, and is not usually identified as a coherent structure.

In Chapter 11 we also mention the work of Aubry and Sanghi, see [25, 323], in which considerably higher-dimensional models for the wall region of the turbulent boundary layer were examined. There it was found that the system's state hovered in the neighborhood, not of a fixed point, but of another invariant set, probably contained in a torus, making occasional excursions from this set along a heteroclinic orbit, and back to another location near the invariant set. Again, the invariant set corresponds roughly to the coherent structures observed (which now have non-trivial temporal as well as spatial behavior), while the trips along the heteroclinic orbit correspond to bursting events.

It is the purpose of this book to explore these deeper identifications of coherent structures with features in the phase spaces of such (fairly) low-dimensional dynamical systems. More generally, we shall argue that the geometrical view of phase space afforded by dynamical systems theory can provide new and useful insights into the spatio-temporal dynamics of turbulent flows containing coherent structures. The presence of such structures implies that a limited set of localized spatial and temporal patterns continually recurs in the flow, albeit irregularly in space and time. It is our task to show how suitable bases or modal representations of velocity fields can be derived from this observation and to show how one can deduce, from the governing Navier–Stokes equations, relatively simple sets of ordinary differential equations (ODEs) which describe the interaction of these modes and which complement the order-of-magnitude estimates and deductions of Sections 2.4 and 2.5 above. The fact that coherent structures dominate the turbulent kinetic energy in turn implies that they correspond to solutions on low-dimensional attractors in these ODE models. The heteroclinic attractors of the models developed by Aubry *et al.* [22] and Aubry and Sanghi [25, 323], mentioned above, provide merely one class of examples, and we feel that our strategy, which effectively produces a map relating physical space to phase space in a manner adapted to coherent structures, permits analysis of a much wider range of dynamical mechanisms underlying the production of turbulence.

3

Proper orthogonal decomposition

The proper orthogonal decomposition (POD) provides a basis for the modal decomposition of an ensemble of functions, such as data obtained in the course of experiments. Its properties suggest that it is the preferred basis to use in various applications. The most striking of these is *optimality*: it provides the most efficient way of capturing the dominant components of an infinite-dimensional process with only finitely many, and often surprisingly few, "modes."

The POD was introduced in the context of turbulence by Lumley in [220]. In other disciplines the same procedure goes by the names: Karhunen–Loève decomposition, principal components analysis, singular systems analysis, and singular value decomposition. The basis functions it yields are variously called: empirical eigenfunctions, empirical basis functions, and empirical orthogonal functions. According to Yaglom (see [221]), the POD was introduced independently by numerous people at different times, including Kosambi [197], Loève [215], Karhunen [183], Pougachev [285], and Obukhov [272]. Lorenz [216], whose name we have already met in another context, suggested its use in weather prediction. The procedure has been used in various disciplines other than fluid mechanics, including random variables [275], image processing [313], signal analysis [5], data compression [7], process identification and control in chemical engineering [118,119], and oceanography [286]. Computational packages based on the POD are now readily available (an early example appeared in [11]).

In the bulk of these applications, the POD is used to analyze experimental data with a view to extracting dominant features and trends – in particular coherent structures. In the context of turbulence and other complex spatio-temporal fields, these will typically be patterns in space *and* time. However, our goal is somewhat different. As outlined in the introductory chapter, we wish to use the POD to provide a "relevant" set of basis functions with which we can identify a low-dimensional subspace on which to construct a model by projection of the governing equations. The POD will produce the key *spatial* ingredients, from which our models will *dynamically* recreate the coherent structures as time-dependent *mixtures* of POD modes. In the following sections we review the theory behind the POD and present some new results and extensions which are useful in the context of low-dimensional models. The POD is an especially important tool in this connection, since it yields an increasing sequence of finite-dimensional subspaces of the full phase space, chosen to resolve those parts of physical and phase space which contain the dominant dynamics.

Although we shall almost exclusively apply it to nonlinear problems, it is important to recognize that the POD is a linear procedure, and the nested sequence of subspaces referred to above are *linear* spaces, even if the dynamical systems ultimately to be defined in them, and the source of the data that generates them, are nonlinear. Linearity is the source of the method's strengths as well as its limitations: appealing to results from the theory of linear operators, we can give a fairly complete account of the properties of representations via the POD, but in stating optimality results, for example, the reader must remember that we imply optimality only with respect to other linear representations. By a linear representation, we mean the superposition by a finite or an infinite sum of modal functions multiplied by appropriate coefficients, such as a Fourier series. These representations in terms of basis functions chosen a priori or by the POD are blind to the origin of the functions they are called upon to represent, which of course may, and in our case do, derive from nonlinear dynamical processes.

We start by introducing the POD in the simple context of scalar fields. The succeeding sections describe the properties of representations using empirical basis functions, precisely characterizing the fields they can reproduce and the sense in which they are optimal, and explaining how symmetries in the data sets affect them. We also discuss the relationship between the use of empirical basis functions and the structure of attractors in phase space. The chapter ends with some comments on the relation of the POD to other techniques for the statistical analysis of data. Basic derivations and descriptions of the POD can be found in various books, most notably in the context of turbulence in Lumley ([221], Section 3.5ff.), but a number of the results we shall find useful for low-dimensional models do not appear well known. We have therefore included statements and proofs of them, relegating some of the more technical derivations and discussions to the Appendix (Section 3.8). The mathematical basis for the POD is the spectral theory of compact, self-adjoint operators and we sketch some of the relevant background of this in the Appendix also, along with those elements of measure theory needed for the definition of averages. Some of the material of this chapter appeared previously in [45].

3.1 Introduction

The fundamental idea behind the POD is straightforward. Suppose that we have an ensemble $\{u^k\}$ of scalar fields, each being a function $u = u(x)$ defined on the domain $0 \le x \le 1$. Ultimately we want to use the theory in combination with Fourier decompositions, and so we allow u be complex valued. In seeking good representations of members of $\{u^k\}$, we need to project each u onto candidate basis functions, and so we assume that the functions belong to an inner product space: for example, the linear, infinite-dimensional Hilbert space $L^2([0, 1])$, of square integrable (complex-valued) functions with inner product

$$(f, g) = \int_0^1 f(x)g^*(x)dx. \tag{3.1}$$

(More information on L^2 spaces was given in Section 1.4.) In this context, we want to find a basis $\{\varphi_j(x)\}_{j=1}^{\infty}$ for L^2 that is optimal for the data set in the sense that finite-dimensional representations of the form

$$u_N(x) = \sum_{j=1}^{N} a_j \varphi_j(x) \tag{3.2}$$

describe typical members of the ensemble better than representations of the same dimension in *any* other basis. The notion of "typical" implies use of an averaging operation, which we denote by $\langle \cdot \rangle$ and which is assumed to commute with the spatial integral (3.1) of the L^2 inner product. Averaging is discussed in more detail in the next section and in the Appendix, but it is sufficient for now to imagine an ensemble average over a number of separate experiments forming $\{u^k\}$, or a time average over an ensemble with members $u^k(x) = u(x, t_k)$, obtained from successive measurements during a single run.

We derive the POD in the context of a general Hilbert space $\mathcal{H}$, with inner product $(\cdot, \cdot)$; that is, we do not require the specific definition (3.1). While this approach may seem overly abstract, it allows us to readily specialize the results to a number of cases of interest, such as functions of more than one variable and vector-valued functions including the three-dimensional velocity fields $\mathbf{u}(\mathbf{x}, t)$ of turbulence. If the abstract formulation is off-putting, it may help to think of the space $\mathcal{H}$ as the space of flowfields at a given time instant, so that an element of $\mathcal{H}$ is a snapshot of the flow.

The mathematical statement of optimality is that we should choose φ such that the average (squared) error between u and its projection onto φ is minimized:

$$\min_{\varphi \in \mathcal{H}} \left\langle \left\| u - \frac{(u, \varphi)}{\|\varphi\|^2} \varphi \right\|^2 \right\rangle, \tag{3.3}$$

where $\| \cdot \|$ is the *induced norm*

$$\|f\| = (f, f)^{1/2}.$$

This is equivalent to maximizing the averaged projection of u onto φ, suitably normalized:

$$\max_{\varphi \in \mathcal{H}} \frac{\langle |(u, \varphi)|^2 \rangle}{\|\varphi\|^2}, \tag{3.4}$$

where $| \cdot |$ denotes the absolute value. Solution of (3.4) as stated would yield only the best approximation to the ensemble members by a *single* function, but the other critical points of this functional are also physically significant, for they correspond to an entire set of functions which, taken together, provide the desired basis.

We now have a problem in the calculus of variations: to extremize $\langle |(u, \varphi)|^2 \rangle$ subject to the constraint $\|\varphi\|^2 = 1$. The corresponding functional for this constrained variational problem is

$$J[\varphi] = \langle |(u, \varphi)|^2 \rangle - \lambda(\|\varphi\|^2 - 1), \tag{3.5}$$

and a necessary condition for extrema is that the functional derivative vanish for all variations $\varphi + \delta\psi \in \mathcal{H}, \delta \in \mathbb{R}$:

$$\frac{d}{d\delta} J[\varphi + \delta\psi] \,|_{\delta=0} = 0. \tag{3.6}$$

From (3.5) we have

$$\frac{d}{d\delta} J[\varphi + \delta\psi]\Big|_{\delta=0}$$
$$= \frac{d}{d\delta}\Big[\langle (u, \varphi + \delta\psi)(\varphi + \delta\psi, u)\rangle - \lambda(\varphi + \delta\psi, \varphi + \delta\psi)\Big]\Big|_{\delta=0}$$
$$= 2\,\mathrm{Re}\Big[\langle (u, \psi)(\varphi, u)\rangle - \lambda(\varphi, \psi)\Big] = 0,$$

where we use the property of the inner product $(f, g) = (g, f)^*$. Interchanging the order of the averaging operation and the inner product, the quantity in brackets may be written

$$\Big(\langle (\varphi, u)u\rangle, \psi\Big) - \lambda(\varphi, \psi), \tag{3.7}$$

or simply

$$\big(\mathcal{R}\varphi - \lambda\varphi, \psi\big), \tag{3.8}$$

where the linear operator $\mathcal{R}$ is defined by

$$\mathcal{R}\varphi = \langle (\varphi, u)u\rangle. \tag{3.9}$$

Finally, since ψ is an arbitrary variation, our condition reduces to the eigenvalue problem

$$\mathcal{R}\varphi = \lambda\varphi. \tag{3.10}$$

Thus the optimal basis is given by the eigenfunctions $\{\varphi_j\}$ of the operator $\mathcal{R}$ that is defined from the empirical data u. They are consequently sometimes called *empirical eigenfunctions*, or *POD modes*.

We shall shortly describe some properties of the POD, but first we illustrate with some examples.

3.1.1 Finite dimensional spaces

Let $\mathcal{H} = \mathbb{R}^N$, with the standard inner product $(\mathbf{x}, \mathbf{y}) = \mathbf{y}^T\mathbf{x}$. Here the "data" are a collection of M vectors $\mathbf{u}^k \in \mathbb{R}^N$, and the average $\langle \cdot \rangle$ is just an arithmetic mean over the ensemble. The operator $\mathcal{R}$ in (3.9) then becomes

$$\mathcal{R}\mathbf{x} = \frac{1}{M}\sum_{k=1}^{M}(\mathbf{u}^k)^T\mathbf{x}\mathbf{u}^k \tag{3.11}$$

or, equivalently, $\mathcal{R}$ is given by

$$\mathcal{R} = \frac{1}{M}\sum_{k=1}^{M}\mathbf{u}^k(\mathbf{u}^k)^T, \qquad R_{ij} = \frac{1}{M}\sum_{k=1}^{M}u_i^k u_j^k, \tag{3.12}$$

or simply the correlation matrix $\langle \mathbf{u}\mathbf{u}^T \rangle$. Thus, $\mathcal{R}$ is a real, symmetric $N \times N$ matrix, and Equation (3.10) is a standard matrix eigenvalue problem on $\mathbb{R}^N$. There is a nice geometrical interpretation in this case: the eigenvectors are simply the principal axes of the cloud of data points $\{\mathbf{u}^k\}$ in the N-dimensional vector space. This idea is discussed further in Section 3.4.2.

3.1.2 Scalar-valued functions

Let $\mathcal{H} = L^2([0, 1])$, with the inner product (3.1). Then for an ensemble of functions $u(x)$, Equation (3.10) becomes

$$\mathcal{R}\varphi(x) = \int_0^1 \langle u(x)u^*(x') \rangle \varphi(x')dx' = \lambda\varphi(x); \qquad (3.13)$$

the kernel of this integral equation is the averaged *autocorrelation function* $R(x, x') = \langle u(x)u^*(x') \rangle$.

3.1.3 Vector-valued functions

Now let $\mathcal{H} = C(\Omega, V)$ denote the space of continuous functions from some (spatial) domain $\Omega \subset \mathbb{R}^3$ to a vector space $V = \mathbb{C}^3$ (e.g., velocity vectors). Define an inner product on $\mathcal{H}$ by

$$\langle \mathbf{u}, \mathbf{v} \rangle = \int_\Omega \mathbf{v}^\star(\mathbf{x}) Q\mathbf{u}(\mathbf{x}) \, dx, \qquad (3.14)$$

where $\star$ denotes the complex-conjugate transpose, and $Q \in \mathbb{C}^{3\times3}$ is a positive-definite Hermitian matrix. (Often we take $Q = I$, the identity matrix, but we shall see later that other inner products may sometimes be useful.) Then the eigenvalue problem (3.10) becomes

$$\mathcal{R}\varphi(\mathbf{x}) = \int_\Omega \langle \mathbf{u}(\mathbf{x})\mathbf{u}^\star(\mathbf{y}) \rangle Q\varphi(\mathbf{y}) \, dy = \lambda\varphi(\mathbf{x}). \qquad (3.15)$$

3.1.4 Technical properties of the POD

Henceforth we focus on the case $\mathcal{H} = L^2([0, 1])$, with inner product (3.1). In the Appendix we give conditions under which $\mathcal{R}$ is a compact self-adjoint operator, in which case spectral theory [304] guarantees that the maximum in (3.4) exists and is equal to the largest eigenvalue of the integral equation (3.13). Moreover, Hilbert–Schmidt theory assures us that there is a countable infinity of eigenvalues and eigenfunctions that provides a diagonal decomposition of the averaged autocorrelation function:

$$R(x, x') = \sum_{j=1}^{\infty} \lambda_j \varphi_j(x) \varphi_j^*(x'), \qquad (3.16)$$

and that the eigenfunctions φ_j are mutually orthogonal in L^2. This derivation is stronger than the variational argument given above, since we are assured a maximum for (3.4) rather than merely a critical point: solution of (3.10, 3.13) is both necessary *and* sufficient.

We can order the eigenvalues by $\lambda_j \geq \lambda_{j+1}$, and observe that the fact that the averaged autocorrelation $R(x, x') = \langle u(x)u^*(x')\rangle$ is non-negative definite implies that the integral operator $\mathcal{R}$ is non-negative definite. This ensures that $\lambda_j \geq 0$ for all j. As we shall see in Section 3.3, almost every member (in a measure or probabilistic sense) of the ensemble used in the averaging $\langle \cdot \rangle$ leading to $R(x, x')$ can be reproduced by a modal decomposition based on the eigenfunctions $\{\varphi_j\}_{j=1}^{\infty}$:

$$u(x) = \sum_{j=1}^{\infty} a_j \varphi_j(x). \tag{3.17}$$

(The proper orthogonal decomposition is literally this equation.) Also, the diagonal representation (3.16) of the two point correlation tensor implies that

$$\langle a_j a_k^* \rangle = \delta_{jk} \lambda_j, \tag{3.18}$$

so that the (random) modal coefficients of the representation are uncorrelated on average. If $u(x)$ is a turbulent velocity field, then the eigenvalues λ_j represent twice the average kinetic energy in each mode φ_j, and thus, picking the subspace spanned by the modes $\{\varphi_j\}_{j=1}^{N}$ corresponding to the first (= largest) N eigenvalues, the representation (3.2) reproduces the most energetic disturbances in the field, as claimed at the outset. The λ_j are called *empirical eigenvalues*.

Thus far we have considered only functions defined on a bounded interval. The unbounded case, which is more natural in the context of open fluid flows, can be dealt with in the same way provided that the inner product (now an infinite integral) is well defined, and that the space of functions still has a countable basis. See the next section for more remarks on this. In dealing with unbounded domains in practice we either select a finite subdomain and use periodic boundary conditions (see Section 3.3.3) or we are concerned with functions which decay to zero rapidly outside a finite region.

Another important point is implicit in Equations (3.17) and (3.18). We have remarked that non-negative definiteness of $R(x, x')$ implies that the empirical eigenvalues λ_j are non-negative themselves, but in general they are not *all* strictly positive. To produce a *complete* basis for L^2 we must include all those "additional" eigenfunctions φ_j with zero eigenvalues, although, in view of (3.18), they carry no information on the original data ensemble. It is therefore often advantageous to consider representations in terms of only those empirical eigenfunctions with non-zero eigenvalues. We shall return to this shortly in Section 3.3.

3.2 On domains and averaging

In our introductory discussion we focused on fields depending on a single variable x. In dealing with turbulence, the fields depend on four variables, three spatial and one temporal. There is no reason a priori to distinguish between space and time and the multi-dimensional theory does not enforce such a distinction: for experiments performed in the one-dimensional spatial domain $0 \leq x \leq 1$ over times of duration T, one simply measures correlations with time lags as well as spatial separations and works in the space $L^2([0, 1] \times [0, T])$, in which case the inner product becomes a double integral over x and t. However, in view of our intended use of the POD in the derivation of low-dimensional models, we generally seek only spatial basis functions. The time-dependent modal coefficients $a_k(t)$ in representations of the form

$$u(x, t) = \sum_k a_k(t)\varphi_k(x) \tag{3.19}$$

and their multi-dimensional analogs, are determined subsequently via projection of the governing equations.

Depending on the physical situation, one has to decide if the problem at hand should be treated as stationary or non-stationary in time. If one seeks purely spatial representations of a space–time field $u(x, t)$ in a statistically stationary problem, the correlation functions between pairs of points in physical space must be measured with no time lag. Assuming ergodicity, time may be used to increase the ensemble size by including measurements taken at appropriately separated intervals during a single experimental run, in which case the ensemble members may be defined as $u^k(x) = u(x, t_k)$ and $\langle \cdot \rangle$ effectively becomes a time average. In contrast, if one wishes to represent a non-stationary field as a function over (x, t)-space, correlations with time lags as well as spatial separations must be measured and multiple runs of the experiment undertaken to increase the ensemble size. In such a case one could still derive purely spatial representations from the zero time lag correlation averaged over ensembles chosen from different experiments of the same "age," for example, or by appeal to some other condition. We refer the reader to [368] for a discussion of such issues. In our applications we shall generally assume stationarity in time.

In fluid mechanical applications involving a well-defined (finite) domain in physical space with well-defined boundary conditions, examination of the mathematical details associated with the averaging operator is usually unnecessary. However, it is important to appreciate that, when dealing with an infinite domain, difficulties may arise. The appropriate function space for time-stationary problems is $L^2(\Omega)$ where Ω is the (three-dimensional) spatial domain of the experiment, analogous to the interval $[0, 1]$ in Section 3.1. The appropriate function space for time-dependent problems, for which there is no a priori upper bound on duration, is $L^2_{\text{loc}}(\Omega \times [0, \infty))$, where the subscript loc implies that the L^2-norm is finite on finite closed intervals in time (the second variable). For problems in fluid mechanics a finite L^2-norm is a reasonable assumption since it corresponds to finite kinetic energy (although if infinite-time experiments were possible then the integrated kinetic energy might become unbounded). Here we merely caution the reader that subtleties arise in the time-dependent case due, for example, to the fact that $L^2_{\text{loc}}(\Omega \times [0, \infty))$ is not a separable space [304] and so does not admit a countable set of basis functions. Most of the analysis in this chapter assumes countable bases.

3.3 Properties of the POD

In much of the literature cited thus far in this chapter the POD is primarily regarded as a tool for analysis of experimental data. We now wish to view it in a more dynamical context. Throughout the remainder of the chapter the reader should imagine that the ensembles from which the autocorrelation function and empirical bases are generated originate from solutions belonging to the *attractor* of a dynamical system such as the Navier–Stokes equations, realized either by a physical or a numerical experiment. Attractors are discussed in more detail in Chapter 6; for the moment it is enough to think of a set in phase space to which all solutions starting sufficiently close approach as time increases. In the best case, the attractor is *ergodic*, which means that time averages and averages over the part of phase space containing the attractor coincide. In this case the initial conditions are "forgotten" as time proceeds, much as in the common hypothesis that certain turbulent flows relax in physical space to "universal equilibrium states" (see Narasimha [256]).

The following subsections describe properties that are especially important in our use of the POD to derive low-dimensional models. In the first two we characterize the classes of functions that can be represented by empirical bases and explain precisely how such representations preserve properties of the observations from which they are derived and how they are optimal. We then consider symmetries, showing that in the case of translation invariance (homogeneous directions), the empirical eigenfunctions are simply Fourier modes, and we obtain results on ergodic attractors invariant under more general symmetry groups. We then show how the rate of decay of the empirical eigenvalues determines geometrical properties of the attractor and how theoretical results on the regularity of solutions of the governing evolution equations are related to this.

3.3.1 Span of the empirical basis

The first step in understanding what can be done with representations using empirical eigenfunctions is in characterizing the class of functions which can be accurately represented by the "relevant" elements of the basis: those containing spatial structures having finite energy on average. This is the set $S = \{\sum a_j \varphi_j | \sum |a_j|^2 < \infty, \lambda_j > 0\}$, or $\text{span}\{\varphi_j | j = 1, \ldots, \infty, \lambda_j > 0\}$. In this section, equality of functions is interpreted as almost everywhere in the spatial domain Ω with respect to Lebesgue measure: two functions f and g are equal in this sense if

$$\int_\Omega |f - g|^2 dx = 0; \tag{3.20}$$

this is the mathematical definition of "accurately." We also frequently use a second notion of *almost every member of an ensemble with respect to the probability measure underlying the averaging operation* $\langle \cdot \rangle$. This is denoted by "a.e." In applications this average is typically a finite sum over a set of realizations or an integral over a finite-time experimental run, but the theory is developed in the ideal case of infinite data sets.

A standing assumption in this section is that the averaged autocorrelation $R(x, x')$ is a continuous function. Discontinuities in R can lead to negative values in the power spectrum – the Fourier transform of R – and negative energies are unreasonable on physical grounds. See both Section 3.10 and Appendix 3.13 of Lumley [221] for further details.

We first show that the empirical basis can reconstruct any function that is indistinguishable in the sense of (3.20) from a member of the original ensemble $\{u^k\}$. Let $u \in L^2(\Omega)$ be any such function and $\{\varphi_j\}$ be the orthonormal sequence of empirical eigenfunctions. The reconstruction of u is a function $u_s(u) = \sum_j (u, \varphi_j)\varphi_j$, belonging to S. (By Parseval's inequality we know that $\sum_j |(u, \varphi_j)|^2$ converges.) We need to show that for a.e. u with respect to the ensemble average, we have $u = u_s(u)$; that is:

$$\langle \|u - u_s\|^2 \rangle = 0. \tag{3.21}$$

From the real valuedness of u, we have

$$\langle \|u(x) - u_s(x)\|^2 \rangle = \langle (u - u_s, u - u_s) \rangle$$
$$= \langle (u, u) - 2(u, u_s) + (u_s, u_s) \rangle. \tag{3.22}$$

Since the functions u are the members of the original ensemble, we have

$$\langle (u, u) \rangle = \left\langle \int_\Omega u(x)u^*(x)dx \right\rangle = \int_\Omega R(x, x)dx, \tag{3.23}$$

and also

$$\langle -2(u, u_s) \rangle = -2 \left\langle \int_\Omega u(x) \left[\sum_j (u^*, \varphi_j^*)\varphi_j^*(x) \right] dx \right\rangle$$

$$= -2 \left\langle \int_\Omega u(x) \sum_j \left[\int_\Omega u^*(x')\varphi_j(x')dx' \right] \varphi_j^*(x)dx \right\rangle$$

$$= -2 \int_\Omega \sum_j \left[\int_\Omega \langle u(x)u^*(x') \rangle \varphi_j(x')dx' \right] \varphi_j^*(x)dx$$

$$= -2 \int_\Omega \left(\sum_j \mathcal{R}\varphi_j \right)\varphi_j^* = -2 \int_\Omega \sum_j \lambda_j \varphi_j(x)\varphi_j^*(x)dx$$

$$= -2 \int_\Omega R(x, x)dx. \tag{3.24}$$

The third term of (3.22) is:

$$\langle (u_s, u_s) \rangle = \left\langle \int_\Omega \left[\sum_i (u, \varphi_i)\varphi_i(x) \right] \left[\sum_j (u^*, \varphi_j^*)\varphi_j^*(x) \right] dx \right\rangle$$

$$= \left\langle \sum_{i,j} (u, \varphi_i)(u^*, \varphi_j^*) \int_\Omega \varphi_i(x)\varphi_j^*(x)dx \right\rangle$$

$$= \left\langle \sum_j \int_\Omega u(x)\varphi_j^*(x)dx \int_\Omega u^*(x')\varphi_j(x')dx' \right\rangle$$

$$= \sum_j \int_\Omega \left[\int_\Omega \langle u(x)u^*(x') \rangle \varphi_j(x')dx' \right] \varphi_j^*(x)dx$$

$$= \sum_j \int_\Omega \lambda_j \varphi_j(x)\varphi_j^*(x)dx.$$

Using the continuity of R we can apply Mercer's theorem for the uniform convergence of the series expression for R and interchange summation and integration, obtaining

$$\int_\Omega \sum_j \lambda_j \varphi_j(x)\varphi_j^*(x)dx = \int_\Omega R(x,x)dx. \tag{3.25}$$

Combining (3.23), (3.24), and (3.25) we obtain (3.21).

We have shown that almost every member of the original ensemble can be reconstructed as a linear combination of empirical eigenfunctions having strictly positive eigenvalues. Now we want to show the converse: that each such eigenfunction can be expressed as a linear combination of observations. This will imply that any property of the ensemble members that is preserved under linear combination is inherited by the empirical basis functions and hence by elements of S.

Let X denote the set of functions (of full measure with respect to the averaging operation) for which reconstructions satisfying (3.21) are possible, and let θ be any function in S. We claim that there is a sequence $\{b_j\}_{j=1}^\infty$ of real numbers and a set of functions $u^j(x) \in X$ for $j = 1, \ldots, \infty$ such that

$$\theta(x) = \sum_{j=1}^\infty b_j u^j(x). \tag{3.26}$$

It immediately follows from (3.26) that, if $\mathcal{P}$ is a closed linear property of a subset of functions in $L^2(\Omega)$ (i.e. all $u \in L^2(\Omega)$ with property $\mathcal{P}$ form a closed linear subspace) and all the ensemble members u^k share that property, then the eigenfunctions of the POD also share the property. The converse holds too. Equation (3.26) and this remark characterize the "empirical subspace" S.

It remains to justify Equation (3.26). Let S' denote the set of *all* functions in $L^2(\Omega)$ with representations $\sum_i b_i u^i$ with $u^i \in X$. We can show that $S'^\perp = S^\perp$, from which it follows that $S' = S$, and so the equation indeed holds.

Now $S^\perp$ is exactly the set of functions θ such that $(\theta, \varphi_i) = 0$ for every φ_i with eigenvalue $\lambda_i > 0$. From the first result of this section we have $u(x) = \sum b_i \varphi_i(x)$ where $\lambda_i > 0$, for a.e. (almost every) u. Thus we have $(\theta, u) = 0$ a.e. and so $(\theta, \sum_i b_i u^i) = 0$. This shows that $S^\perp \subset S'^\perp$.

To show $S'^\perp \subset S^\perp$ and hence conclude $S^\perp = S'^\perp$, assume that $(\theta, u) = \int_\Omega \theta(x')u^*(x')dx' = 0$ for a.e. u. Therefore for a.e. u we have $u(x) \int_\Omega u^*(x')\theta(x')dx' = 0$, and taking the average we get

$$\int_\Omega \langle u(x,t)u^*(x',t)\rangle \theta(x')dx' = 0,$$

which, from the eigenvalue Equation (3.13), implies that $(\theta, \varphi_i) = 0$ for every i such that $\lambda_i > 0$.

The classic example of a property which passes from the data ensemble to the empirical basis is incompressibility. If the autocorrelation tensor $\mathbf{R} = \langle \mathbf{u}(\mathbf{x}, t) \otimes \mathbf{u}^*(\mathbf{x}', t)\rangle$ is formed from realizations of divergence-free vector fields $\mathbf{u}$, then the empirical eigenfunctions $\boldsymbol{\varphi}_j(\mathbf{x})$ are also divergence-free. This is very useful when we project the Navier–Stokes equations onto a subspace spanned by a collection of these eigenfunctions in the next chapter. Other important properties inherited by the eigenfunctions include those of satisfying linear boundary conditions, such as no-slip or no-penetration conditions on fixed surfaces.

We now have a characterization of the span of the eigenfunctions with strictly positive eigenvalues. This linear space S exactly coincides with that spanned by all realizations $u^k(x)$ of the original ensemble a.e. with respect to the measure induced by the averaging operation. (A special case of this result, when the average is a sum on a finite number of points, as would be the case in a computer experiment, was observed in [20].) From this we see that the set of empirical eigenfunctions $\{\varphi_j | \lambda_j > 0\}$ need *not* form a complete basis for $L^2(\Omega)$. While S may be infinite-dimensional, it is generally only a subset of the "big" space $L^2(\Omega)$ in which we are working. It is complete only if one includes the kernel of the operator $\mathcal{R}$ – all the (generalized) eigenfunctions with zero eigenvalues – but in doing so one loses the major advantage of the POD, for in many applications one can argue on physical grounds that the realizations $u(x,t)$ do not and should not span $L^2(\Omega)$. (Many strange things may happen in turbulence, but not *everything*.) In such cases the discussion of this section highlights a strong property of the POD. Its use limits the space studied to the smallest linear subspace that is sufficient to describe the observed phenomena. For our models, the moral is that "you can only describe what you've seen already." As we shall see later in the book, this can lead to interesting paradoxes and problems as well as to significant economies.

3.3.2 Optimality

Suppose we have a decomposition of a time-dependent, statistically stationary signal $u(x,t)$ with respect to *any* orthonormal basis $\{\psi_j(x)\}_{j=1}^{\infty}$:

$$u(x,t) = \sum_j b_j(t)\psi_j(x). \tag{3.27}$$

If the $\psi_j(x)$ are dimensionless, then the coefficients $b_j(t)$ carry the dimension of the quantity u. If $u(x,t)$ is a velocity and $\langle \cdot \rangle$ is a time average, the average kinetic energy per unit mass over the experiment is given by

$$\frac{1}{2}\left\langle \int_{\Omega} u(x,t)u^*(x,t)dx \right\rangle = \frac{1}{2}\left\langle \sum_{ij} b_i(t)b_j^*(t) \int_{\Omega} \psi_i(x)\psi_j^*(x)dx \right\rangle$$

$$= \frac{1}{2}\sum_i \langle b_i(t)b_i^*(t) \rangle,$$

and so the average kinetic energy in the ith mode is given by $\frac{1}{2}\langle b_i(t)b_i^*(t) \rangle$ (no summation implied).

We can now precisely state optimality for the POD. Suppose that we have a stationary random field $u(x,t)$ in $L^2(\Omega)$ and that $\{\varphi_i, \lambda_i | i = 1, \ldots, \infty; \ \lambda_i \geq \lambda_{i-1} > 0\}$ is the set of orthonormal empirical eigenfunctions with their associated eigenvalues obtained from time averages of $u(x,t)$. Let

$$u(x, t) = \sum_i a_i(t)\varphi_i(x) \tag{3.28}$$

be the decomposition with respect to this basis and let $\{\psi_i(x)\}_{i=1}^{\infty}$ be any other arbitrary orthonormal set such that

$$u(x, t) = \sum_i b_i(t)\psi_i(x). \tag{3.29}$$

Then the following hold:

1. $\langle a_i(t)a_j^*(t)\rangle = \delta_{ij}\lambda_i$; i.e. the POD random coefficients are uncorrelated.
2. For every n we have

$$\sum_{i=1}^{n} \langle a_i(t)a_i^*(t)\rangle = \sum_{i=1}^{n} \lambda_i \geq \sum_{i=1}^{n} \langle b_i(t)b_i^*(t)\rangle;$$

i.e. the POD is optimal on average in the class of representations by linear superposition: the first n POD basis functions capture more energy on average than the first n functions of any other basis.

The first assertion derives from the representation of $R(x, x')$, given in Equation (3.16):

$$R(x, x') = \langle u(x, t)u^*(x', t)\rangle = \left\langle \sum_i a_i(t)\varphi_i(x) \sum_j a_j^*(t)\varphi_j^*(x') \right\rangle$$

$$= \sum_{ij} \langle a_i(t)a_j^*(t)\rangle \varphi_i(x)\varphi_j^*(x').$$

But we know that

$$R(x, x') = \sum_i \lambda_i \varphi_i(x)\varphi_i^*(x'),$$

and so, since the $\varphi_i^*(x)$ are an orthonormal family in $L^2(\Omega)$, we see that $\langle a_i(t)a_j^*(t)\rangle = \delta_{ij}\lambda_i$.

The second assertion relies on a result on linear operators. Let $\{\psi_j(x)\}_{j=1}^{n}$ be n arbitrary orthonormal vectors in $L^2(\Omega)$ that may be completed to form an orthonormal basis. Let Q denote projection onto span$\{\psi_1, \ldots, \psi_n\}$. We can express the kernel R in terms of $\{\psi_j\}_{j=1}^{\infty}$ as

$$R(x, x') = \left\langle \sum_i b_i(t)\psi_i(x) \sum_j b_j^*(t)\psi_j^*(x') \right\rangle = \sum_{ij} \langle b_i b_j^* \rangle \psi_i \psi_j^*. \tag{3.30}$$

We can then write R in operator matrix notation as

$$\begin{bmatrix} \langle b_1 b_1^* \rangle & \langle b_1 b_2^* \rangle & \langle b_1 b_3^* \rangle & \cdots \\ \langle b_2 b_1^* \rangle & \langle b_2 b_2^* \rangle & \cdots & \cdots \\ \langle b_3 b_1^* \rangle & \cdots & \cdots & \cdots \\ \vdots & \vdots & \vdots & \vdots \end{bmatrix}$$

and the product $R \circ Q$ yields

$$\begin{bmatrix} \langle b_1 b_1^* \rangle & \langle b_1 b_2^* \rangle & \cdots & \langle b_1 b_n^* \rangle & 0 \cdots & 0 \cdots \\ \langle b_2 b_1^* \rangle & \langle b_2 b_2^* \rangle & \cdots & \langle b_2 b_n^* \rangle & 0 \cdots & 0 \cdots \\ \vdots & \vdots & \vdots & \vdots & \vdots & \vdots \\ \langle b_n b_1^* \rangle & \langle b_n b_2^* \rangle & \cdots & \langle b_n b_n^* \rangle & 0 \cdots & 0 \cdots \\ 0 \cdots & 0 \cdots & \cdots & 0 \cdots & 0 \cdots & 0 \cdots \\ \vdots & \vdots & \vdots & \vdots & \vdots & \vdots \end{bmatrix}.$$

The proof is now completed by appeal to Remark 1.3 in Section V.1.2 (p. 260) of Temam [367] (cf. Riesz and Nagy [304]), which states that the sum of the first n eigenvalues of a self-adjoint operator is greater than or equal to the sum of the diagonal terms in any n-dimensional projection of it:

$$\sum_{i=1}^{n} \lambda_i \geq \mathrm{Tr}(R \circ Q) = \sum_{i=1}^{n} \langle b_i b_i^* \rangle. \tag{3.31}$$

This characterization is the basis for the claim that the POD is optimal for modeling or reconstructing a signal $u(x, t)$. It implies that, among all linear decompositions, the POD is the most efficient in the sense that for a given number of modes, n, the projection on the subspace spanned by the leading n empirical eigenfunctions contains the greatest possible kinetic energy on average. Moreover, the time series of the coefficients $a_i(t)$ are linearly uncorrelated.

3.3.3 Symmetry

We start by describing a particular kind of symmetry, called homogeneity in the turbulence literature. We say that the averaged two point correlation $R(x, x')$ is *homogeneous* if $R(x, x') = R(x - x')$, i.e. R depends only on the difference of the two coordinates: it is *translation invariant*. In general, homogeneity of a system is defined through multipoint moments. Here we need only second order moments, but it is important to note that, while the ensemble of realizations $\{u^k\}$ may be translation invariant on average, individual realizations typically are not. Homogeneity occurs in both spatially unbounded systems and systems with periodic boundary conditions. In either case we may develop R in a Fourier representation. In the case of a finite domain, we have the series

$$R(x - x') = \sum c_k e^{2\pi i k(x - x')}. \tag{3.32}$$

We can then solve the eigenvalue problem (3.13) by substituting the (unique) representation

$$R(x, x') = \sum c_k e^{2\pi i k x} e^{-2\pi i k x'}, \tag{3.33}$$

which implies that $\{e^{2\pi i k x}\}$ are exactly the eigenfunctions with eigenvalues c_k. Conversely, if the eigenfunctions are Fourier modes we can write (3.33), which implies (3.32). However, while homogeneity completely determines the form of the empirical eigenfunctions, the numerical values and ordering of the *eigenvalues* depend upon the Fourier spectrum of the particular data set involved. In summary we can state:

- If $R(x, x') = R(x - x')$ is homogeneous, then the eigenfunctions of the operator $\mathcal{R} = \int_\Omega R(x, x') \cdot dx'$ are Fourier modes, and vice versa.

This is especially useful in systems where the domain Ω is of higher dimension. For example, if $\Omega \subset R^2$ and the x_1-direction is homogeneous, the problem of finding eigenfunctions in the two-dimensional domain is decoupled into a set of one-dimensional problems by writing

$$R(x_1, x_1', x_2, x_2') = R(x_1 - x_1', x_2, x_2')$$

and performing the same procedure as above, yielding an eigenvalue problem for each Fourier wavenumber. In the decomposition of the boundary layer used in later chapters we appeal to homogeneity in the spanwise (x_3) and streamwise (x_1) directions, appropriate to a fully developed flow in, say, a channel or a pipe. Selecting the (finite) domain $[0, L_1] \times [0, L_3]$ in these variables, we may then use a mixed discrete Fourier-empirical decomposition of the form

$$\mathbf{u}(\mathbf{x}, t) = \sum_{k_1} \sum_{k_3} \sum_{n} a_{k_1, k_3, n}(t) e^{2\pi i \left(\frac{k_1 x_1}{L_1} + \frac{k_3 x_3}{L_3} \right)} \boldsymbol{\varphi}_n(k_1, k_3; x_2). \tag{3.34}$$

The vector-valued eigenfunctions $\boldsymbol{\varphi}_n(k_1, k_3; x_2)$ in (3.34) are obtained by solving an operator equation analogous to (3.10) in which the kernel $\mathbf{R}(x_1 - x_1', x_2, x_2', x_3 - x_3')$ is replaced by its Fourier transform in the x_1- and x_3-directions. More details are given in Chapter 10; also see [155, 221, 244]. In the turbulent boundary layer, as in other problems with one or more homogeneous directions, this decomposition produces structures that are not localized in spanwise and streamwise extent, unlike the (instantaneous) events observed. We return to this issue in Section 3.6, where we show how one can retrieve local structures from the statistics in certain cases.

Homogeneity or translation invariance is only one of many types of symmetry that physical systems may exhibit. It is an example of a *continuous* symmetry group, for we may make translations by any distance. Discrete groups are also common: in the case of a boundary layer over a surface treated with riblets – strakes parallel to the mean flow direction – one would have spanwise symmetry only with respect to translations through multiples of the riblet spacing. A jet with a lobed mixer would similarly exhibit symmetry under discrete rotations. However, we stress that while a physical system or a model of it in the form of a dynamical system may well admit such symmetries, we cannot expect either individual observations (solutions) or even ensembles of them to share the full underlying symmetry group. In mathematical terms, the system will generally not be ergodic. A simple example of this appears in Chapters 6 and 7, in a reflection symmetric two-dimensional ODE possessing two stable fixed points (Figure 6.9). Any given solution can only approach one of these, and so to reveal the full structure of the attractor, one must average over a set of initial conditions chosen in light of the symmetry. More generally, if a system has several distinct attractors, the time average of a single solution will reproduce just one of these and so the empirical eigenfunctions generated by time averaging from one experimental run will enjoy less symmetry than the problem as a whole.

To make a precise statement characterizing the relation between underlying symmetries and subspaces spanned by the empirical eigenfunctions, we need the notions of equivariant dynamical systems and invariant subspaces, which are discussed (in the finite dimensional context) in Chapters 6 and 7. We therefore relegate it to the Appendix where it appears as Proposition 3. Even to make the informal statement below, we need a few preliminaries, which we give in the context of finite-dimensional ODEs. Let

$$\dot{\mathbf{x}} = \mathbf{f}(\mathbf{x}) \tag{3.35}$$

be an n-dimensional system and Γ be a symmetry group acting on the phase space $\mathbb{R}^n$: the elements $\boldsymbol{\gamma}$ of Γ being $n \times n$ matrices. To say that (3.35) is *equivariant* under Γ means that the equation

$$\boldsymbol{\gamma}\mathbf{f}(\mathbf{x}) = \mathbf{f}(\boldsymbol{\gamma}\mathbf{x}) \tag{3.36}$$

holds for every $\boldsymbol{\gamma} \in \Gamma$. As described in more detail in Chapter 7, this implies that if $\mathbf{x}(t)$ is a solution of (3.35), then so is $\boldsymbol{\gamma}\mathbf{x}(t)$ (think of the two fixed points of the reflection symmetric system mentioned above). Equivariance also typically implies that eigenvalues come in multiples, for if $\mathbf{A}$ denotes the linearization of $\mathbf{f}$, then we have from (3.36)

$$\boldsymbol{\gamma}\mathbf{A}\mathbf{v} = \boldsymbol{\gamma}\lambda\mathbf{v} = \lambda\boldsymbol{\gamma}\mathbf{v}$$

and also

$$\mathbf{A}\boldsymbol{\gamma}\mathbf{v} = \lambda\boldsymbol{\gamma}\mathbf{v},$$

implying that if $\mathbf{v}$ is an eigenvector of $\mathbf{A}$ with eigenvalue λ, then so is $\boldsymbol{\gamma}\mathbf{v}$. The eigenspaces of the operator $\mathcal{R}$ are similarly constrained by symmetries, so that one typically expects several distinct eigenfunctions for a given eigenvalue, corresponding to the same structure differently oriented or located in physical space.

After this preamble, we can now give the gist of the result:

- If φ_j and λ_j are the empirical eigenfunctions and eigenvalues generated from a set of solutions (= experiments) $\{u^k\}$ of a dynamical system equivariant under a group Γ, then a necessary condition for the flow generating $\{u^k\}$ to be ergodic is that each of the finite-dimensional eigenspaces corresponding to a given empirical eigenvalue be invariant under Γ.

The way one might check this condition experimentally would be to:

1. Perform the experiment to measure $R(x, x')$.
2. Decompose $R(x, x')$ using the POD: $R(x, x') = \sum \lambda_i \varphi_i(x)\varphi_i^*(x')$.
3. Check that for every eigenfunction $\varphi_j \in N_{\lambda_m} = \text{span}\{\varphi_j | \lambda_j = \lambda_m\}$ and every $\boldsymbol{\gamma} \in \Gamma$, $\boldsymbol{\gamma}(\varphi_j) \in N_{\lambda_m}$.

Aubry et al. [23] computed POD bases from numerical integrations of the Kuramoto–Sivashinsky equation (see Chapter 8) and concluded that for certain values of the bifurcation parameter the system is not ergodic. On the other hand, if one assumes that a system *is* ergodic one may use its symmetries to increase the size of the ensemble, generating additional data sets $\{\gamma u^k\}$ from a set of observations $\{u^k\}$ which represent only a limited region of the full attracting set. This approach has been advocated by Sirovich in [330] and adopted in many studies. However, one should be cautious, as there are examples for which the partition into ergodic components is finer than the partition into symmetric components. In this case the image of the basis obtained by one experiment under the symmetry group will not produce the basis obtained by the ensemble average measure. See [48] for further discussion of this point.

The ergodicity assumption is questionable, particularly in cases of "small" systems or special geometries. For example in a square Rayleigh–Bénard cell there is a possibility that the rotation direction of the single roll selected at the time of onset may never change throughout the life of the experiment. This indicates that there are at least two distinct and disjoint parts for the support of the invariant measure, each associated with a rotation direction, much as in Figure 6.9; hence it is not ergodic. Similar phenomena appear to occur in the minimal flow unit of the channel flow simulations by Jiménez and Moin [175].

3.3.4 Attractors

If the observations $\{u^k(x)\}$ from which the POD is generated come from a solution (or solutions) $u(x, t)$ of a dynamical system, then the empirical eigenfunctions and eigenvalues contain information on the attractor(s) of that system. We have already discussed symmetries. In this subsection we develop this observation in several other ways. We first give a probabilistic-geometric interpretation of the location of the dynamics in phase space using Chebyshev's inequality (Feller [108]):

Chebyshev's inequality *Let* $\mathbf{x}$ *be a vector-valued random variable, with mean* $\langle \mathbf{x} \rangle$ *and variance* $\sigma^2 = \mathrm{var}\,(\mathbf{x}) = \langle |\mathbf{x} - \langle \mathbf{x} \rangle|^2 \rangle$. *Then for any* $\epsilon > 0$

$$P\{|\mathbf{x} - \langle \mathbf{x} \rangle| \geq \epsilon\} \leq \frac{\sigma^2}{\epsilon^2},$$

where P denotes probability.

Chebyshev's inequality expresses the notion that the variance says something about the frequency of departures from the mean. In our case we define

$$S_n(\epsilon) = \left\{ u \in L^2(\Omega) | \sum_{m=n+1}^{\infty} |(u, \varphi_m)|^2 < \epsilon \right\},$$

and

$$W_n(\epsilon) = L^2(\Omega) \setminus S_n(\epsilon).$$

Here $S_n(\epsilon)$ is a slab of thickness 2ϵ around the finite-dimensional space span$\{\varphi_1, \ldots, \varphi_n\}$ and $W_n(\epsilon)$ is the rest of phase space, outside this slab. Note that $S_n(\epsilon)$ is infinite-dimensional. We use Chebyshev's inequality to estimate the fraction of the time spent by solutions $u(x, t)$ in $S_n(\epsilon)$. Denote by $\mathbf{x}_n$ the vector-valued random variable

$$\mathbf{x}_n(t) = \{(u, \varphi_m)\}_{m=n+1}^{\infty}$$

representing the infinite "tail" of the process. Then we have $\langle \mathbf{x}_n \rangle = 0$ and $\sigma^2(\mathbf{x}_n) = \sum_{m=n+1}^{\infty} \lambda_m$, and therefore, by Chebyshev's inequality,

$$P\{u \in W_n(\epsilon)\} = P\{|\mathbf{x}_n| \geq \epsilon\} \leq \frac{\sum_{m=n+1}^{\infty} \lambda_m}{\epsilon^2}. \tag{3.37}$$

To obtain a useful result, one lets $\epsilon \to 0$ for fixed n via a subsequence $\epsilon_n \to 0$ satisfying

$$\frac{\sum_{m=n+1}^{\infty} \lambda_m}{\epsilon_n^2} \to 0; \tag{3.38}$$

in other words, the ϵ_ns are chosen such that their squares go to zero slower than the decay of the norm in the residual modes. This gives a series of slabs whose thicknesses go to zero while the probability of solutions being in those slabs goes to one.

The problem now becomes that of estimating the rate of decay of the residual eigenvalues $\sum_{m=n+1}^{\infty} \lambda_m$ in the tail. The analytical evidence outlined below suggests that, when the POD basis is used for turbulent flows, this residual decays at least exponentially fast asymptotically. This enables us to take a series $\epsilon_n^2 \to 0$, with a slightly smaller exponent. The result is a sequence of slabs with thicknesses going exponentially to zero, while the probability of solutions being in a slab goes exponentially to one. This creates a picture in which an attractor is very thin, although possibly infinite-dimensional. It is reasonable to postulate, even in this case, that the essential dynamics are controlled by a finite number of modes. There is even a technical refinement to this, given as Proposition 4 in the Appendix; also see [112].

In the dynamical systems literature a lot of effort has gone into developing methods for estimation of dimension of attractors (e.g. [84, 320, 367]). The underlying idea is that, if the attractor of an infinite-dimensional evolution equation is finite-dimensional, then it should be possible to extract a finite-dimensional model in the form of a set of ODEs of comparable dimension. It is natural to try to relate the POD to such ideas, and even to define a dimension using empirical eigenvalues. Perhaps the most obvious thing to do is to define a (Karhunen–Loève) dimension as the number of non-zero eigenvalues in the POD, as suggested by Aubry *et al.* [20]. However, the number of non-zero eigenvalues is the dimension of the smallest *linear* subspace containing the dynamics, and it consequently provides only a very crude upper bound for the (Hausdorff) dimension of the attractor. Even if the attractor dimension is finite the linear subspace may not be finite-dimensional. It is easy to construct an example having a limit cycle (of dimension 1) which "twists" around in infinitely many directions in the function space and so generates data having an infinite number of non-zero eigenvalues. Indeed, as Sirovich realised in [331], this naive definition is not very useful. He suggested the following working definition: " ... introduce $\dim_{KL}$ ..., the number of actual eigenfunctions required so that the captured energy is at least 90% of the total ..., and that no neglected mode, on the average, contains more than 1% of the energy contained in the principal eigenfunction mode."

Note that the above comments are consistent with the celebrated embedding results of Takens [365] and with the observation that the set of projections of a compact set of Hausdorff dimension k on to a $(2k + 1)$-dimensional subspace is a residual set [234]. In spite of the finite (and possibly low) dimension of the attractor and the set containing it, the

POD will generally have positive eigenvalues with eigenfunctions in the complementary subspace, unless the set is "flat" and entirely contained in the $(2k + 1)$-dimensional linear subspace. As suggested by Equation (3.37) and the example mentioned above, we can only expect this to hold probabilistically and in an asymptotic sense.

In connection with Takens's theorem, which is generally applied to the study of phase space reconstructions effected by delay maps, we note that delay maps are highly nonlinear (as maps from "true" phase space into the embedding space). The POD spectrum is *not* invariant under nonlinear coordinate changes and one cannot expect the dimension in the delay map to be related to the number of non-zero POD eigenvalues in any simple way. Berkooz [37] has suggested an approach to this problem via a "conditional POD," but a discussion of this would take us too far afield.

The remark following Equations (3.37) and (3.38) depends upon exponential decay of empirical eigenvalues in the infinite tail. Is this a reasonable expectation? In the case of turbulent flows the answer appears to be yes both on physical (Tennekes and Lumley [368]) and on mathematical grounds (Promislow [289], Foias *et al.* [112]).

From a physical viewpoint, exponential decay of the spectrum holds only in the far dissipative range, in which length scales are smaller than the Kolmogorov microscale (Equation (2.4)). This should not be confused with the power-law decay of the intermediate inertial range. If asymptotic decay were substantially slower than exponential, high order spatial derivatives of the velocity field $\mathbf{u}(\mathbf{x}, t)$ might not exist. Most fluid mechanicians believe this to be unreasonable in a description of continuum matter.

The relevant mathematical concept is regularity of solutions, which describes the rate of decay of the tail of the spectrum of instantaneous solutions of a PDE in wavenumber space. To get a feel for this, consider the linear heat equation with a spatio-temporal "forcing function":

$$u_t = u_{xx} + f(x, t); \quad u(x, 0) = u_0(x).$$

Solution of this equation effectively requires one to integrate twice in space (and once in time), thereby smoothing the function $f(x, t)$, and the initial data $u_0(x)$. The effect is, of course, more marked the higher the spatial wavenumber. See Section 4.2 for explicit examples. In the Navier–Stokes equation the dissipative term $\nu\mathbf{\Delta u}$ plays a similar rôle to u_{xx}, its major influence also being in the far dissipative range.

In the simplest fluid mechanical situation, where the domain is a rectangular box with periodic boundary conditions and the solutions may be expressed in a (triple) Fourier series with time-dependent coefficients, *Gevrey regularity of class s* is the statement that the Fourier coefficients $a_\mathbf{k}$ decay at a rate such that the sum

$$\sum_\mathbf{k} a_\mathbf{k} e^{\tau |\mathbf{k}|^{2s}} e^{i\mathbf{k}\cdot\mathbf{x}}$$

converges for all τ. In this case the energy spectrum also decays exponentially fast at sufficiently high wavenumbers. In the context of turbulent flows this applies to the dissipation region of the spectrum. For an arbitrary (finite) domain one defines regularity in terms of the decay of the modal coefficients associated with the eigenfunctions of the Stokes operator in that domain. The asymptotics of the eigenvalues of the Stokes operator are the same for all "reasonable" domains [83], hence asymptotic results obtained from analysis of the periodic case remain valid. Rigorous regularity results for the Navier–Stokes equations in two-dimensional domains are given in [116]. In three dimensions there are no complete results, but if one *assumes* that vorticity is bounded above uniformly throughout the flow domain and in time, one can bypass the blow-up problem for solutions and obtain Gevrey regularity in this case too, possibly after solutions have evolved for some time [116].

Equipped with regularity of individual solutions, one still needs to show that the appropriate averages used in the POD are also uniformly exponentially bounded. This can be done [37] and the end result is that Gevrey regularity of solutions of the governing evolution equations implies exponential decay of the empirical eigenvalues. However, as remarked in [112], such exponential decay results apply only to the far dissipative range of turbulence and so are not directly relevant to the low-dimensional models of interest to us, in which we truncate far below that range. For more details see Berkooz [37].

For the sake of completeness we mention an interesting result due to Sirovich and Knight [336] which also concerns the structure of the POD at small spatial scales (high wavenumbers). Foias *et al.* [112] conclude from the results of [336] that, under certain conditions, the asymptotic form of the empirical eigenfunctions is that of Fourier modes. However, the results of Section 3.3.1 on the span of the eigenfunctions show that this cannot be true in full generality. Consider an ensemble of realizations that all vanish identically (or fall below experimental error) on a specific part of the domain. By Proposition 4 of the Appendix, the eigenfunctions themselves also have to be zero on that part of the domain and so cannot be asymptotically close to Fourier modes in that region.

3.4 Further results

In this section we briefly describe some recent extensions of, and other developments related to, the POD. We first discuss an idea which can reduce computational effort and then briefly remark on the "robustness" of the POD with respect to changes in conditions under which the data ensemble is generated and on the relation between the POD and probability density functions.

3.4.1 Method of snapshots

The *method of snapshots* was suggested by Sirovich in [330]. It is a numerical procedure that can save time in solving the eigenvalue problem (3.10) necessary for the POD. The idea is as follows: suppose one performs a numerical simulation on a large number of gridpoints N, that the number of ensemble members deemed adequate for a description of the process is M, and $N \gg M$. (The fundamental question of determining M is not addressed.) In general, since the data functions u^k and eigenfunctions φ_j have been replaced by N-vectors, the eigenfunction computation would become an $N \times N$ eigenvalue problem, as described in Section 3.1.1. Sirovich observed that this can be reduced to an $M \times M$ problem as follows: suppose that $\{\mathbf{u}^i\}_{i=1}^M$ are the realizations of the field and that the inner product on the N-dimensional vector space of realizations is denoted by $(\cdot, \cdot)$; this is the discretized version of the inner product in $L^2(\Omega)$. From the result on the span of the eigenfunctions in Section 3.3.1 we know that if φ is an eigenvector then

$$\varphi = \sum_{k=1}^{M} a_k \mathbf{u}^k, \tag{3.39}$$

where the coefficients a_k remain to be determined. Following (3.12), the N-dimensional eigenfunction problem may then be written

$$\left(\frac{1}{M} \sum_{i=1}^{M} \mathbf{u}^i (\mathbf{u}^i)^{\mathrm{T}}\right) \sum_{k=1}^{M} a_k \mathbf{u}^k = \lambda \sum_{k=1}^{M} a_k \mathbf{u}^k. \tag{3.40}$$

The left-hand side may be rearranged to give

$$\sum_{i=1}^{M} \left[\sum_{k=1}^{M} \frac{1}{M} (\mathbf{u}^k, \mathbf{u}^i) a_k\right] \mathbf{u}^i,$$

and we conclude that a sufficient condition for the solution of (3.10) will be to find coefficients a_k such that

$$\sum_{k=1}^{M} \frac{1}{M} (\mathbf{u}^k, \mathbf{u}^i) a_k = \lambda a_i \; ; \quad i = 1, \ldots, M. \tag{3.41}$$

This is now (one row of) an $M \times M$ eigenvalue problem. Note that in order for (3.41) to be a necessary condition, one needs to assume that the observations $\{\mathbf{u}^i\}_{i=1}^{M}$ are linearly independent.

3.4.2 Relationship to singular value decomposition

In the finite-dimensional case, the POD reduces to a singular decomposition of the given dataset. To see this connection, we first stack the snapshots $\mathbf{u}^k$ into an $N \times M$ *data matrix*

$$\mathbf{X} = \begin{bmatrix} \mathbf{u}^1 & \cdots & \mathbf{u}^M \end{bmatrix}, \tag{3.42}$$

after which the $N \times N$ eigenvalue problem (3.10) may be written

$$\frac{1}{M} \mathbf{X}\mathbf{X}^T \varphi = \lambda \varphi, \tag{3.43}$$

and the expansion (3.39) becomes $\varphi = \mathbf{X}\mathbf{a}$, where $\mathbf{a} = (a_1, \ldots, a_M)$. The eigenvalue problem (3.41) for the method of snapshots then reduces to an M-dimensional eigenvalue problem:

$$\frac{1}{M} \mathbf{X}^T \mathbf{X}\mathbf{a} = \lambda \mathbf{a}. \tag{3.44}$$

Recall that the *singular value decomposition* (SVD) of a real $N \times M$ matrix is given by

$$\mathbf{X} = \mathbf{U}\mathbf{\Sigma}\mathbf{V}^{\mathrm{T}} = \sum_{j=1}^{r} \sigma_j \varphi_j \mathbf{v}_j^{\mathrm{T}}, \tag{3.45}$$

where $\mathbf{U} = [\boldsymbol{\varphi}_1 \;\cdots\; \boldsymbol{\varphi}_N]$ and $\mathbf{V} = [\mathbf{v}_1 \;\cdots\; \mathbf{v}_M]$ are orthogonal matrices ($\mathbf{U}^\mathsf{T}\mathbf{U} = \mathbf{I}_{N\times N}$ and $\mathbf{V}^\mathsf{T}\mathbf{V} = \mathbf{I}_{M\times M}$), r is the rank of $\mathbf{X}$, and the $N \times M$ matrix $\boldsymbol{\Sigma}$ is of the form

$$\boldsymbol{\Sigma} = \begin{bmatrix} \boldsymbol{\Sigma}_1 \\ \mathbf{0} \end{bmatrix}, \quad N \geq M, \qquad \boldsymbol{\Sigma} = \begin{bmatrix} \boldsymbol{\Sigma}_1 & \mathbf{0} \end{bmatrix}, \quad N \leq M, \tag{3.46}$$

where $\boldsymbol{\Sigma}_1$ is a diagonal matrix of real, non-negative singular values σ_j, arranged in descending order. A straightforward calculation reveals that if (3.45) is the singular value decomposition of the data matrix $\mathbf{X}$, then the POD modes are the columns of $\mathbf{U}$, and the eigenvalues of (3.43) are $\lambda_i = \sigma_i^2/M$:

$$\frac{1}{M}\mathbf{X}\mathbf{X}^\mathsf{T}\boldsymbol{\varphi}_i = \frac{1}{M}\sum_{j=1}^{r}\sum_{k=1}^{r}\sigma_j\sigma_k\boldsymbol{\varphi}_j\mathbf{v}_j^\mathsf{T}\mathbf{v}_k\boldsymbol{\varphi}_k^\mathsf{T}\boldsymbol{\varphi}_i = \frac{1}{M}\sigma_i^2\boldsymbol{\varphi}_i, \tag{3.47}$$

since $\{\boldsymbol{\varphi}_j\} \subset \mathbb{R}^N$ and $\{\mathbf{v}_j\} \subset \mathbb{R}^M$ are orthonormal sets. Thus, the columns $\boldsymbol{\varphi}_i$ of U (the *left singular vectors* of $\mathbf{X}$) are the eigenfunctions in (3.43) (the POD modes), with empirical eigenvalues σ_i^2/M.

Also note that the right singular vectors $\mathbf{v}_j$ are related to the temporal coefficients a_j in (3.17), and property (3.18) that these are uncorrelated is reflected in the orthogonality of $\mathbf{v}_j$.

3.4.3 On inner products for compressible flows

For incompressible flows, typically we are interested in velocity fields $\mathbf{u} = (u, v, w)$, and the standard inner product on L^2 is a natural choice, since its induced norm corresponds to kinetic energy, as discussed in Section 1.4. For compressible flows, however, the situation is more complicated. In this case, the vector of quantities of interest contains not only kinematic variables (velocities), but also thermodynamic variables: density, pressure, enthalpy, etc. Here, the standard L^2 inner product may not even make sense. For instance, if we use the flow variables $\mathbf{q} = (\rho, u, v, p)$, defined on a spatial domain Ω, the L^2 inner product is

$$\langle \mathbf{q}_1, \mathbf{q}_2 \rangle = \int_\Omega (\rho_1\rho_2 + u_1u_2 + v_1v_2 + p_1p_2)\, d\mathbf{x}, \tag{3.48}$$

which is not dimensionally consistent, since one should not add a velocity and a pressure.

The simplest approach is perhaps to consider the kinematic variables and thermodynamic variables separately, obtaining separate sets of POD modes. However, this approach yields many more POD modes, and Galerkin models formed using this approach have been shown not to perform as well as when a single set of vector-valued modes is used [316].

One approach, as taken in [228], is normalization: if the quantities (ρ, u, v, p) are scaled by nominal values so that each component is non-dimensional, then combining them as in (3.48) does make sense. This is perhaps the most versatile approach, because one can adopt different scalings for different problems. However, there can be too much freedom here, for it may not be clear which scaling is appropriate.

Another approach is to seek an inner product whose induced norm is also a form of energy, consistent with the notion of the norm in incompressible flows. For instance, the *stagnation enthalpy* of the flow is given by

$$h_0 = h + \frac{1}{2}(u^2 + v^2 + w^2), \tag{3.49}$$

where h is the static enthalpy, and (u, v, w) are velocities. The *stagnation energy* is defined analogously, with enthalpy h replaced by the internal energy per unit mass, given for an ideal gas by $E = h/\gamma$, where γ is the ratio of specific heats. Seeking an energy-based inner product, in [316] an induced norm of the form

$$\frac{1}{2}\|\mathbf{q}\|_\alpha^2 = \int_\Omega \left(\alpha h + \frac{1}{2}(u^2 + v^2 + w^2)\right) d\mathbf{x} \tag{3.50}$$

was proposed, where $\mathbf{q}$ is the vector of flow variables, and $\alpha > 0$ is a constant. The integrand is not quadratic, however, because h appears linearly, so it may not be clear how to write a corresponding inner product. This situation was remedied in [316] by transforming to the flow variables $\mathbf{q} = (u, v, w, a)$, where a is the local *speed of sound*, which for an ideal gas satisfies $a^2 = (\gamma - 1)h$. A corresponding family of inner products can then be defined as

$$\langle \mathbf{q}_1, \mathbf{q}_2 \rangle_\alpha = \int_\Omega \left(u_1 u_2 + v_1 v_2 + w_1 w_2 + \frac{2\alpha}{\gamma - 1} a_1 a_2\right) d\mathbf{x}, \tag{3.51}$$

and parameterized by α, where $\alpha = 1$ corresponds to using the integral of stagnation enthalpy as the norm, and taking $\alpha = 1/\gamma$ corresponds to using stagnation energy. In the case of a nearly isentropic flow, the choice of flow variables $\mathbf{q} = (u, v, w, a)$ also leads to a particularly simple (quadratic) form of the compressible Navier–Stokes equations, given in [314] by

$$\mathbf{u}_t + \mathbf{u} \cdot \nabla \mathbf{u} + \frac{2}{\gamma - 1} a \nabla a = \nu \Delta \mathbf{u},$$

$$a_t + \mathbf{u} \cdot \nabla a + \frac{\gamma - 1}{2} a \nabla \cdot \mathbf{u} = 0.$$

Note, however that neither the integral of the stagnation enthalpy nor the stagnation energy is actually a conserved quantity. The conserved quantity is the *total energy*, given by

$$\int_\Omega \left(\rho E + \frac{1}{2}\rho(u^2 + v^2 + w^2)\right) d\mathbf{x}. \tag{3.52}$$

Although this norm would perhaps be the most natural from an energetic viewpoint, it is not obvious how to choose physically meaningful flow variables for which the equations of motion are reasonably simple. However, it is shown in [314] that in the case of an isentropic flow, choosing $\alpha = 1/(\gamma - 1)$ in (3.51) yields an induced norm that is indeed a conserved quantity.

3.4.4 On using an empirical basis over a parameter range

As we have remarked several times, and made clear in the results of Section 3.3.1, the ability of the POD to represent data is entirely dependent on the ensemble of observations, be they the result of physical experiments or numerical simulations, that goes into the averaging process. A natural question, then, is the "robustness" of the POD under changes in experimental conditions or parameter settings. Suppose, for example, that empirical eigenfunctions are computed for a channel flow at a Reynolds number (based on centerline velocity) of 6000 and one wishes to use them to represent flows in the same channel at $R_e = 8000$. How well can they be expected to perform? There is no firm evidence here, although some indications are encouraging.

Rodriguez and Sirovich [308, 339], working with numerical simulations of the one-space-dimensional Ginzburg–Landau equation in a (temporally) chaotic regime, constructed empirical eigenfunctions and used them to produce low-dimensional projections. They then produced bifurcation diagrams (see Chapter 6) by varying control parameters in the resulting low-dimensional ODEs and compared the bifurcation values with those revealed by the "full" simulations of the original PDE, finding agreement within a few percent over a wide range of parameter variation. Another model problem, the Kuramoto–Sivashisky equation (cf. Chapter 8) was considered in [344]. In neither of these cases, however, were solutions particularly rich spatially, and the empirical eigenfunctions are rather similar to Fourier modes. Perhaps more remarkably, in experiments on channel flows, Adrian and his colleagues [214] found evidence of simple scaling of eigenfunctions with Reynolds number, much as they found earlier in numerical simulations of a randomly forced one-dimensional Burgers equation [72]. In Section 12.4 we discuss some other instances involving direct numerical simulations of flows in complex geometries.

Recently Noack and colleagues have proposed the use of parameterized empirical bases $\{\boldsymbol{\phi}_j^{\kappa}(\mathbf{x})\}$ to encompass wider operating ranges: an issue of particular importance in control applications, in which actuators can change flow patterns around bodies. Examples appear in [207, 248, 249, 329], and in a recent book [267]. Also see [177].

This problem can be avoided if one chooses scaling appropriate to the physical phenomenon under study. For example, as we shall see in the boundary layer models of Chapter 10, if scaling based on wall variables is employed, the models are unaffected by changes in Reynolds number.

We close this section by mentioning a connection between the POD and the probability density function (PDF) in phase (functional) space. The invariant measure associated with solutions of the Navier–Stokes equations in functional space is an object of great interest; if one could obtain it explicitly one would have "a solution to turbulence," since all multipoint (single time) statistics would then be available. From this point of view the POD can be seen as the linear change of basis that turns the phase space coordinates into uncorrelated (although probably dependent) random variables. As shown by Hopf's theory of turbulence, the characteristic functional of the PDF in functional space may be obtained by multipoint correlations [164, 353]. This leads us to propose a very simple model for the PDF in functional space. Using the representation

$$u(x, t) = \sum_k a_k(t)\varphi_k(x), \tag{3.53}$$

we assume that the a_ks are independent and normally distributed with variance λ_k, $a_k \sim N(0, \lambda_k)$, where $N(\mu, \sigma^2)$ denotes the Gaussian distribution with mean μ and variance σ. While this is consistent with the picture the POD gives of the flow in that the coefficients *are* uncorrelated and the spectrum *is* correct, it clearly implies a strong assumption on the modal dynamics. Nonetheless, in the next section we see that this model is closely related to another statistical approach to coherent structures in turbulence.

3.5 Stochastic estimation

In this section we comment on the connection between the POD and linear stochastic estimation, as applied by Adrian and co-workers in [1–3,243]. This picks up on the relation between the POD and probability density functions already mentioned in Section 3.4.4.

We remark that the formulation of the POD and the results developed earlier apply to space, time, or mixed space-time analyses, all depending on the choice of the averaging operator (or equivalently, the measure), as long as the assumptions of Proposition 1 in the Appendix are satisfied. In this regard, an appropriate choice of averaging (i.e. a measure concentrated on a finite number of points, as would be encountered in a computer simulation) will produce the "bi-orthogonal" decomposition of Aubry *et al.* [20]. (Note that this notion of bi-orthogonality, in both space and time, is distinct from the notion of a bi-orthogonal set, considered in Chapter 5.)

Stochastic estimation is a method for predicting the conditional probability density function (CPDF) of a field $u(x)$, given observations $u(x')$ at other points, or possibly a vector of events at several points. There are many reasons to seek a CPDF, including its use in closure models [284] and in order to "produce" coherent structures. We shall outline the method of linear stochastic estimation since it is simple and enlightening, but for the sake of simplicity, we limit ourselves to scalar fields and single point vector events and we consider only estimates linear in $u(x')$, instead of the full CPDF.

Given $u(x')$, we seek an estimate for $u(x)$ of the form $A(x, x')u(x')$ by requiring the function $A(x, x')$ to minimize the expression

$$\langle |u(x) - A(x, x')u(x')|^2 \rangle, \tag{3.54}$$

where $\langle \cdot \rangle$ denotes an average over an ensemble of realizations, as in the rest of this chapter. As in the derivation in Section 3.1, a necessary condition will be that for any $V(x, x')$

$$\frac{d}{d\delta} \langle |u(x) - [A(x, x') + \delta V(x, x')]u(x')|^2 \rangle \bigg|_{\delta=0} = 0. \tag{3.55}$$

But the expression inside the averaging brackets is equal to:

$$\{u(x) - [A(x, x') + \delta V(x, x')]u(x')\} \cdot$$
$$\{u^*(x) - [A^*(x, x') + \delta V^*(x, x')]u^*(x')\},$$

so that, after differentiating with respect to δ, evaluating at $\delta = 0$ and equating to zero we get:

$$2\text{Re}[V^*(x, x')\langle u(x)u^*(x') \rangle] = 2\text{Re}[V^*(x, x')\langle A(x, x')u(x')u^*(x') \rangle]. \tag{3.56}$$

Since $V(x, x')$ is an arbitrary variation this implies that

$$\langle u(x)u^*(x')\rangle = A(x, x')\langle u(x')u^*(x')\rangle \qquad (3.57)$$

and we therefore take

$$A(x, x') = \frac{\langle u(x)u^*(x')\rangle}{\langle u(x')u^*(x')\rangle}. \qquad (3.58)$$

The fluctuations typically present in a turbulent system make the assumption that $\langle u(x')u^*(x')\rangle$ is invertible at every point x' a reasonable one, so that (3.58) will be well defined. We have found that, to provide the "best" linear estimate, $A(x, x')$ should be precisely the averaged two point correlation function $R(x, x')$, suitably normalized. Results of Adrian [1] show that the corrections to the CPDF due to higher order nonlinear terms in $u(x')$ are small, at least for homogeneous turbulence.

We can now introduce the representation (3.16) in terms of the POD to rewrite (3.58) as:

$$A(x, x') = \frac{\sum_{i=1}^{\infty} \lambda_i \varphi_i(x)\varphi_i^*(x')}{\sum_{j=1}^{\infty} \lambda_j |\varphi_j(x')|^2} \stackrel{\text{def}}{=} \sum_{i=1}^{\infty} \varphi_i(x) f_i(x'), \qquad (3.59)$$

where $f_i(x') = \lambda_i \varphi_i^*(x')/\sum_{j=1}^{\infty} \lambda_j |\varphi_j(x')|^2$. We may therefore interpret $f_i(x')$ as the relative contribution of the mode φ_i to $u(x')$ on the average. We conclude that linear stochastic estimation is equivalent to assuming that the estimated value of the POD coefficient of the ith mode, given the velocity at x', is the average contribution of the ith mode to the velocity at x' multiplied by the given velocity.

We now show that we recover exactly the same result from the simplified PDF model based on the POD introduced in Section 3.4, Equation (3.53). There we assumed that $a_i \sim N(0, \mu_i)$ and that these normally distributed coefficients a_is were independent. Let us compute the estimator $\langle u(x)|u(x')\rangle$. Since we have an expression for the PDF we can compute this explicitly. Recall from the formula for the conditional expectation of joint normal variables in probability theory (see [108]) that if $x_i \sim N(0, \sigma_i^2)$ for $i = 1, \dots, m$ then

$$\left\langle x_i \,\middle|\, \sum_{j=1}^{m} x_j = C \right\rangle = \frac{\sigma_i^2 C}{\sum_{j=1}^{m} \sigma_j^2}. \qquad (3.60)$$

Using (3.60), we have

$$\left\langle a_i \varphi_i(x') \,\middle|\, \sum_{j=1}^{\infty} a_j \varphi_j(x') = u(x') \right\rangle = \frac{\lambda_i |\varphi_i(x')|^2 u(x')}{\sum_{j=1}^{\infty} \lambda_j |\varphi_j(x')|^2},$$

which gives

$$\langle u(x)|u(x')\rangle = \frac{\sum_{i=1}^{\infty} \lambda_i |\varphi_i(x')|^2 u(x')\varphi_i(x)/\varphi_i(x')}{\sum_{j=1}^{\infty} \lambda_j |\varphi_i(x')|^2}$$

$$= \frac{\sum_{i=1}^{\infty} \lambda_i \varphi_i^*(x')\varphi_i(x)u(x')}{\sum_{j=1}^{\infty} \lambda_j \, |\varphi_j(x')|^2}. \qquad (3.61)$$

Since we estimated $\langle u(x)|u(x')\rangle$ by $A(x, x')u(x')$, this coincides with the result obtained from linear stochastic estimation, as can be seen by reference to Equation (3.59).

We conclude that the simple PDF model suggested in the previous section results in the best linear estimator of the conditional PDF of velocity, and that linear stochastic estimation may be viewed as a result of the simple PDF model based on the POD. This reveals a fundamental connection between the POD and linear stochastic estimation. In addition we can make the following technical observations based on our previous results:

1. All fields generated by linear stochastic estimation (LSE) possess any closed linear property that all ensemble members share.
2. Suitable averages of LSE events produce the POD eigenfunctions.
3. All LSE events are linear combinations of POD eigenfunctions.

We remark that one can apply the geometric (Chebyshev) result of Section 3.3.4 to obtain bounds on the probability of rare LSE events. Bonnet and co-workers [54] have also studied the relationship between stochastic estimation and the POD, with a view to efficient estimation of instantaneous velocity fields and to developing methods for validating low-dimensional models.

3.6 Coherent structures and homogeneity

The quest for an unbiased descriptor of coherent structures led us to consider the POD as a possible tool for extraction of structure from the statistics of an ensemble of observations on a physical system or solutions of a dynamical system. However, as we saw in Section 3.3.3, the existence of a homogeneous direction in physical space yields empirical eigenfunctions which are simply Fourier modes. Superficially, this does not appear to agree with the observation of localized coherent structures, but one can, of course, produce a localized structure with a suitable *combination* of Fourier modes with the appropriate complex coefficients. The spectrum of empirical eigenvalues supplies the moduli (Equation (3.33)), but one still has to determine phase relationships that yield instantaneous events.

In the work described subsequently in this book, low-dimensional models themselves determine the time-varying amplitudes and phases of the Fourier modes, and so this question is not particularly relevant for our purposes. However, one can ask for ensemble averaged coherent structures which reflect the actual Fourier combinations present, in which case phase information must be estimated from the statistics of the data. The experimentally founded intuition that coherent structures occur randomly in space and time is the basis for the treatment in this section, which follows Lumley's application of the shot noise decomposition [221, 223, 303]. We describe only the simplest case of a scalar field with a single space variable. At the end of this section we briefly discuss a connection to pattern analysis techniques such as [109, 359, 372].

Imagine a "building block" $f(x)$, the basic coherent structure concentrated near 0, and suppose that the process $u(x)$ is generated by randomly sprinkling such blocks on the real line or on a subinterval $[0, L]$. To move the structure so its reference point is at y, we perform the convolution

$$u(x) = \int \delta(\xi - y) f(x - \xi) d\xi,$$

where $\delta(\xi - y)$ is the Dirac delta function based at y. This prompts the following:

Definition *A convolution of the type*

$$u(x, t) = \int g(\xi) f(x - \xi) d\xi, \tag{3.62}$$

where $g(\xi)$ is a random process in the space of generalized functions, will be called a shot noise decomposition of $u(x)$.

The goal of "extracting a coherent structure" implies that one wishes to reconstruct the function f from statistics of $u(x, t)$. With the above definition, we see that a shot noise decomposition is always possible, and that it is moreover far from unique, since one has freedom in choosing f and g. In fact if $\hat{u}(k)$, $\hat{g}(k)$, and $\hat{f}(k)$ are the Fourier transforms of u, g, and f respectively, then $\hat{u} = \hat{g}\hat{f}$. To remove some ambiguity in the decomposition and to formalize the notion that g "randomly" sprinkles the deterministic blocks f, we assume that the random process g is uncorrelated in non-overlapping intervals; i.e.

$$\langle g(x)g^*(x') \rangle = \eta(x - x'), \tag{3.63}$$

where $\eta(x - x')$ satisfies

$$\int \int h(x, x') \eta(x - x') dx \, dx' = \int h(x, x) dx, \tag{3.64}$$

for any continuous function $h(x, x')$.

We can now partially characterize the power spectrum of the function f:

- If $\hat{R}(k)$ is the Fourier transform of the averaged two point correlation for a homogeneous process, then $\hat{R}(k) = |\hat{f}(k)|^2$.

To derive this result, note that

$$\langle u(x)u^*(x') \rangle = R(x - x')$$

$$= \left\langle \int g(\xi) f(x - \xi) \, d\xi \int g^*(\xi') f^*(x' - \xi') \, d\xi' \right\rangle$$

$$= \int \int f(x - \xi) f^*(x' - \xi') \langle g(\xi) g^*(\xi') \rangle \, d\xi \, d\xi'$$

$$= \int f(x - y) f^*(x' - y) \, dy.$$

The last equality comes from (3.64). Changing variables to $s = x - y$, we obtain

$$R(x - x') = \int f(s) f(x' - x + s) \, ds,$$

from which the result follows from well-known convolution equalities.

Our assumption has effectively specified the modulus of the power spectrum of the building block; the phase angles are yet to be determined. This approach formalizes rather well the stochastic sprinkling of deterministic structures in physical space. An extension of the shot noise formalism to include stochastic sprinkling in time is also possible, the *ansatz* being extended to uncorrelatedness in time as well as space. However, coherent structures are typically thought of as having a characteristic "life cycle," which does not necessarily sit well with lack of correlation in time. In this respect the assumption of a single building block is also restrictive, since we expect to meet more than one spatial form of coherent structure during the life cycle. These deficiencies reappear when we try to retrieve the phase information for $\hat{f}$.

One possible way to obtain the phase information is from the bi-spectrum [59, 212]; that is, we want to find a second function $\theta(c)$ such that

$$\hat{f}(c) = \hat{R}(c)^{1/2} e^{2\pi i \theta(c)}. \tag{3.65}$$

Consider the triple correlation:

$$
\begin{aligned}
&\langle u(x)u(x + r_1)u(x + r_2)\rangle \\
&= \left\langle \int g(\xi)f(x - \xi)\, d\xi \int g(\xi')f(x + r_1 - \xi')\, d\xi' \int g(\xi'')f(x + r_2 - \xi'')\, d\xi'' \right\rangle \\
&= \int\!\!\int\!\!\int f(x - \xi)f(x + r_1 - \xi')f(x + r_2 - \xi'')\langle g(\xi)g(\xi') \cdot g(\xi'')\rangle\, d\xi\, d\xi'\, d\xi''.
\end{aligned}
\tag{3.66}
$$

We now extend the assumption on g to be triply uncorrelated on non-overlapping intervals, in which case the expression of (3.66) becomes

$$\int f(x)f(x + r_1)f(x + r_2)\, dx = B(r_1, r_2), \tag{3.67}$$

which serves as the definition of $B(r_1, r_2)$. Upon taking the Fourier transform of B we obtain

$$
\begin{aligned}
\hat{B}(c_1, c_2) &= \int\!\!\int B(r_1, r_2) e^{-2\pi i(c_1 r_1 + c_2 r_2)}\, dr_1\, dr_2 = \hat{f}(c_1)\hat{f}(c_2)\hat{f}^*(c_1 + c_2) \\
&= \hat{R}^{1/2}(c_1)\hat{R}^{1/2}(c_2)\hat{R}^{1/2}(c_1 + c_2) \cdot e^{2\pi i[\theta(c_1) + \theta(c_2) - \theta(c_1 + c_2)]}.
\end{aligned}
\tag{3.68}
$$

In Equation (3.68) the "known" quantities are $\hat{B}(c_1, c_2)$ and $\hat{R}(c)$. As observed in [221] (see also [244]) this problem is, in general, not exactly solvable, since $\hat{B}(c_1, c_2)$ may not be factorizable as the right-hand side prescribes. Moin and Moser [244] observe that this problem is encountered in other disciplines as well, see [31, 237], and ad hoc procedures must be invoked. This is exactly where our assumptions on g come back to haunt us. The lack of an exact solution to the bi-spectrum equation indicates that our assumptions are too simple, either on the existence of a single building block, or on the statistical behavior of g.

We end with a brief and speculative discussion indicating a potential connection between the POD and pattern recognition techniques. With the advent of digital image processing, pattern recognition has become a vast field [313]. We limit ourselves to the relatively basic procedures used in [109, 359, 372]. Coherent structures were originally found by direct observations of flow visualizations. The quest for a quantitative procedure for extracting coherent structures and their dynamics is still a subject of research. Pattern recognition techniques are designed to mimic the human capability of detecting patterns in a noisy medium and thus they may be helpful in identifying such flow structures.

In the following discussion the reader should imagine a set of (two-) dimensional images. The basic procedure is as follows: one wants to identify a recurrent pattern in a noisy medium. First pick a template size and fill it with what is conjectured to look like the coherent structure: this is the first member of the ensemble. The template is then moved around in each frame of the data set and after each movement a correlation is computed. Every time the correlation attains a local maximum the corresponding pattern is added to the ensemble, which is averaged to produce a new reference template. This process is repeated until the template undergoes no further change. The final template is supposed to be the coherent structure. Once the coherent structure is deduced, one attempts to find regions in space well correlated with this structure and to study their contributions to various statistics.

This again, is a subjective procedure, although [359] suggests it is a robust one, with the final template being practically independent of the initial conjectured structure. Our mathematical understanding of the POD may contribute to a better understanding of the results of pattern recognition applications. Observe the similarity in mission between the pattern recognition technique and the shot noise expansion. Both attempt to decompose the flow into building blocks (although in pattern recognition we concentrate on regions of the flow with higher correlation with the template). This suggests caution in interpretation of the resultant template, since, as we saw earlier in this section, *any* template with a suit- able power spectrum might decompose the flow, with an appropriate sprinkling function. This is accentuated by the fact that the experiments of [359], for instance, show a median correlation of about 0.3. Based on the shot noise decomposition, one can propose a test for the objectivity of this method, seeing how well the basic building block is reproduced. Lumley's example [223] would be a good starting point for such a quest.

3.7 Some applications

We close this chapter with a brief survey of applications of the POD in analysis of exper- imental and numerically simulated data from various turbulent flows, and from numerical simulations of model problems. Work with a more dynamical emphasis, including the derivation and analysis of low-dimensional models, is described in Chapter 12. Neither there nor in the present section can we give a complete survey of this rapidly evolving field, and we apologize to those authors whose work has been omitted for reasons of space, or our own ignorance.

3.7.1 Wall bounded flows

In an early use of the POD in turbulence, Bakewell and Lumley [28] measured two point correlations of a single velocity component in the wall region of a fully devel- oped pipe flow. They constructed the autocorrelation tensor using incompressibility and a closure assumption. The flow was approximately homogeneous in spanwise and streamwise directions and, in addition to the mean, they computed only a single eigenfunction corresponding to the zero wavenumber Fourier mode in the streamwise direction. The coherent structure was then reconstructed under the assumption of zero phase shift among the spanwise Fourier modes. This yielded a pair of counter-rotating streamwise rolls much like those pictured in Figures 2.15 and 2.16.

Using the same facility as [28] – a tunnel running almost pure glycerine as the working fluid – Herzog [155] performed a fully three-dimensional study of the wall layer of turbulent pipe flow at a Reynolds number of 8750 based on centerline velocity. He measured streamwise and spanwise velocity components simultaneously with hot film probes and computed the third component from incompressibility. He had rather low spatial resolution (six points in the normal and spanwise directions, and seven in the streamwise direction), but, by averaging over long periods, obtained well-converged statistics which enabled him to compute the three leading eigenfunctions over a fairly wide range of spanwise and streamwise Fourier wavenumbers. Herzog also measured correlation functions with time lags, but only processed the zero time lag data to produce purely spatial eigenfunctions. He did not attempt to reconstruct phase relationships, but rather produced "raw" Fourier-empirical data: eigenfunctions $\varphi_n(k_1, k_3; x_2)$ of the type needed for representation of Equation (3.34). Indeed, it was these data that Aubry *et al.* [22] used in constructing low-dimensional models, as described in Chapter 10. Herzog's was the first full two point data set to be measured.

Moin and Moser [244] used the direct numerical simulation of a turbulent channel flow by Kim *et al.* [190] at a centerline Reynolds number of 3200 to perform a similar analysis. Although it is questionable that the statistics are fully converged (fewer than 200 realizations were included in the ensemble), the spatial resolution is excellent and, moreover, extends across the whole channel, taking in the outer part of the boundary layer as well as the wall region. The computations employed periodic boundary conditions (Fourier modes) in the streamwise and spanwise directions, and their raw decomposition took the same form as Herzog's. However, they also studied various phase reconstruction methods. In addition to using the bi-spectrum in the shot noise decomposition of Section 3.6, they proposed two further techniques, one based on spatial compactness and the other on continuity of the eigenfunctions with respect to changes in spanwise and streamwise wavenumber. With the help of these criteria they extracted the "characteristic structures" which dominate the turbulent kinetic energy in both the wall region and the wall-to-channel center domain.

Also working with a direct numerical simulation of channel flow, but at centerline Reynolds numbers of 1500 and 3–4000, Sirovich *et al.* [332, 333] and Ball *et al.* [29] computed a similar Fourier-empirical decomposition across the full channel. Rather than attempting to estimate phase relations among the Fourier modes, they extracted the temporal behavior of the modal coefficients in an expansion of the type (3.34) directly, by projecting realizations of the solution on to specific modes. They found strong intermittency, as one would expect from experimental observations of the burst-sweep process, such as those described in Section 2.5. They also found evidence that plane waves propagating obliquely can act as triggers for bursting.

In the latter studies when the full channel is taken as the spatial domain over which the optimal basis is to be computed, a relatively large number of eigenfunctions is required to capture, say, 90% of the turbulent kinetic energy on average. Ball *et al.* [29] give a figure of about 500, in reasonable agreement with the Liapunov dimension calculations of Keefe *et al.* [186]. In quoting this figure, we count each triple consisting of a pair of span and streamwise wavenumbers and an eigenfunction number as a separate "mode"; the reader should note that some authors lump Fourier modes together by integrating or summing, and count only empirical mode numbers. In contrast, when the wall layer is the domain of interest, significantly fewer modes are necessary to reproduce the same fraction of average energy. See Figures 2 and 3 in [244] and compare with Figures 10 and 12 of [241]. In all cases the Fourier-empirical representations display an initial convergence rate considerably faster than the Fourier-Chebyshev bases used for the numerical computations, although over the last few percent in energy the empirical basis functions lose much of their advantage. Moin and Moser [244] note that the first eigenfunction (summed over all Fourier wavenumbers) captures 23% compared to only 4% for the lowest Chebyshev polynomial, but that if 90% is to be reproduced, 10 empirical and 12 Chebyshev modes respectively are required.

Finally we mention the work of Rempfer and Fasel on transition in flat plate boundary layers. Like Moin and Moser [244], they used a database from a direct numerical simulation; specifically, one due to Rist and Fasel [305], who employed a finite difference scheme with "buffer" regions at inlet and outlet to simulate a developing layer. In these simulations, disturbances are triggered by perturbations symmetric about a mid-plane in the streamwise/normal direction. This was chosen to match the experiments of Kachanov *et al.* [180], and has the effect of breaking the spanwise translation/reflection symmetry that one might otherwise expect in a computation with periodic boundary conditions in that direction. Since the developing flow is now inhomogeneous in all spatial directions, Fourier decompositions cannot be used, and fully three-dimensional POD computations must be carried out.

In the first part of their work, [297, 299, 302], Rempfer and Fasel use the POD to probe the spatio-temporal evolution of structures. They find that empirical eigenvalues and eigenfunctions occur almost in pairs, reflecting the approximate streamwise translation invariance of relatively slowly growing structures. They identify the leading (pair of) eigenfunction(s) with the dominant physically observed structure and demonstrate its similarity to the lambda vortices revealed in flow visualization. They divide the flow into three regions at different streamwise locations, thereby investigating the changes in the structures as they evolve and propagate downstream. Near the inlet, the leading structure is primarily two-dimensional; moving downstream, three-dimensional effects increase. Time histories of the modal coefficients reveal that, while the leading (first and second) order structures oscillate almost sinusoidally, higher order structures, which typically carry less than 5% of the energy, display less regular, spiky behavior. This is consistent with a propagating Tollmien–Schlichting wave on which "second order" instabilities are developing.

In the second phase of this work, [296, 300, 301], the authors develop low-dimensional models similar to those of Aubry *et al.* [22], and study the energy flow among the various modal components. We describe this in greater depth in Section 12.2.

3.7.2 Free shear flows

Among the earliest studies of free shear flows by the POD were those of Glauser *et al.* [121–123], who considered a jet. They did not account for the growth of the shear layer but assumed approximate homogeneity in the streamwise direction. They found that most of the turbulent kinetic energy resided in a "ridge" of wavenumber modes and proposed a dynamical mechanism for turbulence production. See Section 12.1 for more details.

Sirovich and various co-workers also studied jet flows, using both experimental results and numerically generated databases. In [335], conditionally sampled realizations of concentration fields from a gas jet at Reynolds number 1150 based on exit velocity and nozzle diameter were used to generate a POD that reflects the large-scale structures in a frame moving with the convected velocity. The conditioning breaks streamwise homogeneity and yields a two-dimensional eigenvalue problem. The resulting eigenfunctions, as maps over the streamwise and radial coordinates, clearly display the lobes responsible for entrainment and mixing. Kirby *et al.* [192] performed a similar analysis using data from a large eddy simulation, and Winter *et al.* [388] examined mixing of a jet exhausting into a turbulent cross-flow.

Kirby *et al.* [193] studied two-dimensional pressure and momentum density fields from a numerically simulated supersonic shear layer, also using conditional sampling to "freeze" the convecting structures. They focused on the data compression afforded by use of empirical eigenfunctions.

Glezer *et al.* [130] studied a periodically excited plane mixing layer using an extended POD. Also see Section 12.3.

3.7.3 Rayleigh–Bénard convection

Numerically simulated Rayleigh–Bénard convection problems were studied by Sirovich *et al.* [92, 276, 334, 338]. In the first pair of papers the authors made extensive use of discrete symmetries of the domain, a rectangular parallelepiped, to increase ensemble sizes as suggested in Section 3.3.3. They also used these symmetries to simplify their computations by selecting even or odd parities for eigenfunction components. Here such simplifications are especially important, for there are no homogeneous directions, and fully three-dimensional empirical eigenfunctions must be found. The latter papers focus on scaling properties and the computation of Liapunov and Karhunen–Loève dimensions (see Section 3.3.4). In [337] low-dimensional dynamics and data compression issues were considered.

3.7.4 Model problems

The POD has been used to analyze numerical solutions of several one- and two-space-dimensional model equations. Sirovich and Rodriguez [308, 339, cf. 331] studied the one-space-dimensional complex Ginzburg–Landau equation, investigating the ability of Galerkin projections onto an empirical subspace computed at one parameter value to reproduce dynamical and bifurcation behavior over a fairly wide parameter range.

Chambers *et al.* [72] analyzed a simulation of the viscous Burgers equation defined on the interval [0, 1] with zero boundary conditions and a random forcing term. They showed that the empirical eigenfunctions exhibit "viscous" boundary layers near the ends of the interval and an "outer" region in the interior of the interval essentially independent of the viscosity parameter.

Kirby and Armbruster [191] applied a conditional POD and a moving POD to study bifurcation in the Kuramoto–Sivashinsky equation, thereby identifying traveling wave structures. Their work is similar in spirit to the pattern recognition studies outlined at the end of Section 3.6. Aubry *et al.* [20, 21] studied the same problem, paying particular attention to the relation between the POD and symmetry groups (Section 3.3.3). See Section 7.5 below for more on traveling modes.

Armbruster *et al.* [12, 13] studied the two-dimensional Kolmogorov flow: the Navier–Stokes equations in a rectangular domain subject to a spatially periodic steady body force. They worked with both stream function and vorticity data and investigated how the POD highlights different flow structures when optimizing the L^2-norms of these quantities. In the first paper they argue that the stream function eigenfunctions demonstrate low-dimensional dynamics while the vorticity eigenfunctions display an enstrophy cascade. The second paper is more in the spirit of [21]: discrete and continuous symmetries are used to help identify local and global (homoclinic and heteroclinic) bifurcations. Platt *et al.* [279] studied the same flow and, while they did not use the POD, their dynamical investigations of intermittency share some common ground with [13].

3.8 Appendix: some foundations

The POD relies on the two mathematical concepts treated in this section. The first is averaging, primarily insofar as it is needed to justify interchange of averaging and the inner product operations. The second is the compactness of the operator $\mathcal{R}$ which, via Hilbert–Schmidt theory, ensures the discrete spectrum and orthonormal eigenfunctions. (It is clear from its definition that $\mathcal{R}$ is self-adjoint.) In this technical appendix we state and prove the relevant results for these and some other properties used in the body of this chapter.

3.8.1 Probability measures

We first consider averaging. An averaging operation corresponds to a probability measure, μ, on $L^2(\Omega)$, which, in the time stationary case, satisfies

$$\mu(A) = \mu(S_t^{-1}(A)). \tag{3.69}$$

Here S_t is the solution semi-group or *flow map* of the dynamical system in question and A is a subset in phase space. We discuss flow maps for finite-dimensional processes in Chapter 6; in the present context the reader should think of the mapping induced on an infinite-dimensional phase space by the solutions of the underlying governing equations. The existence of a measure corresponding to the physical ensemble average is an assumption which we phrase as follows:

Ansatz 1 *There exists an S_t invariant probability measure μ on $L^2(\Omega)$ such that for every (Borel) set $A \subset L^2(\Omega)$*

$$\mu(A) = \lim_{N \to \infty} \frac{1}{N} \sum_{k=0}^{N-1} 1_A(u^k), \tag{3.70}$$

where $\{u^k\}_{k=0}^{\infty}$ is a sequence of physically determined states $u^k \in L^2(\Omega)$.

The task of supplying an ensemble of physically determined states $\{u^k\}$ is left to the experimentalist. The states should represent instantaneous velocity fields over the full flow domain which are characteristic of the persistent turbulent activity after transients have decayed: the "statistical equilibrium states" of the flow [256]. They must be representative of the system's behavior in the sense that all statistical moments can, in principle, be derived from them. In dynamical systems terms, $\mu(A)$ encodes the average frequency with which typical orbits visit the region A of phase space. In the time-stationary case, the measure μ defined by (3.70) is invariant under the evolution operator S_t of the system. A simple one-dimensional example is given towards the end of Section 6.5.

We can now give conditions under which our standing assumption, that averaging and spatial integration commute, will hold.

Proposition 1 *Let μ be a probability measure on $L^2(\Omega) \cap C(\Omega)$, where $C(\Omega)$ is the set of continuous functions on Ω, such that for every $x \in \Omega$ we have $\langle |u|^2(x) \rangle < \infty$ and $\langle |u|^2(x) \rangle^{1/2} \in L^2(\Omega)$. Then:*

1. *$\langle u(x_1)u^*(x_2) \rangle$ exists for every $(x_1, x_2) \in \Omega \times \Omega$.*
2. *$\langle u(x_1)u^*(x_2) \rangle$ is in $L^2(\Omega \times \Omega)$.*
3. *The following equation holds (interchange of averaging and integration):*

$$\left\langle \int_0^1 \int_0^1 u(x)u^*(x')\varphi(x')\psi^*(x)\,dx\,dx' \right\rangle = \int_0^1 \left[\int_0^1 \langle u(x)u^*(x') \rangle \varphi(x')\,dx' \right] \psi^*(x)\,dx. \tag{3.71}$$

Proof To show property (1), define $f_{x_i} : L^2(\Omega) \cap C(\Omega) \to \mathbb{R}$ by $f_{x_i}(u) = u(x_i)$. Since we are using continuous functions, the notion of point value is well defined (this is the reason for working in $C(\Omega)$). We have to show that the average

$$\langle u(x_1)u^*(x_2) \rangle = \int_{L^2(\Omega)} f_{x_1}(u) f_{x_2}^*(u)\,d\mu(u)$$

exists. The following upper bound establishes this:

$$\int_{L^2(\Omega)} |f_{x_1}(u) f_{x_2}^*(u)|\,d\mu(u) \leq \left[\int_{L^2(\Omega)} |f_{x_1}|^2 d\mu(u) \right]^{1/2} \left[\int_{L^2(\Omega)} |f_{x_2}|^2 d\mu(u) \right]^{1/2}$$
$$= \langle |u|^2(x_1) \rangle^{1/2} \langle |u|^2(x_2) \rangle^{1/2}.$$

To show property (2), we observe that

$$|\langle u(x_1)u^*(x_2) \rangle \langle u^*(x_1)u(x_2) \rangle| \leq \langle |u|^2(x_1) \rangle \langle |u|^2(x_2) \rangle,$$

and therefore

$$\left| \int_\Omega \int_\Omega \langle u(x_1) u^*(x_2) \rangle \langle u^*(x_1) u(x_2) \rangle dx_1 dx_2 \right| \leq \left[\int_\Omega \langle |u|^2(x) \rangle dx \right]^2 < \infty.$$

Finally, to show that Equation (3.71) holds we use Tonelli's theorem, which permits change of order of integration [319]. It is sufficient to show that the integral

$$\int_\Omega \int_\Omega \langle u(x_1) u^*(x_2) \rangle \varphi(x_2) \psi^*(x_1) dx_2 dx_1$$

exists. But we have

$$\left| \int_\Omega \int_\Omega \langle u(x_1) u^*(x_2) \rangle \varphi(x_2) \psi^*(x_1) dx_2 dx_1 \right| \leq$$

$$\left[\int_\Omega \int_\Omega |\langle u(x_1) u^*(x_2) \rangle|^2 dx_2 dx_1 \right]^{1/2} \left[\int_\Omega \int_\Omega |\varphi(x_2) \psi^*(x_1)|^2 dx_2 dx_1 \right]^{1/2}$$

$$< \infty.$$

□

The requirement that $\langle |u|^2(x) \rangle^{1/2} \in L^2(\Omega)$ is physically reasonable: its interpretation is that kinetic energy must be finite on average in a (finite) spatial domain. The requirement that for every x we have $\langle |u|^2(x) \rangle < \infty$ is also physical and corresponds to finite velocity at any point, on average. We take the hypotheses of Proposition 1 as standing assumptions and so throughout we may freely interchange the order of the operations $\langle \cdot \rangle$ and $\int_\Omega \cdot dx$.

3.8.2 Compactness of $\mathcal{R}$

We now address the issue of compactness of the operator $\mathcal{R}$ introduced in Section 3.1.

Proposition 2   *Under the assumptions on u in Proposition 1, and if additionally* supp(μ) *is precompact, the operator* $\mathcal{R} : L^2(\Omega) \rightarrow L^2(\Omega)$ *defined by*

$$\mathcal{R}\psi = \int_\Omega \langle u(x_1) u^*(x_2) \rangle \psi(x_2) dx_2 \tag{3.72}$$

is compact.

Proof   Let $\psi_n \rightharpoonup \psi$ be a weakly convergent sequence. We have

$$\|\mathcal{R}(\psi_n - \psi)\| = \left\| \int_\Omega \langle u(x_1) u^*(x_2) \rangle [\psi_n(x_2) - \psi(x_2)] dx_2 \right\|$$

$$= \left\| \left\langle u(x_1) \int_\Omega u^*(x_2) [\psi_n(x_2) - \psi(x_2)] dx_2 \right\rangle \right\|.$$

$$\tag{3.73}$$

Given $\delta > 0$ and using the precompactness of supp(μ), we can choose $u_1, \ldots, u_M$ to be a $\delta > 0$ spanning set in supp(μ), i.e. for every $u \in$ supp(μ) there exists an $1 \leq i \leq M$ such that $\|u - u_i\| < \delta$. Let N be such that for all $n > N$ and $i = 1, \ldots, M$:

$$\left\| \int_\Omega u_i^*(x_2)[\psi_n(x_2) - \psi(x_2)]dx_2 \right\| < \delta.$$

For $u \in \text{supp}(\mu)$ let $\text{Cl}(u)$ be the u_i, $1 \le i \le M$ closest to u, then $\|\text{Cl}(u) - u\| < \delta$, and we may bound the right-hand side of (3.73) by

$$\left\| \left\langle u(x_1) \int_\Omega [u^*(x_2) - \text{Cl}(u^*)(x_2)][\psi_n(x_2) - \psi(x_2)]dx_2 \right\rangle \right\|$$

$$+ \left\| \left\langle u(x_1) \int_\Omega \text{Cl}(u^*)(x_2)[\psi_n(x_2) - \psi(x_2)]dx_2 \right\rangle \right\|$$

$$\le \left\| \left\langle u(x_1) \int_\Omega \delta C \right\rangle \right\| + \|\langle u(x_1)\delta \rangle\|.$$

Choosing $\delta > 0$ small enough we have shown that $\mathcal{R}$ takes weakly convergent sequences to strongly convergent ones. The operator is therefore compact. $\square$

We see that the assumption that the support of the invariant measure be contained in a compact set in phase space suffices to show that the operator $\mathcal{R}$ is compact. Compactness of the attractor in general and the support of the invariant measure in particular has been established for many of the dissipative systems of interest; see Temam [367]. However, for the three-dimensional Navier–Stokes equations, this still requires the assumption of global existence of solutions.

3.8.3 Symmetry and invariant subspaces

We now state and prove the result underlying the discussion of symmetries at the end of Section 3.3.3. It is given in terms of the probability measure μ corresponding to the average $\langle \cdot \rangle$, introduced at the beginning of this appendix.

Proposition 3 *Suppose that we have a semi-group or flow map $S_t : L^2(\Omega) \to L^2(\Omega)$ for a dynamical system which is equivariant under a linear symmetry group Γ, i.e. every element $\gamma \in \Gamma$ is a linear transformation of $L^2(\Omega)$, such that $\gamma \circ S_t = S_t \circ \gamma$. Suppose further that S_t is stationary and let μ be the invariant measure associated with time averages $\langle \cdot \rangle$. Then a necessary condition for μ to be ergodic is that, for almost every experiment, each of the finite-dimensional eigenspaces N_λ corresponding to a given empirical eigenvalue λ be invariant under Γ.*

Proof By ergodicity of μ, the averaged autocorrelation

$$R(x, x') = \langle u(x, t)u^*(x', t) \rangle$$

is independent of the initial condition $u(x, 0)$ for every initial condition in a set of full measure $X \subset L^2(\Omega)$. We can write

$$R(x, x') = \sum_{i=1}^\infty \lambda_i \varphi_i(x)\varphi_i^*(x').$$

Let $N_\lambda = \text{span}\{\varphi_i | \lambda_i = \lambda\}$. We want to show that if S_t is ergodic then, for every $\gamma \in \Gamma$ we have $\gamma(N_\lambda) = N_\lambda$. It is enough to show $\gamma(N_\lambda) \subset N_\lambda$ for every γ, since $\gamma^{-1} \in \Gamma$ and this implies $\gamma^{-1}(N_\lambda) \subset N_\lambda$, showing that $\gamma(N_\lambda) = N_\lambda$. Let $\{u(x,t)|t \in [0,\infty), u(x,0) \in X\}$ be an experiment, then

$$u(x,t) = \sum_{i=1}^{\infty} a_i(t)\varphi_i(x),$$

where $\langle a_i a_i^* \rangle = \lambda_i$. Thus, for every $\gamma \in \Gamma$,

$$\tilde{u}(x,t) = \gamma(u(x,t)) = \sum_{i=1}^{\infty} a_i(t)\gamma(\varphi_i(x))$$

is also an experiment (with probability 1 we have $\tilde{u}(x,0) \in X$). From the independence of the representation of $R(x,x')$ on the initial condition and the uncorrelatedness of the a_is, we deduce that

$$R(x,x') = \sum_i \lambda_i \gamma(\varphi_i(x))\gamma(\varphi_i(x'))^*$$

is a diagonal representation of $R(x,x')$. In particular, the eigenspaces are orthogonal and are spanned by the functions $\gamma(\varphi_i(x))$, so

$$\gamma(\varphi_i(x)) \in N_{\lambda_i},$$

which verifies the claim.   □

3.8.4 Spectral decay and approximate compactness

In Section 3.3.4, we used Chebyshev's inequality to show that, if the tail of the spectrum of empirical eigenfunctions decays sufficiently fast, then the probability is high that solutions of the underlying dynamical system belong to a thin set around a finite-dimensional linear subspace in phase space. However, this set $S_n(\epsilon)$ is not necessarily compact. In Proposition 2 we saw that precompactness of the support of the invariant measure is sufficient to obtain compactness of the operator $\mathcal{R}$. Here we obtain a partial converse of this result by showing that, if the empirical eigenvalues decay rapidly enough, then practically all the support of the invariant measure is contained in a compact set.

Proposition 4 *Let S_t be the flow of a dynamical system in $L^2(\Omega) \cap C(\Omega)$ with an invariant measure μ. Denote by $\langle \cdot \rangle$ the average with respect to this measure. Under the assumptions of Proposition 1, if*

$$\langle u(x)u^*(x') \rangle = \sum_i \lambda_i \varphi_i(x)\varphi_i^*(x') \quad \text{with } \lambda_n = o(\exp(-cn))$$

for some $c > 0$, then for any $\epsilon > 0$ there exists a compact set B_ϵ such that $\mu(B_\epsilon) > 1 - \epsilon$.

Proof Take a sequence $\epsilon_n \to 0$ such that

$$\frac{\sum_{m=n+1}^{\infty} \lambda_m}{\epsilon_n^2} \to 0 \text{ and } \sum_{n=1}^{\infty} \frac{\sum_{m=n+1}^{\infty} \lambda_m}{\epsilon_n^2} < \epsilon \text{ and } \sum \epsilon_n = r^2 < \infty,$$

for some constant r. This is possible since $\lambda_n = o(\exp(-cn))$, and the head of the sequence $\{\lambda_n\}$ is of no consequence. Recall from Section 3.3.4 that $S_n(\epsilon_n)$ is a slab of thickness ϵ_n centered on the finite-dimensional subspace spanned by the (generalized) eigenvectors belonging to the first n eigenvalues. Since $\sum \epsilon_n < \infty$, the closure of $\bigcap_{n=1}^{\infty} S_n(\epsilon_n)$ is a compact set in $L^2(\Omega)$. Indeed, we have

$$\bigcap_{n=1}^{\infty} S_n(\epsilon_n) \subset \{u \in L^2 : \sum_{n=1}^{\infty} \sum_{m \geq n} |(u, \phi_m)|^2 \leq \sum_{n=1}^{\infty} \epsilon_n = r^2 < \infty\}.$$

Changing the order of summation in this expression leads to

$$\sum_{n=1}^{\infty} \sum_{m \geq n} |(u, \phi_m)|^2 = \sum_{n=1}^{\infty} n|(u, \phi_n)|^2,$$

so that we have

$$\bigcap_{n=1}^{\infty} S_n(\epsilon_n) \subset \{u \in L^2 : \sum_{n=1}^{\infty} n|(u, \phi_n)|^2 \leq \sum_{n=1}^{\infty} \epsilon_n = r^2 < \infty\}.$$

This defines a precompact set in L^2 because the right-hand side set is the image of the open ball of L^2 of radius r under the compact linear map T defined by: $(Tv, \phi_k) = k^{-1/2}(v, \phi_k)$ for every $k = 1, 2, \ldots$

Then, setting

$$B_\epsilon = \overline{\bigcap_{n=1}^{\infty} S_n(\epsilon_n)},$$

(the closure of the infinite intersection), we see that

$$\mu(B_\epsilon) = \mu(\bigcap_{n=1}^{\infty} S_n(\epsilon_n)) = 1 - \mu(\bigcup_{n=1}^{\infty} W_n(\epsilon_n))$$

$$\geq 1 - \sum_{n=1}^{\infty} \frac{\sum_{m=n+1}^{\infty} \lambda_m}{\epsilon_n^2} > 1 - \epsilon.$$

$\square$

This final result has interesting implications. If we perform a POD on a system which is not known a priori to have precompact support for its invariant measure and we obtain a discrete POD with some regularity on the rate of decay of the spectrum, we may conclude that most of the measure is concentrated on a compact set. This implies that, in a probabilistic sense, we can well approximate the "full" attractor by a compact attractor, albeit possibly an infinite-dimensional one. It is surprising that such fundamental information may be obtained from a simple procedure.

4

Galerkin projection

In numerical simulations of turbulence, one can only integrate a finite set of differential equations or, equivalently, seek solutions on a finite spatial grid. One method that converts an infinite-dimensional evolution equation or partial differential equation into a finite set of ordinary differential equations is that of Galerkin projection. In this procedure the functions defining the original equation are projected onto a finite-dimensional subspace of the full phase space. In deriving low-dimensional models we shall ultimately wish to use subspaces spanned by (small) sets of empirical eigenfunctions, as described in the previous chapter. However, Galerkin projection can be used in conjunction with any suitable set of basis functions, and so we discuss it first in a general context.

After a brief description of the method in Section 4.1, we apply it in Section 4.2 to a simple problem: the linear, constant-coefficient heat equation in both one- and two-space-dimensions. We recover the classical solutions, which are often obtained by separation of variables and Fourier series methods in introductory applied mathematics courses. We then consider an equation with a quadratic nonlinearity, Burgers' equation, which was originally introduced as a model to illustrate some of the features of turbulence [65]. The remainder of the chapter is devoted to the Navier–Stokes equations. In Section 4.3 we describe Fourier mode projections for fluid flows in simple domains with periodic boundary conditions, paying particular attention to the way in which the incompressibility condition is addressed. The final Section 4.4 focuses on the use of empirical eigenfunctions and introduces some issues that arise in making the drastic truncations necessary to obtain low-dimensional models. We pick up these difficulties again in Chapters 10 and 11, where we discuss models of the wall region of a turbulent boundary layer as an example of the procedure, and propose various "modeling" strategies for overcoming them.

4.1 Introduction

The Galerkin method is most conveniently introduced in the context of the modern theory of evolution equations, which we will briefly sketch. For more information, see Temam [367], Henry [154], or, in the specific context of the Navier–Stokes equations, Doering and Gibbon [96]. The reader is perhaps used to thinking of ordinary differential equations

(ODEs) and partial differential equations (PDEs) as quite different beasts. However, the distinction becomes blurred when one considers systems of the form:

$$\frac{du}{dt} = F(u), \tag{4.1}$$

where $F(\cdot)$ represents a nonlinear operator that may involve spatial derivatives and/or integrals. Just as in the simpler ODEs with finite-dimensional phase spaces considered in Chapter 6, at each (fixed) time t the solution u belongs to a suitable phase or state space, which we denote by X. Thus if $u = u(x, t)$ is a function defined on a spatial domain Ω, X should be a space of functions defined over the domain Ω, which is flexible enough to include all possible solutions of (4.1). The L^2 spaces introduced in Section 1.4 and used extensively in the previous chapter are good candidates. However, if $F(\cdot)$ is a differential operator we may wish to restrict ourselves to subspaces of functions in $L^2(\Omega)$ whose derivatives also belong to L^2. In the case of PDEs the boundary conditions are also incorporated into the definition of X, so that any candidate for a solution to an initial-boundary-value problem, $u \in X$, automatically satisfies the boundary conditions. Once the phase space has been chosen, it is customary to drop explicit reference to the spatial variable x and simply write $u = u(t)$.

Given a finite-dimensional linear subspace S of X, we wish to determine a dynamical system that evolves on S and approximates (4.1) in some sense. This dynamical system may be denoted

$$\frac{du}{dt} = F_S(u), \tag{4.2}$$

where $u(t) \in S$ and F_S is a vector field on S. *Galerkin projection* specifies this vector field by

$$F_S(u) = PF(u), \tag{4.3}$$

where $P : X \to S$ is the orthogonal projection onto S. To apply this method explicitly, we expand $u(t)$ as

$$u(t) = \sum_{j=1}^{n} a_j(t)\varphi_j, \tag{4.4}$$

where φ_j are basis functions that span the subspace S (for instance, these may be Fourier modes, or the POD modes of the preceding chapter). If these basis functions are orthonormal, the projection P is given by

$$Pu = \sum_{j=1}^{n} (u, \varphi_j)\varphi_j, \tag{4.5}$$

so the projected dynamics (4.2) become

$$\sum_{j=1}^{n} \dot{a}_j(t)\varphi_j = \sum_{j=1}^{n} \left(F(u(t)), \varphi_j\right)\varphi_j. \tag{4.6}$$

Since the basis functions φ_j are linearly independent, this simplifies to

$$\dot{a}_j(t) = \big(F(u(t)), \varphi_j\big), \qquad j = 1, \dots, n : \tag{4.7}$$

a system of n ODEs for the coefficients a_k.

To give an idea of how a simple PDE is treated under this approach, consider the scalar reaction-diffusion (heat) equation defined on the spatial interval $0 \le x \le 1$:

$$u_t = u_{xx} + g(u), \tag{4.8}$$

with boundary and initial conditions

$$u(0, t) = u(1, t) = 0; \quad u(x, 0) = u_0(x). \tag{4.9}$$

Here $g(u)$ could be almost any nonlinear function, such as $g(u) = u - u^3$. The appropriate phase space X is H_0^1, the Sobolev space of functions $u(x)$ that, along with their first derivatives, belong to $L^2([0, 1])$, and that satisfy the boundary conditions $u(0) = u(1) = 0$. In short:

$$X = H_0^1 = \{u \in L^2([0, 1]) | u_x \in L^2([0, 1]); \ u(0) = u(1) = 0\}.$$

In the notation H_0^1, the subscript refers to the boundary conditions and the superscript to the degree of differentiability assumed. The space H_0^1 is appropriate for this problem, since analysis of the equation draws heavily on examination of the inner product (u_{xx}, u), which (via integration by parts) is well defined if $\int |u_x|^2 dx$ is finite. See the remarks at the end of this section. Within this set-up, we can rewrite (4.8) as

$$\frac{du}{dt} = Au + g(u); \quad u \in H_0^1, \ u(0) = u_0(x), \tag{4.10}$$

where the linear operator A is the projection of the Laplacian $\partial^2/\partial x^2$ on H_0^1. We give explicit examples in the next section. The reader may object that in (4.8) we take two space derivatives and indeed the *domain* $D(A)$ of the Laplacian operator consists of functions in H^2 whose *second* derivatives belong to L^2, but the well-known smoothing action of the (linear) heat equation implies that solutions starting in H_0^1 rapidly enter H^2 in any case. Also see the remarks at the end of this section.

In order to carry out Galerkin projection, the phase space X must be an inner product space spanned by a suitable set of basis functions. All the examples we shall meet are subspaces of $L^2(\Omega)$ and so will automatically be endowed with the L^2 inner product. Suitable bases include Fourier modes, Chebyshev or other sets of orthogonal polynomials, and wavelets [90], as well as the empirical bases of Chapter 3. Note that the basis functions must satisfy the boundary conditions implicit in the definition of X.

Prior to projection we write the evolution Equation (4.10) or its analog in the form

$$\frac{du}{dt} - Au - g(u) = 0. \tag{4.11}$$

We then express u as a finite sum of time-dependent modal coefficients multiplied by elements of a suitable basis of X:

$$u(x, t) = \sum_{j=1}^{J} a_j(t)\varphi_j(x). \tag{4.12}$$

The basis elements $\varphi_j(x)$ are usually ordered according to some physically motivated criterion, such as increasing spatial wavenumber (decreasing spatial scales). In the case of empirical eigenfunctions, as we saw in Chapter 3, it is decreasing average energy. We then substitute (4.12) into the left-hand side of (4.11), take the inner product of this with each basis function φ_k, $k = 1, 2, \ldots, J$ in turn and equate each expression to zero. The resulting spatial integration removes the x-dependence and we are left with a set of J coupled ODEs for the amplitude coefficients $a_k(t)$. If the basis $\{\varphi_j\}$ is orthonormal then the first term coming from (4.11) is particularly simple:

$$\left(\frac{d}{dt} \sum_{j=1}^{J} a_j \varphi_j, \varphi_k \right) = \sum_{j=1}^{J} \dot{a}_j \left(\varphi_j, \varphi_k \right) = \dot{a}_k, \tag{4.13}$$

where $(\dot{})$ denotes the derivative with respect to time. In the next section we address an explicit example which can be completely solved, but in general an evolution equation of the form (4.10), projected onto a subspace of X spanned by J mutually orthogonal basis functions, results in a set of J first order ODEs of the form

$$\dot{a}_k = \left(\left[\sum_{j=1}^{J} (A\varphi_j) a_j + g\left(\sum_{j=1}^{J} a_j \varphi_j \right) \right], \varphi_k \right). \tag{4.14}$$

The error incurred in approximation of the infinite-dimensional evolution Equation (4.10) by (4.14) is determined by the sum $\sum_{j=J+1}^{\infty} a_j \varphi_j$ of the residuals. A suitable choice of truncation order J depends upon the properties of the original equation. If the linear part contains a dissipative term, which tends to suppress modes with high spatial wavenumbers, as do the examples considered below, including the Navier–Stokes equations, then we can expect to obtain reliable results if J is taken sufficiently large so that the modes included cover the dissipative range. In fact one method used to prove existence and uniqueness of solutions for the Navier–Stokes equations involves taking J-dimensional Galerkin projections and letting J approach infinity; see Ladyzhenskaya [201], Temam [367]. However, convergence is not assured in all cases and depends upon the specific problem at hand. In Section 4.4 we discuss this issue at somewhat greater length.

At this point we remark that, if A is a differential operator such as the Laplacian in (4.8), the inner products

$$(A\varphi_j, \varphi_k) = (\varphi_{j,xx}, \varphi_k) \tag{4.15}$$

in (4.14) can be evaluated by integrating by parts to give

$$\varphi_{j,x} \varphi_k \Big|_0^1 - (\varphi_{j,x}, \varphi_{k,x}) = -(\varphi_{j,x}, \varphi_{k,x}). \tag{4.16}$$

To eliminate the boundary term we appeal to the fact that the basis functions also satisfy the conditions $\varphi_j(0) = \varphi_j(1) = 0$. Note that integration by parts allows us to use the space H_0^1, in which we do not need bounds on second space derivatives. This is referred to as working with weak solutions.

4.2 Some simple PDEs revisited

In this section we use the Galerkin method to solve the linear heat equation. We start with the single space variable problem in the form

$$u_t = u_{xx}, \tag{4.17}$$

$$u_x(0, t) = u_x(1, t) = 0, \tag{4.18}$$

$$u(x, 0) = u_0(x), \tag{4.19}$$

which describes the temperature field in a bar with insulated ends (no heat flux). As for Equation (4.8), the phase space is a subset of H^1, but now incorporating different boundary conditions denoted by the subscript 1:

$$X = H_1^1 = \{ f \in L^2([0, 1]) | f_x \in L^2([0, 1]); \ f_x(0) = f_x(1) = 0 \}.$$

The Fourier cosine modes $\{ \varphi_j(x) \}_{j=0}^{\infty} = \{ \cos(j\pi x) \}_{j=0}^{\infty}$ provide a basis for this space, orthogonal with respect to the L^2 inner product, which can be normalized if desired by multiplication by $\sqrt{2}$. Here we deal only with real-valued functions and so the L^2 inner product is simply

$$(f, g) = \int_0^1 f(x)g(x)dx. \tag{4.20}$$

Following the prescription sketched above we rewrite Equation (4.17) as

$$\left[\frac{d}{dt} - A \right] (u) = 0; \quad A = \partial^2/\partial x^2. \tag{4.21}$$

Substituting $u(x, t) = \sum_j a_j(t)\varphi_j(x)$ into this expression and taking the inner product with each basis function in turn we obtain

$$\left(\left[\frac{d}{dt} - A \right] \sum_j a_j \varphi_j, \varphi_k \right) = 0; \quad k = 1, 2, \ldots \tag{4.22}$$

In replacing the original PDE (4.17)–(4.19) by (4.22), we are actually solving an equation of the form

$$u_t = P u_{xx},$$

where P denotes projection onto the subspace spanned by the Fourier cosine modes. This choice of basis has the convenient property that the projection of A on any finite-dimensional subspace spanned by these functions is exactly $\partial^2/\partial x^2$ and so P is effectively the identity. This is because the second spatial derivative takes cosines to cosines, leaving each subspace invariant. Had our equation contained a first derivative, the projection process would have been significantly complicated: one would effectively have to represent sine functions in terms of a (finite) Fourier cosine series.

Evaluating the expression in (4.22), integrating by parts or using

$$A\varphi_j = \frac{\partial^2}{\partial x^2}[\cos(j\pi x)] = -j^2\pi^2 \cos(j\pi x),$$

we obtain

$$\left(\sum_j \left[\dot{a}_j \varphi_j + j^2\pi^2 a_j \varphi_j\right], \varphi_k\right) = \frac{1}{2}(\dot{a}_k + k^2\pi^2 a_k) = 0. \tag{4.23}$$

Orthogonality of the basis functions makes the final result particularly simple. We have replaced the PDE by a set of uncoupled ODEs for the modal amplitude coefficients:

$$\dot{a}_k = -k^2\pi^2 a_k. \tag{4.24}$$

These may be solved immediately to yield

$$a_k(t) = a_k(0)e^{-k^2\pi^2 t}, \tag{4.25}$$

in which the coefficients $a_k(0)$ are determined from the Fourier series for the initial data (4.19):

$$u_0(x) = \sum_{k=0}^{\infty} a_k(0)\cos(k\pi x) \tag{4.26}$$

or

$$a_0(0) = \int_0^1 u_0(x)dx \; , \; a_k(0) = 2\int_0^1 u_0(x)\cos(k\pi x)dx \; ; k = 1, 2, \ldots \tag{4.27}$$

We note that all the non-zero Fourier modes in the solution (4.25) decay exponentially as time increases; the decay rate is $k^2\pi^2$ and so the higher modes disappear first. This is an explicit expression of the smoothing effect of the Laplacian operator. In the end only the mean temperature is left, uniformly distributed along the bar:

$$u(x, t) \to a_0 \text{ as } t \to \infty. \tag{4.28}$$

In this simple linear problem, as we have pointed out, the Fourier cosine basis has been chosen to diagonalize the Laplacian operator as well as the time derivative term, so that, no matter what the dimension J of the projected system, we obtain uncoupled ODEs of identical form to (4.24). When $J = \infty$ we have, of course, simply reproduced the solution that could have been found by an elementary application of separation of variables. (The basis functions $\{\cos(j\pi x)\}_{j=0}^{\infty}$ are precisely the eigenfunctions of the Sturm–Liouville problem

$$\varphi_{xx} + \lambda\varphi = 0; \quad \varphi_x(0) = \varphi_x(1) = 0$$

appearing in that method, λ being the separation constant.) If we choose a finite order truncation $J < \infty$ there will generally be an error in our representation of the true initial condition, since high order modes are excluded. However, as the solution evolves, this error will decay at a rate $J^2\pi^2 t$ due to the smoothing effect described above.

Inhomogeneous boundary conditions may arise in problems with time-dependent control actuation or external forcing (e.g., blowing or suction on a wall). These can often be dealt with by decomposing the state variable as a superposition of a particular solution $u_p(x, t)$ that satisfies the boundary conditions, and one having homogeneous boundary conditions:

$$u(x, t) = u_p(x, t) + u_h(x, t). \tag{4.29}$$

For example, suppose that the boundary conditions (4.18) are replaced by the time-dependent temperatures

$$u(0, t) = f(t), \ u(1, t) = 0. \tag{4.30}$$

If we choose $u_p(x, t) = (1 - x) f(t)$, which automatically satisfies (4.30), and substitute (4.29) into (4.17), we obtain a problem with homogeneous boundary conditions in which forcing at the boundary has been replaced by a space- and time-varying body or volume force:

$$u_{h,t} = u_{h,xx} - (1 - x) \dot{f}(t), \tag{4.31}$$
$$u_h(0, t) = u_h(1, t) = 0. \tag{4.32}$$

(Here the choice of linear spatial dependence in u_p implies that $u_{p,xx} \equiv 0$, but other functions may be used as long as the boundary conditions are satisfied.)

We may now employ a standard basis for the homogeneous (Dirichlet) problem such as $\{\varphi_j(x)\}_{j=0}^\infty = \{\sin(j\pi x)\}_{j=0}^\infty$, write $u_h(x, t) = \sum_j a_j \varphi_j$, and proceed as above to obtain a set of ODEs, still uncoupled, of the form

$$\dot{a}_k = -k^2 \pi^2 a_k + \dot{f}_j(t), \ \text{where } \dot{f}_j(t) = \dot{f}(t) \int_0^1 (1 - x) \varphi_j(x) \, dx. \tag{4.33}$$

In [157] this simple idea is significantly extended and applied under the name of "lifting" to the control of instabilities in plane channel flow. Using a pair of particular solutions for each combination of streamwise and spanwise wavenumbers, it is shown that exponentially-decaying "feedback kernels" can be truncated to have compact support without degrading control performance. Further applications to channel flow are developed in [170, Chapter 5], and actuation modes have been used in several cylinder wake examples [35, 139, 184, 218, 269, 385].

Before going on to a nonlinear problem, for which separation of variables fails but to which the Galerkin method can still be applied, we consider the two-space-dimensional heat equation:

$$u_t = \Delta u = \left(\frac{\partial^2}{\partial x_1^2} + \frac{\partial^2}{\partial x_2^2} \right) u \tag{4.34}$$

defined on the rectangular domain $0 \leq x_1 \leq L_1; \ 0 \leq x_2 \leq L_2$ with boundary conditions

$$u(0, x_2) = u(L_1, x_2) = u(x_1, 0) = u(x_1, L_2) = 0, \tag{4.35}$$

corresponding to constant temperature (heat sinks). Here the appropriate phase space is $X = H_0^1(\Omega)$, where $\Omega = [0, L_1] \times [0, L_2]$ and the inner product is the double integral

$$(f, g) = \int_0^{L_2} \int_0^{L_1} f(x_1, x_2) g(x_1, x_2) \, dx_1 \, dx_2. \tag{4.36}$$

A suitable basis for $H_0^1(\Omega)$ is provided by the doubly indexed set of Fourier sine modes

$$\varphi_{jk}(x_1, x_2) = \sin\left(\frac{j\pi x}{L_1} \right) \sin\left(\frac{k\pi x}{L_2} \right), \tag{4.37}$$

which, as the reader can check, are also mutually orthogonal with respect to the inner product (4.36). In fact we have

$$(\varphi_{jk}, \varphi_{lm}) = \begin{cases} 1/4 & \text{if } j = l \text{ and } k = m, \\ 0 & \text{otherwise.} \end{cases} \tag{4.38}$$

In conjunction with evaluation of the Laplacian or integration by parts, (4.38) implies that the right-hand side of (4.34) also diagonalizes, leading to the uncoupled ODEs

$$\dot{a}_{lm} = -\left(\frac{l^2\pi^2}{L_1^2} + \frac{m^2\pi^2}{L_2^2}\right) a_{lm}, \tag{4.39}$$

whence the solution may be reconstructed from the modal coefficients via the representation

$$u(x_1, x_2, t) = \sum_l \sum_m a_{lm}(t)\varphi_{lm}(x_1, x_2). \tag{4.40}$$

As in the earlier example, an explicit solution can easily be found, which again coincides with the exact solution found by separation of variables.

For our third example we return to a scalar equation, but now with a nonlinear term. Specifically, we take

$$u_t = v u_{xx} - u u_x; \quad u(0, t) = u(1, t) = 0. \tag{4.41}$$

This is Burgers' equation, a one-dimensional "cartoon" of the Navier–Stokes equations that combines diffusion with the convective nonlinearity typical in fluid mechanics. (One should note, however, that its solutions are typically collections of waves strengthening to N-wave shocks and decaying slowly for small v: quite different from real turbulence.) It is exactly soluble in closed form via the Cole–Hopf transformations (cf. Whitham [386]), but here we are mainly interested in illustrating the Galerkin procedure. We use the fact that, for $v > 0$ the "viscous dissipation" due to the Laplacian damps out high modes just as it does in the linear case, so that a finite-dimensional truncation is reasonable.

Using the Fourier sine basis, we expand u in the form

$$u(x, t) = \sum_j a_j(t)\sqrt{2}\sin(j\pi x) \tag{4.42}$$

and project onto span$\{\sqrt{2}\sin(j\pi x); \; j = 1, \ldots, J\}$. (Here we use normalized basis functions, to avoid spurious factors in the energy expressions below.) The linear terms in (4.41) give essentially the same expression as in (4.23), after insertion of the coefficient v. Projecting onto the lth mode, the nonlinear term is

$$\int_0^1 \sum_j a_j\sqrt{2}\sin(j\pi x) \sum_k k\pi a_k\sqrt{2}\cos(k\pi x)\sqrt{2}\sin(l\pi x)\,dx$$

$$= \sum_j \sum_k 2\sqrt{2}k\pi a_j a_k \int_0^1 \sin(j\pi x)\cos(k\pi x)\sin(l\pi x)\,dx, \tag{4.43}$$

and we can use elementary trignometric identities to reduce the integral to

$$\int_0^1 \frac{1}{4}\{\cos[(j-l+k)\pi x] + \cos[(j-l-k)\pi x]$$
$$- \cos[(j+l+k)\pi x] - \cos[(j+l-k)\pi x]\}\,dx. \tag{4.44}$$

Thus only those terms of the double sum in (4.43) survive for which the relevant combinations of integers j, k, l sum to zero, leading to the single sum

$$\sum_k \frac{k\pi}{\sqrt{2}}(a_{(l-k)} + a_{(l+k)} - a_{-(l+k)} - a_{-(l-k)})a_k. \tag{4.45}$$

The projected ODEs therefore take the form (after cancelling a factor of $1/2$):

$$\dot{a}_l = -l^2\pi^2 v a_l - \frac{\pi}{\sqrt{2}}\sum_k k(a_{(l-k)} + a_{(l+k)} - a_{-(l+k)} - a_{-(l-k)})a_k. \tag{4.46}$$

The celebrated Lorenz equations, discussed in Section 6.5, were derived from the Boussinesq problem involving the coupled Navier–Stokes and heat equations for convection in a two-dimensional layer in much the same fashion as this [217].

To give a more explicit idea of the sort of system one obtains, we display the Equations (4.46) for a truncation of order three, in which one retains only those terms in the sum containing modal coefficients with indices 1, 2, or 3:

$$\left.\begin{aligned}
\dot{a}_1 &= -\pi^2 v a_1 + \frac{\pi}{\sqrt{2}}(a_1 a_2 + a_2 a_3) \\
\dot{a}_2 &= -4\pi^2 v a_2 - \frac{\pi}{\sqrt{2}}(a_1^2 - 2a_1 a_3) \\
\dot{a}_3 &= -9\pi^2 v a_3 - \frac{\pi}{\sqrt{2}}3a_1 a_2.
\end{aligned}\right\} \tag{4.47}$$

Specific truncations such as (4.47) or the general system (4.46) of J equations provide finite-dimensional approximations to the true dynamics of (4.41), more-or-less accurate depending on truncation order and the size of the parameter v. All such projections do, however, preserve a key property of (4.41): the fact that the nonlinear interaction term conserves energy. The analogous term in the Navier–Stokes equations also conserves energy and so it is worth exploring this feature. For the scalar Burgers' equation, the kinetic energy is

$$\mathcal{E} = \frac{1}{2}\int_0^1 |u(x,t)|^2 dx = \frac{1}{2}\|u\|^2, \tag{4.48}$$

and differentiating with respect to t we have

$$\frac{d\mathcal{E}}{dt} = \int_0^1 u u_t dx = (u, u_t). \tag{4.49}$$

Substituting for u_t from the evolution equation, we find the expression governing the energy change along solutions:

$$\frac{d\mathcal{E}}{dt} = (u, v u_{xx} - u u_x) = v(u, u_{xx}) - (u, u u_x). \tag{4.50}$$

The first inner product can be integrated by parts to yield $-\nu(u_x, u_x)$, the boundary term vanishing due to the boundary conditions; the second term in (4.50) also vanishes identically, since it is an exact differential:

$$\int_0^1 u^2 u_x \, dx = \int_0^1 \left(\frac{u^3}{3}\right)_x dx = \frac{u^3}{3}\bigg|_0^1 = 0. \tag{4.51}$$

In summary, we have

$$\frac{d\mathcal{E}}{dt} = -\nu \|u_x\|^2, \tag{4.52}$$

showing that energy decays at a rate proportional to the viscosity ν. Thus, when $\nu = 0$ the "inviscid" Burgers' equation is conservative.

The analog of kinetic energy in the finite-dimensional projections is simply the sum of squares of the modal energies

$$\mathcal{E}_J = \frac{1}{2} \sum_{j=1}^J a_j^2, \tag{4.53}$$

as one can check by substituting (4.42) into (4.48) and using orthogonality. Differentiating (4.53) with respect to t and substituting for $\dot{a}_l$ from the projected ODEs, one can also check that for any finite-dimensional truncation the cubic terms analogous to (4.51) vanish identically, leaving

$$\frac{d\mathcal{E}_J}{dt} = -\nu \sum_{j=1}^J j^2 \pi^2 a_j^2. \tag{4.54}$$

This is the analog of (4.52) in terms of Fourier modes.

In the examples above, the ODEs given by (4.7) could be determined analytically, in terms of the coefficients a_k in an expansion in basis functions. This is useful both for analysis of the resulting ODEs, and for numerical simulations, as the inner product in (4.7) does not need to be computed at every timestep of the simulation. For general nonlinearities, the coefficients in the ODE may not be able to be determined a priori: for instance, in (4.14), the nonlinear term $g(\cdot)$ needs to be evaluated in the high-dimensional space X. However, many of the examples we are interested in have only *quadratic* nonlinearities – for instance, this is true for Burgers' equation discussed above, and for the Navier–Stokes equations – and for quadratic systems, the coefficients in the ODEs *can* be determined analytically, as we now indicate.

Consider a quadratic partial differential equation described by

$$u_t = F(u) = L(u) + Q(u, u), \tag{4.55}$$

where F is a differential operator on a function space X, $L : X \rightarrow X$ is a linear operator, and $Q : X \times X \rightarrow X$ is bilinear (linear in each argument). We may expand $u(t) \in X$ either as (4.4), or more generally, with an offset b:

$$u(t) = b + \sum_j a_j(t)\varphi_j, \tag{4.56}$$

where $b, \varphi_j \in X$. The projected dynamics (4.7) then become

$$\dot{a}_l(t) = \left(L(u) + Q(u, u), \varphi_l\right)$$

$$= \left(L\left(b + \sum_j a_j(t)\varphi_j\right) + Q\left(b + \sum_j a_j(t)\varphi_j, b + \sum_k a_k(t)\varphi_k\right), \varphi_l\right)$$

$$= b_l + \sum_j L_l^j a_j(t) + \sum_{j,k} Q_l^{jk} a_j(t) a_k(t),$$

where the quantities

$$b_l = \left(L(b) + Q(b, b), \varphi_l\right),$$

$$L_l^j = \left(L(\varphi_j) + Q(b, \varphi_j) + Q(\varphi_j, b), \varphi_l\right),$$

$$Q_l^{jk} = \left(Q(\varphi_j, \varphi_k), \varphi_l\right),$$

are constants (independent of t) that may be determined before integrating the ODEs. For instance, for Burgers' equation (4.41), we have $L(u) = \nu u_{xx}$ and $Q(u, v) = -u v_x$. For the basis functions $\varphi_j(x) = \sqrt{2}\sin(j\pi x)$ as in (4.42), and taking $b = 0$, one readily computes

$$L_l^j = -2(j\pi)^2\nu \int_0^1 \sin(j\pi x)\sin(l\pi x)\, dx = -l^2\pi^2\nu\delta_{jl}, \tag{4.57}$$

which gives the linear terms in (4.47). For the nonlinear term, we have

$$Q_l^{jk} = -2\sqrt{2}k\pi \int_0^1 \sin(j\pi x)\cos(k\pi x)\sin(l\pi x)\, dx, \tag{4.58}$$

as in (4.43).

4.3 The Navier–Stokes equations

We now describe the reduction of the Navier–Stokes equations to a set of ordinary differential equations, which at this stage may still be a (countably) infinite set. As we saw above, the choices of function space and basis used in the projection are an essential part of the process and the physical boundary conditions are built into the space in an integral fashion. In the following discussion the boundary conditions are periodic in all three space dimensions and for simplicity we take the domain to be $[0, 2\pi] \times [0, 2\pi] \times [0, 2\pi]$, although arbitrary parallelepipeds can be dealt with by simple rescaling of independent variables. As in Chapter 2, we use Einstein notation throughout this section.

The Navier–Stokes equations may be written as

$$u_{i,t} + u_j u_{i,j} = -p_{,i} + \frac{1}{R_e} u_{i,jj} + f_i, \tag{4.59}$$

$$u_{i,i} = 0, \tag{4.60}$$

$$\mathbf{u} \text{ periodic in } \Omega = [0, 2\pi] \times [0, 2\pi] \times [0, 2\pi], \tag{4.61}$$

where (u_1, u_2, u_3) are the three components of velocity, which depend on space (x_1, x_2, x_3) and time t. For the sake of simplicity we assume that $\mathbf{f} = (f_1, f_2, f_3)$ is a constant body force driving the flow, much the same as a constant mean pressure gradient. We discuss two approaches projecting the Navier–Stokes equations which differ in the basis used, although both approaches employ Fourier modes in their construction.

Before doing this we mention two other forms in which the equations frequently appear. If we wish to consider the behavior of perturbations riding on a given (usually steady) solution U_i of the Navier–Stokes equations, we let $\tilde{u}_i$, $\tilde{p}$ denote the full velocity and pressure fields, write $\tilde{u}_i = U_i + u_i$, $\tilde{p} = P + p$, and use the fact that U_i and P themselves solve the original equation to obtain:

$$u_{i,t} + U_j u_{i,j} + u_j U_{i,j} + u_j u_{i,j} = -p_{,i} + \frac{1}{R_e} u_{i,jj}. \qquad (4.62)$$

As pointed out in Section 2.1, however, there are so few exact solutions known that this form is generally useful only when considering stability and evolution of perturbations on simple flows such as plane Couette or Poiseuille flows. A more common strategy in turbulence is a decomposition of $\tilde{u}_i$, $\tilde{p}$ in which U_i, P are the *mean* velocity and pressure fields with respect to a suitable (ensemble) average $\langle \cdot \rangle$ and $\langle u_i \rangle = \langle p_i \rangle = 0$, as in Sections 2.4 and 2.5. In this case U_i and P satisfy the *averaged* Navier–Stokes equations:

$$U_j U_{i,j} + \langle u_j u_{i,j} \rangle = -P_{,i} + \frac{1}{R_e} U_{i,jj} + f_i, \qquad (4.63)$$

and, using (4.63), the evolution equation for the fluctuations u_i, p becomes

$$u_{i,t} + U_j u_{i,j} + u_j U_{i,j} + u_j u_{i,j} - \langle u_j u_{i,j} \rangle = -p_{,i} + \frac{1}{R_e} u_{i,jj}. \qquad (4.64)$$

In both (4.62) and (4.64) the terms $U_j u_{i,j} + u_j U_{i,j}$ effectively provide a driving force, extracting energy from the (mean) flow and shear.

Here we describe only projection of the "primitive" form (4.59) of the equations, but it can easily be adapted to these other forms. Equation (4.64) in the context of the boundary layer is considered in Chapter 10.

The first and perhaps most obvious Fourier expansion of the velocity field is to take

$$u_i(\mathbf{x}, t) = a_{i,k_1 k_2 k_3}(t) e^{i(k_1 x_1 + k_2 x_2 + k_3 x_3)}, \qquad i = 1, 2, 3, \qquad (4.65)$$

where the time-dependent modal coefficients a_{i,k_1,k_2,k_3} are complex and the sum over the indices (k_1, k_2, k_3) ranges from $-\infty$ to $+\infty$. Since the physical velocity field is real valued, not all coefficients are independent; for example we may require that $a_{i,-k_1 k_2 k_3} = a^*_{i,k_1 k_2 k_3}$. In the space $L^2(\Omega)$ of vector-valued velocity fields, these basis elements are of the form:

$$\left. \begin{array}{l} \frac{1}{\sqrt{2\pi}}(e^{i(k_1 x_1 + k_2 x_2 + k_3 x_3)}, 0, 0), \\[2mm] \frac{1}{\sqrt{2\pi}}(0, e^{i(k_1 x_1 + k_2 x_2 + k_3 x_3)}, 0), \\[2mm] \frac{1}{\sqrt{2\pi}}(0, 0, e^{i(k_1 x_1 + k_2 x_2 + k_3 x_3)}). \end{array} \right\} \qquad (4.66)$$

They are clearly orthonormal, since the three components are uncoupled and each component is a simple Fourier series. So far we have not specified a basis for the pressure field, nor have we discussed the divergence-free condition (4.60). This latter and the pressure are of course related. The Navier–Stokes equations may be viewed as differential-algebraic equations for the modal amplitudes of the velocity and pressure fields with the algebraic constraint among the modal amplitudes deriving from (4.60). In this sense the pressure acts as a Lagrange multiplier, enabling one to satisfy the momentum equation. The presence of this constraint complicates the projection process.

In this first approach, projection is done so that the divergence-free fields form an invariant subspace of the evolution equation. We first take the divergence of the Navier–Stokes equations (4.59) to yield:

$$\left(u_{i,t} + u_j u_{i,j} + p_{,i} - \frac{1}{R_e} u_{i,jj} - f_i \right)_{,i} = 0.$$

Using the assumption of a constant body force and the divergence-free condition (4.60), this simplifies to

$$u_{i,it} + u_{j,i} u_{i,j} = -p_{,ii}.$$

Therefore, if we start from a divergence-free initial condition and ensure that the pressure satisfies

$$u_{j,i} u_{i,j} = -p_{,ii} \tag{4.67}$$

throughout the evolution of the solution, we are guaranteed that the velocity field will remain divergence-free as it develops. An explicit solution of this *Poisson equation* (4.67) for the pressure field is not feasible in most applications since the boundary conditions for pressure are not generally known. However, here the periodic boundary conditions permit a substantial simplification. If we assume that the mean pressure is given (we may take it zero, for the mean does not enter into the equations) and that the mean pressure gradient is also zero (which is justified since we have assumed a constant body force), then we may invert the Poisson equation and explicitly solve for pressure in terms of velocity components. We find that the (k_1, k_2, k_3)-mode in the Fourier decomposition of the divergence of the momentum equation is given by:

$$\widehat{u_{j,i} u_{i,j}} = ik_i' a_{j,k_1' k_2' k_3'} i(k_j - k_j') a_{i,k_1 - k_1' k_2 - k_2' k_3 - k_3'} = \widehat{-p_{,ii}} \tag{4.68}$$

and

$$\widehat{-p_{,ii}} = k_l k_l \hat{p}, \tag{4.69}$$

where $\widehat{(\cdot)}$ denotes the value of the (k_1, k_2, k_3)-component in the Fourier decomposition of $(\cdot)$. In this discrete version of the convolution, summation on the primed wavenumbers from $-\infty$ to $+\infty$ is understood, as well as summation on the indices i and j which range from 1 to 3. We conclude that

$$\hat{p} = -\frac{1}{k_l k_l} k_i' a_{j,k_1' k_2' k_3'} (k_j - k_j') a_{i,k_1 - k_1' k_2 - k_2' k_3 - k_3'}. \tag{4.70}$$

(Here and above we omit the full calculation of the convolutions. The reader may fill in the details.) We are now in a position to project the Navier–Stokes equations onto the basis (4.66). Note that each component in the equations of motion (4.59) and every choice of a

Table 4.1 *Terms in the projected Navier–Stokes equations.*

Term in Navier–Stokes	Term in projected ODE for choice of (i, k_1, k_2, k_3)
$u_{i,t}$	$\dot{a}_{i,k_1k_2k_3}$
$u_j u_{i,j}$	$a_{j,k'_1k'_2k'_3}\mathrm{i}(k_j - k'_j)a_{i,k_1-k'_1 k_2-k'_2 k_3-k'_3}$
$-p_{,i}$	$\frac{\mathrm{i}k_i}{k_lk_l}k'_m a_{j,k'_1k'_2k'_3}(k_j - k'_j)a_{m,k_1-k'_1 k_2-k'_2 k_3-k'_3}$
$u_{i,jj}$	$-k_lk_l a_{i,k_1k_2k_3}$
f_i	$\hat{f}_i$

wavenumber triplet (k_1, k_2, k_3) results in a first order ODE component. In Table 4.1 we list each term in the equation and its counterpart term in the projection, thereby defining the ODEs. Summation is implied on the repeated, primed indices.

This formulation may be adapted to more general situations and boundary conditions, but substantial difficulties arise in solving the Poisson equation without the simplifying assumption of periodic boundary conditions. Nonetheless, similar spectral decompositions lie at the heart of some of the most efficient and accurate numerical simulations of turbulent flows [70, 138].

We now describe a second approach to projection of the Navier–Stokes equations. By now the reader might expect this to result from choosing a different basis. In this case the Navier–Stokes equations are projected onto a *divergence-free basis*. There are several choices for such a basis. It may be derived from that in (4.66) by defining, say, the first and third velocity components (in the x_1- and x_3-directions) as in (4.66), and deriving the second component from the divergence-free condition (4.60). This yields the elements

$$\left. \begin{array}{l} \frac{1}{\sqrt{2\pi}}(e^{\mathrm{i}(k_1x_1+k_2x_2+k_3x_3)}, -\frac{k_1}{k_2}e^{\mathrm{i}(k_1x_1+k_2x_2+k_3x_3)}, 0) \\[2mm] \frac{1}{\sqrt{2\pi}}(0, -\frac{k_3}{k_2}e^{\mathrm{i}(k_1x_1+k_2x_2+k_3x_3)}, e^{\mathrm{i}(k_1x_1+k_2x_2+k_3x_3)}) \end{array} \right\} \tag{4.71}$$

for $k_2 \neq 0$ and k_1, k_3 arbitrary. For $k_2 = 0$, $k_3 \neq 0$ and k_1 arbitrary we may take

$$\frac{1}{\sqrt{2\pi}}(0, e^{\mathrm{i}(k_1x_1+k_3x_3)}, 0), \quad \frac{1}{\sqrt{2\pi}}(e^{\mathrm{i}(k_1x_1+k_3x_3)}, 0, -\frac{k_1}{k_3}e^{\mathrm{i}(k_1x_1+k_3x_3)}), \tag{4.72}$$

for $k_2 = k_3 = 0$, and $k_1 \neq 0$:

$$\frac{1}{\sqrt{2\pi}}(0, e^{\mathrm{i}k_1x_1}, 0), \quad \frac{1}{\sqrt{2\pi}}(0, 0, e^{\mathrm{i}k_1x_1}); \tag{4.73}$$

and for $k_1 = k_2 = k_3 = 0$,

$$(1, 0, 0), \quad (0, 1, 0), \quad (0, 0, 1). \tag{4.74}$$

Note that, except in the last case $k_i \equiv 0$, there are now only two linearly independent vectors for each choice of wavenumber triplet, reflecting the fact that the scalar constraint equation (4.60) has been automatically satisfied.

There are several other possible choices for divergence-free bases. For an example in cylindrical coordinates, see [210]. In particular the proper orthogonal decomposition yields divergence-free bases that will be used later in this book for the construction of low-dimensional projections.

The next step is to index the elements of the basis, which for simplicity we denote $(\varphi_1^n(\mathbf{x}), \varphi_2^n(\mathbf{x}), \varphi_3^n(\mathbf{x}))$. The subscripts identify the components of each basis vector in Cartesian coordinates, the mode number being now indicated by a superscript. The expansion of the velocity field then takes the form

$$(u_1(\mathbf{x}, t), u_2(\mathbf{x}, t), u_3(\mathbf{x}, t)) = a_n(t)(\varphi_1^n(\mathbf{x}), \varphi_2^n(\mathbf{x}), \varphi_3^n(\mathbf{x})) \tag{4.75}$$

or, in brief, $u_i = a_n \varphi_i^n$. In the examples above, each choice of an integer for the single index n corresponds to selection of a triple (k_1, k_2, k_3); in subsequent applications of empirical basis elements such as those of (3.34) we use a multi-index explicitly.

Following the general approach outlined in the first two sections of this chapter, we substitute the modal expansion into the equations of motion and use the inner product in $L^2(\Omega)$ defined by

$$((u_1(\mathbf{x}), u_2(\mathbf{x}), u_3(\mathbf{x})), (v_1(\mathbf{x}), v_2(\mathbf{x}), v_3(\mathbf{x}))) = \int_\Omega u_i(\mathbf{x}) v_i(\mathbf{x}) d\mathbf{x}, \tag{4.76}$$

(Section 1.4). The specific form of the basis determines whether convenient cancellations and simplifications occur in the projected terms. The only point at which the use of a divergence-free basis differs significantly from the application of the basis (4.66) is in the behavior of the pressure term. This term yields the following contribution to the nth projected ODE:

$$\dot{a}^n = \ldots - \int_\Omega p_{,i} \varphi_i^n d\mathbf{x} + \ldots$$

Using the periodicity of the domain, our assumption of a zero mean pressure gradient, and the divergence-free property of the basis functions, we may integrate by parts to obtain:

$$\int_0^{2\pi} \int_0^{2\pi} \int_0^{2\pi} p_{,i} \varphi_i^n \, dx_1 \, dx_2 \, dx_3 =$$

$$\int_0^{2\pi} \int_0^{2\pi} p\varphi_1^n \, dx_2 \, dx_3 \Big|_{x_1=0}^{2\pi} + \int_0^{2\pi} \int_0^{2\pi} p\varphi_2^n dx_1 dx_3 \Big|_{x_2=0}^{2\pi}$$

$$+ \int_0^{2\pi} \int_0^{2\pi} p\varphi_3^n \, dx_1 \, dx_2 \Big|_{x_3=0}^{2\pi} - \int_0^{2\pi} \int_0^{2\pi} \int_0^{2\pi} p\varphi_{i,i}^n \, dx_1 \, dx_2 \, dx_3 \equiv 0. \tag{4.77}$$

In (4.77) we write out the boundary terms explicitly, since we ultimately need to adapt this calculation to a situation in which non-periodic boundary conditions are taken in one direction, so that some boundary terms do survive. In the present simple case, through the use of a divergence-free basis, the pressure entirely drops out, considerably simplifying the resulting ordinary differential equations.

In Chapter 10 we use an empirical eigenfunction basis of the form (3.34) to provide a hierarchy of nested subspaces in which to project the Navier–Stokes equations defined in the wall region of the boundary layer. As we have remarked several times, we are especially interested in (very) low-dimensional models. We wish to truncate at wavenumbers far below those at which viscous dissipation occurs and so we cannot appeal to a simple decay estimate such as (4.25) above to conclude that errors remain small as the solution evolves. Some additional modeling is needed to properly account for the neglected modes. To this we now turn.

4.4 Towards low-dimensional models

In Section 4.2 above we pointed out that the Laplacian operator in the heat equation leads to exponential decay of high wavenumber modes at a rate proportional to the square of the Fourier wavenumber. Since the Navier–Stokes equations also have such a viscous dissipation term, one might expect a similar decay to enable truncations at relatively low wavenumbers to yield reasonable results. However, at Reynolds numbers sufficiently high to give turbulent flow conditions, and taking into account the three-dimensional character of the flow, one still needs enormous numbers of modes to resolve the energy cascade through the inertial range and the dissipative process itself.

In determining an adequate wavenumber cut-off for a truncation, one asks 'How high is "high enough"?' There are three perspectives from which to view this question: the perspective of rigorous numerical analysis, the engineering/computational fluid dynamics perspective, and the nonlinear analyst's perspective.

From the point of view of numerical analysis, we consider a "full" system of ODEs

$$\dot{a} = F(a), \tag{4.78}$$

which is supposed to be a faithful representation of the Navier–Stokes equations and a system that is a perturbation of (4.78):

$$\dot{b} = \tilde{F}(b). \tag{4.79}$$

The perturbation may be due to the numerical approximations or neglected modes or both. An estimate is required of the difference between the functions $F(a)$ and $\tilde{F}(a)$. Once this is available, an estimate that bounds the distance between solutions of (4.78) and (4.79) may be obtained using Gronwall's inequality ([77] p. 37, [144] p. 169). This states that, for real-valued, non-negative functions $v(s)$, $u(s)$, $c(s)$, if the inequality

$$v(t) \leq c(t) + \int_0^t u(s)v(s)ds \tag{4.80}$$

holds for $t \geq 0$, then we may conclude that

$$v(t) \leq c(0) \exp \int_0^t u(s)ds + \int_0^t c'(s) \left[\exp \int_s^t u(\tau)d\tau \right] ds. \tag{4.81}$$

From (4.78) and (4.79), we form an equation for the error $\epsilon(t) = |a(t) - b(t)|$ and integrate to obtain

$$\epsilon(t) = \epsilon(0) + \left| \int_0^t \left[F(a(s)) - \tilde{F}(b(s)) \right] ds \right|$$

$$\leq \epsilon(0) + \int_0^t \left| F(a(s)) - \tilde{F}(b(s)) \right| ds. \tag{4.82}$$

We must now bound the integrand. Let $\tilde{F} = F + F_1$ (where F_1 represents the neglected modes, for example) and assume that we have a Lipschitz estimate of F of the form

$$|F(a) - F(b)| \leq L|a - b|$$

and a *uniform* bound on the function F_1:

$$|F_1| \leq K.$$

Here L and K are constants independent of a and b. In this case (4.82) may be written

$$\epsilon(t) \leq \epsilon(0) + \int_0^t L\epsilon(s)ds + \int_0^t K ds = \epsilon(0) + Kt + \int_0^t L\epsilon(s)ds,$$

and we deduce from (4.81) that

$$\epsilon(t) \leq \epsilon(0)e^{Lt} + \int_0^t K e^{L(t-s)}ds \leq \left[\epsilon(0) + \frac{K}{L} \right] e^{Lt}. \tag{4.83}$$

This is essentially a statement concerning continuous dependence on initial conditions and parameters. Note that, in general, we cannot claim better than an exponentially diverging error.

This strategy works well in many ODE situations and is used, for example, in proving the averaging theorem in which one shows that the influence of rapidly oscillating modes is small [144, 150]. Unfortunately, for nonlinear evolution equations of the type we are concerned with, such as the Navier–Stokes equation, estimation of the constants K and L is not feasible. Since F is a nonlinear differential operator it can in principle amplify the high wavenumber modes so that they dominate the equation and any approximation of it by lower wavenumbers becomes meaningless. There are no a-priori results that assure that the high wavenumber modes will remain small or even bounded. This is the famous (and still unresolved) question of the possibility of blow-up in the Navier–Stokes equations. Therefore, this approach does not yield practical estimates unless additional assumptions are made about the nature of solutions. In Section 13.2, we make such an assumption and use the Gronwall lemma to deduce a probabilistic error estimate for low-dimensional models.

From the point of view of the engineer, a direct numerical simulation (DNS) of the Navier–Stokes equations contains enough modes if the dissipative range in the spectrum is well resolved. This manifests itself as an exponential decay in the high wavenumbers, which emerges as one includes successively higher wavenumber modes. In contrast, a simulation that is not well resolved will display an accumulation of energy at high wavenumbers that has cascaded down the spectral pipeline. The largest numbers currently (2011) used in simulations of a turbulent channel flow, for example, are 6144 Fourier modes in the streamwise direction, 4608 in the spanwise direction, and 633 gridpoints in the wall-normal direction [165], for a total of about 17.9 billion modes. For homogeneous turbulence, the largest calculation is even greater, using 4096^3 Fourier modes [391]. One can reduce such large figures at the expense of losing high wavenumber detail by replacing the dissipative range of the spectrum by an eddy viscosity model. This is referred to as a Large Eddy Simulation (LES), see [309].

Nonlinear analysis introduces the concept of an attracting manifold in phase space. Under certain assumptions (weaker than those required to lend popular turbulence simulations credibility) one can prove that there exists a finite-dimensional manifold in the infinite-dimensional phase space. This is generally called an inertial manifold. The inertial manifold has the property that it attracts all solutions of the equation. Therefore, a simulation of the dynamics of the inertial manifold will produce the long-term dynamics of the original equations. The inertial manifold is a global version of the center-unstable manifold, which is described in the context of ODEs in Chapters 5 and 8. The dynamics of the inertial manifold are in principle determined by a finite set of ODEs (an inertial form). However, for problems such as turbulence, the number of ODEs is very large. The theory of inertial manifolds is described in [85] and [367], and we return to it in Section 13.4.

With these remarks in mind, let us return to the set of projected ODEs for the modal coefficients, $a_n(t)$, obtained by the Galerkin procedure. This is still, in principle, a countably infinite set of equations. To proceed further one must truncate and the acceptable order of truncation depends on what one's purpose is. As we have pointed out, a DNS must resolve the dissipative scales as well as the largest scales in the flow. Classical estimates ([368] p. 21), give the ratio of the Kolmogorov (dissipative) scale to that of the largest eddies (integral scales) as

$$\frac{\eta}{\ell} = R_\ell^{-3/4}, \tag{4.84}$$

as in Section 2.1. Thus, in three spatial dimensions, one expects to require $\mathcal{O}(R_\ell^{9/4})$ modes to adequately represent the full range of active scales. Here the relevant Reynolds number is $R_\ell = u\ell/\nu$, where u is the root-mean-square magnitude of the turbulent fluctuations, and ℓ is their scale.

For example, for a turbulent channel flow at $R_e = UD/\nu = 10\,000$, based on centerline velocity U and channel width D, we have

$$\frac{u\ell}{\nu} = \frac{u}{U} \cdot \frac{UD}{\nu} \cdot \frac{\ell}{D} \approx \frac{1}{30} \cdot \frac{UD}{\nu} \cdot \frac{1}{3},$$

so that the Reynolds number R_ℓ in (4.84) is approximately

$$\frac{10\,000}{90} \approx 111.$$

Thus the dimension of the ODE system required to resolve the smallest scales in the wall region is $\mathcal{O}(10^4 - 10^5)$. Clearly there is little hope for qualitative understanding of the solutions of systems in phase spaces of such high dimension.

However, if we focus on limited regions in the flow and concentrate on the low-wavenumber end of the spectrum, in which the bulk of the turbulent energy is produced, there is some indication that truncations of far lower dimension may be useful. In flows energetically dominated in this range by coherent structures, such as the mixing layer and boundary layer described in Sections 2.4 and 2.5, it is intuitively clear that relatively few modes can capture the spatial form of the structures, provided they are chosen appropriately. Here the details depend crucially on the particular flow and geometry and so representations by empirical eigenfunctions of the kind described in Chapter 3 may be expected to help.

At this point it is useful to recall the representation (3.34) adapted to situations in which two flow directions are statistically homogeneous (say streamwise and spanwise) and the third is non-homogeneous. This applies to the boundary layer of a fully developed channel flow and to mixing layers which grow slowly in the streamwise direction. Fourier modes are used in the homogeneous directions (being themselves optimal). The expansion used in the boundary layer work (Chapter 10) is of the form

$$\mathbf{u}(\mathbf{x}, t) = \sum_{k_1} \sum_{k_3} \sum_n a_{k_1, k_3, n}(t) e^{2\pi i \left(\frac{k_1 x_1}{L_1} + \frac{k_3 x_3}{L_3} \right)} \boldsymbol{\varphi}_n(k_1, k_3; x_2), \qquad (4.85)$$

where the $\boldsymbol{\varphi}_n(k_1, k_3; x_2)$ are vector-valued, divergence-free functions. In such representations we refer to each choice of n as a *family* of empirical eigenfunctions (other terms are also used: Sirovich, for example, calls n a quantum number). In this terminology, speaking of the energy in the first family of eigenfunctions implies summation (or integration) over the Fourier wavenumbers k_1 and k_3. In some papers this is simply called "the energy in the first eigenfunction."

To give an explicit idea of how well such empirical representations can reproduce a turbulent kinetic energy budget, in Figure 4.1 we show data from a study by Moin [241] (also see [244]) in which a database obtained by LES of a channel flow at a centerline velocity $R_e = 13\,800$ was subjected to the proper orthogonal decomposition. The first three graphs (a)–(c) show the streamwise, wall-normal, and spanwise amounts of (averaged) kinetic energy captured by 1, 10, and 20 families of eigenfunctions, as a function of the distance from the wall ($y/\delta = 0$ corresponds to the centerline). We observe that a single family of eigenfunctions can capture 30–40% of the total energy in the boundary layer itself (say between $y/\delta = -1$ and $y/\delta = -0.6$), but that this drops to an insignificant fraction as one approaches the channel center. This feature is even more pronounced in the near-wall region itself (d). Here, to $y/\delta = -0.94$, a single family captures 90% of the streamwise kinetic energy, although this drops to about 50% at $y/\delta = -0.90$. (Note that the results in (a)–(c) and (d) involve *different* families of eigenfunctions, optimized on the full channel and on the near-wall layer respectively. The point is that one cannot expect a basis optimized on the full channel to do well in any particular subdomain.) Thus, if we are prepared to restrict ourselves in physical space as well as wavenumber space, significant reductions of dimension may be possible. When we take up the boundary layer example at greater length in Chapters 10 and 11, we argue that a truncation retaining only a single family of eigenfunctions, a single ($k_1 = 0$) Fourier mode in the streamwise direction, and as few as five Fourier modes in the spanwise direction, can indeed capture key aspects of the bursting process outlined in Section 2.5.

We shall leave the details to those chapters, but remark that in truncations of such low dimension we typically include only modes that are directly linearly unstable. No attempt is made to retain "stable" modes in the inertial range that are dynamically active,

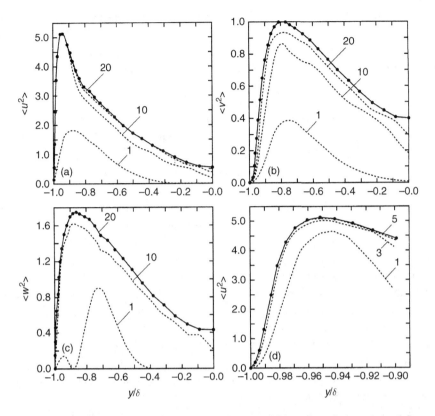

Figure 4.1 Convergence of Karhunen–Loève decompositions of a turbulent channel flow. Panels (a), (b), and (c) show behavior for the three velocity components from wall to centerline; panel (d) shows the near-wall behavior for the streamwise component, based on eigenfunctions computed for the domain $0 \leq x_2^+ \leq 65$. From Moin [241].

being excited in the energy transfer process and which thus act, on average, as energy sinks for the lower modes. It is therefore necessary to account for the effects of these neglected modes on those included in the models. This may be done with a fairly conventional eddy viscosity mechanism of the type originally proposed by Heisenberg for homogeneous turbulence [32]. Details are given in Chapter 10 but we observe here that such a model effectively increases the magnitudes of coefficients in the viscous terms, thereby (partially) stabilizing some of the higher modes in the model. A scalar parameter, α, is introduced to specify the rate of energy extraction due to the bands of neglected modes adjoining in phase space the modes retained. While the model is constructed in such a manner that $\alpha = \mathcal{O}(1)$, we can modestly vary α and hence treat it as a bifurcation parameter, seeking ranges for which the model's behavior appears reasonable. It also provides a simple check on the model. For large values of α, corresponding to unrealistically large dissipation, one expects solutions to decay and all turbulent fluctuations to die out, while unrealistically low values of α should lead to unreasonably large solutions or perhaps even blow-up.

Thus one effect of a low-dimensional truncation is the loss of the natural energy cascade and its replacement by a model. A second effect is more subtle. We have noted that the eigenfunctions of the proper orthogonal decomposition (or indeed *any* divergence-free basis functions such as those of (4.71–4.74)) impose coupling among the individual velocity components. Although these bases are complete, and *any* velocity field can be expressed without error (in the L^2 sense) by an *infinite* sum of their elements, the coupling becomes significant for low-dimensional truncations. In such truncations, one cannot represent all velocity fields with equal accuracy; only those sufficiently close to the spatial structures extracted from the ensemble average will be reproduced well. (This is what is implied in saying that a certain truncation captures 90% of kinetic energy *on average*.)

To take an extreme example, we cannot adequately reproduce purely streamwise perturbations of the form $\mathbf{u} = (u_1, 0, 0)$, or cross-stream ones $\mathbf{u} = (0, u_2, u_3)$ with only one or two families of empirical eigenfunctions derived from a turbulent database, for the eigenfunction structure will reflect the average activity in all three components, implying that a particular spatial dependence in the first component is tied to specific forms in the second and third. Thus the evolution of such "pure" (if artificial) perturbations cannot be studied in the context of projections onto low-dimensional empirical subspaces and one is constrained to look only at "typical" behaviors characteristic of the flows from which the basis elements were derived. This blindness of empirical eigenfunctions to the effects of individual velocity components and the instantaneous effects of coupling among them in the Navier–Stokes equations leads to an interesting feature of the boundary layer models, which we describe in Section 10.7.1.

The fact that empirical eigenfunctions $\boldsymbol{\phi}_j(\mathbf{x})$ can only reproduce mixtures of the data they are constructed from has led to the proposal that *shift modes* be included among the basis functions used in Galerkin projections. In studying low-dimensional models of cylinder wakes, Noack *et al.* found that transient growth of instabilities could not be reproduced using bases derived from numerical simulations of alternating vortex shedding behind a circular cylinder [262]. The mean flow $\mathbf{u}_0(\mathbf{x})$, derived from that database, is significantly different from the unstable laminar flow solution $\mathbf{u}_s(\mathbf{x})$, found by numerical solution of the time-independent Navier–Stokes equations: while both feature symmetric recirculation domains behind the cylinder, that of $\mathbf{u}_s(\mathbf{x})$ extends downstream much further than for $\mathbf{u}_0(\mathbf{x})$ [262, Figure 7]. To include a neighborhood of $\mathbf{u}_s(\mathbf{x})$ in the reduced phase space, the empirical basis was augmented by addition of the shift mode $\mathbf{u}_\Delta(\mathbf{x}) = \mathbf{u}_0(\mathbf{x}) - \mathbf{u}_s(\mathbf{x})$, and the velocity field represented as

$$\mathbf{u}(\mathbf{x}) = \mathbf{u}_0(\mathbf{x}) + a_\Delta(t)\mathbf{u}_\Delta(\mathbf{x}) + \sum_{j=1}^{N} a_i(t)\boldsymbol{\phi}_j(\mathbf{x}). \tag{4.86}$$

During the projection process an $(N + 1)$th ODE is computed for the shift mode coefficient $a_\Delta(t)$. This additional mode contains the unstable laminar equilibrium $\mathbf{u}_s(\mathbf{x})$ and therefore allows the model to reproduce solutions that connect it to the attractor. To ensure orthogonality, $\mathbf{u}_\Delta$ can be replaced by $\mathbf{u}_\Delta - \sum_{j=1}^{N} (\mathbf{u}_\Delta, \boldsymbol{\phi}_j) \, \boldsymbol{\phi}_j$, and the entire basis made orthonormal.

In [262] the transient solutions are in fact shown to evolve near an approximate invariant manifold for the low-dimensional model, related to the Hopf bifurcation in which the laminar flow loses stability, and it is also shown that inclusion of the shift mode allows the model to perform better over a range of Reynolds numbers. A simple 3-dimensional example with a stable limit cycle is constructed to show that naive projection into an empirical basis can produce poor results. This model has been used in numerous flow control applications, e.g. [34, 277]. A more elaborate example in which projection onto a 2-dimensional subspace yields a model with an *unstable* limit cycle can be found in [298]. Other examples of instability control applications include [218, 219] and shift modes are also used in [268].

Other modifications to POD have been proposed. In [179] a "PODS" based on Squire's coordinate system [351] was used in studies of channel turbulence and it was shown that the resulting bases improve convergence properties and better represent modal energetics and Reynolds stresses. A similar idea appeared in [347], where certain empirical eigenfunctions are split into mutually orthogonal pairs and allowed to evolve independently as separate dependent variables. The low-dimensional model that results correctly represents a *stable* laminar solution that coexists with a turbulent attractor in plane Couette flow at a low Reynolds number and allows it to qualitatively reproduce the bifurcation diagram as Reynolds number varies in the range 0–500.

We have introduced some issues related to severe truncation in wavenumber; we now turn to the effects of truncation in the physical domain. In attempting to produce low-dimensional models we do not want to include large populations of coherent structures, for if we do, allowing each to appear independently and allowing enough "ingredients" for the model to evolve reasonably realistically, we will again require too many dependent variables. (Recall Figures 2.14 and 2.17 of Section 2.5.) We need to extract a relatively small spatial subdomain from the large (or notionally infinite) flow region.

In homogeneous directions it seems reasonable to select characteristic lengths L_i sufficient to contain "enough" structures and then apply periodic boundary conditions, much as is routinely done in numerical simulations. In Section 8.4 we discuss this at some length for a simpler scalar one-space dimension evolution equation. However, in non-homogeneous directions, the artificial boundaries introduced by selection of any subdomain within the interior of the flow can present more serious difficulties. In a mixing layer there are two such, above and below, and in a boundary layer one: the eigenfunctions automatically satisfy the no-slip condition at the wall, so this boundary poses no problem. One can of course take a domain large enough so that turbulent fluctuations have effectively decayed (see Sections 2.4 and 2.5) but as the results illustrated in Figure 4.1 indicate, this also demands the inclusion of many modes that may well be redundant for the dynamics of the thin turbulent layer itself.

In the boundary layer model of [22] described in Chapter 10, an upper boundary was selected for the wall layer domain at x_2^+ ($= y^+$) $= 40$ in wall units (see Section 2.5, Equation (2.42)), a choice supported by the observations above and necessitated by the fact that the correlation tensor from which the basis functions were derived had only been measured to this distance [155]. The eigenfunctions do not vanish here and consequently the term

$$\int_0^{L_1} \int_0^{L_3} p\varphi_2^n dx_3 dx_1 \,|_{x_2^+=40},\qquad\qquad(4.87)$$

deriving from integration by parts of the projected pressure term in (4.77) is non-zero (all other terms do vanish, either because the eigenfunctions vanish at the wall $x_2^+ = 0$, or due to the periodic structure in the x_1- and x_3-directions). This term corresponds to a pressure field on the "free" surface $x_2^+ = 40$ and so represents communication between the modeled region and the outer part of the flow. As such, it is a potentially useful feature of the model, since its presence (or absence, if it be set to zero) allows one to distinguish between inner and outer influences on the dynamics in the wall layer. The major practical difficulty is that terms such as (4.87) are not specified within the model itself but must be estimated or supplied from independent computations. One simple idea is to replace these terms by additive random excitations characteristic of disturbances in the outer layer, as we describe in subsequent chapters.

An alternative and more mathematically appealing solution to the problem of selecting a subdomain in non-homogeneous directions was suggested by Zhou and Sirovich [394], who extracted basis functions for a full channel and then took suitable linear combinations of them designed for optimality in the wall region. No boundary terms appear when projection is carried out using an inner product involving integrals across the whole channel, but the model now contains a weighted contribution of the outer layer and, indeed, the whole channel. Models of this type do not provide universal descriptions of the wall layer, as do those of [22]. We return to discuss this in Chapters 10, 12, and 13, but note that there appears to be no obviously correct way to overcome such difficulties. What is mathematically convenient and clean may be physically unrealistic or have unclear implications. Such issues are probably best dealt with on a case by case basis. In any event, if a relatively small spatial subdomain is to be selected, some sort of model to account for missing neighboring domains will certainly be necessary, just as models for neglected wavenumbers are needed.

A final issue arises in the extraction of small spatial domains in the case of a flow driven by mean shear, as are both the boundary and mixing layers. As we pointed out in Chapter 2, the components of the mean profiles U_i are best defined via ensemble averages, although they are frequently determined experimentally by long-term time or space averages. When we decompose the primitive velocity field $\tilde{u}_i = U_i + u_i$ as in Chapter 2, the evolution equation (4.64) for the turbulent fluctuations u_i contains U_i and $U_{i,j}$ as "driving" terms. If we assume that these latter are constant (as they should be in a sufficiently large domain, or over a sufficiently long time), then we may seriously upset the true energy balance in a small subdomain, especially over short time intervals. Initial unpublished work leading to that in [22], in which the mean streamwise velocity and shear were taken constant, revealed that solutions of the low-dimensional projected ODEs typically grew without bound and even appeared to blow up in finite time. (This is not surprising from a dynamical systems viewpoint, for the equations are quadratic and no global trapping region exists.)

What is missing from such a model in physical terms is the local collapse of the shear profile as fluctuations grow. This may be restored by supposing a decomposition in which the "mean" quantities are slowly varying, time-dependent functions such as

$U_1 = U_1(x_2, \epsilon t)$, whose forms may be obtained from the spatially averaged Navier–Stokes equations. We describe how this is done in Section 10.8. The net effect is to replace the temporally constant U_i by two terms: a constant part due to the pressure gradient, which ultimately drives the flow, and a part quadratic in the fluctuations u_i depending upon the Reynolds stress $\langle u_i u_j \rangle$ integrated across the layer. The final projected equations therefore contain a cubic term in addition to the linear and quadratic terms derived directly from the Navier–Stokes equations. In some cases at least, the geometrical effect of this term in phase space is to create a global trapping region: a bounded (but possibly large) domain into which all solutions eventually enter and which contains the attractors. Effectively, the model for the slowly varying mean profile in a local subdomain introduces a stabilizing feedback term that mimics the influence of the missing neighboring spatial domains, much as the Heisenberg model mimics the effect of missing wavenumbers.

5

Balanced proper orthogonal decomposition

As we shall see in Parts III and IV, the techniques of proper orthogonal decomposition and Galerkin projection can be powerful tools for obtaining low-order models that capture the qualitative behavior of complex, high-dimensional systems. However, for certain systems, the resulting models can perform poorly: even if a large fraction of energy (over 99%) is captured by the modes used for projection, the resulting low-order models may still have completely different qualitative behavior. The transients may be poorly captured, and the stability types of equilibria can even be different.

In this chapter, we present a method which can dramatically outperform projection onto traditional energy-based empirical eigenfunctions described in Chapter 3. We focus primarily (though not exclusively) on *linear* systems, for several reasons. Many of the pitfalls of traditional proper orthogonal decomposition can be demonstrated for linear systems, without the additional complexity of nonlinearities. Furthermore, for linear systems, one can use *operator norms* to quantify the difference between a detailed model and its reduced-order approximation. Most importantly, for linear systems, there are established tools for performing model reduction, for instance using balanced truncation, which is described in Section 5.1. In contrast, while some modest extensions to nonlinear systems have been attempted, model reduction of nonlinear systems is still an active area of research.

The techniques described in this chapter also differ from those in Chapters 3 and 4 in that they are formulated for *input–output systems*. The *inputs* represent the external influences on the system, for instance from external disturbances, or from actuators in a flow-control setting. The *outputs* represent the quantities that we wish our reduced-order model to capture accurately. This may be the entire state (all flow variables everywhere in space), or it could be the flow variables in a particular region of space, or it could be a small number of sensor measurements. The modeling procedures discussed in this chapter handle all of these cases, but one must specify the inputs and outputs a priori.

We begin this chapter with an overview of balanced truncation, a model reduction procedure widely used in the control theory community. Balanced truncation is a highly effective model-reduction technique, but requires a coordinate transformation of the entire state, and so is not tractable for systems of very high dimension, as arise in fluid mechanics. In Section 5.2, we present an approximate version of balanced truncation, called balanced proper orthogonal decomposition, that is computationally tractable even for large systems. This method shares many characteristics with the method of snapshots for computing POD modes, discussed in Section 3.4.1. We then describe extensions of the basic approach to systems with large numbers of outputs (Section 5.3) and discuss connections with the traditional proper orthogonal decomposition of Chapter 3 in Section 5.4. Section 5.5 treats extensions to unstable and nonlinear systems, as well as an efficient algorithm for computing reduced-order models that are identical to those produced with balanced POD, without the need for adjoint simulations. Finally, in Section 5.6 we describe several examples illustrating the differences between balanced proper orthogonal decomposition, and the standard POD and Galerkin projection.

5.1 Balanced truncation

Balanced truncation is a model-reduction method introduced by Moore [246] for linear input–output systems. As is standard in the control theory literature, we denote the vector of inputs by $\mathbf{u} \in \mathbb{R}^p$ (not to be confused with fluid flow velocity), and the vector of outputs by $\mathbf{y} \in \mathbb{R}^q$. The linear dynamics then have the form

$$\left. \begin{array}{l} \dot{\mathbf{x}} = \mathbf{Ax} + \mathbf{Bu}, \\ \mathbf{y} = \mathbf{Cx}, \end{array} \right\} \tag{5.1}$$

where $\mathbf{x} \in \mathbb{R}^N$ is the state variable. In this section, we furthermore assume that (5.1) is stable: that is, all the eigenvalues of $\mathbf{A}$ are in the left-half plane (although we will relax this assumption in Section 5.5.1). The goal of model reduction is to find a system of the form (5.1) that approximates the same input–output behavior, but with state dimension smaller than n.

We begin by defining the notions of *controllability* and *observability*, which are the properties that are "balanced" in this approach. Conceptually, the most controllable states are those that are most easily excited by the input $\mathbf{u}$, when the system starts from rest, $\mathbf{x}(0) = 0$. Conversely, the most observable states are those that have the largest effect on future outputs $\mathbf{y}(t)$, in the absence of any input ($\mathbf{u}(t) = 0$). The idea behind balanced truncation is to transform to coordinates in which the states that are the most controllable are also the most observable, and then truncate the states that are least controllable/observable. This is a plausible approach to model reduction, since the states that are truncated are those least affected by the input (least controllable), and also have the least effect on the output (least observable).

These intuitive concepts may be quantified by introducing two $N \times N$ matrices called the *controllability* and *observability Gramians*, given by

$$\mathbf{W}_c = \int_0^\infty e^{\mathbf{A}t} \mathbf{BB}^\star e^{\mathbf{A}^\star t} \, dt, \qquad \mathbf{W}_o = \int_0^\infty e^{\mathbf{A}^\star t} \mathbf{C}^\star \mathbf{C} e^{\mathbf{A}t} \, dt. \tag{5.2}$$

These are symmetric, positive-semidefinite matrices that precisely define the degree of controllability and observability of different states. For instance, if the system has initial condition $\mathbf{x}(0) = \mathbf{x}_0$ and the input $\mathbf{u}(t) = 0$, then the output $\mathbf{y}(t) = \mathbf{C} \exp(\mathbf{A}t)\mathbf{x}_0$, and $\|\mathbf{y}\|^2 = \mathbf{x}_0^\star \mathbf{W}_o \mathbf{x}_0$. In this sense, the observability Gramian quantifies a given state's effect on future outputs: states $\mathbf{x}_0$ that are aligned with eigenvectors corresponding to the largest eigenvalues of $\mathbf{W}_o$ will have the largest future output signals $\mathbf{y}(t)$, and so will be the most observable. We may also regard strongly observable states as states with large *dynamical importance*, since a small perturbation in the direction of a strongly observable state creates a large change in the future dynamics of $\mathbf{y}(t)$. Note that Equations (5.2) measure cumulative (integrated) effects, and that stable dynamics are required for these expressions to be well defined.

Conversely, the states that are aligned with the eigenvectors corresponding to the largest eigenvalues of $\mathbf{W}_c$ are the most controllable states. One can show that the minimal-energy input $\mathbf{u}(t)$ that drives the state to $\mathbf{x}_0$ at $t = 0$ (with $\mathbf{x}(-\infty) = 0$) has energy $\|\mathbf{u}\|^2 = \mathbf{x}_0^\star \mathbf{W}_c^{-1} \mathbf{x}_0$. This quantity may then be used to quantify the degree of controllability: states that may be reached with a small input energy are more controllable than states that require a large input energy.

The Gramians are typically computed by solving the *Liapunov equations*

$$\mathbf{AW}_c + \mathbf{W}_c\mathbf{A}^\star + \mathbf{BB}^\star = 0, \qquad \mathbf{A}^\star\mathbf{W}_o + \mathbf{W}_o\mathbf{A} + \mathbf{C}^\star\mathbf{C} = 0. \tag{5.3}$$

When $\mathbf{A}$ is stable, these linear matrix equations have the unique solutions given by (5.2) (for a derivation of this result, see [98, Theorem 4.1]). However, for fluid flows, the state dimension N is typically on the order of 10^6, and direct solution of these $N \times N$ matrix equations is not feasible with today's computers (2011). It is therefore often preferable to compute $\mathbf{W}_c$ and $\mathbf{W}_o$ directly, by approximating the integrals in (5.2) via quadrature, as done in [202].

The goal of balancing is to transform to coordinates in which the most controllable states coincide with the most observable states. Under a change of coordinates $\mathbf{x} = \mathbf{Tz}$, the Gramians transform according to

$$\mathbf{W}_c \mapsto \mathbf{T}^{-1}\mathbf{W}_c(\mathbf{T}^{-1})^\star, \qquad \mathbf{W}_o \mapsto \mathbf{T}^\star\mathbf{W}_o\mathbf{T}. \tag{5.4}$$

One can show that, as long as the system (5.1) is both controllable and observable (i.e., the Gramians $\mathbf{W}_c$ and $\mathbf{W}_o$ both have rank N), one may find a coordinate system in which the Gramians are *equal* and *diagonal* [98, Proposition 4.7]:

$$\mathbf{T}^{-1}\mathbf{W}_c(\mathbf{T}^{-1})^\star = \mathbf{T}^\star\mathbf{W}_o\mathbf{T} = \mathbf{\Sigma} = \mathrm{diag}(\sigma_1, \ldots, \sigma_n). \tag{5.5}$$

The diagonal elements $\sigma_1 \geq \cdots \geq \sigma_N \geq 0$ are called *Hankel singular values*, and are independent of the coordinate system. Under this coordinate transform, the input–output system (5.1) becomes

$$\left. \begin{aligned} \dot{\mathbf{z}} &= \mathbf{T}^{-1}\mathbf{ATz} + \mathbf{T}^{-1}\mathbf{Bu}, \\ \mathbf{y} &= \mathbf{CTz}. \end{aligned} \right\} \tag{5.6}$$

To obtain a reduced-order model of dimension r, one simply projects (5.6) onto the subspace spanned by the first r components of $\mathbf{z}$.

This procedure of balanced truncation is certainly plausible on intuitive grounds, but it is not an *optimal* method of model reduction. However, it is provably close to optimal, in the following sense. One can show [98, Theorem 4.14] that any reduced-order model $\mathbf{G}_r$ of order r must satisfy

$$\|\mathbf{G} - \mathbf{G}_r\|_\infty > \sigma_{r+1}, \tag{5.7}$$

where $\| \cdot \|_\infty$ is the infinity norm defined in Equation (1.13) of Section 1.4, and σ_{r+1} is the $(r+1)$th Hankel singular value. This provides a lower bound on the achievable accuracy of any reduced-order model.

While balanced truncation is not optimal, one does have an *upper* bound on the error that is close to the lower bound (5.7). In particular, if $\mathbf{G}_r$ is a rank-r model produced by balanced truncation, then

$$\|\mathbf{G} - \mathbf{G}_r\|_\infty < 2 \sum_{j=r+1}^{n} \sigma_j, \tag{5.8}$$

or "twice the sum of the tail." Thus, while balanced truncation is not optimal, it is provably close to optimal. Note that there are other procedures, such as optimal Hankel norm approximation, that have better bounds (half that of balanced truncation), but these methods are more involved, and generally not tractable for very high-dimensional systems such as fluids. For more information, see [342].

5.2 Balanced POD

Balanced truncation is an effective model reduction technique, but it requires a transformation of the full state, and so is not computationally tractable for many fluid systems, which typically have millions of states. However, we may borrow ideas from the proper orthogonal decomposition, in particular the ideas of projecting onto modes and calculating modes using the method of snapshots, in order to form an approximation of balanced truncation that is computationally tractable even for these large systems. We call this procedure *balanced proper orthogonal decomposition*.

First, note that if $\mathbf{x}_i(t)$ denotes the solution of (5.1) with zero initial conditions and an input $\mathbf{u}_i(t) = (0, 0, \dots, \delta(t), \dots, 0)$ (an impulse in the ith component), then $\mathbf{x}_i(t) = \exp(\mathbf{A}t)\mathbf{b}_i$, where $\mathbf{b}_i$ denotes the ith column of $\mathbf{B}$. The controllability Gramian defined in (5.2) may then be written

$$\mathbf{W}_c = \int_0^\infty \left(\mathbf{x}_1(t)\mathbf{x}_1(t)^\star + \cdots + \mathbf{x}_p(t)\mathbf{x}_p(t)^\star \right) dt. \tag{5.9}$$

As in the previous section, we assume that the linear system is stable, so the impulse responses $\mathbf{x}_i(t)$ decay to zero as $t \to \infty$. Note the similarity between this expression and the correlation matrix $\mathcal{R}$ in (3.12). In fact, POD modes of the dataset $\{\mathbf{x}_1(t), \dots, \mathbf{x}_p(t)\}$ are the dominant eigenvectors of $\mathbf{W}_c$ (those corresponding to the largest eigenvalues), or the *most controllable* modes.

The first step in balanced POD is to obtain a quadrature approximation of (5.9) (and of the corresponding observability Gramian), using snapshots from impulse-response simulations. For instance, if we compute M snapshots at times $t_1, \dots, t_M$, and if we have an approximation

$$\int_0^\infty \mathbf{x}_1(t)\mathbf{x}_1(t)^\star \, dt \approx \sum_{j=1}^{M} \mathbf{x}_1(t_j)\mathbf{x}_1(t_j)^\star \delta_j,$$

where δ_j are quadrature coefficients, then we may stack the snapshots as columns of a matrix

$$\mathbf{X} = \left[\mathbf{x}_1(t_1)\sqrt{\delta_1} \cdots \mathbf{x}_1(t_M)\sqrt{\delta_M} \quad \cdots \quad \mathbf{x}_p(t_1)\sqrt{\delta_1} \cdots \mathbf{x}_p(t_M)\sqrt{\delta_M}\right], \tag{5.10}$$

so that the quadrature approximation of (5.9) is

$$\widetilde{\mathbf{W}}_c = \mathbf{X}\mathbf{X}^\star. \tag{5.11}$$

Similarly, we may approximate the observability Gramian using snapshots of an *adjoint system*, defined by

$$\dot{\mathbf{z}} = \mathbf{A}^\star\mathbf{z} + \mathbf{C}^\star\mathbf{v}. \tag{5.12}$$

If we construct impulse responses $\mathbf{z}_1(t), \ldots, \mathbf{z}_q(t)$, where $\mathbf{z}_i(t)$ is the solution of (5.12) with zero initial conditions and an impulse input on the ith component of $\mathbf{v}$, then the observability Gramian defined in (5.2) may be written

$$\mathbf{W}_o = \int_0^\infty \left(\mathbf{z}_1(t)\mathbf{z}_1(t)^\star + \cdots + \mathbf{z}_q(t)\mathbf{z}_q(t)^\star\right)dt. \tag{5.13}$$

As before, if we form a data matrix as in (5.10), then the observability Gramian may be approximated by

$$\widetilde{\mathbf{W}}_o = \mathbf{Y}\mathbf{Y}^\star. \tag{5.14}$$

We now have approximations to the controllability and observability Gramians, based on data from simulations. However, these are full (non-sparse) $N \times N$ matrices, and if the state dimension N is large (millions), it is not feasible to compute them, much less compute a balancing transformation as described in the previous section. However, a similar difficulty arose in computing POD modes (compare (5.11) and (5.14) with (3.43)), and with the method of snapshots, we were able to transform an $N \times N$ eigenvalue problem (3.43) (where N is the number of states) into an $M \times M$ eigenvalue problem (3.44) (where M is the number of snapshots). Here, we may use a related method of snapshots, in which we form the singular value decomposition of the matrix $\mathbf{Y}^\star\mathbf{X}$:

$$\mathbf{Y}^\star\mathbf{X} = \mathbf{U}\mathbf{\Sigma}\mathbf{V}^\star = \begin{bmatrix} \mathbf{U}_1 & \mathbf{U}_2 \end{bmatrix} \begin{bmatrix} \mathbf{\Sigma}_1 & \mathbf{0} \\ \mathbf{0} & \mathbf{0} \end{bmatrix} \begin{bmatrix} \mathbf{V}_1^\star \\ \mathbf{V}_2^\star \end{bmatrix} = \mathbf{U}_1\mathbf{\Sigma}_1\mathbf{V}_1^\star, \tag{5.15}$$

where $\mathbf{\Sigma}_1$ is an invertible, diagonal matrix (whose dimension R is the rank of $\mathbf{Y}^\star\mathbf{X}$), and $\mathbf{U}_1^\star\mathbf{U}_1 = \mathbf{V}_1^\star\mathbf{V}_1 = \mathbf{I}_R$. We then define matrices

$$\mathbf{\Phi} = \mathbf{X}\mathbf{V}_1\mathbf{\Sigma}_1^{-1/2}, \qquad \mathbf{\Psi} = \mathbf{Y}\mathbf{U}_1\mathbf{\Sigma}_1^{-1/2}. \tag{5.16}$$

The columns of $\mathbf{\Phi}$, denoted $\boldsymbol{\varphi}_j$, are called *balancing* or *direct modes*, and the columns of $\mathbf{\Psi}$, denoted $\boldsymbol{\psi}_j$, are the corresponding *adjoint modes*. It follows directly from (5.15) and (5.16) that $\mathbf{\Psi}^\star\mathbf{\Phi} = \mathbf{I}_R$, which implies that the balancing modes and the adjoint modes form a *bi-orthogonal set*: that is,

$$(\boldsymbol{\psi}_j, \boldsymbol{\varphi}_k) = \delta_{jk}. \tag{5.17}$$

We then form a reduced-order model of dimension $r \leq R$ by projecting onto the balancing modes. We approximate $\mathbf{x}(t)$ as

$$\mathbf{x}_r(t) = \sum_{k=1}^{r} a_k(t)\boldsymbol{\varphi}_k, \tag{5.18}$$

insert this expansion into the governing equations (5.1), and take inner products with $\boldsymbol{\psi}_j$, to obtain the reduced-order model

$$\left.\begin{aligned} \dot{a}_j &= \sum_{k=1}^{r} (\boldsymbol{\psi}_j, \mathbf{A}\boldsymbol{\varphi}_k)a_k + (\boldsymbol{\psi}_j, \mathbf{B}\mathbf{u}), \\ y &= \sum_{k=1}^{r} \mathbf{C}\boldsymbol{\varphi}_k a_k. \end{aligned}\right\} \tag{5.19}$$

In state-space form, with $\mathbf{a} = (a_1, \ldots, a_r)$, we have

$$\left.\begin{aligned} \dot{\mathbf{a}} &= \mathbf{A}_r\mathbf{a} + \mathbf{B}_r\mathbf{u}, \\ y &= \mathbf{C}_r\mathbf{a}, \end{aligned}\right\} \tag{5.20}$$

where

$$\mathbf{A}_r = \boldsymbol{\Psi}_r^\star\mathbf{A}\boldsymbol{\Phi}_r, \qquad \mathbf{B}_r = \boldsymbol{\Psi}_r^\star\mathbf{B}, \qquad \mathbf{C}_r = \mathbf{C}\boldsymbol{\Phi}_r, \tag{5.21}$$

where $\boldsymbol{\Phi}_r$ and $\boldsymbol{\Psi}_r$ denote the first r columns of $\boldsymbol{\Phi}$ and $\boldsymbol{\Psi}$.

It is shown in [315] that this procedure perfectly balances the approximate Gramians $\widetilde{\mathbf{W}}_c = \mathbf{X}\mathbf{X}^\star$ and $\widetilde{\mathbf{W}}_o = \mathbf{Y}\mathbf{Y}^\star$. More precisely, if the approximate Gramians are full rank ($R = N$), then $\boldsymbol{\Phi}$ and $\boldsymbol{\Psi}$ are square and invertible, with $\boldsymbol{\Psi} = \boldsymbol{\Phi}^{-1}$, and $\boldsymbol{\Phi}$ is precisely the transformation $\mathbf{T}$ that balances $\widetilde{\mathbf{W}}_c, \widetilde{\mathbf{W}}_o$, as in (5.5). In the more common case that the approximate Gramians are not full rank, $\boldsymbol{\Psi}, \boldsymbol{\Phi}$ still form a balancing transformation, as shown by the following Proposition, proved in [315]:

Proposition 5 *Suppose* $\mathbf{Y}^\star\mathbf{X}$ *has rank* $R < N$. *Then there exist matrices* $\boldsymbol{\Phi}_2, \boldsymbol{\Psi}_2 \in \mathbb{R}^{N\times(N-R)}$ *such that for*

$$\mathbf{T} = \begin{bmatrix} \boldsymbol{\Phi} & \boldsymbol{\Phi}_2 \end{bmatrix}, \qquad \mathbf{S} = \begin{bmatrix} \boldsymbol{\Psi} & \boldsymbol{\Psi}_2 \end{bmatrix},$$

$\mathbf{T}$ *is invertible with* $\mathbf{T}^{-1} = \mathbf{S}$, *and*

$$\mathbf{S}\widetilde{\mathbf{W}}_c\widetilde{\mathbf{W}}_o\mathbf{T} = \begin{bmatrix} \boldsymbol{\Sigma}_1^2 & 0 \\ 0 & 0 \end{bmatrix}, \tag{5.22}$$

and furthermore,

$$\mathbf{S}\widetilde{\mathbf{W}}_c\mathbf{S}^\star = \begin{bmatrix} \boldsymbol{\Sigma}_1 & 0 \\ 0 & \mathbf{M}_1 \end{bmatrix}, \qquad \mathbf{T}^\star\widetilde{\mathbf{W}}_o\mathbf{T} = \begin{bmatrix} \boldsymbol{\Sigma}_1 & 0 \\ 0 & \mathbf{M}_2 \end{bmatrix}. \tag{5.23}$$

Thus, (5.23) shows that in the new coordinates, the first R modes are perfectly balanced, with singular values contained in $\boldsymbol{\Sigma}_1$. Furthermore, (5.22) shows that any additional modes (those not spanned by the columns of $\boldsymbol{\Phi}$) are either uncontrollable (as measured by $\widetilde{\mathbf{W}}_c$) or unobservable (as measured by $\widetilde{\mathbf{W}}_o$).

5.3 Output projection

The procedure described in the previous section involves data from a number of impulse–response simulations of the linear system (5.1) and its adjoint (5.12). In particular, we require one impulse-response simulation for each input (each component of $\mathbf{u}$), and one adjoint simulation for each output (each component of $\mathbf{y}$). The amount of computation is reasonable as long as the number of inputs and outputs is not too large. However, we might be interested in problems where the output $\mathbf{y}$ has large dimension as well. For instance, in a flow control setting, even if the number of sensor measurements is small, the cost function for optimal control may include the kinetic energy in the entire flow field, which would require us to capture the full state $\mathbf{x}$ accurately (so we would want $\mathbf{y} = \mathbf{x}$ in developing our low-dimensional model). Since the state dimension for a fluid problem is of the order of millions, it is clearly not feasible to run millions of adjoint simulations, so a different approach is required.

Here, we describe one such approach, considering the projection of the *output* onto a lower-dimensional subspace. In particular, consider the projected system

$$\left.\begin{array}{l} \dot{\mathbf{x}} = \mathbf{A}\mathbf{x} + \mathbf{B}\mathbf{u}, \\ \mathbf{y} = \mathbf{P}\mathbf{C}\mathbf{x}, \end{array}\right\} \tag{5.24}$$

where $\mathbf{P}$ is an orthogonal projection with rank r_{op}, and the number of outputs is q, as before. Such a projection allows us to compute the approximate observability Gramian $\widetilde{\mathbf{W}}_{\mathrm{o}}$ using only r_{op} simulations of the adjoint system, rather than q simulations. To see this, recall that any orthogonal projection $\mathbf{P}$ may be written as the product $\mathbf{P} = \mathbf{\Theta}\mathbf{\Theta}^{\star}$, where $\mathbf{\Theta}$ is a $q \times r_{\mathrm{op}}$ matrix, with $\mathbf{\Theta}^{\star}\mathbf{\Theta} = \mathbf{I}_{r_{\mathrm{op}}}$. The observability Gramian (5.2) then becomes

$$\mathbf{W}_{\mathrm{o}} = \int_{0}^{\infty} e^{\mathbf{A}^{\star}t}\mathbf{C}^{\star}\mathbf{\Theta}\mathbf{\Theta}^{\star}\mathbf{C}e^{\mathbf{A}t}\,dt, \tag{5.25}$$

and so may be computed from r_{op} simulations of the adjoint system

$$\dot{\mathbf{z}}(t) = \mathbf{A}^{\star}\mathbf{z} + \mathbf{C}^{\star}\mathbf{\Theta}\mathbf{v}, \tag{5.26}$$

where $\mathbf{v} \in \mathbb{R}^{r_{\mathrm{op}}}$. When the number of outputs q is large, the reduction in computational cost is substantial.

We wish to choose $\mathbf{P}$ such that the input–output behavior of (5.24) is as close as possible to that of (5.1). In particular, if we consider $\|\mathbf{G} - \mathbf{P}\mathbf{G}\|_2$, the two-norm on *systems*, defined by Equation (1.14) of Section 1.4, this corresponds to minimizing

$$\int_{0}^{\infty} \|\mathbf{G}(t) - \mathbf{P}\mathbf{G}(t)\|_{F}^{2}\,dt, \tag{5.27}$$

where $\|\mathbf{A}\|_{F} = \mathrm{Tr}(\mathbf{A}^{\star}\mathbf{A})^{1/2}$ is the Frobenius norm on matrices, induced by the inner product $(\mathbf{A}, \mathbf{B}) = \mathrm{Tr}(\mathbf{B}^{\star}\mathbf{A})$. (In obtaining (5.27), we have used Parseval's theorem to convert the integral in (1.14) to the time domain.)

The solution of the minimization problem (5.27) is actually found by the proper orthogonal decomposition described in Chapter 3: the projection $\mathbf{P}$ that minimizes the error (5.27) is precisely the projection onto the first r_{op} POD modes of the dataset $\mathbf{G}(t)$. For instance, if

$$\Theta = \begin{bmatrix} \theta_1 & \cdots & \theta_{r_{op}} \end{bmatrix} \tag{5.28}$$

is a matrix containing the first r_{op} POD modes of $\mathbf{G}(t)$, then $\mathbf{P} = \Theta\Theta^\star$ is the projection that minimizes (5.27).

This result is particularly convenient, since the data required to compute the POD modes described above already need to be computed to find the balanced POD modes in the previous section. Thus, for output projection, we use the impulse-response data twice: first to compute POD modes in order to define the projection $\mathbf{P}$ above, and again to compute the balancing modes as described in Section 5.2.

5.4 Connections with standard POD

As described in Section 3.1.3, balanced POD produces modes that are not orthogonal; rather, it produces two sets of modes (balancing modes φ_j and adjoint modes ψ_j) that together form a bi-orthogonal set. Traditional POD, of course, produces orthogonal modes, but apart from this distinction, the two methods are actually quite closely related. For instance, the expansion (5.18) looks very much like the expansion (4.12) in POD modes, and the reduced-order model (5.19) is precisely analogous to the model (4.14) determined by Galerkin projection onto POD modes, but with the adjoint modes ψ_j used in the projection. The connection between the methods actually goes much deeper than this, however, as we describe in this section.

5.4.1 Non-orthogonal projection

First, we illustrate the geometry of the projection using bi-orthogonal modes. Suppose the modes $\{\varphi_j \mid j = 1, \ldots, r\}$ are columns of a matrix $\mathbf{\Phi}_r$, and the corresponding adjoint modes $\{\psi_j\}$ are columns of a matrix $\mathbf{\Psi}_r$. If the modes form a bi-orthogonal set, then $\mathbf{\Psi}_r^\star \mathbf{\Phi}_r = \mathbf{I}_r$, and the projection onto the modes φ_j is given by

$$\mathbf{P}_r = \mathbf{\Phi}_r \mathbf{\Psi}_r^\star. \tag{5.29}$$

One easily verifies that $\mathbf{P}_r^2 = \mathbf{P}_r$, so $\mathbf{P}_r$ is a projection, and the nullspace of this projection is orthogonal to the space spanned by the columns of $\mathbf{\Psi}_r$. The resulting projection is illustrated in Figure 5.1.

In contrast, the orthogonal projection onto the same subspace is given by

$$\mathbf{P}_r' = \mathbf{\Phi}_r (\mathbf{\Phi}_r^\star \mathbf{\Phi}_r)^{-1} \mathbf{\Phi}_r^\star, \tag{5.30}$$

also illustrated in Figure 5.1. Note that, if the columns of $\mathbf{\Phi}_r$ are orthogonal (as they are for POD modes), then $\mathbf{\Phi}_r^\star \mathbf{\Phi}_r = \mathbf{I}_r$, and this expression reduces to $\mathbf{P}_r' = \mathbf{\Phi}_r \mathbf{\Phi}_r^\star$. The reduced dynamics given by (5.19) are obtained by projecting the full dynamics using $\mathbf{P}_r$, while the Galerkin projection given by (4.14) is obtained by projecting the full dynamics using $\mathbf{P}_r'$. This distinction turns out to be the main difference between models determined by standard

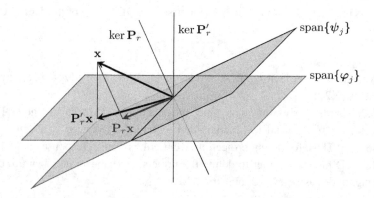

Figure 5.1 Subspaces spanned by balancing modes φ_j and adjoint modes ψ_j, and the difference in orthogonal projection $\mathbf{P}'_r$ onto span$\{\varphi_j\}$ and non-orthogonal projection $\mathbf{P}_r$ onto span$\{\varphi_j\}$, using ψ_j. (Ker denotes Kernel.)

POD and balanced POD: for instance, in the examples in Section 5.6, the balancing modes φ_j are close to the POD modes, but the resulting reduced-order models are very different, because of the difference in projection using the adjoint modes ψ_j.

5.4.2 Observability Gramian as an inner product

The connections between standard POD and balanced POD go significantly beyond this, however. In fact, balanced POD can be viewed as a special case of POD, for a linear system, in which one uses impulse responses for a dataset (i.e., the matrix $\mathbf{X}$ in (5.10)), and uses the *observability Gramian* as an inner product. To see this, define an inner product on $\mathbb{C}^n$ by

$$(\mathbf{a}, \mathbf{b})_{\widetilde{\mathbf{W}}_o} = \mathbf{b}^\star \widetilde{\mathbf{W}}_o \mathbf{a}, \qquad (5.31)$$

where $\widetilde{\mathbf{W}}_o$ is the (approximate) observability Gramian given by (5.14). As described in Section 5.1, $\widetilde{\mathbf{W}}_o$ measures states of large dynamical importance, so this inner product weights dynamically important states more heavily. The POD modes of the dataset $\mathbf{X}$ with respect to this inner product are eigenvectors of $\mathbf{R} = \mathbf{X}\mathbf{X}^\star \widetilde{\mathbf{W}}_o$, as described in Section 3.1.3. Since $\mathbf{R}$ is self-adjoint with respect to the inner product (5.31), these eigenvectors are orthogonal with respect to this inner product (though not with respect to the standard inner product).

Since the dataset $\mathbf{X}$ was produced so that $\mathbf{X}\mathbf{X}^\star = \widetilde{\mathbf{W}}_c$, the POD modes are just the eigenvectors of $\mathbf{R} = \widetilde{\mathbf{W}}_c \widetilde{\mathbf{W}}_o$: in other words, they are the balancing modes, normalized differently. Furthermore, the eigenvalues of $\mathbf{R}$ (the "empirical eigenvalues" arising from POD) are the squares of the Hankel singular values. If we compute the POD modes using the method of snapshots as in Section 3.4.1–2, we form the singular value decomposition $\mathbf{X}^\star \widetilde{\mathbf{W}}_o \mathbf{X} = \mathbf{V}_1 \mathbf{\Sigma}_1^2 \mathbf{V}_1^\star$, and the POD modes are columns of

$$\widetilde{\mathbf{\Phi}} \stackrel{\text{def}}{=} \begin{bmatrix} \widetilde{\varphi}_1 & \cdots & \widetilde{\varphi}_r \end{bmatrix} = \mathbf{X}\mathbf{V}_1\mathbf{\Sigma}_1^{-1}. \qquad (5.32)$$

Note that these modes are the same as the columns of Φ in (5.16), with a different scaling. If we define "adjoint modes" $\tilde{\boldsymbol{\psi}}_j = \widetilde{\mathbf{W}}_0 \tilde{\boldsymbol{\varphi}}_j$, then

$$(\tilde{\boldsymbol{\psi}}_i, \tilde{\boldsymbol{\varphi}}_j) = \tilde{\boldsymbol{\varphi}}_j^* \tilde{\boldsymbol{\psi}}_i = \tilde{\boldsymbol{\varphi}}_j^* \widetilde{\mathbf{W}}_0 \tilde{\boldsymbol{\varphi}}_i = (\tilde{\boldsymbol{\varphi}}_i, \tilde{\boldsymbol{\varphi}}_j)_{\widetilde{\mathbf{W}}_0} = \delta_{ij}, \tag{5.33}$$

so these adjoint modes may be viewed as a bi-orthogonal decomposition with respect to the standard inner product $(\boldsymbol{\psi}, \boldsymbol{\varphi}) = \boldsymbol{\varphi}^* \boldsymbol{\psi}$, as in Section 5.2. These adjoint modes are also rescaled versions of the rows of $\boldsymbol{\Psi}$ in (5.16), since with $\widetilde{\mathbf{W}}_0 = \mathbf{Y}\mathbf{Y}^*$, and $\mathbf{Y}^*\mathbf{X} = \mathbf{U}_1 \Sigma_1 \mathbf{V}_1^*$,

$$\widetilde{\boldsymbol{\Psi}} \stackrel{\text{def}}{=} \begin{bmatrix} \tilde{\boldsymbol{\psi}}_1 & \cdots & \tilde{\boldsymbol{\psi}}_r \end{bmatrix} = \widetilde{\mathbf{W}}_0 \Phi = \mathbf{Y}\mathbf{Y}^*\mathbf{X}\mathbf{V}_1 \Sigma_1^{-1} = \mathbf{Y}\mathbf{U}_1, \tag{5.34}$$

i.e., a rescaling of $\boldsymbol{\Psi}$ in Equation (5.16).

5.4.3 Guaranteed stability

A useful property of balanced truncation is that the resulting reduced-order models are guaranteed to be stable [98]. Unfortunately, Galerkin projection onto POD modes does not have this property, as demonstrated explicitly in Section 5.6.1. However, balanced POD models preserve this property, as long as the approximate observability Gramian $\widetilde{\mathbf{W}}_0 = \mathbf{Y}\mathbf{Y}^\star$ is close enough to the true observability Gramian.

To see this, we first present a property of Galerkin projection, proved in [316]. Consider a nonlinear system with a stable equilibrium at the origin, and suppose that we have an inner product such that $V(\mathbf{x}) = (\mathbf{x}, \mathbf{x})$ is a Liapunov function. (This is referred to in [316] as an *energy-based inner product*, since energy, like a Liapunov function, typically remains constant or decreases as solutions evolve.) Then if we form a reduced-order model by orthogonal projection (with respect to this inner product), the resulting model will also have a stable equilibrium at the origin, no matter what subspace we use for the projection.

In the previous section, we showed that balanced POD is equivalent to standard POD with respect to an inner product (5.31) defined by the observability Gramian. It follows immediately from (5.3) that the function $V(\mathbf{x}) = (\mathbf{x}, \mathbf{x})_{\mathbf{W}_0}$ is a Liapunov function of the linearized system $\dot{\mathbf{x}} = \mathbf{A}\mathbf{x}$, with $\dot{V}(\mathbf{x}) = -\mathbf{C}^*\mathbf{C} \leq 0$. Thus, as long as the approximate observability Gramian $\widetilde{\mathbf{W}}_0$ is close enough to the true observability Gramian that it is still a Liapunov function, the reduced-order model produced by balanced POD will also be stable.

5.5 Extensions of balanced POD

The methods described above are effective, but are limited to stable, linear systems. This restriction may appear necessary, since the snapshot-based procedure is based on impulse response simulations, and the impulse response is unbounded for an unstable system. However, a simple extension to unstable linear systems is possible, by decoupling the stable and unstable subspaces, as we discuss below. Extensions to nonlinear systems are much more challenging, but we sketch a simple approach that is effective as long as the nonlinear dynamics are spanned (at least approximately) by the balancing modes determined from a linearized system. Finally, we present a method for computing reduced-order models that are equivalent to those from balanced POD, without the need for adjoint simulations.

5.5.1 Unstable systems

For an unstable linear system, even the definitions of the controllability and observability Gramians that form the basis for balanced truncation are not immediately applicable, as the integrals in (5.2) do not converge if A has eigenvalues in the right-half plane. A generalization of balanced truncation was proposed in [393], however, using Gramians defined in the frequency domain. One then decouples the system into stable and unstable parts, and retains the unstable portion, performing model reduction on the stable portion. This procedure is plausible as long as the number of unstable states is small, which is the case for many problems of interest.

For large problems, one must find a computationally tractable way of decoupling stable and unstable components. If the number of unstable eigenvalues of A is small, one may in principle calculate the unstable left and right eigenvectors, for instance using an Arnoldi method [373]. Assembling the unstable right eigenvectors as columns of a matrix $\mathbf{\Phi}_u$, and the corresponding left eigenvectors as columns of a matrix $\mathbf{\Psi}_u$ (normalized so that $\mathbf{\Psi}_u^\star \mathbf{\Phi}_u = \mathbf{I}$), one can then define the projection onto the stable subspace as

$$\mathbf{P}_s = \mathbf{I} - \mathbf{\Phi}_u \mathbf{\Psi}_u^\star. \tag{5.35}$$

The dynamics restricted to the stable subspace may then be computed by applying this projection at each timestep of a simulation (or every few timesteps). The corresponding adjoint system restricted to the stable subspace of $\mathbf{A}^\star$ may be computed analogously, and a reduced-order model on the stable subspace may then be computed using balanced POD. If the resulting balancing modes are written as columns of a matrix $\mathbf{\Phi}_s$, and the corresponding adjoint modes are columns of $\mathbf{\Psi}_s$, then one approximates the state $\mathbf{x}$ by

$$\mathbf{x}_r = \mathbf{\Phi}_u \mathbf{a}_u + \mathbf{\Phi}_s \mathbf{a}_s, \tag{5.36}$$

where $\mathbf{a}_u \in \mathbb{R}^{n_u}$ and $\mathbf{a}_s \in \mathbb{R}^r$, where n_u is the number of unstable eigenvectors, and r is the number of balancing modes retained in the reduced-order model. If this expansion is inserted into (5.1), the unstable and stable dynamics decouple (the cross terms such as $\mathbf{\Psi}_s^\star \mathbf{A} \mathbf{\Phi}_u$ all vanish), and one has the reduced-order model

$$\begin{aligned}
\begin{pmatrix} \dot{\mathbf{a}}_u \\ \dot{\mathbf{a}}_s \end{pmatrix} &= \begin{bmatrix} \mathbf{\Lambda}_u & 0 \\ 0 & \mathbf{\Psi}_s^\star \mathbf{A} \mathbf{\Phi}_s \end{bmatrix} \begin{pmatrix} \mathbf{a}_u \\ \mathbf{a}_s \end{pmatrix} + \begin{bmatrix} \mathbf{\Psi}_u^\star \mathbf{B} \\ \mathbf{\Psi}_s^\star \mathbf{B} \end{bmatrix} \mathbf{u}, \\
\mathbf{y} &= \begin{bmatrix} \mathbf{C}\mathbf{\Phi}_u & \mathbf{C}\mathbf{\Phi}_s \end{bmatrix} \begin{pmatrix} \mathbf{a}_u \\ \mathbf{a}_s \end{pmatrix},
\end{aligned} \right\} \tag{5.37}$$

where $\mathbf{\Lambda}_u = \mathbf{\Psi}_u^\star \mathbf{A} \mathbf{\Phi}_u$ is the diagonal matrix of unstable eigenvalues of $\mathbf{A}$. For further details, see [4], in which this method was used to develop a model-based controller to stabilize oscillations in the flow past a flat plate at a large angle of attack.

5.5.2 Nonlinear systems

The methods of this chapter are applicable primarily to linear systems. Several methods for generalizing balanced truncation to nonlinear systems have been proposed [202, 324], but these are not computationally tractable for large-dimensional systems such as fluids.

However, a simple extension to nonlinear systems is possible, using ideas similar to those in [202]. One can first linearize a nonlinear system about an equilibrium point, and obtain balancing modes and adjoint modes $\boldsymbol{\varphi}_j$, $\boldsymbol{\psi}_j$ of the linearized system. One can then project the full nonlinear dynamics $\dot{\mathbf{x}} = \mathbf{f}(\mathbf{x}, \mathbf{u})$ onto these modes, as

$$\dot{a}_j = \left(\boldsymbol{\psi}_j, \mathbf{f}(\mathbf{x}_r, \mathbf{u})\right), \tag{5.38}$$

where $\mathbf{x}_r$ is given by (5.18). In general, one should not expect such a nonlinear model to be valid far from the equilibrium, if the modes are determined from linearized dynamics, but if the modes $\boldsymbol{\varphi}_j$ do approximately span the portion of state space visited by the nonlinear system, then this approach is reasonable. Initial studies have shown that this procedure can work well, at least for a model problem of the complex Ginzburg–Landau equation [171], the linearized version of which is considered in Section 5.6.2.

5.5.3 Adjoint-free balancing

One of the drawbacks of the balanced POD procedure described in Section 5.2 is that it requires information from adjoint simulations, and thus cannot be used with experimental data. However, if all one requires is a reduced-order model of the input–output dynamics, and one does not need the balancing and adjoint modes themselves, then it is possible to obtain the reduced-order model (5.20) directly from impulse-response data, without the need for adjoint simulations.

This method is most easily demonstrated for a discrete-time system, of the form

$$\left.\begin{aligned} \mathbf{x}(k+1) &= \mathbf{A}\mathbf{x}(k) + \mathbf{B}\mathbf{u}(k), \\ \mathbf{y}(k) &= \mathbf{C}\mathbf{x}(k), \end{aligned}\right\} \tag{5.39}$$

where, as in Equation (5.1), $\mathbf{u}(k)$ is the vector of inputs at time $k \in \mathbb{Z}$, $\mathbf{y}(k)$ is the vector of outputs, and $\mathbf{x}(k)$ is the state. To apply balanced POD to such a discrete-time system, the matrix of impulse responses defined by (5.10) becomes

$$\mathbf{X} = \begin{bmatrix} \mathbf{B} & \mathbf{A}\mathbf{B} & \mathbf{A}^2\mathbf{B} & \cdots & \mathbf{A}^M\mathbf{B} \end{bmatrix}, \tag{5.40}$$

and the matrix of impulse responses of the adjoint system becomes

$$\mathbf{Y} = \begin{bmatrix} \mathbf{C}^\star & \mathbf{A}^\star\mathbf{C}^\star & (\mathbf{A}^\star)^2\mathbf{C}^\star & \cdots & (\mathbf{A}^\star)^M\mathbf{C}^\star \end{bmatrix}. \tag{5.41}$$

The matrix $\mathbf{Y}^\star\mathbf{X}$ then has the form

$$\mathbf{H} \overset{\text{def}}{=} \mathbf{Y}^\star\mathbf{X} = \begin{bmatrix} \mathbf{CB} & \mathbf{CAB} & \cdots & \mathbf{CA}^M\mathbf{B} \\ \mathbf{CAB} & \mathbf{CA}^2\mathbf{B} & \cdots & \mathbf{CA}^{M+1}\mathbf{B} \\ \vdots & \vdots & \ddots & \vdots \\ \mathbf{CA}^M\mathbf{B} & \mathbf{CA}^{M+1}\mathbf{B} & \cdots & \mathbf{CA}^{2M}\mathbf{B} \end{bmatrix}; \tag{5.42}$$

this is called a *generalized Hankel matrix* in control theory literature. The terms $\mathbf{CA}^k\mathbf{B}$ are commonly called *Markov parameters*, and we note that they may be determined solely by the input–output dynamics (in particular, the impulse response): they are invariant under coordinate transformations $\mathbf{x}(k) = \mathbf{T}\tilde{\mathbf{x}}(k)$, and no adjoint simulations are required to compute them. Thus, the singular value decomposition in (5.15) may be computed directly from snapshots of the output $\mathbf{y}(k)$ from an impulse response simulation (or experiment).

Note, however, that one cannot compute the modes in (5.16) without the snapshots $\mathbf{x}(k)$ and $\mathbf{z}(k)$, and if adjoint information is not available, then $\mathbf{z}(k)$ are not known. Nevertheless, as demonstrated in [231], one may still calculate the reduced-order model in the form

$$\left.\begin{aligned} \mathbf{x}_r(k+1) &= \mathbf{A}_r\mathbf{x}_r(k) + \mathbf{B}_r\mathbf{x}_r(k), \\ \mathbf{y}(k) &= \mathbf{C}_r\mathbf{x}_r(k), \end{aligned}\right\} \tag{5.43}$$

where

$$\mathbf{A}_r = \mathbf{\Sigma}_r^{-1/2}\mathbf{U}_r^*\mathbf{H}'\mathbf{V}_r\mathbf{\Sigma}_r^{-1/2},$$

$$\mathbf{B}_r = \text{the first } p \text{ columns of } \mathbf{\Sigma}_r^{1/2}\mathbf{V}_1^\star,$$

$$\mathbf{C}_r = \text{the first } q \text{ rows of } \mathbf{U}_r\mathbf{\Sigma}_r^{1/2},$$

and $\mathbf{U}_r, \mathbf{V}_r$ are the first r columns of the matrices $\mathbf{U}, \mathbf{V}$ in the singular value decomposition (5.15), p and q are the numbers of inputs and outputs, $\mathbf{\Sigma}_r$ is the diagonal matrix of the first r singular values, and $\mathbf{H}'$ is given by

$$\mathbf{H}' \overset{\text{def}}{=} \begin{bmatrix} \mathbf{CAB} & \mathbf{CA}^2\mathbf{B} & \cdots & \mathbf{CA}^{M+1}\mathbf{B} \\ \mathbf{CA}^2\mathbf{B} & \mathbf{CA}^3\mathbf{B} & \cdots & \mathbf{CA}^{M+2}\mathbf{B} \\ \vdots & \vdots & \ddots & \vdots \\ \mathbf{CA}^{M+1}\mathbf{B} & \mathbf{CA}^{M+2}\mathbf{B} & \cdots & \mathbf{CA}^{2M+1}\mathbf{B} \end{bmatrix}.$$

Note that $\mathbf{H}'$ contains the same Markov parameters as $\mathbf{H}$ (plus one additional entry), simply rearranged, so little additional computation is required to compute $\mathbf{H}'$.

As explained in [231], the method above is precisely equivalent to a system identification technique called the Eigensystem Realization Algorithm [178]. If all that is desired is a reduced-order model of the form (5.43), then the Eigensystem Realization Algorithm explained in this section is preferable to balanced POD: adjoint simulations are not required, and the method is more computationally efficient, as the direct construction of the Hankel matrix in (5.42) is much faster than the calculation of $\mathbf{Y}^*\mathbf{X}$. However, balanced POD provides modes φ_j, ψ_j, which provide insight into the coherent structures in a system. These modes also may be used to obtain a nonlinear reduced-order model, as explained in the previous section, and furthermore they allow one to retain parameters in a reduced-order model: for instance, if the $\mathbf{A}, \mathbf{B}, \mathbf{C}$ matrices depend on a parameter μ, then a projection of the form (5.19) will give reduced-order $\mathbf{A}_r, \mathbf{B}_r, \mathbf{C}_r$ matrices whose coefficients depend on μ. While there is no guarantee that the reduced-order model will perform well at off-design parameter values, a study in [172] has shown that balanced POD models of linearized channel flow perform well even at Reynolds numbers far from that at which the model was obtained.

5.6 Some examples

We now illustrate the methods of this chapter by means of several examples that highlight the effects of impulsive perturbations. Our aim here is to compare reduced-order models obtained by balanced POD with those obtained by projection onto traditional POD modes, as described in Chapters 3 and 4, and to illustrate the differences and potential pitfalls of POD-based models.

Unlike examples in later chapters, here we focus on *linear* systems. This is partly out of necessity, since balanced POD modes are defined only for linear systems (although one can certainly project a nonlinear system onto modes determined from a linearization, as in Equation (5.38)). But this restriction has its advantages as well: for linear systems, it is straightforward to quantify the difference between models using norms, such as the infinity norm defined by (1.13). For nonlinear systems, it is much more difficult to compare models qualitatively, and hence harder to evaluate the accuracy of a given reduced-order model.

Throughout this section, we focus on *non-normal* linear systems. Recall that a linear operator A is *normal* if it commutes with its adjoint – that is, if $AA^\star = A^\star A$, where $A^\star$ denotes the adjoint of A (as defined in Section 1.4). For instance, self-adjoint, skew-symmetric, and unitary operators are all examples of normal operators. These operators have the property that their eigenfunctions are orthogonal, and as a result, for a stable linear system $\dot{x} = Ax$ (one whose spectrum lies entirely in the left half of the complex plane), the L^2 norm or "energy" $\|x(t)\|^2$ decays monotonically with increasing time t. Non-normal operators do not share these characteristics: their eigenfunctions are not orthogonal, and in fact can be nearly parallel, leading to large transient growth. Even if a non-normal operator A is stable, the energy can grow, sometimes substantially, before eventually decaying. Non-normality and large transient growth occur in a wide variety of problems in fluid mechanics, as discussed in [73, 325]. For these cases, standard POD modes often do not work well, as we shall see, because they focus on the most energetic structures, while low-energy features can be critically important to the dynamics of non-normal systems.

5.6.1 Example 1: a non-normal ODE

Many of the differences between balanced POD and standard POD and Galerkin projection, and the potential pitfalls that can arise for non-normal systems, may be seen in simple examples, such as the linear system (5.1), with

$$\mathbf{A} = \begin{bmatrix} -1 & 0 & 100 \\ 0 & -2 & 100 \\ 0 & 0 & -5 \end{bmatrix}, \qquad \mathbf{B} = \begin{bmatrix} 1 \\ 1 \\ 1 \end{bmatrix}, \qquad \mathbf{C} = \begin{bmatrix} 1 & 1 & 1 \end{bmatrix}. \qquad (5.44)$$

Although the dimension of this system is already small, let us consider the problem of determining a reduced-order model that approximately reproduces its input–output behavior.

The impulse response for this system (or, equivalently, the solution of $\dot{\mathbf{x}} = \mathbf{Ax}$, with $\mathbf{x}(0) = (1, 1, 1)$) can be computed analytically, and is shown in Figure 5.2 (left), which

Table 5.1 *POD and balancing modes for the system* (5.44).

	POD		Balancing		Adjoint	
	θ_1	θ_2	φ_1	φ_2	ψ_1	ψ_2
x_1	−0.8422	0.5386	−3.0018	−3.4900	−0.1685	−0.0170
x_2	−0.5390	−0.8406	−2.0320	−1.4895	−0.1127	−0.1093
x_3	−0.0094	−0.0570	−0.0552	0.1572	−4.8077	4.9485

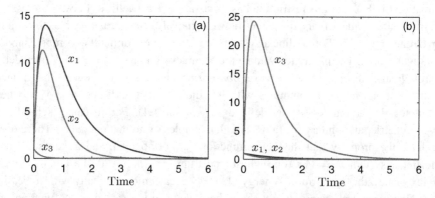

Figure 5.2 (a) Response of the state (x_1, x_2, x_3) to an impulsive input, for the system (5.44); (b) impulse response of the adjoint system.

illustrates the large transient growth. In particular, the first two states, x_1 and x_2, grow significantly before decaying, but x_3 merely decays. This transient growth arises because of the high degree of non-normality: the eigenvectors corresponding to the eigenvalues −1, −2, and −5, are $(1, 0, 0)$, $(0, 1, 0)$, and $(−25, −33.3, 1)$. The first two eigenvectors are orthogonal, but the third is almost parallel to the subspace spanned by the first two, indicating that the system is highly non-normal. As seen in Figure 5.2(left), $x_3(t)$ decays monotonically, but not before strongly exciting x_1 and x_2 via the large entries in column 3 of **A**.

Computing POD modes for this impulse response dataset, one finds the empirical eigenvalues are $\{42.64, 0.996, 0.011\}$. The first two POD modes (denoted θ_1, θ_2 in Table 5.1) capture 99.97% of the energy in this dataset, so one might therefore conclude that a model projected onto these first two modes should be excellent.

However, when we compute a low-dimensional model by Galerkin projection of Equation (5.1) onto span$\{\theta_1, \theta_2\}$, the results are quite disappointing, as demonstrated by the impulse response and frequency response shown in Figure 5.3. The second-order model misses the transient peak in the impulse response, the frequency response looks completely different, both in magnitude and phase, and the model is actually unstable, with eigenvalues $\{0.0019 \pm 1.13i\}$. This poor behavior is particularly alarming, as the first two POD modes captured such a large fraction of the overall energy in the impulse response (over 99.9%).

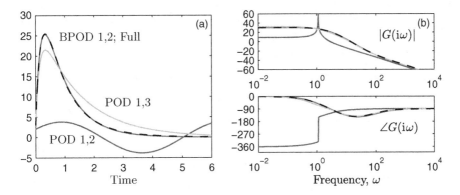

Figure 5.3 Reduced-order models of order 2: impulse response (a), and frequency response (b). Black dashed: full third-order system; black: POD/Galerkin model (modes 1 and 2); gray underneath black dashed: balanced POD model; gray: POD/Galerkin model (modes 1 and 3).

But perhaps we are asking too much of our tools: could it be that it is simply not possible to obtain a reliable second-order approximation for this system?

Let us compare the results when balanced POD is used. Here, we require not just the impulse response, but also the impulse response of the adjoint system, also shown in Figure 5.2(right) (cf. Section 5.2). Note that it is x_3 (rather than x_1) that exhibits large transient growth for the adjoint system, indicating that x_3 is highly *observable*, so small errors in x_3 have a large impact on future dynamics. (This strong sensitivity to x_3 can also be seen from the large coefficients of x_3 in the first two rows of the governing equations.) From these impulse responses, we construct the balancing and adjoint modes and the Hankel singular values as described in Section 5.2. The Hankel singular values are $\{17.5, 1.86, 0.178\}$, and the small value of the last Hankel singular value suggests that it *should* be possible to obtain a good approximation by a second-order system.

Indeed, we find that a second-order model obtained by projection onto the balancing modes, given by Equation (5.20), works quite well, as shown in the impulse response and frequency response in Figure 5.3. The transient response is captured almost perfectly, as are both the magnitude and phase of the frequency response. The eigenvalues of the balanced POD model are $\{-1.23, -5.73\}$, also not far from the first and third eigenvalues of the full system, $\{-1, -2, -5\}$.

How can we explain such a dramatic difference in the behavior of the models? Why do the POD models fail, even though the first two modes capture over 99.9% of the energy, and why do the balanced POD models perform so well?

It is instructive to examine the balancing modes φ_j and corresponding adjoint modes ψ_j, shown in Table 5.1, and compare these to the POD modes also shown in the table. Although the balancing modes are normalized differently than the POD modes, and are not orthogonal to each other, the subspace spanned by them is actually relatively close to that spanned by the first two POD modes. One way of quantifying the closeness of the subspaces is to compute the angle between them: the normal to the subspace spanned by the first two

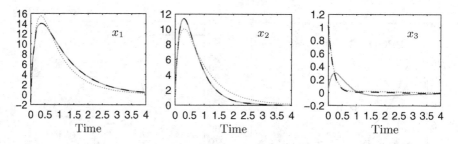

Figure 5.4 Projection of impulse response data onto POD modes and balancing modes. Dashed: impulse response of full system; gray: projection on to POD modes; dotted: projection onto balancing modes.

POD modes is $(0.023, -0.053, 0.998)$, while the normal vector to the subspace spanned by the first two balancing modes is $(0.164, -0.269, 0.949)$, and the angle between them is $15.1°$. Both normal vectors are roughly aligned with the x_3-direction, so the subspaces are approximately the (x_1, x_2) subspace. However, the space spanned by the *adjoint* modes is quite different. The largest component of the adjoint modes is x_3, indicating a strong "weighting" of this component in the projection of dynamics. The normal vector to the space spanned by the adjoint modes is $(0.758, -0.652, -0.011)$, and the angle between this subspace and that spanned by the POD modes is $87.7°$, nearly orthogonal. We can therefore expect the projection of the dynamics onto the balancing modes to be quite different from the orthogonal projection of the dynamics onto POD modes, even though the subspaces spanned by $\{\varphi_j\}$ and $\{\theta_j\}$ are close.

It is also instructive to examine the projection of the impulse response data (generated by the full third-order system) onto the POD modes and the balancing modes, shown in Figure 5.4. Note that the projection of the data $\mathbf{x}(t)$ onto POD modes is given by $\Theta\Theta^T\mathbf{x}(t)$, while the projection onto balancing modes is $\Phi\Psi^T\mathbf{x}(t)$. The POD modes are clearly superior to the balancing modes at representing these data: indeed, this is what the POD modes are designed to capture, and they are optimal in the sense of minimizing the error in this projection. The POD modes capture the largest-energy states, x_1 and x_2, almost perfectly, while the projection onto balancing modes is much less accurate. However, recall that what balancing is trying to reproduce is not the full state $\mathbf{x}$, but rather the output $y = \mathbf{Cx} = x_1 + x_2 + x_3$, and here, because the errors in the projection of x_1 and x_2 have opposite sign, the resulting errors in y are actually quite small. More importantly, the projection onto balancing modes much more faithfully captures the behavior of x_3, which, as we have seen, is the most *dynamically* important state. The balancing procedure incorporates this through the use of adjoint information, while standard POD has no knowledge of which states are more dynamically important.

We have learned that, for the purposes of developing a reduced-order Galerkin model, the most energetic modes are not always the best modes to retain. We might then consider a model obtained by projecting onto POD modes 1 and 3: although mode 3 contains only 0.026% of the energy in the impulse response (about $1/100$ of the energy in mode 2), it is given by $\theta_3 = (0.023, -0.053, 0.998)$, and so it does capture the x_3 component, which we now know is the most dynamically important state. The impulse response and frequency response of the resulting model are also shown in Figure 5.3, and indeed, the model projected onto modes 1 and 3 does perform much better than the model projected onto the two most energetic modes, although not as well as the balanced model.

This example should engender a healthy skepticism of reduced-order models formed from POD modes: for this third-order linear system, the first two POD modes capture 99.97% of the energy, and yet the reduced-order model obtained by projection onto these modes fails dramatically. However, it is possible to obtain an accurate reduced-order model, and in fact a two-dimensional balanced model performs extremely well. The failure of POD in this example is due to the strong non-normality: the state x_3 contains little energy, and yet has a strong influence on the dynamics. POD modes are not well suited to this case, as they focus on the most energetic features, but balanced POD remedies this situation by considering the dynamical importance as well. We shall see in subsequent examples that many of these effects are also observed in fluid systems governed primarily by non-normal linear effects.

5.6.2 Example 2: a one-dimensional PDE

In our next example, we consider the complex Ginzburg–Landau (G–L) equation, a one-dimensional model of spatially developing instabilities in a fluid that describes the evolution of a complex-valued function $q(x, t)$, representing the amplitude and phase of an instability wave. The nonlinear G–L equation is typically used in weakly nonlinear stability studies [74, 88], but here we discuss only the linearized form, given by

$$q_t = -\nu q_x + \mu(x)q + \gamma q_{xx}, \qquad x \in (-\infty, \infty), \tag{5.45}$$

where $\nu = U + 2ic_u$ represents a complex convection velocity, $\gamma = 1 + ic_d$ is a complex diffusion coefficient, and the real parameter $\mu(x)$ is a spatially dependent growth rate, which we take to be

$$\mu(x) = (\mu_0 - c_u^2) + \mu_2 \frac{x^2}{2}, \tag{5.46}$$

as in [88]. For this example, we take $\mu_0 = 0.962 = 0.85\mu_c$ (for which the equilibrium is "locally absolutely unstable," but "globally stable," in the terminology of [166]), and we consider $U = 2.0$, $c_d = -0.1$, $c_u = 0.3$, and $\mu_2 = -0.08$, which determine the convection velocity, dispersion, most unstable wavenumber, and how non-parallel the flow is. For more information about the physical significance of these parameters, see the reviews [27, 73].

We solve the system (5.45) numerically using a spectral method, expanding $q(x, t)$ in terms of Hermite polynomials, as in [27]. We take 100 Hermite polynomials, whose corresponding collocation points form a computational domain $x \in [-30, 30]$. With μ given by Equation (5.46), the region of amplification is $x \in [-x_c, x_c]$, where $x_c = \sqrt{-2\mu_0/\mu_2} \approx 4.9$: inside this region, the most unstable wavenumber of q is amplified, and outside it all wavenumbers are attenuated.

In order to apply the balanced POD method, we must define the inputs and outputs of our system. For inputs, we consider a Gaussian disturbance of the form

$$d(x, t) = e^{-((x+x_c)/1.6)^2} u(t), \tag{5.47}$$

centered about the upstream end of the region of amplification. We allow u to take on complex values, so the system effectively has two input parameters. We are in principle interested in a reduced-order model that captures the behavior of the entire field $q(x, t)$, so we take the output to be $y = q$ (hence the **C** matrix in (5.1) is the identity).

Since the state dimension is relatively small (200 real states), it is more efficient to compute the Gramians by solving the Liapunov equations (5.3), and then taking $X = W_c^{1/2}$, $Y = W_o^{1/2}$, rather than using the snapshot-based quadrature approximation (5.10). The snapshot-based approach gives identical results, however, if enough snapshots are used. Note that here, if the snapshot-based approach is used, the output projection method of Section 5.3 is also needed, or else one requires 200 adjoint simulations (one for each output). The POD modes of the impulse response dataset may also be easily computed as eigenvectors of W_c, as described in Section 5.2.

The absolute value of the first POD mode is shown in Figure 5.5, along with the first balancing mode and the first adjoint mode. The qualitative characteristics seen in this figure are the same for the higher modes as well: the POD modes θ_j and balancing modes ϕ_j both have supports centered near the downstream end of the region of amplification, while the adjoint modes ψ_j are quite different, with supports centered near the *upstream* end of the region of amplification. The first adjoint mode ψ_1 is, in fact, nearly orthogonal to the POD mode θ_1:

$$\frac{(\psi_1, \theta_1)}{\|\psi_1\|\|\theta_1\|} = -0.0149, \tag{5.48}$$

corresponding to an "angle" of 90.9°. This near-orthogonality results from the small extent of spatial overlap between the POD mode and the adjoint mode, and is analogous to the near-orthogonality of the adjoint modes and POD modes in the ODE example of the previous section (Table 5.1).

The first six POD modes contain 99.92% of the energy in the impulse–response dataset, so it is reasonable to expect that a Galerkin projection onto these first six modes would perform well. However, as with the example in the previous section, the resulting Galerkin model actually performs quite poorly, as shown in Figure 5.6: the transient peaks in the

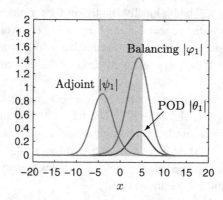

Figure 5.5 Absolute value of modes for the linearized complex Ginzburg–Landau equation. The shaded region denotes the region of amplification, $[-x_c, x_c]$.

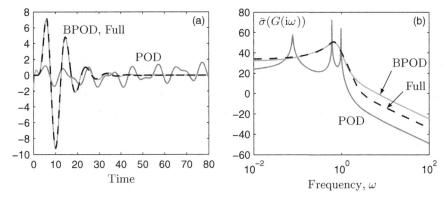

Figure 5.6 Impulse response and singular value plot for linearized complex Ginzburg–Landau equation: full system (black, dashed), and six-mode models from POD (black) and balanced POD (gray underneath black dashed). (a) Real part at $x = 5.23$; (b) maximum singular value of $G(i\omega)$.

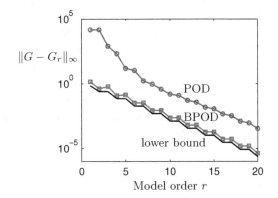

Figure 5.7 Error norms $\|G - G_r\|_\infty$ for reduced-order models of linearized complex Ginzburg–Landau equation: POD models (top ○); balanced POD models (middle □); and lower bound (5.7) for any reduced-order model (bottom —).

impulse response are not captured, and the response decays much too slowly, because of some lightly damped eigenvalues. The frequency response of the POD model is similarly poor, with spurious peaks, again resulting from lightly damped eigenvalues. By contrast, the six-mode projection onto balanced POD modes performs quite well: both the impulse response and frequency response are very close to that of the original 200-dimensional system, the only discernible difference being in the early times of the impulse response, and magnitudes in the high-frequency range of the frequency response function, where the response is small anyway.

A more comprehensive comparison of the models is presented in Figure 5.7, which shows the error norm ($\| \cdot \|_\infty$, given by Equation (1.13)) of the difference between the full system and the reduced-order model, for different orders of models produced by projection onto POD modes and balanced POD modes. The balanced POD models have error norms typically an order of magnitude smaller than the corresponding models using standard POD, and furthermore, the balanced POD models are quite close to the minimum possible error at each rank (computed from (5.7)).

This example reveals that many of the characteristics observed for the simple ODE in the previous section also arise in a PDE model of spatially developing instabilities in fluids. In particular, low-order models based on projection onto POD modes can fail even when over 99.9% of the energy is captured by the modes: the models do not accurately reproduce the transient growth or the frequency response, and they tend to have eigenvalues close to the imaginary axis, making them "fragile" to perturbations. Balanced models perform well, however, and at any given order, the errors are provably close to the best achievable by any linear model.

5.6.3 Example 3: linearized channel flow

The final example we consider is flow in a plane channel. In particular, we consider the flow between two parallel plates, in a domain periodic in both the streamwise and spanwise directions. The laminar solution, a parabolic velocity profile, is stable up to a Reynolds number (based on channel half-width) of 5772, but in experiments, transition to turbulence occurs at much lower Reynolds numbers [97]. It is hypothesized that one of the reasons for this discrepancy is the large transient growth that occurs due to the non-normality of the linearized equations [295].

For this geometry, the flow may be described by only two variables: the wall-normal velocity v, and the wall-normal vorticity η (the remaining flow variables may be calculated using the continuity equation). The Navier–Stokes equations are linearized about the laminar profile $U(y)$ to give

$$\left. \begin{array}{r} \left((\partial_t + U\partial_x)\Delta - U''\partial_x - \dfrac{1}{Re}\Delta^2 \right) v = f(x, y, z, t), \\[2ex] \left(\partial_t + U\partial_x - \dfrac{1}{Re}\Delta \right) \eta = -U'\partial_z v, \end{array} \right\} \tag{5.49}$$

where $f(x, y, z, t)$ is a forcing term specified below, $\Delta = \partial_x^2 + \partial_y^2 + \partial_z^2$ is the Laplacian, and $Re = U_c\delta/\nu$ is the Reynolds number based on the channel half-width δ, the center-line velocity U_c, and the kinematic viscosity ν. We solve this system numerically using a spectral method described in [190], with 32 Fourier modes in the streamwise and spanwise directions, and 65 Chebyshev polynomials in the wall-normal direction, for a Reynolds number of 2000. For more details of the results discussed here, see [172].

As in the previous examples, we wish to obtain a low-dimensional model for the response to an impulsive disturbance, and here we consider a vertical body force in the center of the channel, given by

$$f(x, y, z, t) = \left(1 - \frac{r^2}{\alpha^2} \right) e^{(-r^2/\alpha^2 - y^2/\alpha_y^2)} \big[\cos(\pi y) + 1 \big] u(t), \tag{5.50}$$

with $r^2 = x^2 + y^2$ and with parameters $\alpha = 0.7$, $\alpha_y = 0.6$. Figure 5.8 shows the response to an impulsive input $u(t) = \delta(t)$: the localized disturbance elongates and grows significantly in amplitude, forming streamwise streaks, before eventually decaying.

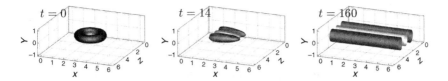

Figure 5.8 Development of an impulsive disturbance for linearized channel flow, showing level sets of wall-normal velocity at three different times. Positive velocity is light, negative velocity is dark, and flow is from left to right. Reprinted with permission from Ilak *et al.* [172]. Copyright 2008, American Institute of Physics.

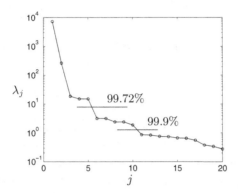

Figure 5.9 Empirical eigenvalues for impulse response of linearized channel flow. Reprinted with permission from Ilak *et al.* [172]. Copyright 2008, American Institute of Physics.

Calculating POD modes from this impulse response data, we find that the first five modes contain 99.72% of the energy, and the first ten modes contain 99.9%, as shown in Figure 5.9. We might therefore expect that a five-mode POD model should perform well, and a ten-mode model should be even more accurate. However, as with the previous examples in this chapter, the behavior of these models is disappointing.

The impulse response of the resulting POD models is shown in Figure 5.10. We see that, while the original system exhibits significant transient growth, the five-mode POD model severely underestimates this transient peak, producing very little transient growth. The ten-mode model captures more transient growth, but its behavior is still far from the true impulse response. Furthermore, both models have very lightly-damped eigenvalues, and produce oscillatory behavior: the spurious second peak around $t = 900$ repeats itself several times before eventually decaying (not shown here). The frequency response for this case is shown in [172], and is qualitatively similar to that of the complex Ginzburg–Landau equation, shown in Figure 5.6, with the POD model displaying spurious resonances.

It is interesting to note that, as with the ODE example of Section 5.6.1, one can dramatically improve the behavior of the models by including selected low-energy modes. We find that a five-mode model containing POD modes 1–3, 10, and 17 captures the transient peak well, and is a dramatic improvement over the five-mode model containing the most energetic modes.

Figure 5.10 also shows the results of balanced POD models for this system, and we find that the balanced POD models perform much better. Even a three-mode balanced model is

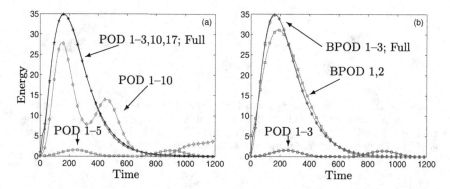

Figure 5.10 Impulse response for POD and balanced POD models of linearized channel flow. (a) POD models with modes 1–5, 1–10, and a five-mode model with modes 1–3, 10, and 17; (b) Balanced POD models with two and three modes, compared with three-mode POD model. Adapted from Ilak *et al.* [172].

able to capture the transient peak well, with the only discernible errors occurring at early times ($t < 40$). For a more detailed study of the application of balanced POD to linear channel flow, see [170, 172].

The reasons for the poor performance of the POD models of linearized channel flow are the same as those in the previous examples: the linearized system is highly non-normal and exhibits large transient growth. For such non-normal systems, low-energy features can have a large effect on the dynamics, and they must therefore be included for a model to capture the behavior of the full system. Since POD modes are based on energy, the resulting models usually neglect these low-energy features, and consequently perform poorly. Careful selection of the modes retained in a model can dramatically improve the performance of the models. By contrast, balanced POD models use adjoint simulations to gain information about the dynamical importance (observability) of different states, and therefore typically perform much better than standard POD models for these highly non-normal systems.

At this point in the book we have provided some background on turbulence itself, described two major tools that provide good basis functions for the description of flows with coherent structures, reviewed the Galerkin projection method, which converts the Navier–Stokes equations into ODEs on subspaces spanned by those functions, and introduced some of the modeling issues that arise in connection with low-dimensional truncations. Assuming that all these steps can be carried out, we can obtain a finite and perhaps fairly small set of coupled ordinary differential equations, that are generally non-linear. But our program is far from complete. We now wish to probe the behavior of this system to determine the global structure of its solutions in phase space and ultimately to relate these to coherent structures and their dynamics in the physical domain of the flow itself. To tackle the first of these, we appeal to developments in dynamical systems theory or what is popularly (and inappropriately) known as "chaos theory." The next four chapters, making up the second part of the book, offer an introduction to this subject, after which we return to turbulence once more.

PART TWO

Dynamical systems

6

Qualitative theory

This chapter and the following one provide a review of some aspects of the qualitative theory of dynamical systems that we need in our analyses of low-dimensional models derived from the Navier–Stokes equations. Dynamical systems theory is a broad and rapidly growing field which, in its more megalomaniacal forms, might be claimed to encompass all of differential equations (ordinary, partial, and functional), iterations of mappings (real and complex), devices such as cellular automata and neural networks, as well as large parts of analysis and differential topology. Here our aim is merely the modest one of introducing, with simple examples, some tools for analysis of nonlinear ordinary differential equations that may not be as familiar as, say, perturbation and asymptotic methods.

The viewpoint of dynamical systems theory is geometric, and invariant manifolds play a central rôle, but we do not assume or require familiarity with differential topology. In the same way, symmetries are crucial in determining the behavior, and permitting the analysis, of the low-dimensional models of interest, but we avoid appeals to the subtleties of group theory in our introduction to symmetric bifurcations. Thus, it should be clear that these two chapters cannot substitute for a serious course (or, more likely, courses) in dynamical systems theory. The makings of such a course can be found in the books of Arnold [15,17], Guckenheimer and Holmes [144], Arrowsmith and Place [18], or Glendinning [129], and in other references cited below. In particular we omit entirely any discussion of partial differential equations, which may seem scandalous, since this book ostensibly treats turbulence as described by the Navier–Stokes equations. However, we recall that we are concerned here only with relatively low spatial wavenumbers and with models, obtained by projection onto low-dimensional but "relevant" subspaces, that take the form of ordinary differential equations. The extension of the considerable theory for finite-dimensional nonlinear systems to infinite-dimensional evolution equations is still an active and changing research field. Readers wishing to learn more might start with the books of Henry [154], Temam [367], Vishik [376], or Doering and Gibbon [96].

Before beginning this chapter we must comment on an important issue. Many readers, acknowledging the impossibility of solving nonlinear differential equations in general, will ask why we do not immediately resort to numerical simulation. Numerical methods are certainly a major tool in determining the behavior of nonlinear systems and their importance and use will continue to grow with the wider introduction of flexible, workstation-based packages. Indeed, it was with one of these, DsTool (Back *et al.* [26]), that many of the illustrations in this part of the book were made. But we believe that analytical methods, used hand in hand with careful simulations, are more useful now than ever. There are three major reasons for this:

(1) One never has enough computer time to produce, or human time to interpret and evaluate, a "complete" picture of the solutions to a system of reasonable dimension (≥ 4 is sufficient) that depends on several parameters (≥ 3 is ample). Preliminary analyses to locate interesting parameter ranges and regions of phase space are crucial. Naturally, numerical simulation can prompt and suggest such analyses, and hence they should proceed hand in hand.

(2) The very nature of chaotic dynamical systems, their sensitive dependence on initial conditions, immediately renders numerical approximations suspect. While rigorous results (shadowing lemmas) do imply that numerical solutions faithfully track true solutions in certain cases, in general we have no such guarantees. Moreover, as we shall see later in this book, the symmetric systems that we deal with are particularly sensitive to symmetry-breaking induced by numerical approximation, and one can obtain quite misleading results, including seemingly clear evidence of chaos, where in truth only regularity reigns.

(3) Numerical simulations and numerical analysis, the mathematical theory underlying them, typically deal with finite time integrations, while the realm of dynamical systems theory includes asymptotic, infinite time properties and phenomena such as attractors. From this viewpoint the two approaches are complementary. (In fact, rigorous numerical estimates coupled with theorems regarding limit sets have been used to prove the existence of limit cycles in planar systems, see, e.g. [143, 233].)

6.1 Linearization and invariant manifolds

It is a pleasant paradox that much of our understanding of nonlinear systems derives from judicious linearization. We shortly come to this but, to begin, we introduce the world in which we shall operate. The *state* of the system is specified by a vector, $\mathbf{x}(t)$, in n-dimensional Euclidean space, $\mathbb{R}^n$, with components $(x_1(t), x_2(t), \ldots, x_n(t))$ representing the dependent variables, t is the independent variable (time) and the dynamics are given by the *vector field* $\mathbf{f}(\mathbf{x})$ with components $(f_1(\mathbf{x}), f_2(\mathbf{x}), \ldots, f_n(\mathbf{x}))$. We consider the autonomous ordinary differential equation (ODE) defined by

$$\dot{\mathbf{x}} = \mathbf{f}(\mathbf{x}). \tag{6.1}$$

Although this is a system of n first order equations, we shall refer to it in the singular. We remark that higher order systems can be put in this form at the expense of introducing additional dependent variables.

At this point it is useful to recall a remark from Chapter 3. The proper orthogonal decomposition takes place in a linear (Hilbert) space of functions, but we apply it to the representation of functions that derive from strongly nonlinear turbulent processes. Similarly, in the theory of dynamical systems the phase spaces are often linear (for us, they are simply

Euclidean real or complex spaces), but the functions defining the ODEs are emphatically nonlinear. There is no contradiction in generating or describing a nonlinear process in a linear phase space.

Rather than attempting to solve an initial value problem by finding (or approximating) the solution or orbit of (6.1) starting at a point $\mathbf{x}(0) = \mathbf{x}_0$, dynamical systems theory attempts to characterize the global structure of *all* solutions in a qualitative fashion. The aim is to produce a *phase portrait*: a description of *phase* or *state space* in which the prominent features are the *attractors*, which "capture" all orbits in their *basins* or *domains of attraction*, the *separatrices*, which divide those basins, and the *invariant* and *non-wandering* sets. Attractors are defined later, but we must define the latter two objects here. An *invariant set* Λ for (6.1) is a subset of the phase space composed of solutions of (6.1): thus if the solution starts at a point in Λ it remains in Λ for all time. A point $\mathbf{x}$ is *non-wandering* if some solution starting arbitrarily close to $\mathbf{x}$ returns arbitrarily close to $\mathbf{x}$ infinitely often.

It may be helpful to imagine the phase space filled with a phase fluid on which the *flow* induced by the vector field $\mathbf{f}$ imprints streak lines. The attractors and other invariant sets can then be seen as (higher-dimensional) generalizations of the sinks, stagnation points, and sources in a real fluid flow.

At the outset we must admit that the global description of phase space for arbitrary vector fields is presently feasible only for two-dimensional systems ($n = 2$), but we remark that much partial information can be gained by intelligent local analyses and the clever use of symmetries, which is the subject of the next chapter. The simplest non-wandering sets are *fixed points* or *equilibria*: points $\mathbf{x}_e \in \mathbb{R}^n$ which satisfy

$$\mathbf{f}(\mathbf{x}_e) = \mathbf{0}, \tag{6.2}$$

and it is wise to seek them first. Of course, solution of the algebraic system (6.2) might be almost as intractable as integration of the differential system (6.1), but, supposing that we have found an equilibrium $\mathbf{x}_e$, we next ask about the behavior of solutions to (6.1) near $\mathbf{x}_e$. Letting

$$\mathbf{x} = \mathbf{x}_e + \boldsymbol{\xi}, \tag{6.3}$$

where $|\boldsymbol{\xi}|$ is taken sufficiently small, we substitute (6.3) into (6.1) and expand in Taylor series to obtain:

$$\dot{\boldsymbol{\xi}} = \mathbf{Df}(\mathbf{x}_e)\boldsymbol{\xi} + \mathcal{O}(|\boldsymbol{\xi}|^2), \tag{6.4}$$

where

$$\mathbf{Df}(\mathbf{x}_e) = \left[\frac{\partial f_i}{\partial x_j}\right]\Bigg|_{\mathbf{x}=\mathbf{x}_e} \tag{6.5}$$

is the (real) $n \times n$ Jacobian matrix of first derivatives evaluated at $\mathbf{x}_e$. Ignoring for the moment the "error" of $\mathcal{O}(|\boldsymbol{\xi}|^2)$, Equation (6.4) is a constant coefficient linear system and can be solved by conventional methods, which we summarize.

Let $\lambda_i, \mathbf{v}_i, i = 1, \ldots, n$ be the eigenvalues and eigenvectors of $\mathbf{Df}(\mathbf{x}_e)$, including generalized eigenvectors if necessary in the event of multiple eigenvalues, and let $\mathbf{X}_e(t)$ be the (real) $n \times n$ matrix with the linearly independent columns $\mathbf{v}_i e^{\lambda_i t}$ ($\mathbf{v}_i t e^{\lambda_i t}$, etc. for generalized eigenvectors). The matrix $\mathbf{X}_e$ is called a *fundamental solution matrix*. If complex eigenvalues $\lambda_\pm = \alpha \pm i\beta$ occur, $\mathbf{X}_e$ can still be chosen real by taking adjacent columns as $e^{\alpha t}[\mathbf{u}\cos(\beta t) - \mathbf{w}\sin(\beta t)]$ and $e^{\alpha t}[\mathbf{w}\cos(\beta t) + \mathbf{u}\sin(\beta t)]$, where $\mathbf{v} = \mathbf{u} \pm i\mathbf{w}$ is the complex conjugate pair of eigenvectors belonging to $\lambda_\pm$. Then the matrix

$$\Phi_e(t) = \mathbf{X}_e(t)\mathbf{X}_e^{-1}(0) \tag{6.6}$$

defines the *flow map* which takes solutions from initial points $\boldsymbol{\xi}(0) = \boldsymbol{\xi}_0$ to their states $\boldsymbol{\xi}(t, \boldsymbol{\xi}_0)$ at time t. For each fixed t, $\Phi_e(t)$ is a linear mapping of n-dimensional Euclidean space. Thus the formula

$$\boldsymbol{\xi}(t, \boldsymbol{\xi}_0) = \Phi_e(t)\boldsymbol{\xi}_0 \tag{6.7}$$

solves the global problem for the linear system

$$\dot{\boldsymbol{\xi}} = \mathbf{Df}(\mathbf{x}_e)\boldsymbol{\xi} \tag{6.8}$$

in the sense that (6.6) and (6.7) define an explicit linear transformation which takes any initial condition $\boldsymbol{\xi}_0$ into the state $\boldsymbol{\xi}(t, \boldsymbol{\xi}_0)$ reached at time t by the solution starting at that point.

Since we shall need it below, we take this opportunity to observe that, provided that $\mathbf{f}$ satisfies the conditions required for local existence and uniqueness of solutions of (6.1) (which it does automatically if it is a differentiable function), then a similar *flow map* $\boldsymbol{\phi}_t$ exists for the nonlinear system, with the following (important) caveat. Unless solutions exist for all time, $\boldsymbol{\phi}_t$ is *not* globally defined and the integration time generally depends on the initial condition. For example, the solution

$$x(t) = \frac{x_0}{1 - tx_0}$$

of the initial value problem

$$\dot{x} = x^2, \; x(0) = x_0$$

reveals that positive solutions are only defined up to time $t = 1/x_0$, when they "blow up" to infinity. Perhaps more practically, there is no general formula analogous to (6.6) to compute the flow map

$$\mathbf{x}(t, \mathbf{x}_0) = \boldsymbol{\phi}_t(\mathbf{x}_0) \tag{6.9}$$

for an arbitrary nonlinear ODE, although the basic existence-uniqueness theory for ODEs does tell us that, like $\Phi_e(t)$, it is a smooth invertible mapping.

The key fact of importance to us here is that the closed form solution (6.6) of (6.8) is a good approximation for solutions of (6.4) as long as all the eigenvalues of $\mathbf{Df}(\mathbf{x}_e)$ have non-zero real parts and $|\boldsymbol{\xi}|$ remains sufficiently small. More precisely, the Hartman–Grobman theorem asserts [18, 144] that there is a continuous change of coordinates transforming solutions of the full nonlinear system (6.1) in a neighborhood $B(\mathbf{x}_e)$ of the fixed point into those of (6.8) in a similar neighborhood of $\boldsymbol{\xi} = \mathbf{0}$. This is called *topological equivalence*. Even better, the decomposition of the phase space $\mathbb{R}^n$ of (6.8) into invariant subspaces spanned by (collections of) eigenvectors $\mathbf{v}_i$ also holds for the nonlinear system in $B(\mathbf{x}_e)$, in that it possesses *invariant manifolds* filled with solutions whose qualitative behavior echoes that of the linearized system. This geometrical fact is perhaps the most important single idea in dynamical systems theory, and to it we now turn.

First, recall that under the flow map $\Phi_e(t)$ any solution initially lying in a k-dimensional subspace $\text{span}\{\mathbf{v}_{i_1}, \ldots, \mathbf{v}_{i_k}\}$, $(k < n)$ remains in that subspace. Thus we can define two distinguished invariant subspaces whose names will reflect the asymptotic properties of solutions belonging to them. Suppose that $\mathbf{Df}(\mathbf{x}_e)$ has s ($\leq n$) and u ($\leq n - s$) eigenvalues with negative and positive real parts respectively and number them $\lambda_1, \ldots, \lambda_s$ and $\lambda_{s+1}, \ldots, \lambda_{s+u}$ (counting multiplicities). Let

$$E^s = \text{span}\{\mathbf{v}_1, \ldots, \mathbf{v}_s\} \quad \text{and} \quad E^u = \text{span}\{\mathbf{v}_{s+1}, \ldots, \mathbf{v}_{s+u}\};$$

then E^s is the *stable subspace* and E^u is the *unstable subspace*. Solutions in E^s decay exponentially and those in E^u grow exponentially as t increases. The stable manifold theorem states that for (6.1) in a neighborhood $B(\mathbf{x}_e)$ of the equilibrium, there exist *local stable and unstable manifolds* $W^s_{\text{loc}}(\mathbf{x}_e)$, $W^u_{\text{loc}}(\mathbf{x}_e)$ of dimensions s and u respectively, tangent at $\mathbf{x}_e$ to E^s and E^u, and characterized as follows, using the flow map (6.9):

$$W^s_{\text{loc}}(\mathbf{x}_e) = \left\{ \mathbf{x} \in B(\mathbf{x}_e) \mid \boldsymbol{\phi}_t(\mathbf{x}) \to \mathbf{x}_e \text{ as } t \to +\infty \right.$$
$$\left. \text{and } \boldsymbol{\phi}_t(\mathbf{x}) \in B(\mathbf{x}_e) \text{ for all } t \geq 0 \right\};$$

$$W^u_{\text{loc}}(\mathbf{x}_e) = \left\{ \mathbf{x} \in B(\mathbf{x}_e) \mid \boldsymbol{\phi}_t(\mathbf{x}) \to \mathbf{x}_e \text{ as } t \to -\infty \right.$$
$$\left. \text{and } \boldsymbol{\phi}_t(\mathbf{x}) \in B(\mathbf{x}_e) \text{ for all } t \leq 0 \right\}.$$

In words, the local stable manifold consists of all solutions that start and remain near the equilibrium for all future time and approach it as time tends to infinity, and the unstable manifold is defined similarly, with the substitutions "past time" and "minus infinity."

These manifolds are smooth, curved surfaces, which locally look like the linear subspaces E^s and E^u (as the surface of the earth looks locally like a plane). We can make this more precise as follows. In terms of the local coordinates $\boldsymbol{\xi}$ near $\mathbf{x}_e$ the smooth manifolds $W^s_{\text{loc}}(\mathbf{x}_e)$ and $W^u_{\text{loc}}(\mathbf{x}_e)$ can be expressed as graphs over E^s and E^u respectively. Picking an eigenvector basis and letting $E^{s\perp}$ denote the $(n - s)$-dimensional orthogonal complement to E^s and $\mathbf{y} \in E^s$, $\mathbf{z} \in E^{s\perp}$ be local coordinates, we can write

$$W^s_{\text{loc}}(\mathbf{x}_e) = \{(\mathbf{y}, \mathbf{z}) | (\mathbf{y}, \mathbf{z}) \in B(\mathbf{0}) \text{ and } \mathbf{z} = \mathbf{g}(\mathbf{y})\}$$

for some smooth function $\mathbf{g} : E^s \to E^{s\perp}$. We cannot generally compute $\mathbf{g}$ (if we could we would have found solutions of (6.1)), but we can approximate it, as we see below.

Note: Here the acute reader may wonder what has happened to the subspace E^c spanned by eigenvectors whose eigenvalues have zero real part, if any such exist. This *center subspace* comes into its own in our study of local bifurcations in Section 6.4, but for the moment we observe that there is a *center manifold* $W^c_{\text{loc}}(\mathbf{x}_e)$ tangent to E^c and that it may contain solutions belonging to $W^s_{\text{loc}}(\mathbf{x}_e)$, $W^u_{\text{loc}}(\mathbf{x}_e)$, or to neither of these manifolds (solutions that neither approach $\mathbf{x}_e$ as $t \to +\infty$ nor as $t \to -\infty$). Thus, if E^c is non-empty, the manifolds tangent to E^s, E^u may contain only *parts* of the "full" stable and unstable manifolds. See the example of Equation (6.15), below.

Equipped with the local manifolds we can define the *global* stable and unstable manifolds:

$$W^s(\mathbf{x_e}) = \bigcup_{t \leq 0} \phi_t(W^s_{loc}(\mathbf{x_e})) \; ; \quad W^u(\mathbf{x_e}) = \bigcup_{t \geq 0} \phi_t(W^u_{loc}(\mathbf{x_e})) \; .$$

These are the unions of backwards and forwards images of the local manifolds under the (nonlinear) flow map. Thus $W^s(\mathbf{x_e})$ is the set of *all* points whose orbits approach $\mathbf{x_e}$ as $t \to +\infty$, even if they leave $B(\mathbf{x_e})$ for a while, and $W^u(\mathbf{x_e})$ is the set of all points whose orbits approach $\mathbf{x_e}$ as $t \to -\infty$.

It is time for an example which illustrates these concepts. Consider the two-dimensional system

$$\left. \begin{array}{l} \dot{x}_1 = -x_1, \\ \dot{x}_2 = x_2 - x_1^2, \end{array} \right\} \tag{6.10}$$

which has the unique fixed point $(x_1, x_2) = (0, 0)$. The linearization at that point

$$\begin{pmatrix} \dot{\xi}_1 \\ \dot{\xi}_2 \end{pmatrix} = \begin{bmatrix} -1 & 0 \\ 0 & 1 \end{bmatrix} \begin{pmatrix} \xi_1 \\ \xi_2 \end{pmatrix} \tag{6.11}$$

is already conveniently diagonal, with eigenvalues $\{-1, +1\}$ and eigenvectors $\{(1, 0), (0, 1)\}$ respectively. We therefore have the stable and unstable subspaces:

$$E^s = \text{span}\{(1, 0)\} \; , \quad E^u = \text{span}\{(0, 1)\} \; ,$$

which are just the x_1- and x_2-axes. We therefore expect the stable and unstable manifolds to be curves tangent at the origin to these one-dimensional subspaces. In this case we can solve explicitly for the global stable manifold by eliminating the independent variable, dividing the components of (6.10) to obtain the linear ODE

$$\frac{dx_2}{dx_1} = -\frac{x_2}{x_1} + x_1, \tag{6.12}$$

which has the general solution

$$x_2 = \frac{x_1^2}{3} + \frac{c}{x_1} \stackrel{\text{def}}{=} g(x_1, c). \tag{6.13}$$

The only member of the one parameter family of curves $x_2 = g(x_1, c)$ tangent to E^s at $(0, 0)$ is

$$g(x_1, 0) = \frac{x_1^2}{3}, \tag{6.14}$$

and so we conclude that the global stable manifold is given by

$$W^s(0, 0) = \left\{ (x_1, x_2) \middle| x_2 = \frac{x_1^2}{3} \right\}.$$

We next observe that if $x_1(0) = 0$ then the first component $x_1(t)$ of the solution of (6.10) remains zero for all time. The subspace E^u is therefore invariant for the full nonlinear system and, in this case, coincides with the global unstable manifold:

$$W^u(0, 0) = \{(x_1, x_2) | x_1 = 0\}.$$

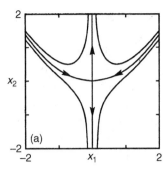
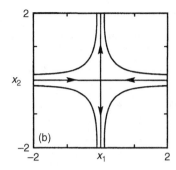

Figure 6.1 Phase portraits of (a) the full system (6.10), and (b) its linearization (6.11). Arrows indicate the sense of increasing time.

The topological equivalence of general solutions $\xi_1 \xi_2 = c$ of the linearized system (6.11) with those of the full system given by (6.13) is made clear in Figure 6.1, which also shows the eigenspaces and invariant manifolds. We see a "nonlinear saddle point."

This example can serve one more purpose. Note that solutions of the nonlinear equation are invariant under the change of coordinates $(x_1, x_2) \mapsto (-x_1, x_2)$: reflection about the x_2-axis. The linearized system (6.11) shares this property and in addition has a reflection symmetry about the x_1-axis. Linearization cannot destroy symmetries, but, as here, it typically increases them. We pick up this theme in the next chapter.

In discussing stable and unstable manifolds we did not assume that all n eigenvalues of $\mathbf{Df}(\mathbf{x}_e)$ have non-zero real parts. If they do, so that $s + u = n$ and E^s, E^u together span $\mathbb{R}^n$, we call the fixed point *hyperbolic* and its stability type is completely determined by the eigenvalues. The name derives from the solutions of the linearized flow near a two-dimensional saddle point, whose graphs are hyperbolae: Figure 6.1(b). More generally, the topological equivalence of the Hartman–Grobman theorem implies that

- If all eigenvalues of $\mathbf{Df}(\mathbf{x}_e)$ have strictly negative real parts then $\mathbf{x}_e$ is an asymptotically stable fixed point. If at least one eigenvalue has a strictly positive real part, then $\mathbf{x}_e$ is unstable.

An asymptotically stable fixed point is called a *sink* and if all eigenvalues have positive real parts it is called a *source*. Otherwise we typically have a *saddle point*. The former two terms are borrowed from potential flow problems in fluid mechanics. Note that to conclude stability from the linearization we require hyperbolicity, while detection of instability merely requires a positive eigenvalue.

When the real parts of one or more eigenvalues are zero, things are not so easy and we must examine the nonlinear terms. The reader can verify that the example

$$\left. \begin{array}{l} \dot{x}_1 = -x_1 \\ \dot{x}_2 = \alpha x_2^3 \end{array} \right\} \tag{6.15}$$

has a stable fixed point at $(0, 0)$ if $\alpha \leq 0$ and an unstable fixed point if $\alpha > 0$, its solutions being given by

$$(x_1(t), x_2(t)) = (x_1(0)e^{-t}, \pm|x_2(0)|[1 - 2\alpha(x_2(0))^2 t]^{-1/2}). \tag{6.16}$$

In connection with the note on page 159 above, we observe that, while the center sub-space *and* the center manifold for the linearized system corresponding to (6.15) are the x_2-axis ($W^c(0, 0) = \mathrm{span}\{(0, 1)\}$), for $\alpha < 0$ the center manifold is contained in the stable manifold, which is the whole plane, since all solutions decay and approach $(0, 0)$ in this case. On the other hand, if $\alpha > 0$ the center manifold is actually also the unstable manifold. Thus in general we cannot attach a name denoting stability type to the center manifold; the asymptotic properties of solutions in it are controlled by nonlinear terms (here, by αx_2^3). In Section 6.4 we focus on "degenerate" cases such as this and their relevance in bifurcations, but before that we briefly turn to the next simplest non-wandering sets: periodic orbits.

6.2 Periodic orbits and Poincaré maps

The reader may have met the van der Pol oscillator, originally proposed to model electronic diodes with nonlinear resistive characteristics (van der Pol [375]). It can be written as the first order system:

$$\left.\begin{aligned} \dot{x}_1 &= x_2 - (\tfrac{1}{3}x_1^3 - x_1), \\ \dot{x}_2 &= -x_1. \end{aligned}\right\} \tag{6.17}$$

The origin is the only fixed point and it is an unstable source, its eigenvalues being $(1 \pm i\sqrt{3})/2$, but solutions do not escape to infinity; rather, laborious estimates show that there is an annular region $\mathcal{B}$ surrounding the origin into which the vector field defined by the right-hand side of (6.17) is directed. The Poincaré–Bendixson theorem [129, 156] then implies that a *limit cycle* lies within this *trapping region* $\mathcal{B}$. (This theorem is beyond the scope of our outline and is, in any case, limited to two-dimensional systems.) Numerical simulations reveal the phase portrait of Figure 6.2.

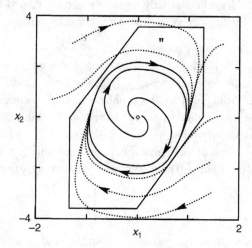

Figure 6.2 The trapping region and the limit cycle of the van der Pol equation (6.17).

The limit cycle of Figure 6.2 is an example of an asymptotically stable or attracting, periodic orbit. Periodic orbits can, of course, occur in differential equations of any dimension $n \geq 2$, but finding them analytically is much more difficult than finding fixed points, since one must effectively solve the ODE to do it. It is still useful, however, to develop some analytical tools. The Poincaré map is one such tool which allows us to reduce the (local) analysis of solutions near a periodic orbit γ to the study of a mapping of dimension $n - 1$, one lower than that of the original phase space. One defines a (small) piece Σ of an $(n - 1)$-dimensional manifold (a hyper-plane often suffices), pierced by γ at a point $\mathbf{p}$ and transverse to the flow in the sense that the component of the vector field $\mathbf{f}$ normal to Σ does not vanish at any point in Σ. The (hyper-) surface Σ is called a *cross section* or *Poincaré section* for the flow and solutions cross it, all in the same direction, with non-zero speed. By continuity, solutions starting at points $\mathbf{q} \in \Sigma$ near $\mathbf{p}$ will follow γ sufficiently closely to intersect Σ again at, say $\mathbf{q}'$, thus implicitly defining a map on Σ:

$$\mathbf{q} \to \mathbf{q}' = \mathbf{P}(\mathbf{q}), \tag{6.18}$$

where $\mathbf{P}$ is called the *Poincaré map*, or *first-return map*. Note that the point $\mathbf{p}$ at which the periodic orbit intersects the cross section Σ is a fixed point of $\mathbf{P}$.

In the example (6.16) above, the positive x_1-axis

$$\Sigma = \{(x_1, x_2)|x_1 > 0, x_2 = 0\}$$

is a suitable cross section, and while we cannot integrate (6.17) explicitly to obtain a formula for the one-dimensional map P analogous to (6.7), the instability of the fixed point $(0, 0)$ and attractivity of the trapping region $\mathcal{B}$ from outside implies that P takes the form sketched in Figure 6.3. It is clear that the continuous function P must intersect the diagonal in at least one point $p > 0$, so that $p = P(p)$ is a fixed point corresponding to the limit cycle. In fact it is true, although not easy to prove, that p is unique and that the linearized

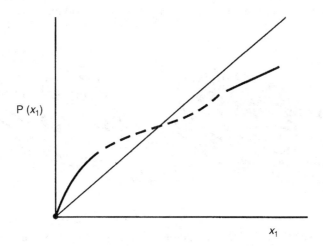

Figure 6.3 The Poincaré map for Equation (6.17).

map satisfies $(dP/dx_1)|_{x_1=p} = \lambda_p < 1$, implying asymptotic stability of both the fixed point and the corresponding periodic orbit.

More generally, from the theory of iterated matrices, if $\mathbf{DP(p)}$, an $(n-1)$-dimensional Poincaré map linearized at a fixed point $\mathbf{p}$, has only eigenvalues of modulus strictly less than one, then $\mathbf{p}$ and the associated periodic orbit are asymptotically stable. If at least one eigenvalue has modulus greater than one, $\mathbf{p}$ and the periodic orbit are unstable. The degenerate case, with one or more eigenvalues of unit modulus, is dealt with rather like the analogous non-hyperbolic fixed point, and we return to it in Section 6.4.

As in the case of flows generated by solutions of differential equations, along with the (discrete) orbits of iterated mappings come invariant manifolds, tangent to the appropriate stable and unstable subspaces of the linearized maps. Suppose that $\mathbf{DP(p)}$ has $s \le n-1$ eigenvalues of modulus less than one and $u = n - 1 - s$ of modulus greater than one (so that, for simplicity, $\mathbf{p}$ is hyperbolic). Then $\mathbf{DP(p)}$ has, on Σ, s- and u-dimensional stable and unstable subspaces $E^s(\mathbf{p})$, $E^u(\mathbf{p})$ respectively. The stable manifold theorem for maps then yields local manifolds $W^s_{\text{loc}}(\mathbf{p})$, $W^u_{\text{loc}}(\mathbf{p})$, tangent to $E^s(\mathbf{p})$, $E^u(\mathbf{p})$ at $\mathbf{p}$, just as for flows. A solution to the ODE started on Σ in $W^s_{\text{loc}}(\mathbf{p})$ will circulate near γ and return to a point of $W^s_{\text{loc}}(\mathbf{p})$ closer to $\mathbf{p}$ at its next intersection with Σ. All such solutions started in a neighborhood $B(\mathbf{p}) \subset \Sigma$ form the $(s+1)$-dimensional local stable manifold of the periodic orbit γ, $W^s_{\text{loc}}(\gamma)$, while solutions started at points of $W^u_{\text{loc}}(\mathbf{p})$ form the $(u+1)$-dimensional local unstable manifold of γ, $W^s_{\text{loc}}(\gamma)$. Figure 6.4 shows two possible structures in the three-dimensional case in which $s = u = 1$. In case (a) the real eigenvalues of $\mathbf{DP(p)}$ satisfy $0 < \lambda_1 < 1 < \lambda_2$; in case (b) $\lambda_2 < -1 < \lambda_1 < 0$. In the latter case it is easy to see that the (diagonalized) matrix

$$\begin{bmatrix} \lambda_1 & 0 \\ 0 & \lambda_2 \end{bmatrix}$$

of $\mathbf{DP(p)}$, which takes $(x_1, x_2) \in \Sigma$ to $(\lambda_1 x_1, \lambda_2 x_2)$, rotates the vector connecting $(0, 0)$ to (x_1, x_2) by (approximately) π. As solutions circulate in a tubular neighborhood of γ, therefore, they pick up an odd number of half twists. In case (a) the number of half twists

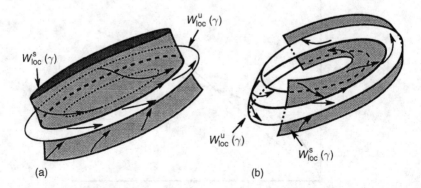

$W^u_{\text{loc}}(\gamma)$

$W^s_{\text{loc}}(\gamma)$

$W^u_{\text{loc}}(\gamma)$

$W^u_{\text{loc}}(\gamma)$

$W^s_{\text{loc}}(\gamma)$

(a)   (b)

Figure 6.4 Periodic orbits in a three-dimensional flow with (a) orientable, and (b) non-orientable stable and unstable manifolds. The stable manifolds are lightly shaded, and in (b) part of $W^s_{\text{loc}}(\gamma)$ is removed for clarity.

is even. Figure 6.4 depicts the simplest cases of 0 and 1 half twists respectively. In the latter case the local stable and unstable manifolds are Möbius bands. Starting on the right of the periodic orbit, say, and following it on either manifold for one circuit, one returns on its left.

6.3 Structural stability and genericity

The notions of structural stability and generic properties have been important stimuli in the development of the abstract theory of dynamical systems and it is worth giving a brief introduction to them here. As we have seen, instead of seeking an exact or approximate solution of an initial value problem in closed form, the qualitative theory poses less detailed questions about *all* solutions of a given ODE. In the same vein, rather than focusing on a specific ODE, we can ask about "typical" properties of all or "most" ODEs in some class.

We must first extend the notion of topological equivalence, already introduced in the context of linearization and the Hartman–Grobman theorem. Let $\boldsymbol{\phi}_t^f$ and $\boldsymbol{\phi}_t^g$ be the (nonlinear) flow maps of the two n-dimensional ODEs

$$\dot{\mathbf{x}} = \mathbf{f}(\mathbf{x}) \quad \text{and} \quad \dot{\mathbf{x}} = \mathbf{g}(\mathbf{x}) \tag{6.19}$$

respectively. We say that the systems are *topologically equivalent* if there is a continuous mapping or change of coordinates $\mathbf{h}$ such that $\mathbf{h}(\boldsymbol{\phi}_{t_f}^f(\mathbf{x})) = \boldsymbol{\phi}_{t_g}^g(\mathbf{h}(\mathbf{x}))$. Thus $\mathbf{h}$ maps the phase portrait of the $\mathbf{f}$ vector field onto that of the $\mathbf{g}$ vector field, taking solutions of one equation to solutions of the other. Note that the elapsed time t_g of the $\mathbf{g}$ flow generally differs from t_f, that of the $\mathbf{f}$ flow: $\mathbf{h}$ preserves the sense but not the exact parameterization by time. This makes precise the notion of two phase portraits "looking the same."

To discuss structural stability we also need the notion of a *space of systems*, equipped with a suitable norm or notion of "closeness." To illustrate, consider the class of linear ODEs of the form

$$\dot{\mathbf{x}} = \mathbf{A}\mathbf{x}, \tag{6.20}$$

where $\mathbf{A}$ is a constant coefficient $n \times n$ matrix, possibly obtained as in (6.4 and 6.5) by linearization. Equation (6.20) is completely specified by the n^2 coefficients a_{ij} of $\mathbf{A}$ and thus the space of systems is equivalent to $\mathbb{R}^{n^2}$ and the Euclidean or any other convenient norm can be used to measure the distance between two systems. So we can say that two systems of the form (6.20), defined by matrices $\mathbf{A}$ and $\mathbf{B}$, are ϵ-close if $|a_{ij} - b_{ij}| < \epsilon$ for all $1 \le i, j \le n$. For brevity, we write $\|\mathbf{A} - \mathbf{B}\| < \epsilon$, $\|\cdot\|$ denoting the norm. We can then state the definition:

- The linear system $\dot{\mathbf{x}} = \mathbf{A}\mathbf{x}$ is *structurally stable* if all sufficiently close systems $\dot{\mathbf{x}} = \mathbf{B}\mathbf{x}$ are topologically equivalent to $\dot{\mathbf{x}} = \mathbf{A}\mathbf{x}$.

This is nicely illustrated by the special case of two-dimensional linear systems of the form

$$\left. \begin{array}{l} \dot{x}_1 = a_{11}x_1 + a_{12}x_2, \\ \dot{x}_2 = a_{21}x_1 + a_{22}x_2. \end{array} \right\} \tag{6.21}$$

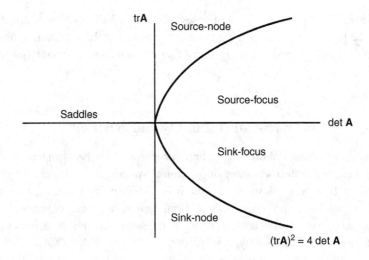

Figure 6.5 Stability types of the fixed point at the origin for Equation (6.21).

The eigenvalues of the matrix $\mathbf{A} = [a_{ij}]$ are given by the roots of the characteristic equation

$$\lambda^2 - \lambda \, \mathrm{tr} \, \mathbf{A} + \det \mathbf{A} = 0,$$

where $\mathrm{tr}\, \mathbf{A} = a_{11} + a_{22}$ and $\det \mathbf{A} = a_{11}a_{22} - a_{12}a_{21}$. Thus, if $\det \mathbf{A} < 0$ then $(0, 0)$ is a saddle point, while if $\det \mathbf{A} > 0$ and $\mathrm{tr}\, \mathbf{A} < 0$ it is a sink and if $\det \mathbf{A} > 0$ and $\mathrm{tr}\, \mathbf{A} > 0$ it is a source. The diagram of Figure 6.5 summarizes the qualitative behavior. On the diagram we also indicate the parabola $(\mathrm{tr}\, \mathbf{A})^2 = 4 \det \mathbf{A}$ on which the eigenvalues change from real to complex, but this does not correspond to a *topological* change in the character of solutions: a sink is a sink, whether solutions approach it directly or in spirals.

Here the space of systems is effectively the plane with $\det \mathbf{A}$, $\mathrm{tr}\, \mathbf{A}$ as coordinates and any system not on the $\mathrm{tr}\, \mathbf{A}$-axis or the positive $\det \mathbf{A}$-axis is structurally stable, since all systems in a sufficiently small neighborhood of it share its behavior. On the positive $\det \mathbf{A}$-axis, however, we have purely imaginary eigenvalues and the ODE behaves like an undamped harmonic oscillator. Such systems are structurally unstable. Moving off the axis – adding damping, no matter how small – yields qualitatively different behavior. The periodic orbits are all destroyed and replaced by solutions spiralling into a sink, or out from a source.

The attentive reader will note that we have somehow characterized a four parameter problem (the four coefficients of $\mathbf{A}$) by only two parameters: $\det \mathbf{A}$ and $\mathrm{tr}\, \mathbf{A}$. However, recall that Figure 6.5 only indicates stability types, and does not distinguish among finer properties of the phase portraits. For these, eigenvectors as well as eigenvalues must be computed, requiring knowledge of all four matrix coefficients.

For nonlinear systems we are faced with the fact that the space of n-dimensional systems is itself infinite-dimensional (think of describing an analytic function f by all the derivatives $d^n f/df^n$ which appear in its Taylor series). But fortunately, in checking closeness we need only include first derivatives:

- Two *n*-dimensional vector fields **f** and **g** are $\epsilon - C^1$ close if for all $\mathbf{x} \in K \subset \mathbb{R}^n$ (*K* is a bounded set) $\|\mathbf{f} - \mathbf{g}\| < \epsilon$ and $\|\mathbf{Df} - \mathbf{Dg}\| < \epsilon$, where $\| \cdot \|$ denote any convenient norms on $\mathbb{R}^n$, $\mathbb{R}^{n^2}$ respectively.

We cannot require that **f** and **g** are $\epsilon - C^1$ close at *all* points $\mathbf{x} \in \mathbb{R}^n$. The problem is that for nonlinear systems the derivatives as well as the values typically become unbounded as $|\mathbf{x}| \to \infty$. For example, the real-valued functions

$$f = 2x \text{ and } g = 2x + \alpha x^2$$

are not ϵ-close on the domain $\mathbb{R}$ for any α. But if we restrict ourselves to the interval $K = [-1, 1]$, say, then f and g are $2\alpha - C^1$ close. Hence the need for the bounded set K in the definition.

We can now say that:

- The nonlinear system $\dot{\mathbf{x}} = \mathbf{f}(\mathbf{x})$ is *structurally stable* if all sufficiently $\epsilon - C^1$ close systems $\dot{\mathbf{x}} = \mathbf{g}(\mathbf{x})$ are topologically equivalent to $\dot{\mathbf{x}} = \mathbf{f}(\mathbf{x})$.

A nice example of structural stability in the class of linear systems (6.20) relevant to our earlier discussions in Section 6.1, and which expands on the example (6.21) above, is provided by those systems whose defining matrices have only eigenvalues with non-zero real parts. The eigenvalues of a matrix depend continuously on the matrix coefficients and so small perturbations cannot move eigenvalues with non-zero real parts onto the imaginary axis. Indeed, in general we have the necessary condition:

- If a nonlinear system (6.1) is structurally stable, then all its equilibria and periodic orbits must be hyperbolic.

However, as we shall see in the next section, in dealing with nonlinear systems this is not enough. Structural stability can fail even if all equilibria and periodic orbits are hyperbolic, due to the "fragile" behavior of solutions leaving and returning to saddle points. The global behavior of such homoclinic and heteroclinic orbits is especially important in what follows. Structural stability is a property possessed by the system *as a whole* and it must not be confused with the more usual notion of (asymptotic) stability of an individual fixed point or periodic orbit.

Structural stability is important in our context for the following reason. The (polynomial) ODEs produced by projection onto a subspace spanned by empirical eigenfunctions have coefficients obtained by (numerical) integration of products of those eigenfunctions and their derivatives. Such coefficients are necessarily subject to uncertainty and one hopes that (small) changes in them will not cause changes in qualitative behavior of the ODEs. Thus, our systems should at least be structurally stable. However, as we see in Chapter 7, one must be careful in discussing families of systems depending on parameters or systems with symmetry. The key is in one's definition of the *space of systems* in which to work. We return to this in Section 6.4 and in Chapter 7, where we discuss symmetric systems.

The other notion, generic properties, is more difficult to define and one requires a deeper knowledge of function spaces and their topology to appreciate it. A property possessed by a system is said to be *generic* if it is shared by an open, dense set of systems in the appropriate class. Loosely speaking, this means that almost all systems enjoy the property in question: it is "typical." To give an idea of this, we return to the linear ODEs of the form (6.20). Among them, those having a single zero eigenvalue correspond to a $(n^2 - 1)$-dimensional hypersurface in $\mathbb{R}^{n^2}$ given by $\det \mathbf{A} = 0$. The complement is clearly an open set, and it is dense since points on $\det \mathbf{A} = 0$ can be approximated arbitrarily closely by points with $\det \mathbf{A} \neq 0$. Moreover, linear systems having several eigenvalues with zero real parts correspond to points in the intersections of similar hypersurfaces and so hyperbolicity of equilibria is a generic property for linear, and indeed nonlinear, systems. In fact we can say that structural stability is a generic property for linear systems. Many of the deeper wonders and complexities of dynamical systems theory arise from the fact that this is *not* true for nonlinear ODEs, due to the global behaviors that such systems can exhibit. One can find examples in which structural *in*stability is persistent as one moves about in the space of systems. The Lorenz equation, considered in Section 6.5 below, is one of them. Again, homoclinic orbits are a major culprit.

6.4 Bifurcations local and global

The differential equations of interest in this book contain coefficients which correspond to physical quantities which an experimentalist could change, or to adjustable parameters chosen in modeling assumptions. We therefore need some elements of *bifurcation theory*: the study of how system behavior changes as *control parameters* are varied. Our world is now that of *families* of systems:

$$\dot{\mathbf{x}} = \mathbf{f}(\mathbf{x}; \boldsymbol{\mu}) \ \text{ or } \ \dot{\mathbf{x}} = \mathbf{f}_{\boldsymbol{\mu}}(\mathbf{x}), \tag{6.22}$$

where $\boldsymbol{\mu} \in \mathbb{R}^m$ is a (vector of) parameter(s). Often all but one are (temporarily) fixed and we consider a one-parameter family, which corresponds to a path in the appropriate space of systems.

The example of the one-parameter, one-dimensional family

$$\dot{x} = \mu - x^2 \tag{6.23}$$

serves to introduce local bifurcation theory. Equation (6.23) is soluble in closed form, but here we need only observe that, for $\mu < 0$ it has no equilibria and all solutions approach $-\infty$ as t increases and $+\infty$ as t decreases, while for $\mu > 0$ there are two hyperbolic equilibria $x_{e_1} = +\sqrt{\mu}$ and $x_{e_2} = -\sqrt{\mu}$, the former being an asymptotically stable sink and the latter an unstable source, and that all solutions starting at $x(0) > -\sqrt{\mu}$ approach the sink as $t \to +\infty$. At $\mu = 0$ the single equilibrium $x_e = 0$ is non-hyperbolic and the linearization $\dot{\xi} = 2x_e\xi = 0$ gives no information on stability. For all $\mu \neq 0$ (6.23) is, in fact, structurally stable: its behavior changes as it passes through the point $\mu = 0$ of structural instability. This information can be conveniently pictured on a *bifurcation diagram*: a plot of the branches of equilibria versus the parameter, see Figure 6.6. This is an example of a *saddle–node bifurcation*.

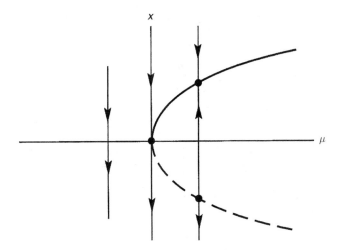

Figure 6.6 The bifurcation diagram for Equation (6.23). Vertical lines represent phase portraits of individual systems, with arrows corresponding to time and thus displaying stability; —— represents a branch of sinks, - - - - a branch of sources.

- More generally, a value μ_0 of μ for which Equation (6.22) is structurally unstable is a *bifurcation value*.

The bifurcation at $\mu = 0$ for (6.23) is called *local* because for $\mu \approx 0$ the qualitative change in the phase portrait is confined to a neighborhood of $x = 0$. In fact when such bifurcations occur in more complex, higher-dimensional systems, they can be analyzed by local techniques: primarily the examination of truncated Taylor series expansions near the degenerate (non-hyperbolic) equilibrium or periodic orbit involved in the bifurcation.

If we add several uncoupled equations involving non-degenerate behavior to (6.23) the picture remains unchanged. For example, for the ODE

$$\left.\begin{aligned}
\dot{x}_1 &= -x_1, \\
\dot{x}_2 &= \mu - x_2^2, \\
\dot{x}_3 &= 2x_3,
\end{aligned}\right\} \tag{6.24}$$

"nothing happens" in the x_1- and x_3-directions as μ varies. At $\mu = 0$ the degenerate equilibrium $\mathbf{x} = \mathbf{0}$ is hyperbolic in those two directions and they can effectively be ignored. It is a remarkable fact that this remains true even when there is nonlinear coupling between the hyperbolic and non-hyperbolic directions. This is the burden of the center manifold theorem ([187, 280], cf. Guckenheimer and Holmes [144]), which supplements the stable manifold theorem in the case of a non-hyperbolic equilibrium $\mathbf{x}_e$, by asserting the existence of a smooth local *center manifold* $W^c_{\text{loc}}(\mathbf{x}_e)$, tangent at $\mathbf{x}_e$ to E^c, the eigenspace spanned by eigenvectors belonging to those eigenvalues having zero real parts. Solutions starting in $W^c_{\text{loc}}(\mathbf{x}_e)$ stay in it as long as they remain in a neighborhood of $\mathbf{x}_e$, so this manifold is locally invariant.

Figure 6.7 provides a schematic picture of the phase space near a degenerate equilibrium with s negative, u positive, and c zero (real part) eigenvalues. As the picture suggests,

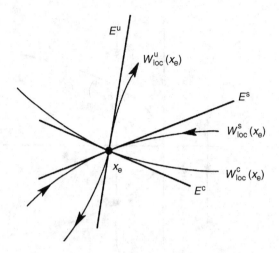

Figure 6.7 The stable, center, and unstable manifolds.

the "nonlinear coordinates" implicit in the definitions of W_{loc}^s, W_{loc}^u, and W_{loc}^c allow us to separate stable, unstable and non-hyperbolic or bifurcation behaviors and so to *reduce* our analysis to that of a c-dimensional system restricted to the center manifold.

To describe the reduction process we assume that coordinates have been chosen with the degenerate equilibrium $\mathbf{x}_e$ at the origin and such that the matrix $\mathbf{Df}(\mathbf{x}_e)$ is block diagonalized. Thus (6.22) can be written in the form:

$$\left.\begin{aligned}\dot{\mathbf{x}} &= \mathbf{Ax} + \mathbf{f}(\mathbf{x}, \mathbf{y}), \\ \dot{\mathbf{y}} &= \mathbf{By} + \mathbf{g}(\mathbf{x}, \mathbf{y}),\end{aligned}\right\} \tag{6.25}$$

where $\mathbf{x}$ belongs to the center subspace E^c and $\mathbf{y}$ belongs to the stable and unstable subspaces. We drop the explicit reference to parameter dependence. Here the $c \times c$ matrix $\mathbf{A}$ has all its eigenvalues with zero real part and the $(n - c) \times (n - c) = (s + u) \times (s + u)$ matrix $\mathbf{B}$ has only eigenvalues with non-zero real parts. The center manifold can now be expressed as a graph $\mathbf{h}$ over E^c:

$$\mathbf{y} = \mathbf{h}(\mathbf{x}). \tag{6.26}$$

This implies that, as long as solutions remain on the center manifold, the full state $(\mathbf{x}, \mathbf{y})$ of the system can be specified, via (6.26), by the state of the $\mathbf{x}$ variables alone. The *reduced system* is then defined as

$$\dot{\mathbf{x}} = \mathbf{Ax} + \mathbf{f}(\mathbf{x}, \mathbf{h}(\mathbf{x})) \tag{6.27}$$

and the stability and bifurcation behavior near the degenerate equilibrium can be deduced from it (Carr [71], Guckenheimer and Holmes [144]).

As an example, consider the two-dimensional system

$$\left.\begin{aligned}\dot{x}_1 &= \mu x_1 - x_1 x_2, \\ \dot{x}_2 &= -x_2 + x_1^2,\end{aligned}\right\} \tag{6.28}$$

whose linear part is already diagonalized. At $\mu = 0$ the eigenvalues are 0 and -1 respectively and $E^c = \text{span}\{(1, 0)\}$, $E^s = \text{span}\{(0, 1)\}$, so we seek the center manifold as a function $x_2 = h(x_1)$. This function will be a particular solution of the ODE

$$\frac{dx_2}{dx_1} = \frac{-x_2 + x_1^2}{-x_1 x_2} = \frac{1}{x_1} - \frac{x_1}{x_2}, \tag{6.29}$$

which cannot be solved explicitly. Here and in general we must resort to approximating the graph $h(x_1)$ as a Taylor series, which, in view of the local nature of our study, will be adequate provided that certain leading terms do not vanish.

Returning to the general problem (6.25), we substitute $\mathbf{y} = \mathbf{h}(\mathbf{x})$ into the second component and use the chain rule and the first component to obtain a partial differential equation for the function $\mathbf{h}(\mathbf{x})$:

$$\mathbf{Dh}(\mathbf{x})[\mathbf{Ax} + \mathbf{f}(\mathbf{x}, \mathbf{h}(\mathbf{x}))] = \mathbf{Bh}(\mathbf{x}) + \mathbf{g}(\mathbf{x}, \mathbf{h}(\mathbf{x})) \tag{6.30}$$

with the "boundary conditions"

$$\mathbf{h}(\mathbf{0}) = \mathbf{0}, \quad \mathbf{Dh}(\mathbf{0}) = \mathbf{0}, \tag{6.31}$$

which result from the tangency of W_{loc}^c to E^c at $\mathbf{x}_e = \mathbf{0}$. We shall seek an approximate solution of (6.30) and (6.31) in the form of a Taylor series. For our example (6.28) at $\mu = 0$, therefore, we set

$$x_2 = h(x_1) = a_2 x_1^2 + a_3 x_1^3 + \mathcal{O}(|x_1|^4)$$
$$\implies Dh(x_1) = 2a_2 x_1 + 3a_3 x_1^2 + \mathcal{O}(|x_1|^3); \tag{6.32}$$

due to the boundary conditions, the Taylor series necessarily starts at second order. Equation (6.30) becomes, in this case,

$$[2a_2 x_1 + 3a_3 x_1^2 + \mathcal{O}(|x_1|^3)]\{-x_1[a_2 x_1^2 + a_3 x_1^3 + \mathcal{O}(|x_1|^4)]\}$$
$$= -[a_2 x_1^2 + a_3 x_1^3 + \mathcal{O}(|x_1|^4)] + x_1^2. \tag{6.33}$$

Equating terms of comparable orders in $|x_1|$ yields:

$$\left.\begin{array}{ll} \mathcal{O}(|x_1|^2): & 0 = -a_2 + 1 \\ \mathcal{O}(|x_1|^3): & 0 = a_3 \end{array}\right\} \tag{6.34}$$

and we conclude that the function h may be written

$$x_2 = h(x_1) = x_1^2 + \mathcal{O}(|x_1|^4). \tag{6.35}$$

At $\mu = 0$ the reduced equation is therefore

$$\dot{x}_1 = -x_1[x_1^2 + \mathcal{O}(|x_1|^4)] = -x_1^3 + \mathcal{O}(|x_1|^5), \tag{6.36}$$

the behavior of which is dominated, for small $|x_1|$, by the negative cubic term which pushes solutions towards $x_1 = 0$. The degenerate equilibrium is evidently stable at the bifurcation point, as well as for all $\mu < 0$. This is not directly obvious from (6.28): the linearization

$\dot{x}_1 = \mu x_1$ of the first component tells us nothing at $\mu = 0$ and the reader can check, repeating the calculations above, that a change in sign in the term x_1^2 in the second component turns stability on its head!

Calculations in higher-dimensional cases employ vector-valued Taylor series and are more awkward, but they proceed in essentially the same manner. Computer algebra can be very helpful in managing such cases; see Rand and Armbruster [294], for example.

To include parameter variation and obtain a family of reduced systems valid *near* as well as *at* the bifurcation value μ_0, we augment (6.25) with the (trivial) equation $\dot{\mu} = \mathbf{0}$:

$$
\left.
\begin{aligned}
\dot{\mathbf{x}} &= \mathbf{A}(\mu)\mathbf{x} + \mathbf{f}(\mathbf{x}, \mathbf{y}; \mu), \\
\dot{\mathbf{y}} &= \mathbf{B}(\mu)\mathbf{y} + \mathbf{g}(\mathbf{x}, \mathbf{y}; \mu), \\
\dot{\mu} &= \mathbf{0},
\end{aligned}
\right\}
\tag{6.37}
$$

and consider an augmented $(c + m)$-dimensional center manifold $\mathbf{y} = \mathbf{h}(\mathbf{x}; \mu)$, tangent to the subspace spanned by $\mathbf{y}$ and μ. In our example we set

$$
x_2 = h(x_1; \mu) = a_2 x_1^2 + b_2 x_1 \mu + c_2 \mu^2 + \mathcal{O}(3)
$$

$$
\implies \quad \frac{\partial h}{\partial x_1} = 2a_2 x_1 + b_2 \mu + \mathcal{O}(2)
$$

$$
\frac{\partial h}{\partial \mu} = b_2 x_1 + 2c_2 \mu + \mathcal{O}(2)
\tag{6.38}
$$

and (6.30) becomes

$$
\begin{aligned}
\big[2a_2 x_1 + b_2 \mu + \mathcal{O}(2) \quad & b_2 x_1 + 2c_2 \mu + \mathcal{O}(2) \big] \\
& \cdot \begin{pmatrix} \mu x_1 - x_1 [a_2 x_1^2 + b_2 x_1 \mu + c_2 \mu^2 + \mathcal{O}(3)] \\ 0 \end{pmatrix} \\
& = -[a_2 x_1^2 + b_2 x_1 \mu + c_2 \mu^2 + \mathcal{O}(3)] + x_1^2,
\end{aligned}
\tag{6.39}
$$

which is still solved by

$$
h(x_1, \mu) = x_1^2 + \mathcal{O}(3)
\tag{6.40}
$$

(parameter dependence in h enters only at higher order). The reduced family is then given by

$$
\dot{x}_1 = \mu x_1 - x_1^3 + \mathcal{O}(4).
\tag{6.41}
$$

In the above, $\mathcal{O}(n)$ denotes terms of order n jointly in x_1 and μ. The bifurcation diagram for the reduced system is given in Figure 6.8. It is a *pitchfork bifurcation* and representative phase portraits of the full system for $\mu < 0$, $\mu = 0$ and $\mu > 0$ appear in Figure 6.9.

In the general case, the reduced family is given by

$$
\dot{\mathbf{x}} = \mathbf{A}(\mu)\mathbf{x} + \mathbf{f}(\mathbf{x}, \mathbf{h}(\mathbf{x}; \mu); \mu).
\tag{6.42}
$$

Several more examples of this procedure are given by Guckenheimer and Holmes ([144], Section 3.2), and we use it again in a higher-dimensional context in Chapter 8. As we have noted, computer algebra systems such as MACSYMA or MAPLE can be very useful

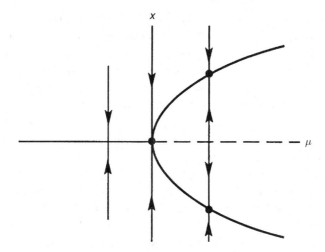

Figure 6.8 The bifurcation diagram for Equation (6.28); —— represents a branch of sinks, - - - - a branch of saddle points.

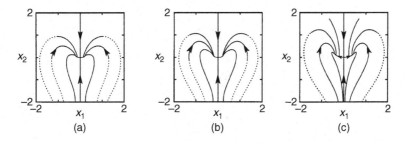

Figure 6.9 Phase portraits of the two-dimensional system (6.28): (a) $\mu = -0.2$, (b) $\mu = 0$, (c) $\mu = 0.2$.

here, especially if the original phase space dimension n is large, because $\mathbf{h}(\mathbf{x}; \boldsymbol{\mu})$ is an m-parameter family of $(n - c)$ functions each of c variables, see Rand [293] and Rand and Armbruster [294].

Invariant manifolds and their close relatives inertial manifolds (see Temam [367], and Chapter 13 below) can be seen as geometric constructions which provide a rigorous basis for the idea of eddy viscosity or the modeling of energy losses to neglected modes. The latter idea was originally suggested for homogeneous turbulence by Heisenberg; see Tennekes and Lumley [368] and Batchelor [32]. Returning to Equation (6.25), one interprets $\mathbf{x}$ as the resolved modes and $\mathbf{y}$ as the neglected modes. Thus the $\mathbf{x}$-subspace contains all the unstable and neutral or critical modes and some weakly stable modes, while the complementary $\mathbf{y}$-subspace contains strongly stable (dissipative) modes. The function $\mathbf{y} = \mathbf{h}(\mathbf{x})$, determined by (6.30), then accounts for the effects of nonlinear coupling due to the functions $\mathbf{f}$ and $\mathbf{g}$ of (6.25): in fact it takes care of energy transfer in *both* directions.

However, as we mentioned in the introductory chapter, the assumption of an invariant manifold that can be expressed as a (single-valued) function implies that we have resolved *all* dynamically relevant spatial scales and neglected only those modes that are slaved to them. In other words, the state $(\mathbf{x}, \mathbf{y} = \mathbf{h}(\mathbf{x}))$ of the full system is supposed to be completely determined by that of the $\mathbf{x}$-subsystem alone. As pointed out in Section 4.4, this implies that the dimension of $\mathbf{x}$ is likely to be very high for truly turbulent processes and, as we briefly describe in Chapter 13, a probabilistic interpretation is likely to be necessary if invariant manifold theory is to play a central rôle in derivation of low-dimensional models. In the following we use it primarily as a tool in the analysis of low-dimensional models derived by other methods, but in Chapter 8 it is used directly to obtain such models from a "simple" partial differential equation.

Having reduced the dimension of our problem, we next attempt to simplify it by coordinate changes. Much as similarity transformations can remove off-diagonal terms in linear ODEs, suitable *nonlinear* coordinate changes can remove many (but usually not all) higher order terms in nonlinear equations. We include the following description for completeness, but we do not explicitly use it below, although we shall rely on its implications. The reader may prefer to skim and pick up the story on the following page.

We write (6.42) or (6.27) as a truncated Taylor series, recalling that we only know $\mathbf{h}(\mathbf{x}; \boldsymbol{\mu})$ approximately in any case. Suppressing the parameter $\boldsymbol{\mu}$, we have:

$$\dot{\mathbf{x}} = \mathbf{A}\mathbf{x} + \sum_{k=2}^{N} \mathbf{f}^k(\mathbf{x}) \tag{6.43}$$

and we seek a near-identity coordinate change of the form

$$\mathbf{x} = \mathbf{z} + \sum_{k=2}^{N} \mathbf{j}^k(\mathbf{z}), \tag{6.44}$$

where $\mathbf{f}^k$ and $\mathbf{j}^k$ are homogeneous polynomials of order k. Substitution of (6.44) into (6.43) and use of the chain rule yields

$$\dot{\mathbf{z}} = \left[\mathbf{I} + \sum_{k=2}^{N} \mathbf{D}\mathbf{j}^k(\mathbf{z}) \right]^{-1} \left[\mathbf{A} \left(\mathbf{z} + \sum_{k=2}^{N} \mathbf{j}^k(\mathbf{z}) \right) + \mathbf{f}^k \left(\mathbf{z} + \sum_{k=2}^{N} \mathbf{j}^k(\mathbf{z}) \right) \right],$$

or, at order $|\mathbf{z}|^2$:

$$\dot{\mathbf{z}} = \mathbf{A}\mathbf{z} + \mathbf{A}\mathbf{j}^2(\mathbf{z}) - \mathbf{D}\mathbf{j}^2(\mathbf{z})\mathbf{A}\mathbf{z} + \mathbf{f}^2(\mathbf{z}) + \mathcal{O}(|\mathbf{z}|^3). \tag{6.45}$$

So if $\mathbf{j}^2$ can be chosen so that the commutator or *Lie bracket* $\mathbf{A}\mathbf{j}^2(\mathbf{z}) - \mathbf{D}\mathbf{j}^2(\mathbf{z})\mathbf{A}\mathbf{z} = [\mathbf{A}\mathbf{z}, \mathbf{j}^2(\mathbf{z})]$ cancels every term in $\mathbf{f}^2$, we have transformed all quadratic terms in (6.44) into cubic and higher order terms. Usually we cannot do quite this well, but the idea is to choose $\mathbf{j}^2$ to remove as many "non-resonant" terms as possible and then proceed to the next order, choosing $\mathbf{j}^3$ appropriately. The terms which cannot be removed at each step are determined by the (degenerate) matrix $\mathbf{A}(\boldsymbol{\mu}_0)$. There are several more or less elegant ways of keeping track of the process: Guckenheimer and Holmes ([144], Section 3.3) provide a relatively simple introduction; a deeper view can be found in [17] or [62]. As in so much else, the ideas go back to Poincaré [282].

The object resulting from this procedure is called a *normal form*. Obviously many different ODEs, subjected to suitable transformations, yield the same normal form and so the emphasis in dynamical systems theory is on classifying and describing the behaviors of all possible normal forms of dimensions 1, 2 . . ., subject to suitable non-degeneracy conditions such as non-vanishing of the leading nonlinear terms, etc. (Generic properties again!) For dimensions 1 and 2 and reasonably low orders of degeneracy this is well in hand, see [17] and [144].

The one-dimensional case is simple: the single eigenvalue passes through zero at the bifurcation point and the degenerate normal form is the scalar ODE

$$\dot{x} = a_2 x^2 + a_3 x^3 + \dots \tag{6.46}$$

Since the degenerate "matrix" is just the scalar 0, *no* terms can be removed by nonlinear transformation. If no special symmetries exist, we expect that the leading coefficient $a_2 \neq 0$ and so a further scale change and truncation yields the two cases

$$\dot{x} = \pm x^2. \tag{6.47}$$

In studying local bifurcations we need to know the kinds of behavior that occur for parameter values near the bifurcation point as well as at it. In cataloging the possible behaviors it is therefore necessary to unfold the degenerate system. By an *unfolding* of (6.47) we mean a parameterized family of ODEs containing (6.47) and also *all possible* nearby vector fields of equal or lower degeneracy, including, of course, the structurally stable ones. This usually means that we add only lower order terms, and then only the minimum number required. Here it is not hard to see that

$$\dot{x} = \mu \pm x^2 \tag{6.48}$$

provides such an unfolding, since the degenerate zero of the function x^2 can be removed entirely or split into two simple zeros by suitable variation of μ. This is the *saddle–node bifurcation*, which we have already met and illustrated in Figure 6.6.

If, due perhaps to some physical constraint, the branch of fixed points persists in a full neighborhood of the bifurcation point, then the appropriate unfolding is the *transcritical bifurcation*:

$$\dot{x} = \mu x \pm x^2. \tag{6.49}$$

In the case of reflection symmetry the vector field is given by an odd function and we have

$$\dot{x} = \mu x \pm x^3. \tag{6.50}$$

This is the *pitchfork bifurcation*, which we met earlier in Equation (6.41) and Figure 6.8, and which we meet again in our discussion of local bifurcation with symmetry in Section 7.2.

The analysis of fixed points and stability for the scalar ODEs (6.48)–(6.50) is easy. They can all be integrated explicitly, but pictorial *bifurcation diagrams* are of more interest to us. We have already shown the saddle-node and pitchfork cases in Figures 6.6 and 6.8, and Figure 6.10 shows the remaining case of Equation (6.49). In all three figures we have illustrated only the "−" case in "±."

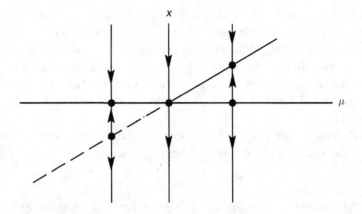

Figure 6.10 The transcritical bifurcation diagram: "−" case.

For these one-dimensional cases it is easy to see that addition of higher order terms will not qualitatively change the local behavior near the bifurcation point $x = 0$, $\mu = 0$; for example, any additional equilibria will appear far from $x = 0$. These pictures faithfully capture all possible new solution types that can bifurcate from $x = 0$ as μ varies near 0.

The simplest two-dimensional case is the *Hopf bifurcation*, first described by Poincaré and Andronov (see Andronov *et al.* [8]). Here a pair of complex conjugate eigenvalues crosses the imaginary axis with non-zero imaginary parts. The normal form can be written

$$\begin{bmatrix} \dot{x}_1 \\ \dot{x}_1 \end{bmatrix} = \begin{bmatrix} 0 & -\omega \\ \omega & 0 \end{bmatrix} \begin{bmatrix} x_1 \\ x_2 \end{bmatrix} + a_3 r^2 \begin{bmatrix} x_1 \\ x_2 \end{bmatrix} + b_3 r^2 \begin{bmatrix} -x_2 \\ x_1 \end{bmatrix} + \mathcal{O}(|\mathbf{x}|^5), \qquad (6.51)$$

where $r^2 = x_1^2 + x_2^2$ is a radial coordinate. Note that the normal form transformation has eliminated all six quadratic terms (x_1^2, $x_1 x_2$, x_2^2 in each row) and six of the eight possible cubic terms. Explicit computation of the transformation yields formulae, in terms of the original coefficients, for the new coefficients a_3 and b_3. The former of these determines the direction of bifurcation and the stability of the resulting periodic orbits, as is most easily seen by passing to the unfolding in polar coordinates:

$$\left.\begin{aligned} \dot{r} &= \mu r + a_3 r^3 + \mathcal{O}(|r|^5), \\ \dot{\theta} &= \omega + b_3 r^2 + \mathcal{O}(|r|^4). \end{aligned}\right\} \qquad (6.52)$$

To any finite order in r, the angular component decouples and the analysis essentially collapses to that of the one-dimensional pitchfork (6.50) above. A fixed point $r = \sqrt{-\mu/a_3}$ exists when $\mu/a_3 < 0$, corresponding to a stable limit cycle if $a_3 < 0$ and an unstable one if $a_3 > 0$, see Figure 6.11. Here, since the reduced phase space is two-dimensional, one plots $|\mathbf{x}|$ versus μ.

The Hopf bifurcation provides a simple way of showing analytically that a given non-linear system has periodic orbits. It is very rare that one can find such orbits explicitly, but the Hopf theorem yields them via local analysis, see [144] for more details.

Bifurcations of periodic orbits themselves can be analyzed via the associated Poincaré maps. This requires a version of normal form theory for maps, which would take us too far

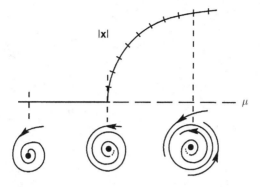

Figure 6.11 The Hopf bifurcation for the case $a_3 < 0$; —— indicates a stable equilibrium, - - - - an unstable equilibrium and +-+-+- a stable periodic orbit (limit cycle). Phase portraits for $\mu < 0$, $\mu = 0$, and $\mu > 0$ are shown below the bifurcation diagram.

afield, so we shall merely give a cursory description. At a saddle–node bifurcation the map has an isolated fixed point **p** and the linearization **DP(p)** has a single simple eigenvalue $\lambda = 1$. The unfolding is analogous to that for an ODE with a non-hyperbolic fixed point with eigenvalue $\lambda = 0$, and as in the bifurcation diagram of Figure 6.6, one finds no equilibria on one "side" of the bifurcation value and two on the other. These fixed points of **P** correspond to periodic orbits of the underlying flow that coalesce into a non-hyperbolic orbit and disappear at the bifurcation point. A simple eigenvalue of a map can also exit the unit circle via $\lambda = -1$, which corresponds to the *period-doubling* or *flip bifurcation* in which the original orbit changes stability type but persists while an orbit of twice its period appears. Finally, a complex conjugate pair of eigenvalues can exit the unit circle at points $\mu_0 = \cos\theta \pm i\sin\theta$, for any $\theta \in (0, \pi)$. Apart from some degenerate resonant values ($\theta = \pi/3$, $\pi/4$), this leads to the creation of an invariant torus which surrounds the original periodic orbit and typically carries quasiperiodic solutions. This is the *Hopf* or *Neimark–Sacker bifurcation* for a map. We do not give specific normal forms for these three cases, but merely indicate the behavior in the bifurcation diagrams of Figure 6.12; more details can be found in [144]. It is clear that the last two cases described above can occur only in ODEs of dimension ≥ 3.

The discussion above is far from a complete treatment of local bifurcations, but we have included all the *codimension-one bifurcations*: roughly speaking, these are the ones whose unfoldings require addition of lower order terms depending upon a single parameter. They are those which generically occur in one-parameter families of ODEs. The classification of such bifurcations is complete and more detail can be found in the texts cited earlier.

Codimension-two bifurcations of fixed points and periodic orbits are almost completely classified, although some technical problems remain. Here the behaviors are much richer and the unfoldings reveal *global* bifurcations in embryo. These involve the "large-scale" rearrangement in state space of solutions such as saddle separatrices. One of the most interesting (and common) is the creation of a saddle loop or *homoclinic orbit*: an orbit which is both forward and backward asymptotic to the same fixed point. Figure 6.13 shows

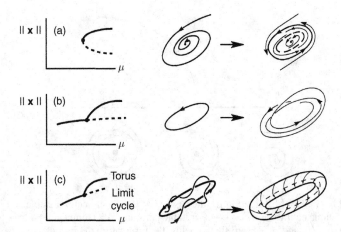

Figure 6.12 (a) The saddle-node, (b) period-doubling, and (c) Hopf bifurcations for a periodic orbit.

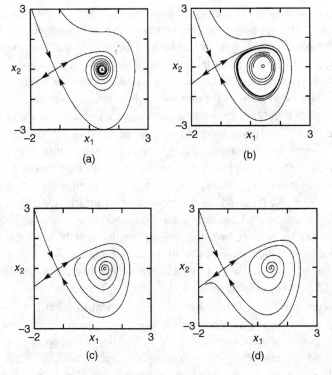

Figure 6.13 A global bifurcation: the saddle-loop or homoclinic orbit for Equation (6.53).
(a) $\mu = -1.1$, (b) $\mu = -0.9$, (c) $\mu = -0.7435$, (d) $\mu = -0.65$.

an example in which an attracting periodic orbit grows and forms the "infinite period" loop before vanishing as a parameter is varied. The specific system is

$$\left. \begin{aligned} \dot{x}_1 &= x_2, \\ \dot{x}_2 &= 1 - x_1^2 + \mu x_2 + x_1 x_2, \end{aligned} \right\} \tag{6.53}$$

which has two fixed points $(x_1, x_2) = (1, 0)$ and $(-1, 0)$. The latter is always a saddle point, but the former is a sink for $\mu < -1$ and a source for $\mu > -1$; it undergoes a Hopf bifurcation at $\mu = -1$ in which an attracting periodic orbit appears. This orbit grows with increasing μ to form the homoclinic loop at $\mu \approx -5/7$. The existence of the homoclinic loop cannot be found by local (Taylor series) methods, as in the case of local bifurcations, but certain classes of system are amenable to perturbation analyses such as Melnikov's method, see Guckenheimer and Holmes ([144], Sections 4.5 and 7.3ff.). It is such an analysis that yields the estimate $\mu \approx -5/7 = -0.7143\ldots$ for the homoclinic bifurcation point. The numerical simulations for several values of μ illustrated in Figure 6.13 suggest the value ≈ -0.7435.

More generally, if $\mathbf{p}_1$ and $\mathbf{p}_2$ are (distinct) fixed points and there is an orbit $\mathbf{q}(t)$ which satisfies

$$\lim_{t \to -\infty} \mathbf{q}(t) = \mathbf{p}_1, \qquad \lim_{t \to +\infty} \mathbf{q}(t) = \mathbf{p}_2, \qquad (6.54)$$

we call $\mathbf{q}$ *heteroclinic* from $\mathbf{p}_1$ to $\mathbf{p}_2$. If $\mathbf{p}_1 = \mathbf{p}_2 = \mathbf{p}$ then $\mathbf{q}$ is said to be *homoclinic* to $\mathbf{p}$. In terms of the stable and unstable manifold notation, in the first case we have $\mathbf{q}(t) \in W^u(\mathbf{p}_1) \cap W^s(\mathbf{p}_2)$ and in the second $\mathbf{q}(t) \in W^u(\mathbf{p}) \cap W^s(\mathbf{p})$. The dimensions of these manifolds must satisfy $\dim(W^u(\mathbf{p})) + \dim(W^s(\mathbf{p})) \le n$, where n is the dimension of the state space itself, and the two manifolds must intersect along the one-dimensional solution curve $\mathbf{q}(t)$ itself if a homoclinic orbit is to exist. This implies that a homoclinic orbit cannot be a structurally stable phenomenon. The case of the two one-dimensional manifolds (separatrices) of the planar saddle point in Figure 6.13 is a good example: any small perturbation, such as a change in the parameter μ, destroys the loop and the two branches of the manifolds "miss each other." But this example also shows that, just as non-hyperbolic equilibria can occur "stably" at a bifurcation point in a parameterized family, so can these homoclinic loops. And as we see in the next chapter, symmetries can utterly change this state of affairs.

6.5 Attractors simple and strange

As we have seen, dynamical systems theory addresses the behavior of all solutions of a given system or family of systems of ODEs. It is particularly concerned with long time behavior: questions such as: "Where do most solutions go in phase space and what do they do as time increases?" are paramount. Such questions lead naturally to the study of *attractors*. We work up to this notion in stages.

Let us recall the idea of an *invariant set* introduced in Section 6.1. A set Λ is invariant for the flow $\boldsymbol{\phi}_t$ generated by an ODE if, when the initial condition $\mathbf{x}(0)$ lies in Λ, then so does the solution $\mathbf{x}(t) = \boldsymbol{\phi}_t(\mathbf{x}(0))$ for all t. Once in, you are in for the long haul. Equipped with this idea, we can define an *attracting set* $\mathcal{A}$ as an invariant set which attracts all solutions starting in some (open) neighborhood of $\mathcal{A}$. We then have our definition:

- An *attractor* is an attracting set which contains a dense orbit.

An orbit is *dense* in a set Λ if it goes arbitrarily close to every point in Λ. The requirement of a dense orbit guarantees that almost all solutions in an attractor display the "typical" behavior of that attractor, they go everywhere in it instead of getting hung up in one corner. This implies ergodicity.

The classical attractors are stable fixed points (sinks) and periodic orbits (limit cycles), such as that in the van der Pol equation, Figure 6.2. The dynamical behaviors associated with these attractors are a little dull: either everything dies in a sink, or everything repeats *ad infinitum*. Somewhat more exotic, but still exhibiting fairly regular behavior, are $(d \geq 2)$-dimensional tori carrying quasiperiodic motions with incommensurate frequencies, see Figure 6.12(c) for an example. As we remarked in Chapter 1, quasiperiodic motions with increasingly many frequencies were supposed, in the theories of Landau and Hopf, to model the transition to turbulence. One of the major contributions of dynamical systems theory was the realization that other, more complex, attractors and attracting sets not only exist, but can also be structurally stable. These attractors are now called "strange," although they are far from unusual: they occur quite frequently. In fact Ruelle and Takens [322] argued that quasiperiodic motions of the types proposed by Landau and Hopf, having more than three independent frequencies, were structurally *un*stable and therefore probably did *not* appear in practice, while certain strange attractors, being structurally stable, might be more likely to appear in specific examples. The attractors considered by Ruelle and Takens were abstract mathematical constructions and it is still not clear how best to define strange attractors in the context of physical problems (see [144, 320]). The definitions we adopt here are therefore not universally held. In particular, some authors ([96, 367]) do not require a dense orbit; in our terminology, their attractors are merely attracting sets.

For us, a *strange attractor* is an attractor that contains a chaotic invariant set. In turn, a *chaotic invariant set* is an invariant set Λ having the property that almost all pairs of solutions $\mathbf{x}_a(t)$, $\mathbf{x}_b(t)$ started arbitrarily $\mathcal{O}(\epsilon)$ close in Λ, diverge by an $\mathcal{O}(1)$ amount after a finite time. If Λ is hyperbolic (a generalization of the notion of hyperbolicity for fixed points and periodic orbits), then divergence proceeds at an exponential rate, so that the estimate

$$|\mathbf{x}_a(t) - \mathbf{x}_b(t)| \approx c|\mathbf{x}_a(0) - \mathbf{x}_b(0)|e^{\lambda t} \tag{6.55}$$

holds for some constants $c > 0$ and $\lambda > 0$. Of course, in an attractor solutions are bounded and so this divergence is only local in time; in the future the solutions may find themselves close again, but they will separate again, and so on. The stretching and folding or cutting implicit in this is the qualitative signature of a strange attractor.

Since it also contains a dense orbit, local (exponential) separation implies that almost all orbits in a strange attractor, or asymptotic to one, display *sensitive dependence on initial conditions*. We illustrate this and the other rather abstract concepts above in a moment, but we first note that, while strange attracting sets can be shown to exist in many specific differential equations, it is usually much harder to show that we have a true attractor. Many examples, including that of Lorenz, are in addition structurally unstable, although this seems not to matter as much as one might fear.

How does one find attracting sets and attractors more complicated than simple sinks and limit cycles? One approach is to first find a *trapping region*. As in the two-dimensional example of the van der Pol equation in Section 6.2 (Figure 6.2), this is a closed, bounded region $\mathcal{B}$ in phase space into which all (forward) orbits are directed: a positively invariant domain. Once in, orbits are literally trapped and so they must have some sort of limiting behavior. One defines the attracting set $\mathcal{A}$ as the intersection of all forward images of $\mathcal{B}$ under the flow:

$$\mathcal{A} = \bigcap_{t \geq 0} \phi_t(\mathcal{B}).$$

It is to the Lorenz equation that we turn for our main example of a strange attractor and our outline of an important method – symbolic dynamics – by which chaotic solutions may be analyzed and understood. While Lorenz derived the following three-dimensional ODE as a drastic truncation of a Galerkin projection of the coupled Navier–Stokes and heat equations modeling the Rayleigh–Bénard convection problem, the behavior of interest to us occurs in a parameter range far outside that in which the truncation is justifiable, and we do not imply any fluid-mechanical relevance. (We observe, however, that the equation *does* provide a reasonable model of the thermosiphon: convection in a thin toroidal loop filled with fluid, oriented in a vertical plane and heated from below; see, e.g. [137, 387] and [383, 384].)

The ODE is [217]:

$$\begin{aligned} \dot{x}_1 &= \sigma(x_2 - x_1), \\ \dot{x}_2 &= \rho x_1 - x_2 - x_1 x_3, \\ \dot{x}_3 &= -\beta x_3 + x_1 x_2, \end{aligned} \right\} \tag{6.56}$$

where σ, β, $\rho > 0$ are parameters. The origin is always a fixed point and for $\rho > 1$ it is a saddle point with one positive and two negative (real) eigenvalues and there are two additional fixed points:

$$\mathbf{q}_\pm = (\pm\sqrt{\beta(\rho - 1)}, \pm\sqrt{\beta(\rho - 1)}, (\rho - 1)). \tag{6.57}$$

For $\sigma = 10$, $\rho = 28$, $\beta = 8/3$ (the values chosen by Lorenz), these are also saddle points, but with one negative eigenvalue and two complex conjugate eigenvalues with positive real parts. In addition, one can find a large trapping region, so that the equation evidently has an attracting set which contains all three unstable fixed points (see [96, 350]). This object appears to be a true attractor for σ and β as specified above and all ρ in a range starting a little below

$$\rho_H = \sigma(\sigma + \beta + 3)/(\sigma - \beta - 1) \approx 23.74,$$

at which parameter value Hopf bifurcations occur in which unstable, saddle type periodic orbits coalesce with $\mathbf{q}_\pm$, and extending to $\rho \approx 30$. The fixed points $\mathbf{q}_\pm$ are sinks for $\rho < \rho_H$.

Figure 6.14(a) shows a three-dimensional view of a single solution of (6.56), started near the saddle point at $(0, 0, 0)$. It winds back and forth about the other equilibria $\mathbf{q}_\pm$ in an irregular manner, approaching the attractor, which we now describe.

To be more precise, we describe a class of *geometric Lorenz attractors* [142, 147], which seem to share all the important features of the "real" Lorenz attractors but which have only

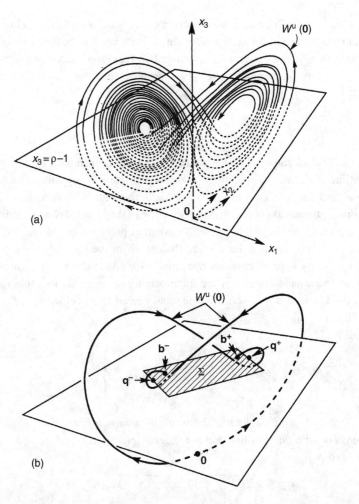

Figure 6.14 (a) A numerical solution of the Lorenz equation (6.56), after Lanford [205]
and Guckenheimer and Holmes [144]. (b) The cross section and Poincaré map. See text for
a description of Σ, etc.

been rigorously derived from specific Lorenz-like ODEs in certain limiting cases [306]
(but see [239]). We define a cross section Σ to the flow whose boundary contains the fixed
points $\mathbf{q}_\pm$ and their local stable manifolds. The section Σ is tangent to (but not identical
to) the plane $z = \rho - 1$, see Figure 6.14(b). Numerical simulation reveals that solutions
asymptotic to the attractor repeatedly pierce Σ as they circulate, thus inducing a two-
dimensional Poincaré or return map $\mathbf{F}$ on Σ. Actually, $\mathbf{F}$ is not defined everywhere on Σ;
implicit in our definition is the fact that points on the boundary of Σ belonging to $W^s(\mathbf{q}_\pm)$
flow into $\mathbf{q}_\pm$ and $W^s(\mathbf{0})$ also intersects the interior of Σ in a curve containing the point
$(0, 0, \rho - 1)$. Solutions starting on this curve flow into $\mathbf{0}$ and so never return to Σ.

 We next assume that (nonlinear) coordinates $(u, v) \in \Sigma$ can be chosen so that the return
map $\mathbf{F}$ may be written in the form

$$\mathbf{F}(u, v) = (f(u), g(u, v)), \tag{6.58}$$

and that the boundaries of Σ are given by $|u| = 1$, $|v| = 1$. More precisely, $\mathbf{F}$ is defined on the open rectangles $\Sigma^- = (-1, 0) \times (-1, 1)$ and $\Sigma^+ = (0, 1) \times (-1, 1)$; points of the form $(0, v) \in \Sigma$ flow into $\mathbf{0}$. We also assume that $\mathbf{F}(-u, -v) = -\mathbf{F}(u, v)$, in concert with the symmetry $(x_1, x_2, x_3) \to (-x_1, -x_2, x_3)$ of (6.56). Regarding the individual components f and g we assume:

- $f(\pm 1) = \pm 1, 0 < \lim_{u \to 0} |f(u)| < 1, df/du(u) > \sqrt{2}$ for $u \neq 0$, and $df/du(u) \to \infty$ as $u \to 0$;
- $g(\pm 1, \mp 1/2) = 0, 0 < \partial g/\partial v < c < 1$ for $u \neq 0$, and $\partial g/\partial v \to 0$ as $u \to 0$.

Many of these assumptions derive directly from the behavior of solutions passing near the saddle point $\mathbf{0}$, which has eigenvalues satisfying $0 < -\lambda_1 < \lambda_2 < -\lambda_3$, implying a strong contracting direction "transverse" to the attractor, with a moderate expanding direction and a weak contracting direction "in" the attractor. A careful study of the linearized system

$$\left. \begin{array}{l} \dot{z} = \lambda_1 z, \\ \dot{u} = \lambda_2 u, \\ \dot{v} = \lambda_3 v, \end{array} \right\} \tag{6.59}$$

which uses the new coordinates u, v aligned with the eigenvectors belonging to λ_2 and λ_3, shows that solutions passing arbitrarily close to $\mathbf{0}$, and so starting on Σ near $u = 0$, are pinched together in the v-direction and next intersect Σ near the points where $W^u(\mathbf{q}_+)$ or $W^u(\mathbf{q}_-)$ first intersect Σ. The contraction properties of $g(u, v)$ in the v-direction follow from this. In contrast, the expansive property of $f(u)$ derives from the unstable spiraling behavior near the saddles $\mathbf{q}_\pm$ and the (moderate) expansion due to λ_2 near $\mathbf{0}$. The key assumption that does not easily follow is that an invariant *stable foliation* exists: a continuous family of curves $\mathcal{F}$ (here, the vertical line segments $u = $ const.) such that if $\mathcal{C} \in \mathcal{F}$ then $\mathcal{F}(\mathcal{C}) \in \mathcal{F}$. This enables us to write the map (6.58) so that the u component decouples. Our assumptions imply that $\mathbf{F}$ acts on Σ as pictured in Figure 6.15.

The image $\mathbf{F}(\Sigma^-)$ of the left-hand rectangle $(-1, 0) \times (-1, 1)$ is a strip of length $\geq \sqrt{2}$ extending entirely across Σ^- into Σ^+, pinched to a point $\lim_{u \to 0^-} \mathbf{F}(u, v) \overset{\text{def}}{=} (r^+, s^+)$

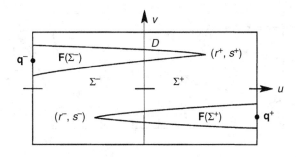

Figure 6.15 The cross section $\Sigma^\pm$ and the images $\mathbf{F}(\Sigma^\pm)$. After [144].

at its right-hand end and of vertical width everywhere less than $2c$. Similarly $\mathbf{F}(\Sigma^+)$ is pinched at its left-hand end $(r^-, s^-) = (-r^+, -s^+)$. (We observe that the points $(r^\pm, s^\pm)$ belong to $W^u(0)$: this will be useful later.) The two fixed points $\mathbf{q}_\pm$ lie in the left- and right-hand boundaries $u = -1, +1$ of Σ^- and Σ^+ respectively. Note that the images $\mathbf{F}(\Sigma^-)$ and $\mathbf{F}(\Sigma^+)$ cannot intersect, for that would violate uniqueness of solutions of the ODE (running backwards). Thus $\mathbf{F}(\Sigma)$ necessarily has a smaller area than Σ. Successive iteration further decreases area and the attractor (of the Poincaré map), which may loosely be defined as the intersection of all forward images:

$$\mathcal{A} = \bigcap_{n \geq 0} \mathbf{F}^n(\Sigma),$$

has zero area. It is, however, a very complicated set, as readers will appreciate if they continue the iteration process begun in Figure 6.15. At the second step, one gets four thinner strips, two each inside $\mathbf{F}(\Sigma^-)$ and $\mathbf{F}(\Sigma^+)$, then eight strips, and so on. Rather than doing this, we make use of the fact, already noted, that the expansive u dynamics can be uncoupled and study the one-dimensional map f defined on the interval $[-1, 1]$. This map is sketched in Figure 6.16(b).

We first observe that, apart from the unstable fixed points $u = \pm 1$, all orbits of f eventually enter and remain in the subinterval $[r^-, r^+]$, to which corresponds a subrectangle $V = [r^-, r^+] \times [-1, 1] \subset \Sigma$ for the two-dimensional map $\mathbf{F}$. We show V in Figure 6.16(a). Thus we clearly have an attracting set; moreover, almost all orbits in it are chaotic, since, unless their u coordinates coincide, the expansive nature of the map $(f' > \sqrt{2})$ implies that (6.55) is satisfied for the flow with $\lambda = \sqrt{2}$. It takes a little more work (done in Guckenheimer and Holmes [144], Section 5.7) to show that there is a dense orbit and hence that $\mathcal{A}$ is really an attractor. (Here the lower bound of $\sqrt{2}$ on stretching is important; for chaos alone, any constant bigger than 1 would do.)

The essence of the dynamical behavior is captured by f, but to fully explore even this (deceptively) simple one-dimensional map would take us into the intricacies of *kneading theory* (see [82] and [144]). To avoid this, we instead give a complete and elementary analysis of a somewhat similar piecewise linear map defined on the interval $[0, 1]$ by

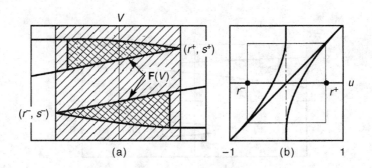

Figure 6.16 More on the maps $\mathbf{F}$ and f. (a) $\mathbf{F}(V) \subset V$; (b) the graph of $f(u)$.

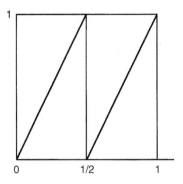

Figure 6.17 The linear doubling map h.

$$h(x) = \begin{cases} 2x & \text{if } 0 \le x < \frac{1}{2} \\ 2x - 1 & \text{if } \frac{1}{2} < x \le 1 \end{cases} \tag{6.60}$$

and illustrated in Figure 6.17. An orbit of h is the sequence $\{x_n\}_{n=0}^{\infty}$ obtained by successively doubling the initial value x_0 and subtracting the integer part at each step; for example:

$$0.3653 \mapsto 0.7306 \mapsto 1.4612 = 0.4612 \mapsto 0.9224$$
$$\mapsto 1.8448 = 0.8448 \mapsto \cdots$$

To understand the sensitive dependence on initial conditions and its consequences, it is best to represent the numbers in [0, 1] (the phase space of this little dynamical system) in binary form. This is the idea of *symbolic dynamics*. We let

$$x_0 = \frac{a_1}{2} + \frac{a_2}{2^2} + \frac{a_3}{2^3} + \cdots + \frac{a_k}{2^k} + \cdots, \tag{6.61}$$

where each coefficient a_j takes the value 0 or 1. It is then easy to see that

$$x_1 = h(x_0) = 2 \left(\sum_{j=1}^{\infty} \frac{a_j}{2^j} \right) = a_1 + \frac{a_2}{2} + \frac{a_3}{2^2} + \frac{a_4}{2^3} + \cdots$$

Actually, this is incorrect: we must remove the integer part and so, regardless of whether $a_1 = 0$ or $a_1 = 1$, we have

$$x_1 = \sum_{j=2}^{\infty} \frac{a_j}{2^{j-1}},$$

and in general

$$x_k = h^k(x_0) = \sum_{j=k+1}^{\infty} \frac{a_j}{2^{j-k}}. \tag{6.62}$$

Thus, applying h is equivalent to shifting the "binary point" and dropping the leading coefficient in the binary representation:

$$(a_1 a_2 a_3 a_4 \ldots) \mapsto (a_2 a_3 a_4 \ldots), \tag{6.63}$$

much as multiplication by 10 shifts the more familiar decimal point. In this operation the leading symbol is removed and *information is lost*. If we knew x_0 to infinite accuracy (so we had all the a_ks for k between 1 and infinity), this would not matter, for at each step we would still know the current state x_k exactly. But alas, our minds and our computers are finite and so we can only store, say, the first N binary places $(a_1, a_2, \ldots, a_N)$. The result is that, after N iterations, we cannot even say whether the state x_{N+1} lies above or below $1/2$! Moreover, even if two initial conditions differ only at the Nth binary place, so that they lie within distance $1/2^{N-1}$ of one another, after N iterations they lie on opposite sides of $1/2$ and thereafter they behave essentially independently. Here is the sensitive dependence: the dynamics amplifies small errors.

Worse is yet to come. The binary number construction shows that to each infinite sequence of 0s and 1s there corresponds a number between zero and one, and vice versa. (The correspondence is not unique, but we can remedy that by identifying pairs of sequences whose heads coincide but whose tails are $1000 \ldots 0 \ldots$ and $0111 \ldots 1 \ldots$ respectively, so that, for example, $001111 \ldots$ and $01000 \ldots$ both represent $1/4$.) This implies that, given any random sequence (generated by tossing a coin and assigning heads = 0, tails = 1 for instance), there is an initial state x_0 such that the orbit $h^k(x_0)$ realizes that sequence. Hence our system has infinitely many such random or chaotic orbits. There are also infinitely many periodic orbits, corresponding to periodic sequences such as $001001001 \ldots$, but since the irrational numbers form a set of full measure, the generic behavior is chaotic rather than periodic.

One can also use the binary representation to show that a dense orbit exists. Consider the sequence $\mathbf{a}^*$ formed by concatenating all possible sequences of lengths $1, 2, 3, \ldots$ end to end: $\mathbf{a}^* = 0 \ 1 \ 00 \ 01 \ 10 \ 11 \ 000 \ 001 \ldots$. As one iterates and drops leading symbols, every possible sub-sequence of any given length appears at the head. This implies that the orbit of the point x^* corresponding to $\mathbf{a}^*$ contains points which approximate, to any desired accuracy, *every* point in the interval $[0, 1]$ and so $h^k(x^*)$ is our dense orbit.

The symbol sequences also allow one to enumerate the periodic orbits simply by listing all distinct (i.e. non shift-equivalent) sequences of lengths $1, 2, 3, \ldots$ which do not contain lower period sub-sequences, see Table 6.1. Asymptotically there are $\approx 2^N/N$ orbits of period N.

Here all the complication comes from the expansive properties of h, the doubling operation, coupled with the fact that orbits cannot escape the interval $[0, 1]$ but are reinjected due to the removal of the integer part at each step. Small errors are magnified: no matter how small ϵ, for some N, $2^N \epsilon$ is large. The Lorenz map f shares these features, although the expansion rate is not constant, being merely bounded below by $\sqrt{2}$, and more significantly, not all binary sequences correspond to orbits of f: in particular those starting with arbitrarily long strings of 0s or 1s do not occur. Examination of Figure 6.16(b) shows that an orbit starting at r^- (or r^+) takes only finitely many iterations to cross $u = 0$ into the other half of the interval $[r^-, r^+]$.

The representation and analysis of orbits of dynamical systems by symbol sequences is far more generally applicable than the happy coincidence of doubling and binary sequences might suggest. Given a (smooth) map $\mathbf{F}$ defined on (a subset of) $\mathbb{R}^n$, if one can find a *Markov partition* $\mathcal{R} \subset \mathbb{R}^n$ – a disjoint collection of "rectangles" R_j, $0 \leq j \leq M$, each of

Table 6.1 *Enumerating periodic orbits for h.*

Length	Sequences	Number of orbits
1	0, 1	2
2	01 (=10)	1
3	001, 011	2
4	0001, 0011, 0111	3
5	00001, 00011, 00101, 00111, etc.	6
6	... ...	9
.	... ...	
.	... ...	.
25	... ...	1 342 176
.	... ...	.

which is mapped over (some of) the others by **F** – and if one can show that the map **DF**(**x**) linearized at each point **x** ∈ $\mathcal{R}$ has eigenvalues bounded away from $|\lambda| = 1$ (uniform hyperbolicity), then there is a one to one correspondence between orbits of **F** and sequences of the symbols $j = 0, 1, \ldots, M$. For the one-dimensional map h above, the rectangles are just the two intervals $(0, \frac{1}{2})$ and $(\frac{1}{2}, 0)$. As in the Lorenz case, certain sequences might be inadmissible and rather than a *full shift on two symbols*, as in the doubling map h, one obtains a *subshift of finite type* or something similar. But the symbol sequence associated with a point x_0 still describes the *itinerary* of the orbit $\mathbf{F}^n(\mathbf{x}_0)$ as it moves about in phase space.

The piecewise linear map h of Figure 6.17 also serves to introduce another important characterization of dynamical systems: that of an *invariant measure*. We have seen via symbolic dynamics that the generic behavior of orbits of h is chaotic and essentially as random as tossing a fair coin. It is therefore natural to seek a statistical description of orbits. One can ask, for example, what is the probability of finding a point on a "typical" orbit at a particular location in the interval $[0, 1]$. The answer to this question is provided by the invariant measure associated with the dynamics.

We describe the situation for maps of the unit interval $I = [0, 1]$, but it readily generalizes to other phase spaces, and to flows associated with ODEs. Let $g : I \to I$ be a mapping and $\mu(x)$ denote a probability density on I, suitably normalized: in this case $\int_0^1 \mu(x)dx = 1$. The measure μ is *invariant* for g if, for every set $A \subset I$,

$$\mu(g^{-1}(A)) = \mu(A). \tag{6.64}$$

Note that the definition employs inverse images; this is to encompass non-invertible maps like f and h above. In such cases, the set $g^{-1}(A)$ must include all pre-images of A. For the map h, there are two pre-images for each set, one in each half of I, and since h doubles length, each pre-image has length precisely half that of A. For example, the subinterval $[\frac{5}{8}, \frac{7}{8}]$ has pre-image $[\frac{5}{16}, \frac{7}{16}] \cup [\frac{13}{16}, \frac{15}{16}]$, as the reader can check by appeal to (6.60), or geometrical construction. This shows that *Lebesgue measure*, here simply length itself, is invariant for h. (Poincaré maps associated with energy-conserving Hamiltonian systems also preserve Lebesgue measure: areas, volumes, or their higher-dimensional generalizations. The famous Poincaré recurrence theorem is based on this fact [16, 86].)

Invariance itself says nothing about chaos. The identity map, under which nothing moves, also preserves Lebesgue measure. Two further notions are important here: *ergodicity* and *mixing*. An invariant measure μ is *ergodic* for g if there are no sets B such that both B and its complement $I \setminus B$ are invariant under g and have positive measure. This rules out trivial behavior such as that of the identity map or simple periodic interchange of regions.

A good example of an ergodic *flow* is provided by the irrational flow on a torus such as typically appears near a Hopf bifurcation for a periodic orbit, like that shown in Figure 6.12(c). In this situation, all solution curves wind densely around the torus, so that no subset of positive measure remains isolated within the dynamics. However, this flow is not chaotic; neighboring solutions flow parallel to one another and do not separate at all, let alone exponentially (Equation (6.55)). Such sensitive behavior is associated with the notion of mixing.

A map $g : I \to I$ is *mixing* with respect to the invariant measure μ if, for any two sets $A_1, A_2 \in I$ of positive measure, the following holds:

$$\lim_{n \to \infty} \mu(g^{-n}(A_1) \cup A_2) = \mu(A_1) \cdot \mu(A_2). \tag{6.65}$$

This equation implies that, over the course of its motion under g, the set A_1 will intersect the (stationary) set A_2 by an amount proportional to the size of A_1. Colloquially: a set of positive measure mixes uniformly in phase space [86]. The doubling map h is a classic example of a mixing dynamical system.

In dissipative systems in which phase space volumes contract and solutions approach an attractor or attracting set, Lebesgue measure clearly cannot be preserved. In such cases one seeks an invariant measure (or measures) supported on the attracting set itself. While there is an extensive theory of such *Sinai–Ruelle–Bowen measures* (see [86, 320]), it would take us far beyond our more modest goals.

Ergodic theory is a substantial field with strong ties to dynamical systems theory, see [86] for a proper (and extensive) treatment. The ergodic theorem of Birkhoff, which states that time and phase space averages with respect to an invariant measure are equivalent, is perhaps the most important result in this field. It is essentially to this and similar results that we appeal in estimating ensemble averages via time averages in applying the POD of Chapter 3. In Chapter 13 we also use averages over suitable invariant measures to obtain results comparing the statistical behavior of low-dimensional models with the full evolution equations.

Before leaving the Lorenz example, we return to the issue of structural stability and point out that attractors of Lorenz type, as defined via $\mathbf{F}$ and its components f and g of (6.58), are not structurally stable. Consider the orbit of the point (r^+, s^+), or equivalently (r^-, s^-). There are basically two cases: either the orbit lands on $u = 0$ after a finite number (N, say) of iterations, or it does not. In the former case, since the points $(r^\pm, s^\pm)$ belong to $W^s(\mathbf{0})$, we have a pair of homoclinic orbits to $\mathbf{0}$. As in Figure 6.13, a small perturbation, such as a change in ρ, will destroy these N circuit loops. However, in contrast to the planar homoclinic bifurcation, in the present case we claim that the members of the Lorenz family of maps possessing homoclinic loops are dense, so that, as soon as a pair of loops is destroyed, it is replaced by another. To see this, consider pre-images under $\mathbf{F}$ of the set $D = \{(u, v) \in \Sigma | u = 0\}$ belonging to $W^s(\mathbf{0})$ (see Figure 6.15). The u coordinates of this set of vertical line segments can be deduced from inverse iteration of f. The first pre-images of 0 are two points, the second pre-images are four points, and in general the nth pre-images of 0 under f are a set of 2^n points. Moreover the density of these pre-images in $[-1, 1]$, which we do not prove here, implies that the pre-images of D form a dense set of vertical line segments in Σ. As we perturb $\mathbf{F}$, remaining in an $\epsilon - C^1$ neighborhood, the points $(r^{\pm}, s^{\pm})$ move by an $\mathcal{O}(\epsilon)$ amount, but no matter how small the change in r^+ and r^- it takes us across a pre-image of D and hence through a system with homoclinic orbits. Thus no system of Lorenz type is structurally stable.

What is happening here is that (unstable) periodic orbits are being destroyed and created as each of these homoclinic connections is made and broken. The systems on either side of such a bifurcation point have different periodic orbits and so cannot be topologically equivalent. But their attractors still exhibit similar behavior, as numerical simulations reveal, and so, while technically not structurally stable, these attractors are still rather "robust."

We have restricted our discussion in this and the following chapter to finite-dimensional dynamical systems, but many of the ideas, definitions, and results generalize to infinite-dimensional evolution equations, including the Navier–Stokes equation, see [84, 85, 96, 154, 367], for example.

7

Symmetry

Physical systems often exhibit symmetry: we have already remarked on the symmetries of spanwise translation and reflection in boundary layers and shear layers and of rotations in circular jets. One could cite many more such cases. Of course, symmetric systems do not always, or even typically, exhibit symmetric behavior, and the study of spontaneous symmetry breaking is an important field in physics. These physical phenomena have their analogs in dynamical systems and in particular in ODEs, as we describe in this chapter.

The theory of symmetric dynamical systems and their bifurcations relies heavily on group theory and especially the notions of invariant functions and equivariant vector fields. The major references are the two volumes by Golubitsky and Schaeffer [134] and Golubitsky *et al.* [136]. In this chapter, as in the last, we attempt to sketch relevant parts of the theory using simple examples and without undue reliance on abstract mathematical ideas.

We have already met symmetric ODEs, in our discussion of the Lorenz equation (6.56) for example. This equation remains unchanged under the transformation $(x_1, x_2, x_3) \mapsto (-x_1, -x_2, x_3)$, which corresponds to rotation of the phase space about the x_3-axis through an angle π. This implies that, when the equilibrium $\mathbf{0}$ changes stability type from sink to saddle as ρ passes through 1, the resulting bifurcation is a pitchfork (see Equation (6.50) and Figure 6.8). Moreover, the global behavior, determined in part by the structure of the unstable manifold $W^u(\mathbf{0})$ as we saw at the close of Chapter 6, is also influenced by symmetry in that the "left" and "right" branches of $W^u(\mathbf{0})$ are mapped into each other by the symmetry. Thus, if one branch makes a homoclinic connection, so does the other. More generally, the return map $\mathbf{F}(u, v)$ of (6.58) inherits this symmetry in that

$$\mathbf{F}(u, v) = -\mathbf{F}(-u, -v).$$

7.1 Equivariant vector fields

As in Chapter 6, we consider (parameterized families of) ordinary differential equations of the form:

$$\dot{\mathbf{x}} = \mathbf{f}_\mu(\mathbf{x}) \quad \text{or} \quad \dot{\mathbf{x}} = \mathbf{f}(\mathbf{x}, \mu). \tag{7.1}$$

Let Γ be a group of transformations on $\mathbb{R}^n$ which describes a symmetry class. Without loss of generality we can take Γ to be a subgroup of the group of $n \times n$ matrices.

Here the terms "group" and "subgroup" mean that the product of any two matrices in the (sub)group is another matrix in that same (sub)group. This is necessary so that one can apply successive transformations. Here are some examples:

The group of reflections about the x_2-axis in the plane $\mathbb{R}^2$ is generated by the matrices

$$\mathbf{I} = \begin{bmatrix} 1 & 0 \\ 0 & 1 \end{bmatrix} \quad \text{and} \quad \mathbf{R}_f = \begin{bmatrix} -1 & 0 \\ 0 & 1 \end{bmatrix}. \tag{7.2}$$

The group D_4 of reflections and rotations about the origin through $\pi/2$, the symmetry group of the square, is generated by the matrices

$$\mathbf{I}, \quad \mathbf{R}_f, \quad \text{and} \quad \mathbf{R}_{\pi/2} = \begin{bmatrix} 0 & -1 \\ 1 & 0 \end{bmatrix}. \tag{7.3}$$

In saying "generated by" we mean that *any* element of the group can be obtained as a finite composition of the generating elements. In the second case, for example, reflection about the x_1-axis is given by:

$$\begin{bmatrix} 1 & 0 \\ 0 & -1 \end{bmatrix} = \mathbf{R}_{\pi/2}^3 \mathbf{R}_f \mathbf{R}_{\pi/2}. \tag{7.4}$$

The above are examples of *discrete* or *finitely generated* symmetry groups. In contrast, the *continuous* group $O(2)$ of reflections and infinitesimal counter-clockwise rotations about the origin in $\mathbb{R}^2$ is generated by

$$\mathbf{R}_f = \begin{bmatrix} -1 & 0 \\ 0 & 1 \end{bmatrix} \quad \text{and} \quad \mathbf{R}_\phi = \begin{bmatrix} \cos\phi & -\sin\phi \\ \sin\phi & \cos\phi \end{bmatrix}, \quad \phi \in [0, 2\pi), \tag{7.5}$$

and the subgroup $SO(2)$ of infinitesimal rotations alone is generated by $\mathbf{R}_\phi$. These two groups, and $O(2)$ in particular, are especially important in this book.

We note that the identity matrix $\mathbf{I}$ ($=\mathbf{R}_{\phi=0}$ in the cases of $O(2)$ and $SO(2)$) is always an element of Γ and that all other elements of Γ can be formed by finite products of the generating matrices. For example, in $O(2)$, reflection about a line at angle α to the x_1-axis is obtained by

$$\mathbf{R}_{\frac{3\pi}{2}+\alpha} \mathbf{R}_f \mathbf{R}_{\frac{\pi}{2}-\alpha}.$$

A set $\mathcal{S} \subset \mathbb{R}^n$ has symmetry Γ if it is *invariant* under every element of the group Γ, i.e. for each $\gamma \in \Gamma$, if $\mathbf{x} \in \mathcal{S}$ then $\gamma(\mathbf{x}) \in \mathcal{S}$. The square in the plane with corners $(\pm 1, \pm 1)$ is invariant under reflections about either axis, and under the larger group D_4, but it is *not* invariant under $O(2)$ or $SO(2)$, since rotation by $\pi/4$, for example, yields a "diamond" with vertices $(0, \pm\sqrt{2})$, $(\pm\sqrt{2}, 0)$.

The idea of an *invariant function* is important. We say that the real- or complex-valued function $f(\mathbf{x})$ is *invariant under the group* Γ, or Γ-invariant, if for all $\gamma \in \Gamma$ and $\mathbf{x} \in \mathbb{R}^n$ the equation

$$f(\gamma(\mathbf{x})) = f(\mathbf{x}) \tag{7.6}$$

is satisfied. Thus, the function $f(x_1, x_2) = x_1^2 + \sin(x_2)\cos(x_1)$ is invariant under the reflection group (7.2), as is any function of two variables provided it is even in the first.

The more stringent requirement of invariance under $SO(2)$ implies that f depends only on the radial coordinate $x_1^2 + x_2^2$. To see this, one can transform to polar coordinates $x_1 = r\cos\theta$, $x_2 = r\sin\theta$ and use the fact that the action of $\mathbf{R}_\phi$ of (7.5) becomes

$$(r, \theta) \xrightarrow{\mathbf{R}_\phi} (r, \theta + \phi): \tag{7.7}$$

thus, $SO(2)$-invariant functions depend on r alone.

In studying ODEs we need the additional notion of *equivariance*. We say that the vector field $\mathbf{f}(\mathbf{x})$ is *equivariant under the group* Γ, or Γ-equivariant, if for every $\boldsymbol{\gamma} \in \Gamma$ the equation

$$\boldsymbol{\gamma}\mathbf{f}(\mathbf{x}) = \mathbf{f}(\boldsymbol{\gamma}\mathbf{x}) \tag{7.8}$$

holds. This has the important consequence that, if a set Λ is invariant for the flow of (7.1), then $\boldsymbol{\gamma}\Lambda$ is an invariant set for (7.1) also. This follows from the fact that the differential equation holds for solutions $\mathbf{x}(t) \in \Lambda$ and so, via (7.8), the equation

$$\boldsymbol{\gamma}\dot{\mathbf{x}} = \boldsymbol{\gamma}\mathbf{f}(\mathbf{x}) = \mathbf{f}(\boldsymbol{\gamma}\mathbf{x}) \tag{7.9}$$

is satisfied for every solution $\boldsymbol{\gamma}\mathbf{x}(t) \in \boldsymbol{\gamma}\Lambda$ as well.

As an example, consider planar ODEs of the form

$$\left.\begin{aligned}\dot{x}_1 &= f_1(x_1, x_2), \\ \dot{x}_2 &= f_2(x_1, x_2),\end{aligned}\right\} \tag{7.10}$$

and suppose that (7.10) is equivariant under reflection about the x_2-axis. Then, for (7.8) to hold we require that

$$\begin{bmatrix} -1 & 0 \\ 0 & 1 \end{bmatrix} \begin{pmatrix} f_1(x_1, x_2) \\ f_2(x_1, x_2) \end{pmatrix} = \begin{pmatrix} f_1(-x_1, x_2) \\ f_2(-x_1, x_2) \end{pmatrix}. \tag{7.11}$$

Therefore f_1 must be odd in x_1 and f_2 even in x_1; no restriction is placed on behavior with respect to x_2. We observe that (provided the vector field is continuous) this necessarily implies that $f_1(0, x_2) = 0$ and so the x_2-axis is an invariant set for (7.10) in this case. We return to this when discussing global behavior.

For a more sophisticated example, suppose (7.10) is equivariant under $O(2)$. Evidently the origin must remain fixed and so we may write (7.10) as

$$\dot{\mathbf{x}} = \mathbf{A}\mathbf{x} + \tilde{\mathbf{f}}(\mathbf{x}), \quad \tilde{\mathbf{f}}(\mathbf{x}) = \mathcal{O}(|\mathbf{x}|^2). \tag{7.12}$$

Consider first the linear part of (7.12), which must commute with the elements of $O(2)$, so that

$$\mathbf{R}_f \mathbf{A} = \mathbf{A}\mathbf{R}_f \quad \text{and} \quad \mathbf{R}_\phi \mathbf{A} = \mathbf{A}\mathbf{R}_\phi \tag{7.13}$$

for every $\phi \in [0, 2\pi)$. The reader can check that the only 2×2 matrices which satisfy (7.13) are multiples of the identity:

$$\begin{bmatrix} \lambda & 0 \\ 0 & \lambda \end{bmatrix}. \tag{7.14}$$

It is more difficult to compute the requisite form for the nonlinear terms, but here invariant functions come to our aid. Let us represent the nonlinear function $\tilde{\mathbf{f}}(\mathbf{x})$ of (7.12) in the form

$$\tilde{\mathbf{f}}(\mathbf{x}) = \tilde{g}(\mathbf{x})\mathbf{B}\mathbf{x}, \tag{7.15}$$

where $\tilde{g}$ is real valued and $\mathbf{B}$ is a 2×2 matrix. Then we have

$$\gamma\tilde{\mathbf{f}}(\mathbf{x}) = \gamma\tilde{g}(\mathbf{x})\mathbf{B}\mathbf{x} = \tilde{g}(\mathbf{x})\gamma\mathbf{B}\mathbf{x} \quad \text{and} \quad \tilde{\mathbf{f}}(\gamma\mathbf{x}) = \tilde{g}(\gamma\mathbf{x})\mathbf{B}\gamma\mathbf{x}. \tag{7.16}$$

From Equation (7.8), equivariance of $\tilde{\mathbf{f}}(\mathbf{x})$ implies that these two expressions must be equal. This holds provided that

$$\tilde{g}(\mathbf{x}) = \tilde{g}(\gamma\mathbf{x}) \quad \text{and} \quad \gamma\mathbf{B}\mathbf{x} = \mathbf{B}\gamma\mathbf{x}, \tag{7.17}$$

i.e. $\tilde{g}$ is Γ-invariant and the linear vector field $\mathbf{B}\mathbf{x}$ is Γ-equivariant. We have already observed that functions depending only on the radial coordinate $x_1^2 + x_2^2$ are $SO(2)$-invariant; in fact they are also $O(2)$-invariant. This, together with our finding on the form of Γ-equivariant 2×2 matrices (7.14), implies that any planar ODE of the form

$$\begin{pmatrix} \dot{x}_1 \\ \dot{x}_2 \end{pmatrix} = [\lambda + \tilde{g}(x_1^2 + x_2^2)] \begin{pmatrix} x_1 \\ x_2 \end{pmatrix} \tag{7.18}$$

is $O(2)$-equivariant. In fact the converse is also true: *any $O(2)$-equivariant planar ODE can be written in the form* (7.18), but the argument establishing this would take us too far into group theory.

If we require equivariance only under the subgroup $SO(2)$ of $O(2)$, then the linear part can take the form

$$\begin{bmatrix} \lambda & -\omega \\ \omega & \lambda \end{bmatrix} \begin{pmatrix} x_1 \\ x_2 \end{pmatrix} = \lambda \begin{pmatrix} x_1 \\ x_2 \end{pmatrix} + \omega \begin{pmatrix} -x_2 \\ x_1 \end{pmatrix}, \tag{7.19}$$

and the analogous equivariant ODE is

$$\begin{pmatrix} \dot{x}_1 \\ \dot{x}_2 \end{pmatrix} = [\lambda + \tilde{g}_r(x_1^2 + x_2^2)] \begin{pmatrix} x_1 \\ x_2 \end{pmatrix} + [\omega + \tilde{g}_\theta(x_1^2 + x_2^2)] \begin{pmatrix} -x_2 \\ x_1 \end{pmatrix}, \tag{7.20}$$

where the subscripts reflect the fact that the invariant function $\tilde{g}_r$ defines a radial vector field while $\tilde{g}_\theta$ defines an azimuthal (purely rotational) vector field. We observe that this is precisely the normal form of the (unfolded) Hopf bifurcation (6.51), but note that in that case the "error" terms $\mathcal{O}(|\mathbf{x}|^5)$ are *not* generally $SO(2)$-equivariant and that the symmetry obtained via normal form transformations is in general only approximate.

The simple examples above show that, in the presence of symmetry, one expects to find invariant linear subspaces of lower dimension than that of the full phase space. We have already remarked that the x_2-axis is invariant for any continuous vector field equivariant under reflection about that axis. The invariant subspaces of the $O(2)$-equivariant vector field (7.18) are far richer: *any* line through the origin is invariant, as is clear from the polar coordinate representation

$$\dot{r} = r(\lambda + \tilde{g}(r^2)),$$
$$\dot{\theta} = 0,$$

since the angular variable $\theta(t) \equiv \theta(0)$ remains constant. (In connection with the discussion preceding Equation (7.9), observe that any radial line is the image of another such line under some rotation matrix $\mathbf{R}_\phi$.) The existence of such lower-dimensional invariant subspaces is very helpful in analysis of systems of dimension $n \geq 3$, as we see when we deal with global behavior in Sections 7.3 and 7.4.

7.2 Local bifurcation with symmetry

In dealing with a family of equivariant vector fields, the center manifold reduction and normal form simplification methods outlined in Section 6.4 can be carried through in a fashion which respects the symmetry class. The reduced system is at least as symmetric as the original one. (We have already remarked how, in the case of the Hopf bifurcation, truncated normal forms can be more symmetric than the original vector fields.) Returning to the example of (6.28), we observe that this ODE:

$$\left.\begin{aligned} \dot{x}_1 &= \mu x_1 - x_1 x_2, \\ \dot{x}_2 &= -x_2 + x_1^2, \end{aligned}\right\} \tag{7.21}$$

is equivariant under reflection about the x_2-axis (cf. (7.2) and (7.11)). The center manifold, given approximately by

$$x_2 = x_1^2 + \ldots \tag{7.22}$$

(Equation (6.35)), is also invariant under such reflections. Knowing this in advance, we can seek a (parameterized) center manifold described by a function $x_2 = h(x_1, \mu)$, even in x_1, which significantly reduces the computational effort. It is also no surprise that the reduced system (6.41) exhibits a pitchfork bifurcation, or that the two equilibria in the full system:

$$(x_1, x_2) = (\pm\sqrt{\mu}, \mu) \tag{7.23}$$

are images of one another under $\mathbf{R}_f$.

A yet more symmetric bifurcation that we meet in our models is the $O(2)$-equivariant pitchfork. Suppose that a one-parameter family of ($n \geq 2$-dimensional) $O(2)$-equivariant vector fields passes through a bifurcation value μ_0 at which the linearization has a zero eigenvalue. Then, in view of the form of $\mathbf{A}$ in the planar case (7.14), the eigenvalue is necessarily of multiplicity two, with a diagonal block

$$\begin{bmatrix} 0 & 0 \\ 0 & 0 \end{bmatrix},$$

and the two-dimensional reduced system takes the form (7.18). Truncating at cubic terms, we obtain

$$\begin{pmatrix} \dot{x}_1 \\ \dot{x}_2 \end{pmatrix} = \left[\mu + a_3(x_1^2 + x_2^2) + \cdots \right] \begin{pmatrix} x_1 \\ x_2 \end{pmatrix}, \tag{7.24}$$

or, in polar coordinates

$$\left.\begin{aligned} \dot{r} &= \mu r + a_3 r^3 + \cdots \\ \dot{\theta} &= 0. \end{aligned}\right\} \tag{7.25}$$

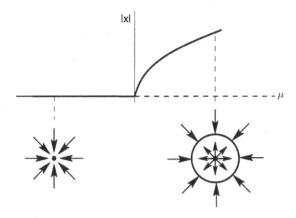

Figure 7.1 The $O(2)$-equivariant pitchfork bifurcation.

Thus, when $a_3 < 0$, a circle of equilibria bifurcates from the origin as μ increases through $\mu_0 = 0$, see Figure 7.1. We remark that the equilibria form an invariant circle and that they are stable only in the weak sense that solutions approach the circle asymptotically; individual equilibria are Liapunov stable, but not asymptotically stable themselves. There is no dynamics "within the circle." This is a rather typical situation for ODEs equivariant under continuous groups such as $O(2)$.

7.3 Global behavior with symmetry

A major consequence of the presence of symmetries is that global structures and in particular homoclinic and heteroclinic orbits can occur in a structurally stable manner. The theory of such global phenomena and their bifurcations is still in its infancy, but here we give three examples which are well understood. The last of these, to be discussed in Section 7.4, reappears in several guises in later chapters.

The first example is that of the reflection symmetric ODE

$$\left.\begin{aligned} \dot{x}_1 &= x_1 x_2, \\ \dot{x}_2 &= 1 - x_2^2 + x_1^2 x_2. \end{aligned}\right\} \tag{7.26}$$

As pointed out in connection with Equation (7.11), for such vector fields the x_2-axis is necessarily invariant. In the present case, this axis also contains the two fixed points

$$(x_1, x_2) = (0, \pm 1) \stackrel{\text{def}}{=} \mathbf{p}_\pm, \tag{7.27}$$

at which the linearizations are, respectively:

$$\begin{bmatrix} 1 & 0 \\ 0 & -2 \end{bmatrix} \quad \text{and} \quad \begin{bmatrix} -1 & 0 \\ 0 & 2 \end{bmatrix}, \tag{7.28}$$

so that both are saddle points. Moreover, the x_2-axis contains the following stable and unstable manifolds:

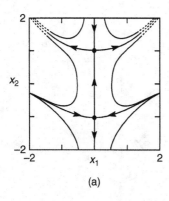 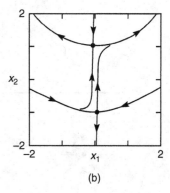

(a)   (b)

Figure 7.2 (a) The heteroclinic connection for Equation (7.26), and (b) the broken connection for Equation (7.29) with $\epsilon = 0.05$.

$$W^s(\mathbf{p}_+) = \{(x_1, x_2) \mid x_1 = 0, \ x_2 > -1\},$$
$$W^u(\mathbf{p}_-) = \{(x_1, x_2) \mid x_1 = 0, \ x_2 < 1\},$$

and any solution starting in the segment $\{(x_1, x_2) | x_1 = 0, -1 < x_2 < 1\}$ is a heteroclinic orbit from $(0, -1)$ to $(0, 1)$. Such orbits persist for all sufficiently small perturbations of (7.26) *sharing the same reflection symmetry*, for such perturbations cannot destroy the hyperbolic equilibria, which necessarily still lie on the invariant x_2-axis, as shown in Figure 7.2(a). Thus the heteroclinic connection is structurally stable in the class of reflection symmetric systems.

Suppose now that (7.26) suffers a small perturbation which breaks the symmetry, for example:

$$\left.\begin{aligned} \dot{x}_1 &= \epsilon + x_1 x_2, \\ \dot{x}_2 &= 1 - x_2^2 + x_1^2 x_2. \end{aligned}\right\} \tag{7.29}$$

The fixed points now move off the x_2-axis (for $\epsilon > 0$ the upper one $\mathbf{p}_+$ to the left and the lower $\mathbf{p}_-$ to the right) and the formerly invariant line is now crossed from left to right by solutions ($\dot{x}_1 = \epsilon$ on $x_1 = 0$). It follows that the upper branch of $W^u(\mathbf{p}_-)$ is contained in the right-hand half plane and the lower branch of $W^s(\mathbf{p}_+)$ in the left-hand half plane. Hence the heteroclinic connection is broken for all $\epsilon > 0$ (and similarly for all $\epsilon < 0$). As we expect, the connection is structurally *unstable* in the larger space of all planar vector fields (recall Figure 6.13).

Our second example, drawn from [145], is of a three-dimensional system, again equivariant under a discrete group, but in this case the 24 element permutation group is generated by the matrices $\mathbf{I}$ and

$$\mathbf{R}_f = \begin{bmatrix} -1 & 0 & 0 \\ 0 & 1 & 0 \\ 0 & 0 & 1 \end{bmatrix}, \qquad \mathbf{R}_p = \begin{bmatrix} 0 & 1 & 0 \\ 0 & 0 & 1 \\ 1 & 0 & 0 \end{bmatrix}. \tag{7.30}$$

The ODE in question is

$$\left.\begin{aligned}
\dot{x}_1 &= x_1(\mu - ax_1^2 - bx_2^2 - cx_3^2), \\
\dot{x}_2 &= x_2(\mu - ax_2^2 - bx_3^2 - cx_1^2), \\
\dot{x}_3 &= x_3(\mu - ax_3^2 - bx_1^2 - cx_2^2),
\end{aligned}\right\} \tag{7.31}$$

where the parameters satisfy $\mu, a, b, c > 0$ and we are only concerned with solutions in the positive octant $x_j \geq 0$.

This system was derived and studied by Busse *et al.* [66, 67] as a model for convection in a Rayleigh–Bénard problem for a layer of fluid heated from below and slowly rotating. The independent variables x_j represent the (non-negative) amplitudes of three convection roll patterns whose orientations differ by 120 degrees.

We first note that each coordinate axis and each coordinate plane is invariant for (7.31), that the origin $(0, 0, 0)$ is a source and that three "pure mode" fixed points,

$$\mathbf{p}_1 = (\sqrt{\mu/a}, 0, 0), \quad \mathbf{p}_2 = (0, \sqrt{\mu/a}, 0), \quad \mathbf{p}_3 = (0, 0, \sqrt{\mu/a}),$$

exist which are permuted one into another by the matrix $\mathbf{R}_p$. Linearizing at $\mathbf{p}_1$, we obtain the diagonal matrix

$$\begin{bmatrix} -2\mu & 0 & 0 \\ 0 & \mu(1 - c/a) & 0 \\ 0 & 0 & \mu(1 - b/a) \end{bmatrix}, \tag{7.32}$$

and, if we additionally assume that $c > a > b$, so that $(1-c/a) < 0$ and $(1-b/a) > 0$, this equilibrium is a saddle point with a two-dimensional stable manifold $W^s(\mathbf{p}_1)$ contained in the plane $\{x_3 = 0\}$ and a one-dimensional unstable manifold $W^u(\mathbf{p}_1)$ in the plane $\{x_2 = 0\}$. Similarly, at $\mathbf{p}_2$ we have the linearization matrix

$$\begin{bmatrix} \mu(1 - b/a) & 0 & 0 \\ 0 & -2\mu & 0 \\ 0 & 0 & \mu(1 - c/a) \end{bmatrix}, \tag{7.33}$$

which is merely a permutation of (7.32) and which shows that the stable manifold $W^s(\mathbf{p}_2)$ is contained in the plane $\{x_1 = 0\}$ and the unstable manifold $W^u(\mathbf{p}_2)$ is contained in the plane $\{x_3 = 0\}$.

We claim that $W^u(\mathbf{p}_2) \subset W^s(\mathbf{p}_1)$, so that there is a heteroclinic orbit from $\mathbf{p}_2$ to $\mathbf{p}_1$. Since both these manifolds lie in the invariant plane $\{x_3 = 0\}$, it is sufficient to consider the two-dimensional subsystem

$$\left.\begin{aligned}
\dot{x}_1 &= x_1(\mu - ax_1^2 - bx_2^2), \\
\dot{x}_2 &= x_2(\mu - ax_2^2 - cx_1^2),
\end{aligned}\right\} \tag{7.34}$$

restricted to that plane. We first observe that no equilibrium exists in the interior of the positive quadrant $\{x_1, x_2 > 0\}$, for if it did it would satisfy

$$ax_1^2 + bx_2^2 = \mu = ax_2^2 + cx_1^2,$$

implying that

$$(a - c)x_1^2 = (a - b)x_2^2,$$

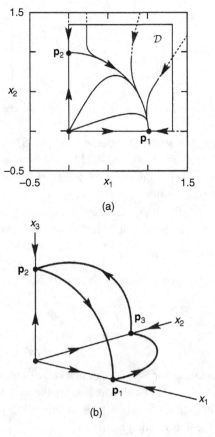

Figure 7.3 (a) A connection from $\mathbf{p}_2$ to $\mathbf{p}_1$. (b) The heteroclinic cycle.

in contradiction of our assumption $c > a > b$. Also, if we take a sufficiently large square $\mathcal{D}$ with outer boundaries $\{x_1 = d; x_2 = d\}$ and inner boundaries $\{x_1 = 0; x_2 = 0\}$ then the vector field of (7.34) is directed inwards on the outer boundary lines since

$$\dot{x}_1|_{x_1=d} = d(\mu - ad^2 - bx_2^2) \le d(\mu - ad^2) \le 0$$

and similarly

$$\dot{x}_2|_{x_2=d} \le d(\mu - ad^2) \le 0.$$

This, together with the facts that the origin is a source and the inner boundaries $\{x_1 = 0; x_2 = 0\}$ are invariant, imply that $W^u(\mathbf{p}_2)$ is trapped inside $\mathcal{D}$ and has nowhere to go but to the fixed point $\mathbf{p}_1$, which, as far as the reduced system (7.34) is concerned, is a sink (see Figure 7.3(a)). (In proving this assertion we must also verify that no other non-wandering sets exist in the positive quadrant, so that orbits in $W^u(\mathbf{p}_2)$ cannot limit on any set other than one on the inner boundary of $\mathcal{D}$. This follows from index theory and the Poincaré–Bendixson theorem, [144, 156].)

Now we obtain two futher heteroclinic connections: $W^u(\mathbf{p}_1) \subset W^s(\mathbf{p}_3)$ and $W^u(\mathbf{p}_3) \subset W^s(\mathbf{p}_2)$ "for free" by simply applying the permutation matrix $\mathbf{R}_p$ to the vector field of

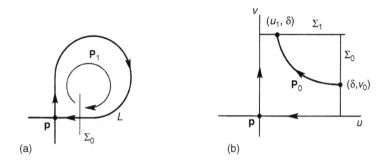

Figure 7.4 (a) A homoclinic loop, and (b) local coordinates near the saddle point.

(7.34). We have found a *heteroclinic cycle* composed of three orbit segments connecting $\mathbf{p}_2$ to $\mathbf{p}_1$, $\mathbf{p}_1$ to $\mathbf{p}_3$, and $\mathbf{p}_3$ back to $\mathbf{p}_2$ (see Figure 7.3(b)). Much as in the previous example, this cycle is structurally stable within the class of three-dimensional permutation symmetric systems, basically because the coordinate planes remain invariant and we simply have a structurally stable saddle to sink connection on each such plane.

We now turn to the question of asymptotic stability of the heteroclinic cycle: does it attract or repel nearby orbits? To develop the idea, we first consider a planar homoclinic loop $\mathcal{L}$ containing a single hyperbolic saddle point as in Figure 7.4(a) (recall the global bifurcation illustrated in Figure 6.13). Solutions approaching $\mathcal{L}$ from inside spend the bulk of their transit time "hesitating" near the saddle point $\mathbf{p}$ and hence their behavior is dominated by that of the flow linearized at $\mathbf{p}$. In local (stable and unstable manifold) coordinates we can write this as

$$\left.\begin{aligned} \dot{u} &= \lambda_s u, \\ \dot{v} &= \lambda_u v, \end{aligned}\right\} \tag{7.35}$$

where λ_s and λ_u are the stable and unstable eigenvalues respectively ($\lambda_s < 0 < \lambda_u$). We wish to approximate the one-dimensional return map P on the local cross section Σ_0 given by $\{u = \delta | \delta \ll 1\}$. This map may be expressed as the composition of a local map $P_0 : \Sigma_0 \to \Sigma_1$, which takes points on Σ_0 past the saddle to a second cross section Σ_1, given by $\{v = \delta\}$, and a global map $P_1 : \Sigma_1 \to \Sigma_0$, which takes points on Σ_1 around near the homoclinic loop and returns them to Σ_0. A Gronwall estimate (see [77], Section 4.1 of [144], and see Section 4.4 above) reveals that this latter map, obtained by integrating the ODE for a finite time interval (assuming no equilibria other than $\mathbf{p}$ lie near $\mathcal{L}$) is a bounded distortion. In contrast, we shall show that the map P_0 has an arbitrarily large effect on solutions starting on Σ_0 sufficiently close to $\mathcal{L}$ and hence that the flow linearized at $\mathbf{p}$ dominates the overall return map, as claimed.

Specifically, for the linearized system (7.35), the time of flight τ for a solution starting at $(u(0), v(0)) = (\delta, v_0) \in \Sigma_0$ to reach $(u_1, \delta) \in \Sigma_1$ is given by integrating the second component:

$$v(\tau) = \delta = v_0 e^{\lambda_u \tau} \Rightarrow \tau = \frac{1}{\lambda_u} \ln \left| \frac{\delta}{v_0} \right|. \tag{7.36}$$

We can then use (7.36) in integration of the first component to obtain

$$u(\tau) = u_1 = \delta e^{\lambda_s \tau} = \delta \exp\left(\frac{\lambda_s}{\lambda_u} \ln\left|\frac{\delta}{v_0}\right|\right)$$

$$= \delta^{1+\lambda_s/\lambda_u}|v_0|^{-\lambda_s/\lambda_u}, \tag{7.37}$$

see Figure 7.4(b). This solution of the linearized equation provides an estimate for P_0 which becomes more accurate the smaller we take δ. The bounded distortion in the global part P_1 of the return map implies that the solution next intersects Σ_0 at a point (δ, v_1) with $c_1 u_1 < v_1 < c_2 u_1$ for two constants c_1, c_2 which depend on δ but not on u_0 or v_1.

Now suppose $-\lambda_s > \lambda_u\ (> 0)$, so that the stable eigenvalue is stronger than the unstable one. Thus $-\lambda_s/\lambda_u > 1$ and we have the estimate:

$$v_1 < c_2 u_1 = c_2 \delta^{1+\lambda_s/\lambda_u}|v_0|^{-\lambda_s/\lambda_u}. \tag{7.38}$$

Since the power to which $|v_0|$ is raised is greater than one, the graph of the return map P is tangent to the v_0-axis at the origin and so $v_1 < v_0$ for v_0 sufficiently small. (Having fixed δ and established error bounds for (7.37) we can always take v_0 small enough to dominate any constant larger than c_2 that such an error bound might demand.) In contrast, if $\lambda_u > -\lambda_s$, we have

$$v_1 > c_1 u_1 = c_1 \delta^{1+\lambda_s/\lambda_u}|v_0|^{-\lambda_s/\lambda_u}, \tag{7.39}$$

and since $0 < -\lambda_s/\lambda_u < 1$ we conclude that $v_1 > v_0$ for v_0 sufficiently small. In the former case, then, the homoclinic loop attracts solutions from the inside while in the latter case it repels solutions.

Returning to the three-dimensional example (7.31), we note that all three saddle points have the same eigenvalues and that an argument analogous to that above applied at any one of them therefore determines stability or instability. For stability it suffices that *both* stable eigenvalues are larger in magnitude than the unstable eigenvalue; i.e. that

$$2\mu > \mu(1 - b/a) \quad \text{and} \quad \mu(c/a - 1) > \mu(1 - b/a). \tag{7.40}$$

Since $1 > -b/a$ is automatically satisfied for $a, b > 0$, we need only examine the second inequality, which holds as long as

$$c + b > 2a. \tag{7.41}$$

So, to summarize, we have shown that, provided $c > a > b > 0$, $c + b > 2a$, and $\mu > 0$, Equation (7.31) has an attracting heteroclinic cycle on the boundary of the positive octant formed by the three equilibria $\mathbf{p}_1, \mathbf{p}_2, \mathbf{p}_3$, and their unstable manifolds.

The behavior of solutions attracted to such a cycle is interesting. They remain almost stationary for relatively long periods while they creep past each saddle point before making rapid transits to the neighborhood of the next saddle in the cycle. The time of flight estimate (7.36) and the rate of attraction estimate (7.38) for a loop with a single saddle imply that successive "waiting times" τ_n near the saddles obey

$$\tau_n = \frac{1}{\lambda_u} \ln\left|\frac{\delta}{v_n}\right| \quad \text{and} \quad \tau_{n+1} = \frac{1}{\lambda_u} \ln\left|\frac{\delta}{v_{n+1}}\right|,$$

with

$$v_{n+1} < c_2\delta^{1+\lambda_s/\lambda_u}|v_n|^{-\lambda_s/\lambda_u}.$$

This leads to the approximate relationship

$$\tau_{n+1} \approx \left(\frac{-\lambda_s}{\lambda_u}\right)\tau_n. \tag{7.42}$$

No numerical simulation can reproduce the resulting exponential increase in waiting times, primarily because finite arithmetic precision (v, say) limits the distance within which solutions can approach the invariant coordinate planes, effectively ensuring that, provided v_n is not rounded down to 0, $\lim_{n\to\infty} v_n \geq v$ and hence that

$$\lim_{n\to\infty} \tau_n \leq \frac{1}{\lambda_u}\ln\left|\frac{\delta}{v}\right|. \tag{7.43}$$

Of course, if v_n is rounded down to zero for some n, then the numerical solution henceforth lies in an invariant plane and approaches one of the saddle points along its unstable manifold. In this case $\tau_n = \infty$.

There is also no guarantee, a priori, that the finite step of a numerical integrator cannot carry a solution *across* an invariant plane. Thus one must interpret numerical work with care; see, e.g. [46]. Nonetheless, we illustrate the behavior of Equation (7.31) with a simulation in Figure 7.5.

While the heteroclinic cycle of this symmetric system is structurally stable within the symmetry class, small perturbations which break that symmetry can be expected to destroy it, as in the former example (Figure 7.2). The addition of small random forcing terms was indeed suggested by Busse and Heikes [67], who argued that they would result in a "statistical limit cycle." If the perturbations are sufficiently small they cannot significantly influence solutions away from the equilibria and so the heteroclinic "jumps" persist. But near the saddle points *any* additive perturbation, however small, can be expected to have a radical effect, especially on the waiting times τ_n. We return to analyze these effects in some detail in Chapter 9, for one class of boundary layer models of interest that contains such quasirandom terms.

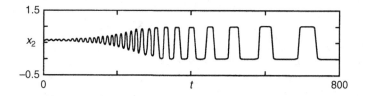

Figure 7.5 A solution of Equation (7.31) with $\mu = a = 1, b = 0.55, c = 1.5; x_1(0) = 0.5, x_2(0) = x_3(0) = 0.499$, obtained by fourth order Runge–Kutta method with step size 0.02.

7.4 An $O(2)$-equivariant ODE

Our third example is so central to later chapters of this book that it deserves a separate section. Here we draw heavily on the paper of Armbruster *et al.* [9], in which more detail can be found. Proctor and Jones [176, 288] analyzed a similar system independently at around the same time, and these papers were the first to give explicit examples of structurally stable heteroclinic cycles in an equation equivariant under a continuous group of symmetries. Since these studies, such "robust" cycles have been observed and proven to occur in numerous symmetric systems, many of them models arising in fluid mechanics: [198] provides a survey.

The ODE we study models the interaction of two Fourier modes whose spatial wavenumbers are in the ratio $1 : 2$. If one considers harmonic waves of wavelengths L and $L/2$ respectively, propagating freely in a one-dimensional continuum, the state of each wave can be specified by its amplitude and phase, or equivalently by a complex number $z_j = x_j + i y_j$, where x_j and y_j are the amplitudes of the cosine and sine components respectively. Thus our model can be written rather compactly as a pair of coupled, complex ODEs. The underlying symmetry in physical space is that of *translations* (waves are supposed not to be pinned and there is no spatial inhomogeneity in the medium) and *reflections* (waves can propagate in either direction with equal ease). Thus, in the physical domain $x \in \mathbb{R}$, the governing equations are unchanged under the transformations

$$T_\alpha : x \mapsto x + \alpha \quad \text{and} \quad R_f : x \mapsto -x. \tag{7.44}$$

More specifically, let us represent a waveform $w(x, t)$ by two complex Fourier modes:

$$w(x, t) = \sum_{j=-2,\ j\neq0}^{2} z_j(t) \exp\left(\frac{2\pi i j x}{L}\right), \tag{7.45}$$

where, for reality of w, we require $z_{-j} = z_j^*$. The asterisk denotes complex conjugation. Applying the translation T_α we obtain

$$T_\alpha w(x) = w(x + \alpha) = \sum z_j \exp\left[\frac{2\pi i j (x + \alpha)}{L}\right]$$
$$= \sum \exp\left(\frac{2\pi i j \alpha}{L}\right) z_j \exp\left(\frac{2\pi i j x}{L}\right), \tag{7.46}$$

and the reflection yields

$$R_f w(x) = w(-x) = \sum z_j \exp\left[\frac{2\pi i j (-x)}{L}\right]$$
$$= \sum z_{-j} \exp\left(\frac{2\pi i j x}{L}\right) = \sum z_j^* \exp\left(\frac{2\pi i j x}{L}\right). \tag{7.47}$$

We therefore see that, in the phase space spanned by the Fourier modes, the representations of translation and reflection in physical space are respectively:

$$\left.\begin{aligned} T_\alpha &: z_j \mapsto \exp\left(\frac{2\pi i j \alpha}{L}\right) z_j, \\ R_f &: z_j \mapsto z_j^*. \end{aligned}\right\} \tag{7.48}$$

We now seek the appropriate $O(2)$-equivariant ODE of the form

$$\left.\begin{aligned}\dot{z}_1 &= f_1(z_1, z_2), \\ \dot{z}_2 &= f_2(z_1, z_2),\end{aligned}\right\} \tag{7.49}$$

which is unchanged under the transformations T_α and R_f of (7.48). In this case the invariant functions turn out to depend only on the three combinations $|z_1|^2$, $|z_2|^2$, and $z_1^2 z_2^* + z_1^{*2} z_2$, and the equivariant vector fields are

$$\begin{pmatrix} z_1 \\ z_2 \end{pmatrix} \quad \text{and} \quad \begin{pmatrix} z_1^* z_2 \\ z_1^2 \end{pmatrix}$$

so that, up to third order, we have the normal form

$$\left.\begin{aligned}\dot{z}_1 &= (\mu_1 + d_{11}|z_1|^2 + d_{12}|z_2|^2)z_1 + c_{12}z_1^* z_2, \\ \dot{z}_2 &= (\mu_2 + d_{21}|z_1|^2 + d_{22}|z_2|^2)z_2 + c_{11}z_1^2,\end{aligned}\right\} \tag{7.50}$$

in which all the coefficients are purely real. Assuming that $c_{12}, c_{11} \neq 0$ we can make a further rescaling, reversing the direction of time if necessary, to obtain:

$$\left.\begin{aligned}\dot{z}_1 &= z_1^* z_2 + (\mu_1 + e_{11}|z_1|^2 + e_{12}|z_2|^2)z_1, \\ \dot{z}_2 &= \pm z_1^2 + (\mu_2 + e_{21}|z_1|^2 + e_{22}|z_2|^2)z_2,\end{aligned}\right\} \tag{7.51}$$

where $e_{11} = d_{11}/|c_{11}c_{12}|$, $e_{12} = d_{12}/c_{12}^2$, $e_{21} = d_{21}/|c_{11}c_{12}|$, and $e_{22} = d_{22}/c_{12}^2$. Such a reduction in the number of parameters is important: a complete analysis of both the "+" and "−" cases of the six-parameter family (7.51) is still formidable enough! Here we shall restrict ourselves to the "−" case and, assuming that $\mu_j > 0 > e_{ij}$, focus on the heteroclinic cycles and bifurcations from them. It is this case which reappears in the boundary layer model of Chapter 11.

Letting $z_j = r_j e^{i\theta_j}$, (7.51) may conveniently be rewritten in polar coordinates as:

$$\left.\begin{aligned}\dot{r}_1 &= r_1 r_2 \cos\phi + r_1(\mu_1 + e_{11}r_1^2 + e_{12}r_2^2), \\ \dot{r}_2 &= -r_1^2 \cos\phi + r_2(\mu_2 + e_{21}r_1^2 + e_{22}r_2^2), \\ \dot{\phi} &= -(2r_2 - r_1^2/r_2)\sin\phi,\end{aligned}\right\} \tag{7.52}$$

where $\phi = 2\theta_1 - \theta_2$. The emergence of the single phase difference ϕ and reduction to three (real) dimensions is a consequence of the $O(2)$ symmetry. Since polar coordinates are singular when $r_1, r_2 = 0$ it will also be necessary to use the real Cartesian form

$$\left.\begin{aligned}\dot{x}_1 &= x_1 x_2 + y_1 y_2 + x_1(\mu_1 + e_{11}r_1^2 + e_{12}r_2^2), \\ \dot{y}_1 &= x_1 y_2 - y_1 x_2 + y_1(\mu_1 + e_{11}r_1^2 + e_{12}r_2^2), \\ \dot{x}_2 &= -(x_1^2 - y_1^2) + x_2(\mu_2 + e_{21}r_1^2 + e_{22}r_2^2), \\ \dot{y}_2 &= -2x_1 y_1 + y_2(\mu_2 + e_{21}r_1^2 + e_{22}r_2^2),\end{aligned}\right\} \tag{7.53}$$

where $r_j^2 = x_j^2 + y_j^2$. This form is especially useful in our discussion of heteroclinic cycles. Reflection symmetry implies that the purely real subspace $\{y_1 = y_2 = 0\}$ is invariant, as is any rotation of it under the first element of (7.48), for example the subspace $\{x_1 = y_2 = 0\}$. This is also used below.

Finally, the scaling $r_j = \epsilon s_j$, $\mu_j = \epsilon^2 v_j$, and $(\dot{\ }) = \epsilon(\ )'$ transforms the polar system (7.52) to

$$
\left.
\begin{aligned}
s_1' &= s_1 s_2 \cos\phi + \epsilon s_1 (v_1 + e_{11} s_1^2 + e_{12} s_2^2), \\
s_2' &= -s_1^2 \cos\phi + \epsilon s_2 (v_2 + e_{21} s_1^2 + e_{22} s_2^2), \\
\phi' &= -(2 s_2 - s_1^2/s_2) \sin\phi,
\end{aligned}
\right\}
\tag{7.54}
$$

and reveals two integrals for the limit $\epsilon \to 0$. These are

$$
E = s_1^2 + s_2^2
\tag{7.55}
$$

and

$$
L = s_1^2 s_2 \sin\phi,
\tag{7.56}
$$

and as the reader can check, they remain constant under the flow of (7.54) when $\epsilon = 0$. This limit is appropriate to the study of the "almost critical" system obtained when $\mu_1, \mu_2 \approx 0$ and one is close to a doubly degenerate equilibrium with two pairs of zero eigenvalues.

Before starting our study of heteroclinic cycles it is worth discussing the various other kinds of solution exhibited by Equations (7.51), (7.52), and (7.54). It is important here to recognize that the individual phase evolution equations

$$
\dot{\theta}_1 = -r_2 \sin\phi \quad \text{and} \quad \dot{\theta}_2 = -(r_1^2/r_2) \sin\phi
\tag{7.57}
$$

have been suppressed in writing (7.52). With this in mind, *steady solutions* or fixed points of (7.51) correspond to fixed points of (7.52) or (7.54) with $\phi = 0$ or π. Among these, in addition to the trivial solution $\{r_1 = r_2 = 0\}$ we have pure modes $\{r_1 = 0, r_2 \neq 0\}$ and mixed modes $\{r_1 \neq 0, r_2 \neq 0\}$. Fixed points of (7.52) or (7.54) with $\phi \neq 0$ or π correspond to *traveling waves*, that is periodic orbits of (7.51) in which the phase difference ϕ remains constant, but θ_1 and θ_2 both increase or decrease linearly with time. Periodic orbits of (7.52) or (7.54) on the subspaces $\phi = 0$, π correspond to *standing waves*, while periodic orbits with $\phi \neq 0$, π correspond to *modulated traveling waves*, that is doubly periodic solutions of (7.51). These quasiperiodic orbits lie on invariant tori, as the simulation shown below in Figure 7.9 illustrates.

In the case of this symmetric system, an important class of periodic orbits, the traveling waves, can be found analytically, since they are simply fixed points with $\phi \neq 0$ or π for the polar coordinate equation (7.52). For such fixed points we require $\dot{\phi} = 0$ and, since $\sin\phi \neq 0$, this implies that $2 r_2^2 = r_1^2$. Using this to eliminate r_1 in the first two components of (7.52) and solving for (r_2, ϕ), we obtain the conditions

$$
r_2^2 = \frac{-(2\mu_1 + \mu_2)}{(4 e_{11} + 2 e_{12} + 2 e_{21} + e_{22})},
\tag{7.58}
$$

$$
r_1^2 = 2 r_2^2,
\tag{7.59}
$$

$$
\cos\phi = \frac{\mu_2 (2 e_{11} + e_{12}) - \mu_1 (2 e_{21} + e_{22})}{[-(2\mu_1 + \mu_2)(4 e_{11} + 2 e_{12} + 2 e_{21} + e_{22})]^{\frac{1}{2}}}.
\tag{7.60}
$$

Note that, if there is one solution to (7.58–7.60) with phase difference $\phi = \hat{\phi}$, then there is typically also a second with $\phi = 2\pi - \hat{\phi}$. The pair of solutions corresponds to waves traveling to the right and to the left. These periodic orbits bifurcate from the circle of mixed mode fixed points $\{r_1 \neq 0, r_2 \neq 0\}$ when $\cos \phi = \pm 1$. They are examples of *relative equilibria*, that is "rigidly rotating" motions which become true equilibria in a suitably rotating coordinate frame. The bifurcations in which they appear are peculiar to $O(2)$-equivariant systems.

In Armbruster *et al.* [9] explicit conditions on the parameters μ_j, e_{ij} are given for the existence and stability of all these various types of solution, but here, as we have remarked, we focus on heteroclinic cycles. We first observe that the pure second mode subspace $\{z_1 = 0\}$ is invariant and, for $\mu_2 > 0 > e_{22}$, it carries the circle of pure mode equilibria

$$|z_2| = \sqrt{-\mu_2/e_{22}}. \tag{7.61}$$

In fact, as μ_2 passes through zero, these equilibria appear in an $O(2)$-equivariant pitchfork bifurcation, as in Figure 7.1. Picking the equilibrium $\mathbf{p}_+ = (0, 0, \sqrt{-\mu_2/e_{22}}, 0)$ and computing the linearization from (7.53), we obtain the diagonal matrix

$$\begin{bmatrix} \lambda_+ & 0 & 0 & 0 \\ 0 & \lambda_- & 0 & 0 \\ 0 & 0 & -2\mu_2 & 0 \\ 0 & 0 & 0 & 0 \end{bmatrix}, \tag{7.62}$$

where

$$\lambda_\pm = \mu_1 - \frac{\mu_2 e_{12}}{e_{22}} \pm \sqrt{-\frac{\mu_2}{e_{22}}}. \tag{7.63}$$

Since any other equilibrium on the circle (7.61) can be obtained by rotation of $\mathbf{p}_+$ under the first component of (7.48), and the corresponding linearization by a similar rotation, all equilibria on the circle share these eigenvalues, although their matrix representations differ and the associated eigenvectors twist as one moves around the circle.

The eigenvector $(0, 0, 0, 1)$ belonging to the zero eigenvalue of $\mathbf{p}_+$ is tangent to the circle of equilibria and, as in (7.24) and (7.25), within the z_2-plane this circle is asymptotically stable in view of the negative eigenvalue $-2\mu_2$ with eigenvector $(0, 0, 1, 0)$. We are concerned with the situation in which the two remaining eigenvalues satisfy $\lambda_+ > 0 > \lambda_-$, so that the circle of equilibria exhibits saddle type behavior in the z_1-subspace orthogonal to z_2. Thus each equilibrium on the circle has a one-dimensional unstable manifold, a two-dimensional stable manifold, and a one-dimensional center manifold, which is the circle itself.

In general it is difficult to establish global behavior in a system of dimension $n \geq 3$. Here, however, the existence of two-dimensional invariant subspaces comes to our aid. In particular we focus on the two equilibria $\mathbf{p}_\pm = (0, 0, \pm\sqrt{-\mu_2/e_{22}}, 0)$ contained in the purely real subspace $\{y_1 = y_2 = 0\}$, which, as the reader can confirm by reference to (7.53), is also an invariant subspace. On that subspace the equation reduces to

$$\left. \begin{aligned} \dot{x}_1 &= x_1 x_2 + x_1(\mu_1 + e_{11}x_1^2 + e_{12}x_2^2), \\ \dot{x}_2 &= -x_1^2 + x_2(\mu_2 + e_{21}x_1^2 + e_{22}x_2^2), \end{aligned} \right\} \tag{7.64}$$

and $\mathbf{p}_+ = (0, +\sqrt{-\mu_2/e_{22}})$ has eigenvalues λ_+ and $-2\mu_2$ with eigenvectors $(1, 0)$ and $(0, 1)$ respectively, while $\mathbf{p}_- = (0, -\sqrt{-\mu_2/e_{22}})$ has eigenvalues λ_- and $-2\mu_2$ with eigenvectors $(1, 0)$ and $(0, 1)$ respectively. Thus, *restricted to this subspace*, $\mathbf{p}_+$ is a saddle and $\mathbf{p}_-$ a sink. As in the previous example (7.31) we first demonstrate that there is an orbit connecting $\mathbf{p}_+$ to $\mathbf{p}_-$ and then use a symmetry transformation to show that it is part of a larger cycle in which $\mathbf{p}_-$ is connected back to $\mathbf{p}_+$.

The linearization of (7.64) reveals that $W^u(\mathbf{p}_+)$ enters the half plane $\{x_1 > 0\}$ (and also $\{x_1 < 0\}$). As in the previous example, we show that there is a trapping region, given here by the interior of a large circle $x_1^2 + x_2^2 = d^2 \gg 1$, so that $W^u(\mathbf{p}_+)$ remains bounded. In fact, differentiating the (Liapunov) function $E = r_1^2 + r_2^2$ (cf. (7.55)) along solutions of (7.51), we obtain

$$\frac{dE}{dt} = \dot{z}_1 z_1^* + z_1 \dot{z}_1^* + \dot{z}_2 z_2^* + z_2 \dot{z}_2^* \qquad (7.65)$$

$$= z_1^*[z_1^* z_2 + z_1(\mu_1 + e_{11}|z_1|^2 + e_{12}|z_2|^2)]$$
$$+ z_1[z_1 z_2^* + z_1^*(\mu_1 + e_{11}|z_1|^2 + e_{12}|z_2|^2)]$$
$$+ z_2^*[-z_1^2 + z_2(\mu_2 + e_{21}|z_1|^2 + e_{22}|z_2|^2)]$$
$$+ z_2[-z_1^{*2} + z_2^*(\mu_2 + e_{21}|z_1|^2 + e_{22}|z_2|^2)]$$
$$= 2[\mu_1|z_1|^2 + \mu_2|z_2|^2 + e_{11}|z_1|^4$$
$$+ (e_{12} + e_{21})|z_1|^2|z_2|^2 + e_{22}|z_2|^4] \qquad (7.66)$$

and, since $e_{ij} < 0 < \mu_j$, we see that $E(t)$ decreases with increasing t provided E is sufficiently large. Thus the full system has a trapping region bounded by the three-dimensional sphere

$$x_1^2 + y_1^2 + x_2^2 + y_2^2 = d^2$$

in the four-dimensional phase space.

Under our assumptions on μ_j and e_{ij}, there are no further fixed points or periodic orbits of (7.64) in the interior of the half planes $\{x_1 \neq 0\}$ and so orbits in $W^u(\mathbf{p}_+)$ have no choice but to approach the sink $\mathbf{p}_-$ as $t \to +\infty$. We have our heteroclinic orbit from $\mathbf{p}_+$ to $\mathbf{p}_-$; in fact we have two, one in each half plane, for (7.64) is also equivariant under reflections about the x_2-axis.

It is now easy to find the returning orbits. Application of the rotation T_α of (7.48) with $\alpha = L/4$ maps the real coordinates as follows:

$$(x_1, y_1, x_2, y_2) \mapsto (y_1, -x_1, -x_2, -y_2),$$

the first mode being rotated by $\pi/2$ and the second by π. Thus the (x_1, x_2)-plane $\{y_1 = y_2 = 0\}$ becomes the $(y_1, -x_2)$-plane $\{x_1 = y_2 = 0\}$, $\mathbf{p}_+$ is taken to $\mathbf{p}_-$ and all solution curves on $\{y_1 = y_2 = 0\}$ are mapped to their analogs on $\{x_1 = y_2 = 0\}$. In particular the two connecting orbits from $\mathbf{p}_+$ to $\mathbf{p}_-$ are mapped to orbits from $\mathbf{p}_-$ to $\mathbf{p}_+$, and we have our heteroclinic cycle.

But we have much more. Applying the rotation T_α of (7.48) to the $\{y_1 = y_2 = 0\}$ and $\{x_1 = y_2 = 0\}$ planes with *any* α yields a pair of invariant subspaces each containing

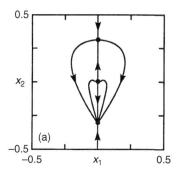 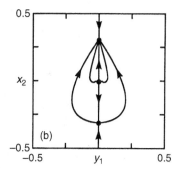

Figure 7.6 Components of the heteroclinic cycle of Equation (7.51) in (a) the (x_1, x_2)-plane, and (b) the (y_1, x_2)-plane. Parameter values $\mu_1 = 0.05$, $\mu_2 = 0.2$, and $e_{11} = -4$, $e_{12} = -1$, $e_{21} = -2$, $e_{22} = -2$, as at point a on Figure 7.8, below.

two diametrically opposite saddle points $\mathbf{p}_\alpha$, $\mathbf{p}_{\alpha+\pi}$ connected by heteroclinic orbits and so we find that *every* pair of diametrically opposite equilibria in the circle $|z_2| = \sqrt{-\mu_2/e_{22}}$ belongs to such a heteroclinic cycle. Figure 7.6 illustrates the cycle containing $\mathbf{p}_-$ and $\mathbf{p}_+$.

Determination of asymptotic stability of the cycles is more subtle than in the previous case and we refer the reader to [9] for specific details and [199] for more general analyses. It turns out that the eigenvalues $-2\mu_2$ and 0, whose eigenvectors span the invariant plane $z_1 = 0$, are irrelevant and stability is determined solely by the ratio $-\lambda_-/\lambda_+$ of the other two eigenvalues having eigenvectors in the "transverse" z_1-subspace. Much as in the two-dimensional case of the loop described in Section 7.3, we find that if $-\lambda_-/\lambda_+ > 1$ the cycle is attracting while if $-\lambda_-/\lambda_+ < 1$ it is unstable. Thus, from (7.63), the bifurcation in which stability changes occurs on the set

$$\mu_1 e_{22} - \mu_2 e_{12} = 0. \tag{7.67}$$

Armbruster *et al.* [9] show that, while the heteroclinic cycles themselves persist through this bifurcation value, an invariant two-dimensional torus filled with quasiperiodic orbits – modulated traveling waves – bifurcates from it. This is done by analysis of the $\epsilon \to 0$ limit in Equation (7.54) and we merely sketch it here. At $\epsilon = 0$, as we have observed, the ODE can be integrated in the sense that, using the relations

$$E = s_1^2 + s_2^2 = \text{const.}, \qquad L = s_1^2 s_2 \sin\phi = \text{const.}, \tag{7.68}$$

two variables can be eliminated and (7.54) reduced to a quadrature. This can be integrated explicitly using Jacobian elliptic functions, but here we are more concerned with the geometry of the three-dimensional (s_1, s_2, ϕ) phase space. Solutions evidently lie on the intersections of surfaces given by (7.68), which yields the structure of Figure 7.7. Here we can see, in embryo, several types of solution. In particular there are two lines of fixed points $\{s_1^2 = 2s_2^2;\ \phi = \pi/2, 3\pi/2\}$, corresponding to traveling waves; a family of periodic orbits depending on the two parameters E and L, corresponding to modulated traveling waves, and the heteroclinic cycles themselves, which appear in this representation as quarter circle arcs on the planes $\phi = 0, \pi$. In the polar coordinate system, a solution reaching $s_2 = 0$ at $s_1 \neq 0$ on $\phi = 0$ instantly reappears at $\phi = \pi$: it is a solution whose z_2-component passes

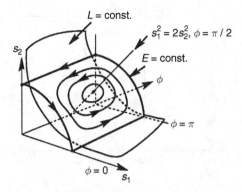

Figure 7.7 Level sets of the integrals E and L, showing a family of unperturbed closed orbits on a surface $E = $ const. We show only the part of phase space $0 \leq \phi \leq \pi$; the part $\pi \leq \phi \leq 2\pi$ is obtained by reflection in the plane $\phi = \pi$.

through the origin while $z_1 \neq 0$. Similarly, the circles of equilibria in the z_2-plane become pairs of isolated equilibria at $\phi = 0$, π on the line $\{r_1 = 0, \ r_2 = $ const.$\}$.

For $\epsilon \neq 0$ small, the quantities E and L no longer remain constant, but they evolve only slowly and one can use averaging and perturbation methods to estimate the changes in them as a solution of Equation (7.54) evolves for a finite time. Using this idea, Armbruster *et al.* estimate the changes ΔE, ΔL as the solution makes one circuit near an unperturbed periodic orbit. Examining the zeroes of ΔE and ΔL as functions of E and L and using properties of elliptic integrals, they show that precisely one periodic orbit is preserved for each set of parameter values satisfying $(4e_{11} + 2e_{12} + 2e_{21} + e_{22}) \neq 0$, $e_{ij} < 0$, $\mu_j > 0$ and lying in the interior of the set bounded by

$$v_2 = v_1 e_{22}/e_{12} \tag{7.69}$$

and

$$v_2 = v_1 \left[1 + \frac{9(e_{22} - e_{12})}{(4e_{11} + 5e_{12} + 2e_{21} - 2e_{22})} \right]. \tag{7.70}$$

This analysis is valid for $\epsilon \neq 0$ and sufficiently small and so, for μ_1 and μ_2 small, we deduce analogs of (7.69) and (7.70) with v_j replaced by μ_j (recall that $\mu_j = \epsilon^2 v_j$ in our rescaling to obtain (7.54)).

When (7.70) is satisfied, a Hopf bifurcation occurs in which families of periodic orbits branch from the traveling wave equilibria and when (7.69) is satisfied the periodic orbits coalesce with the heteroclinic cycle. (Recall that the latter changes stability precisely when $\mu_1 e_{22} - \mu_2 e_{12} = 0$.) The periodic orbits are attracting if

$$1 + \frac{9(e_{22} - e_{12})}{(4e_{11} + 5e_{12} + 2e_{21} - 2e_{22})} < \frac{e_{22}}{e_{12}}, \tag{7.71}$$

and unstable if this inequality is reversed. Figure 7.8 shows a partial bifurcation set in the positive quadrant of the (μ_1, μ_2)-plane with the heteroclinic and Hopf bifurcation curves

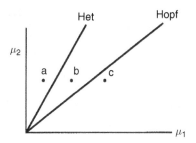

Figure 7.8 A partial bifurcation set for Equation (7.51) with $e_{11} = -4, e_{12} = -1,$
$e_{21} = -2, e_{22} = -2$. The phase portraits in Figure 7.6 correspond to point a, those in
Figures 7.9 and 7.10 to points b and c respectively.

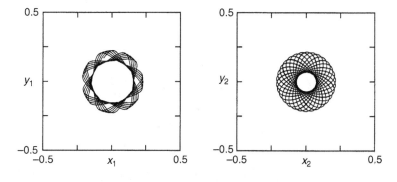

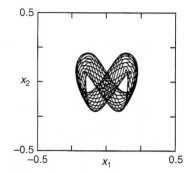

Figure 7.9 Projections of solutions of Equation (7.51) onto three planes for the parameter
values $\mu_1 = 0.135$, $\mu_2 = 0.2$, point b in Figure 7.8, showing quasiperiodic flow on a
torus: a modulated traveling wave.

marked for a case in which stable periodic orbits occur. The corresponding phase portraits for parameter values at the indicated points a,b, and c are shown in Figures 7.6, 7.9
and 7.10.

Figure 7.9 clearly shows that, in the full four-dimensional state space, the periodic orbit
of (7.54) is in fact a solution winding quasiperiodically on a torus. In Figure 7.10 we see

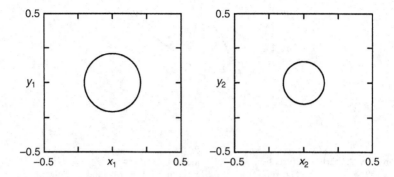

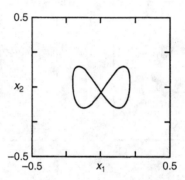

Figure 7.10 Projections of solutions of Equation (7.51) onto three planes for the parameter values $\mu_1 = 0.16$, $\mu_2 = 0.2$, point c in Figure 7.8, showing a periodic orbit: a traveling wave.

that the traveling wave solution describes a circle in each coordinate plane, with frequency ratio $1 : 2$ indicated by the Lissajous figure in the (x_1, x_2)-plane.

To reconstruct the waveform in physical space from the solutions of the ODE, we substitute $z_j(t)$ into (7.45), recalling that $z_{-j} = z_j^*$ and $\theta_{-j}(t) = -\theta_j(t)$:

$$
\begin{aligned}
w(x, t) &= \sum_{j=-2,\ j\neq 0}^{2} z_j(t) \exp\left(\frac{2\pi i j x}{L}\right) \\
&= \sum r_j(t) \exp\left[i\theta_j(t)\right] \exp\left(\frac{2\pi i j x}{L}\right) \\
&= \sum_{j=1}^{2} r_j(t) \left(\exp\left\{-i\left[\frac{2\pi j x}{L} + \theta_j(t)\right]\right\} \right. \\
&\qquad\qquad \left. + \exp\left\{i\left[\frac{2\pi j x}{L} + \theta_j(t)\right]\right\} \right) \\
&= 2 \sum_{j=1}^{2} r_j(t) \cos\left[\frac{2\pi j x}{L} + \theta_j(t)\right].
\end{aligned}
\tag{7.72}
$$

For the traveling wave fixed points we have $r_1^2 = 2r_2^2 \equiv \rho^2$, $\phi = 2\theta_1 - \theta_2 \equiv \pi \pm \beta$, and $2\dot{\theta}_1 = \dot{\theta}_2 = -2r_2 \sin \phi = \pm\sqrt{2}\rho \sin \beta \equiv 4\pi c/L$, so that, up to a phase shift determined by the initial conditions, we can write:

$$\theta_1(t) = \pm\frac{2\pi ct}{L}, \quad \theta_2(t) = \pm\frac{4\pi ct}{L} - (\pi \pm \beta). \tag{7.73}$$

In this case the waveform (7.72) becomes

$$w(x, t) = \rho \left\{ 2\cos\left[\frac{2\pi}{L}(x \pm ct)\right] - \sqrt{2}\cos\left[\frac{4\pi}{L}(x \pm ct) \mp \beta\right] \right\} \tag{7.74}$$

and we explicitly see the pair of left- and right-going waves of constant form.

In this section we have somewhat arbitrarily taken only two Fourier modes to represent a waveform which could, a priori, require arbitrarily many. In the next chapter, we see how such a reduction can sometimes be justified by appeal to the center manifold theorem. The $O(2)$-equivariant ODE (7.50) emerges naturally from a partial differential equation in the neighborhood of a bifurcation point. First, however, we show how translation invariance may be exploited in a different way, which allows more complex waveforms to be represented with a small number of modes, while still retaining the symmetry properties of the governing equations.

7.5 Traveling modes

When a system is translation invariant, the empirical eigenfunctions are Fourier modes, as explained in Section 3.3.3. In the previous section, we exploited this fact, and studied the interactions of two of these Fourier modes. However, solutions of PDEs are not always accurately described by such a small set of Fourier modes. For instance, in the phenomenon of a rotating stall in an axial compressor, the shape of a "stall cell" resembles a square pulse that travels around the annulus of the compressor. This system is translation invariant (and periodic), so the empirical eigenfunctions are Fourier modes. A square wave cannot be represented accurately by a small number of Fourier modes, because of Gibbs phenomena. This difficulty led to the study of expansions in *traveling modes* [127]. Instead of the usual representation (3.19), one considers the expansion

$$u(x, t) = \sum_k a_k(t)\varphi_k(x + c(t)), \tag{7.75}$$

which allows the POD modes to translate by an amount $c(t)$ (possibly with varying speed). This expansion in traveling modes was used earlier in [191] for the Kuramoto–Sivashinsky equation, which we study in the next chapter.

Expansion in traveling modes is an example of *symmetry reduction*, in which one exploits a symmetry by *removing* it. For instance, in the two-body problem of celestial mechanics, one usually begins by using a coordinate system fixed to the center of mass of the two bodies. This is a type of symmetry reduction, since translation invariance of the system has been removed, now that coordinates are relative to its mass center. (One can later reconstruct its position, if desired, but this information is not needed to study the relative motion of the bodies.) When considering traveling wave solutions of PDEs, we can proceed analogously, and shift to a coordinate system that translates with the wave. Translation invariance then implies that the resulting dynamics will be independent of the position of the wave.

In this section, we illustrate the approach for the case of periodic functions on $L^2([0, 2\pi])$, assuming the dynamics are governed by a PDE that is translation invariant. The approach generalizes to other symmetry groups, including systems with self-similar solutions: for a more general treatment, see [317, 318].

The first step is to determine the appropriate reference frame. One approach to this, used in [191], is called *template fitting* or *freezing*. In this approach, one defines a fixed *template function*, and shifts the solution at each time so that it lines up best with the template. The choice of template function is arbitrary, but a common choice is a snapshot from the dataset to be used for POD. Letting $v(x)$ denote the template function, and $u(x)$ denote the solution at a particular time, the minimization problem is then

$$\min_{c\in[0,2\pi]} \int_0^{2\pi} \left[u(x-c) - v(x)\right]^2 dx. \tag{7.76}$$

Defining the *shift operator* on periodic functions

$$S_c[u](x) = u(x+c), \tag{7.77}$$

this minimization problem is equivalent to

$$\max_{c\in[0,2\pi]} \left(u, S_c[v]\right). \tag{7.78}$$

That is, at any time instant, the shift amount $c(t)$ is the value that maximizes the projection of $u(x, t)$ onto the shifted template $v(x + c(t))$. Differentiating (7.78) with respect to c, we find the necessary condition

$$\left(u, S_c[v_x]\right) = 0, \qquad \text{or} \qquad \left(S_{-c}[u], v_x\right) = 0. \tag{7.79}$$

Equation (7.79) has a nice geometric interpretation: it says that the shifted solution $S_{-c}[u]$ is orthogonal to v_x, the spatial derivative of the template function. Thus, the set of all functions that are "correctly shifted" to line up with the template is a linear subspace.

The next step is to determine the dynamics on this space of functions that are aligned with the template. Let $u(x, t) = \hat{u}(x + c(t), t)$, where $\hat{u}$ is aligned with the template (i.e., $(\hat{u}, v_x) = 0$), and expand $\hat{u}$ in terms of POD modes, as

$$\hat{u}(x, t) = \sum_k a_k(t)\varphi_k(x). \tag{7.80}$$

Inserting this expansion into the governing dynamics $u_t = f(u)$, using translation invariance of f (i.e., $f(S_c[u]) = S_c[f(u)]$), and taking inner products with φ_k, one obtains the projected dynamics

$$\dot{a}_k = \left(f(\hat{u}), \varphi_k\right) - \dot{c}\left(\hat{u}_x, \varphi_k\right). \tag{7.81}$$

Note that, while these dynamics do not depend on c, they do depend on $\dot{c}$, so we need an additional equation to determine the translation speed. The approach taken in [318] is to simply differentiate (7.79) with respect to time, and solve for $\dot{c}$, giving

$$\dot{c} = \frac{(f(\hat{u}), v_x)}{(\hat{u}_x, v_x)}. \tag{7.82}$$

This equation is called a *reconstruction equation*, and is precisely analogous to the problem of reconstructing the "phase" (here the translation amount $c(t)$) in more general symmetry reduction problems [236].

The overall procedure is then as follows:

1. Select a template function $v(x)$.
2. Given a set of snapshots $\{u^k(x)\}$, obtain a set of shifted snapshots $\hat{u}^k(x) = u^k(x - c^k)$, where c_k is chosen such that $(\hat{u}^k, v_x) = 0$.
3. Calculate POD modes for the shifted data $\{\hat{u}^k\}$.
4. The resulting Galerkin model is given by (7.81) and (7.82).

Galerkin models formed using this procedure can perform much better than those using the standard approach with Fourier modes. For instance, in a study of the Kuramoto–Sivashinsky equation (with a parameter value $L = 29.30$, for which modulated traveling waves occur), a Galerkin model with three traveling (real) modes reproduced the dynamics more faithfully than a standard Galerkin model projected onto eight complex Fourier modes (16 real ODEs) [318]. Other examples in which this approach has been applied include self-similar solutions of Burgers' equation [317], and spiral waves in excitable media [50].

8

One-dimensional "turbulence"

In the previous two chapters we assembled many of the tools of dynamical systems theory that we shall need. We now apply them to a relatively simple problem: the Kuramoto–Sivashinsky partial differential equation:

$$u_t + u_{xx} + u_{xxxx} + \frac{1}{2}(u_x)^2 = 0, \quad 0 \le x \le L, \tag{8.1}$$

with periodic boundary conditions

$$u(0, t) = u(L, t), \quad u_x(0, t) = u_x(L, t), \dots \tag{8.2}$$

This equation was originally proposed as a model for instabilities on interfaces and flame fronts [341] and "phase turbulence" in chemical reactions [200].* It and related equations have also been used to model directional solidification [270, 271] and, in multiple spatial dimensions, weak fluid turbulence [169,259]. Here we consider only one spatial dimension and make no claims for *direct* relevance to turbulence. We are concerned primarily with illustrating aspects of our general approach on a tractable example.

This example is appealing in that it exhibits interesting dynamical behavior at low "Reynolds numbers": a parameter range in which few modes are active and projections onto low-dimensional subspaces do not require additional modeling to account for neglected modes and so forth. (The modeling is effectively done by the center manifold theorem.) It therefore more clearly displays the main principles of our approach than applications to truly turbulent systems.

In the first section to follow, we describe Galerkin projection onto a subspace spanned by Fourier modes, which are the optimal eigenfunctions for this system. Sections 8.2 and 8.3 contain local bifurcation analyzes which illustrate use of center manifold and normal form methods, and lead to a four-dimensional reduced system which is a special case of the one analyzed in Section 7.4. As we show, this system captures key qualitative and quantitative aspects of the dynamical and bifurcation behavior for lengths $L \approx 4\pi$. In particular it

---

* A version of Eq. (8.1) appeared earlier in the context of free surface flows (G. M. Homsy. Model equations for wavy viscous film flow. In A. C. Newell (ed.), *Nonlinear wave motion*, Lectures in Applied Mathematics, Vol 15, American Mathematical Society, Providence, R. I., 1974). It was also derived as a reduced model in multidimensional plasma dynamics (B. I. Cohen, J. A. Krommes, W. M. Tang, and M. N. Rosenbluth. Nonlinear saturation of the dissipative trapped-ion mode by mode coupling, *Nuclear Fusion*, **16**, 971–92, 1976).

exhibits $O(2)$-equivariant heteroclinic cycles and modulated traveling waves. However, this behavior is not strictly speaking temporally chaotic – the heteroclinic attractor is not strange – and moreover it involves only a few spatial modes.

In the final section we consider a very different parameter range, $L \approx 400$, analogous to high Reynolds numbers, in which many modes are active and one sees spatio-temporal behavior and coherent structures somewhat more reminiscent of Navier–Stokes turbulence. We use this régime to illustrate that the selection of small spatial subdomains and suitable truncation in wavenumber space effectively extracts a class of low-dimensional models which are identical in form to the reduced models of Section 8.3. In this example the heteroclinic behavior can therefore be interpreted as a model for localized coherent structure interactions: a theme which reappears in our treatment of the boundary layer model in Chapters 10 and 11.

The Kuramoto–Sivashinsky equation shares two general features with the Navier–Stokes equations. The term u_{xx} in (8.1) is destabilizing and acts as an energy source while u_{xxxx} is a stabilizing or energy sink term whose effect increases with (spatial) wavenumber. The Fourier transform of the linear part of (8.1) with respect to x,

$$\hat{u}_t(k, t) = k^2(1 - k^2)\hat{u}(k, t), \tag{8.3}$$

makes this clear, exhibiting a range of unstable wavenumbers $|k| < 1$ with a peak at $|k| = 1/\sqrt{2}$ corresponding to the wavenumber of the most rapidly growing linear mode. We return to this in Section 8.4, but for the moment we observe that the terms u_{xx} and u_{xxxx} in (8.1) provide "one-dimensional" analogs to the energy production and dissipation terms $u_j U_{i,j}$ and $u_{i,jj}$ in the Navier–Stokes equations (4.62) or (4.64).

Perhaps more significantly, the translation and reflection symmetries of the Kuramoto–Sivashinsky equation with periodic boundary conditions are the same as the spanwise symmetries of the Navier–Stokes equations in the boundary layer. Hence the Fourier series representations and the resulting modal interactions are similar in the two problems. In fact, when only a single eigenfunction family and just the $k_1 = 0$ streamwise modes are included in the boundary layer model, the structure of the linear and quadratic terms in the projected equations is *identical*: only the numerical values of the coefficients differ. The Kuramoto–Sivashinsky equation is therefore doubly relevant to our purpose.

8.1 Projection onto Fourier modes

In this and the next two sections we draw heavily upon another paper by Armbruster *et al.* [10], to which the reader should refer for more details. We treat the period L in Equations (8.1) and (8.2) as a bifurcation parameter and concentrate for the most part on "small" values of L, for which, as we shall see, only few spatial modes are active.

We first observe that, because only even orders of spatial derivatives appear in each term and we have constant coefficients and periodic boundary conditions, the system is equivariant under spatial translations and reflections:

$$T_\alpha : x \mapsto x + \alpha \quad \text{and} \quad R_f : x \mapsto -x. \tag{8.4}$$

This is exactly the situation envisaged in our introduction to Section 7.4. Moreover, as pointed out in Chapter 3, translation invariance implies that the empirical eigenfunctions obtained from appropriate ensemble averages are Fourier modes. It is therefore reasonable to examine sets of ODEs obtained by projection into subspaces spanned by the basis functions

$$\phi_k(x) = \exp\left(\frac{2\pi i k x}{L}\right). \tag{8.5}$$

We see that such a choice also diagonalizes the linear part of (8.1), significantly simplifying our calculations. We therefore represent the dependent variable $u = u(x, t)$ as

$$u(x, t) = \sum_k a_k(t)\phi_k(x). \tag{8.6}$$

Here the a_k are complex modal coefficients and, as in Section 7.4, reality of the scalar field u implies that

$$a_{-k}(t) = a_k^*(t), \tag{8.7}$$

where a_k^* denotes the complex conjugate of a_k.

The Fourier modes (8.5) form an orthogonal basis with respect to the L^2 inner product

$$(f, g) = \int_0^L f(x)g^*(x)dx; \tag{8.8}$$

in fact we have

$$(\phi_k, \phi_l) = \int_0^L \exp\left[\frac{2\pi i(k-l)x}{L}\right] dx = \delta_{kl} L. \tag{8.9}$$

Now, using the relationships

$$(\phi_k)_x = \frac{2\pi i k}{L}\phi_k, \quad (\phi_k)_{xx} = -\left(\frac{2\pi k}{L}\right)^2 \phi_k, \quad (\phi_k)_{xxxx} = \left(\frac{2\pi k}{L}\right)^4 \phi_k, \tag{8.10}$$

application of the Galerkin procedure to (8.1) yields

$$\int_0^L \left\{ \sum_k \left[\dot{a}_k\phi_k - \left(\frac{2\pi k}{L}\right)^2 a_k\phi_k + \left(\frac{2\pi k}{L}\right)^4 a_k\phi_k \right. \right.$$

$$\left. \left. + \frac{1}{2}\left(\frac{2\pi i k}{L}a_k\phi_k\right)\sum_j\left(\frac{2\pi i j}{L}a_j\phi_j\right) \right]\phi_{-l} \right\} dx = 0. \tag{8.11}$$

Use of (8.9) reduces (8.11) to

$$L\left[\dot{a}_l - \left(\frac{2\pi l}{L}\right)^2 a_l + \left(\frac{2\pi l}{L}\right)^4 a_l - \frac{1}{2}\left(\frac{2\pi}{L}\right)^2 \sum_j j(l-j)a_j a_{l-j} \right] = 0, \tag{8.12}$$

since in the integrals involving the quadratic terms, we have

$$\int_0^L \exp\left[\frac{2\pi i(k+j-l)x}{L}\right] dx = 0,$$

unless $k + j = l$. Finally, rescaling time by the multiplicative factor $(2\pi/L)^2$, we can rewrite (8.12) in the more compact form:

$$\dot{a}_l = l^2 \left[1 - \left(\frac{2\pi l}{L} \right)^2 \right] a_l + \frac{1}{2} \sum_j j(l - j) a_j a_{l-j}. \tag{8.13}$$

So far we have been reticent about the sum appearing in the representation (8.6) and in the ODE (8.13). In fact the form (8.13) applies to *any* truncation order K originally assumed in (8.6) and so in general (8.13) defines an ODE on complex $(2K+1)$-dimensional space. (If we interpret sequences $\{a_k\}$ as belonging to the space l^2 of infinite sequences having "finite energy" $\sum_{k=-\infty}^{\infty} |a_k|^2 < \infty$, we may even let $K = \infty$.) We can modestly reduce the dimension of (8.13) by noting that the equation for a_0, the spatial average, uncouples and is merely driven by the other modes according to

$$\dot{a}_0 = -\frac{1}{2} \sum_j j^2 |a_j|^2; \tag{8.14}$$

a_0 does not enter the equations for the other components. Moreover, the reality condition (8.7) implies that we need only consider the equations for a_l with $l \geq 1$. Thus, for a system truncated at order K, we have only to solve K complex or $2K$ real, first order ODEs.

To give a feeling for the structure of the system, and for our future convenience in center manifold calculations, we display the equations obtained by truncation at order $K = 4$:

$$\left.\begin{aligned}
\dot{a}_1 &= [1 - (2\pi/L)^2]a_1 - 2a_1^* a_2 - 6a_2^* a_3 - 12a_3^* a_4, \\
\dot{a}_2 &= 4[1 - (4\pi/L)^2]a_2 + \tfrac{1}{2}a_1^2 - 3a_1^* a_3 - 8a_2^* a_4, \\
\dot{a}_3 &= 9[1 - (6\pi/L)^2]a_3 + 2a_1 a_2 - 4a_1^* a_4, \\
\dot{a}_4 &= 16[1 - (8\pi/L)^2]a_4 + 2a_2^2 + 3a_1 a_3.
\end{aligned}\right\} \tag{8.15}$$

As we expect from the discussion of symmetries and equivariance of Fourier representations in Section 7.4, we observe that (8.13) in general, and (8.15) in particular, are equivariant under the group $O(2)$ with representation

$$\left.\begin{aligned}
T_\alpha &: a_l \mapsto \exp(2\pi i l\alpha/L)a_l, \\
R_f &: a_l \mapsto a_l^*.
\end{aligned}\right\} \tag{8.16}$$

8.2 Local bifurcations from $u = 0$

The stability characteristics of the equilibrium at the origin ($a_l = 0$ for all l) can be determined directly from the linearization

$$\dot{a}_l = l^2 \left[1 - \left(\frac{2\pi l}{L} \right)^2 \right] a_l, \quad l = 1, 2, \ldots \tag{8.17}$$

in which all modes decouple and each one is simply governed by the two real equations

$$\left.\begin{aligned}\dot{x}_l &= l^2 \left[1 - (2\pi l/L)^2\right] x_l, \\ \dot{y}_l &= l^2 \left[1 - (2\pi l/L)^2\right] y_l. \end{aligned}\right\} \tag{8.18}$$

Every eigenvalue, given by

$$\lambda_l = l^2 \left[1 - \left(\frac{2\pi l}{L}\right)^2\right], \tag{8.19}$$

has multiplicity two. Evidently the origin is asymptotically stable if $L < 2\pi$ and unstable if $L > 2\pi$. At $L = 2\pi$ we have a double zero eigenvalue and a bifurcation occurs: a situation repeated at $L = L_m \overset{\text{def}}{=} 2\pi m$ for all integers $m = 1, 2, 3, \ldots$ From the discussion in Chapter 7, we expect these bifurcations to be $O(2)$-equivariant pitchforks, and we shall now show that this is indeed the case.

At $L = L_m$ the center eigenspace is the complex a_m-plane and so we seek a one (complex-) dimensional center manifold given by a set of complex-valued functions of the form

$$a_l = h_l(a_m, a_m^*), \quad l \neq m, -m, \quad h_l = \mathcal{O}(|a_m|^2), \tag{8.20}$$

satisfying $h_{-l} = h_l^*$. In adapting the form for the center manifold in the real case to complex coordinates, we must allow dependence on both a_m and a_m^*. The reduced system is then given, as usual, by substitution of (8.20) into the mth component of (8.13), and it provides an accurate picture of solutions on the center manifold for $L \approx L_m = 2\pi m$. After this substitution, most of the nonlinear terms $a_j a_{l-j}$ in (8.13) are of order four or higher; only those involving a_m or $a_{-m} = a_m^*$ contain cubic terms. We therefore explicitly separate the latter, corresponding to $j = -m$ and $j = 2m$ (the $j = 0$, $j = m$ terms vanish identically):

$$\dot{a}_m = m^2 \left[1 - \left(\frac{2\pi m}{L}\right)^2\right] a_m - 2m^2 a_m^* h_{2m}(a_m, a_m^*) + \mathcal{O}(|a_m|^4). \tag{8.21}$$

Note that only the single function h_{2m} enters at third order.

To compute the leading (quadratic) term in h_{2m} we apply the procedure of Section 6.4, modestly generalized to accommodate complex functions. We augment each complex ODE with its conjugate:

$$\dot{a}_l^* = l^2 \left[1 - \left(\frac{2\pi l}{L}\right)^2\right] a_l^* + \frac{1}{2} \sum_j j(l - j) a_j^* a_{l-j}^* \tag{8.22}$$

and write the analog of Equation (6.30) for the function h_{2m} as

$$\left(\frac{\partial h_{2m}}{\partial a_m}, \frac{\partial h_{2m}}{\partial a_m^*}\right) \cdot \begin{pmatrix} \dot{a}_m \\ \dot{a}_m^* \end{pmatrix} = (2m)^2 \left[1 - \left(\frac{4\pi m}{L}\right)^2\right] h_{2m}$$

$$+ \frac{1}{2} m^2 a_m^2 + \frac{1}{2} \sum_{j \neq m} j(2m - j) h_j h_{2m-j}. \tag{8.23}$$

At $L = 2\pi m$ the linear terms in $\dot{a}_m$ and $\dot{a}_m^*$ vanish and the left-hand side of (8.23) is consequently of $\mathcal{O}(|a_m|^3)$. The term on the extreme right is also of $\mathcal{O}(|a_m|^3)$ and so, balancing the quadratic terms, we obtain

$$h_{2m} = \frac{1}{24}a_m^2 + \mathcal{O}(|a_m|^3). \tag{8.24}$$

Substitution of this into the reduced system (8.21) yields

$$\dot{a}_m = m^2\left[1 - \left(\frac{2\pi m}{L}\right)^2\right]a_m - \frac{1}{12}m^2|a_m|^2a_m + \mathcal{O}(|a_m|^4), \tag{8.25}$$

and finally, letting $L = 2\pi m(1 + \mu)$, with $\mu \ll 1$, we obtain the bifurcation equation

$$\dot{a}_m = m^2 a_m\left(\mu - \frac{1}{12}|a_m|^2\right) + \mathcal{O}(|a_m|^4, \mu^2, \dots). \tag{8.26}$$

In Cartesian coordinates on each (real) modal plane, this is simply

$$\begin{pmatrix} \dot{x}_m \\ \dot{y}_m \end{pmatrix} = m^2\left[\mu - \frac{1}{12}(x_m^2 + y_m^2)\right]\begin{pmatrix} x_m \\ y_m \end{pmatrix} + \cdots, \tag{8.27}$$

which, truncated at cubic order, is precisely the normal form of the $O(2)$-equivariant pitchfork bifurcation (7.24), as claimed. We conclude that, as L increases through each bifurcation value $L_m = 2\pi m$, a circle of equilibria branches from the origin and that, for $L - L_m > 0$ sufficiently small, this circle lies on an invariant two-(real-) dimensional manifold tangent to the a_m-plane and given by (8.24). (We note that Equation (3.3) in Armbruster *et al.* [10] – the analog of (8.21) in the present treatment – contains the incorrect numerical factor -4 in place of -2. Thus the cubic term of the reduced equation in that paper is also in error by a factor of 2.)

As we remarked in passing in the introduction to this chapter, the center manifold effectively models the neglected stable modes via the algebraic relationship it establishes among these and the critical modes included in the reduced system, as in Equation (8.24). Geometrically, one is replacing the flat subspace of resolved modes in phase space by a curved manifold which in general imposes coupling among all modes and attracts nearby solutions. One might loosely think of this as a nonlinear eddy viscosity model.

The local bifurcation analysis is also valid for the original PDE. The fact that the Fourier representation diagonalizes the self-adjoint linear operator of (8.1) implies that the (unscaled) eigenvalues $\lambda_l = (2\pi l/L)^2 \times [1 - (2\pi l/L)^2]$ of the linearized ODE coincide with the eigenvalues of the differential operator, which are also each of multiplicity two, with eigenfunctions

$$\cos\left(\frac{2\pi l x}{L}\right), \quad \sin\left(\frac{2\pi l x}{L}\right).$$

The center manifold theorem for evolution equations [71] then guarantees the existence, for $L = L_m$, of a two-dimensional invariant manifold tangent to the center eigenspace

$$\text{span}\left\{\cos\left(\frac{2\pi l x}{L}\right), \sin\left(\frac{2\pi l x}{L}\right)\right\}$$

in the space of L-periodic functions, which can be extended to a family of center manifolds for $L \approx L_m$ that carry the circles of equilibria for $L - L_m > 0$ sufficiently small.

8.3 The second bifurcation point

The circle of equilibria created in the first $O(2)$-equivariant pitchfork bifurcation at $L_1 = 2\pi$ is asymptotically stable for $L - L_1 > 0$ and sufficiently small. This follows from (8.26) with $m = 1$ and the fact that all the other eigenvalues of the equation linearized at the origin are strictly negative (see (8.19)). Near the second bifurcation point $L_2 = 4\pi$, we therefore have a two-dimensional unstable manifold tangent to the a_1-plane and a two-dimensional center manifold tangent to the a_2-plane at the origin. The other directions, tangent to the subspaces spanned by the Fourier modes ϕ_l, $l \geq 3$, all lie in the stable manifold. It therefore seems reasonable to seek a reduced system restricted to a four-dimensional *center-unstable* manifold if we wish to study the dynamics for parameter values $L \approx L_2$. We first define a new (small) parameter:

$$\mu = \left(\frac{2\pi}{L_2}\right)^2 - \left(\frac{2\pi}{L}\right)^2 = \frac{1}{4} - \left(\frac{2\pi}{L}\right)^2, \tag{8.28}$$

so that

$$\left[1 - \left(\frac{2\pi}{L}\right)^2\right] = \frac{3}{4} + \mu \ \text{ and } \ 4\left[1 - \left(\frac{4\pi}{L}\right)^2\right] = 16\mu. \tag{8.29}$$

We next follow the procedure of Section 6.4, Equations (6.37–6.42), seeking the parameterized family of center manifolds as graphs of the form:

$$a_l = h_l(a_1, a_1^*, a_2, a_2^*, \mu), \ l = 3, 4, 5, \ldots \tag{8.30}$$

Much as in our derivation of the reduced system (8.26) on the two-dimensional center manifold in the preceding section, it turns out that we need explicitly compute only the leading terms of the functions h_l for $l = 3$ and 4 to obtain accuracy up to $\mathcal{O}(|a_m|^4)$. The reduced system in this case takes the form

$$\left.\begin{array}{l} \dot{a}_1 = (\frac{3}{4} + \mu)a_1 - 2a_1^* a_2 - 6a_2^* h_3 - 12h_3^* h_4 - \sum_{k \geq 4} k(k+1)h_k^* h_{k+1}, \\ \dot{a}_2 = 16\mu a_2 + \frac{1}{2}a_1^2 - 3a_1^* h_3 - 8a_2^* h_4 - \sum_{k \geq 3} k(k+2)h_k^* h_{k+2}. \end{array}\right\} \tag{8.31}$$

We claim that $h_l = \mathcal{O}(|a_l|^3)$ for all $l \geq 5$, implying that the terms in the sums on the extreme right of (8.31) are all of $\mathcal{O}(|a_m|^5)$ and so will not appear in the reduced system approximated up to $\mathcal{O}(|a_m|^4)$.

To verify this claim, we consider the components of Equation (6.30), which in this case are each of the form

$$\frac{\partial h_l}{\partial a_1}\dot{a}_1 + \frac{\partial h_l}{\partial a_1^*}\dot{a}_1^* + \frac{\partial h_l}{\partial a_2}\dot{a}_2 + \frac{\partial h_l}{\partial a_2^*}\dot{a}_2^* = l^2\left[1 - \left(\frac{2\pi l}{L}\right)^2\right]h_l$$

$$+ \frac{1}{2}\sum_{j=1}^{l-1} j(l-j)a_j a_{l-j} - \sum_{j \geq 1} j(j+l)a_j^* a_{j+l}, \quad l \geq 3, \tag{8.32}$$

where we understand that $a_l = h_l(a_1, a_1^*, a_2, a_2^*, \mu)$ for $l \geq 3$ and we have rewritten the nonlinear terms so that only positive indices appear. We know a priori that $h_l = \mathcal{O}(|a_m|^2)$ and, from (8.31), that $\dot{a}_1, \dot{a}_2 = \mathcal{O}(|a_1|), \mathcal{O}(\mu|a_2|)$. One order is lost in differentiating h_l. Thus, for $l = 5$, the left-hand side of (8.32) is $\mathcal{O}(|h_5|)$ while the three terms on the right-hand side are respectively $\mathcal{O}(|h_5|)$, $\mathcal{O}(|a_1h_4|, |a_2h_3|)$, and $\mathcal{O}(|a_1h_6|, |a_2h_7|, |h_3h_8|, \ldots, |h_kh_{k+5}|, \ldots)$. Balancing the lowest order terms we find that

$$h_5 = \mathcal{O}(|a_m|^3), \tag{8.33}$$

and similarly that

$$\left.\begin{array}{l} h_6 = \mathcal{O}(|a_1h_5|, |a_2h_4|, |h_3|^2) + \mathcal{O}(|a_1h_7|, |a_2h_8|, \ldots) = \mathcal{O}(|a_m|^3), \\ h_7 = \mathcal{O}(|a_1h_6|, |a_2h_5|, |h_3h_4|) + \mathcal{O}(|a_1h_8|, |a_2h_9|, \ldots) = \mathcal{O}(|a_m|^4). \end{array}\right\} \tag{8.34}$$

In general we have $h_l = \mathcal{O}(|a_m|^{[(l+1)/2]})$, where the square brackets denote the integer part. Our claim is justified.

To compute h_3 and h_4 up to and including terms of $\mathcal{O}(|a_m|^3)$ and $\mathcal{O}(\mu|a_m|^2)$, we return to Equation (6.30) with the specific components indicated in (8.32). Neglecting terms of $\mathcal{O}(|a_m|^4)$ and $\mathcal{O}(\mu^2|a_m|^2)$, we have

$$\begin{bmatrix} \dfrac{\partial h_3}{\partial a_1} & \dfrac{\partial h_3}{\partial a_1^*} & \dfrac{\partial h_3}{\partial a_2} & \dfrac{\partial h_3}{\partial a_2^*} \\ \dfrac{\partial h_4}{\partial a_1} & \dfrac{\partial h_4}{\partial a_1^*} & \dfrac{\partial h_4}{\partial a_2} & \dfrac{\partial h_4}{\partial a_2^*} \end{bmatrix} \cdot \begin{pmatrix} (\frac{3}{4} + \mu)a_1 - 2a_1^*a_2 + \cdots \\ (\frac{3}{4} + \mu)a_1^* - 2a_1a_2^* + \cdots \\ 16\mu a_2 + \frac{1}{2}a_1^2 + \cdots \\ 16\mu a_2^* + \frac{1}{2}a_1^{*2} + \cdots \end{pmatrix}$$
$$= \begin{pmatrix} 9(-\frac{5}{4} + 9\mu)h_3 + 2a_1a_2 - 4a_1^*h_4 + \cdots \\ 16(-3 + 16\mu)h_4 + 2a_2^2 + 3a_1h_3 + \cdots \end{pmatrix}. \tag{8.35}$$

This system of two simultaneous complex differential equations is solved, up to the order required, by

$$\left.\begin{array}{l} h_3 = \frac{1}{6}(1 + \frac{16}{3}\mu)a_1a_2 - \frac{1}{162}a_1^3 + \frac{1}{72}a_1^*a_2^2 + \cdots \\ h_4 = \frac{1}{24}(1 + \frac{14}{3}\mu)a_2^2 + \frac{1}{108}a_1^2a_2 + \cdots \end{array}\right\} \tag{8.36}$$

We observe that these expressions satisfy

$$h_l(e^{i\beta}a_1, e^{-i\beta}a_1^*, e^{2i\beta}a_2, e^{-2i\beta}a_2^*, \mu) = e^{il\beta}h_l(a_1, a_1^*, a_2, a_2^*, \mu)$$
$$\text{and} \quad h_l^* = h_{-l}, \tag{8.37}$$

so that our approximation to the center-unstable manifold is also invariant under the action of the group $O(2)$. Finally, substituting the expressions of Equations (8.36) into (8.31), we obtain the reduced system

$$\left.\begin{array}{l} \dot{a}_1 = (\frac{3}{4} + \mu)a_1 - 2a_1^*a_2 - (1 + \frac{16}{3}\mu)a_1|a_2|^2 + \frac{1}{27}a_1^3a_2^* \\ \qquad - \frac{1}{6}a_1^*a_2|a_2|^2 + \cdots, \\ \dot{a}_2 = 16\mu a_2 + \frac{1}{2}a_1^2 - \frac{1}{2}(1 + \frac{16}{3}\mu)a_2|a_1|^2 - \frac{1}{3}(1 + \frac{14}{3}\mu)a_2|a_2|^2 \\ \qquad - \frac{1}{54}a_1^2|a_1|^2 - \frac{1}{24}a_1^{*2}a_2^2 - \frac{2}{27}a_1^2|a_2|^2 + \cdots, \end{array}\right\} \tag{8.38}$$

in which the errors are of $\mathcal{O}(|a_m|^5)$ and $\mathcal{O}(\mu^2|a_m|^3)$.

Truncating at cubic order we have

$$\left.\begin{array}{l} \dot{a}_1 = (\frac{3}{4} + \mu)a_1 - 2a_1^*a_2 - a_1|a_2|^2, \\ \dot{a}_2 = 16\mu a_2 + \frac{1}{2}a_1^2 - a_2(\frac{1}{2}|a_1|^2 + \frac{1}{3}|a_2|^2), \end{array}\right\} \tag{8.39}$$

which is a special case of the $O(2)$-equivariant normal form (7.50). In fact, making use of the change of variables $-2a_2 \mapsto a_2$, we obtain the form (7.51):

$$\left.\begin{array}{l} \dot{a}_1 = (\frac{3}{4} + \mu)a_1 + a_1^* a_2 - \frac{1}{4}a_1|a_2|^2, \\[2mm] \dot{a}_2 = 16\mu a_2 - a_1^2 - a_2(\frac{1}{2}|a_1|^2 + \frac{1}{12}|a_2|^2), \end{array}\right\} \tag{8.40}$$

and we can read off the behavior directly from our earlier analysis.

As in the simpler bifurcation analysis of Section 8.2, the center manifold given by (8.36) provides an algebraic relation expressing the neglected modes a_3, a_4 in terms of the modes a_1, a_2 of the reduced system.

We are especially interested in dynamical behavior and bifurcations which occur near $\mu = 0$ ($L = L_2 = 4\pi$) in the neighborhood of $a_1 = a_2 = 0$. For $\mu > 0$ the circle of fixed points $\{a_1 = 0; |a_2| = 8\sqrt{3\mu}\}$ has eigenvalues with eigenvectors in the a_1-plane given by

$$\lambda_{\pm} = \frac{3}{4} - 47\mu \pm 8\sqrt{3\mu} \tag{8.41}$$

(from Equations (7.61–7.63)). Thus these fixed points are saddles with one-dimensional unstable manifolds for all μ satisfying

$$0.002183\ldots < \mu < 0.116649\ldots; \tag{8.42}$$

these extreme μ values correspond to $\lambda_- = 0$ and $\lambda_+ = 0$ respectively. Moreover, since $(3/4 + \mu) > 0$ if $\mu > 0$, we also have heteroclinic cycles in this μ range. From the discussion in Section 7.4, these cycles are asymptotically stable provided $-\lambda_-/\lambda_+ > 1$, or

$$\mu > \mu_{\text{Het}} = \frac{3}{4 \times 47} = 0.015957\ldots, \tag{8.43}$$

and at $\mu = \mu_{\text{Het}}$ a branch of quasiperiodic modulated traveling waves emerges. This family of invariant tori in turn coalesces with the traveling wave solutions at

$$16\mu = \left(\frac{3}{4} + \mu\right)\left(\frac{7}{25}\right) \quad \text{or} \quad \mu = \mu_{\text{Hopf}} = 0.013359\ldots, \tag{8.44}$$

from Equation (7.70). Since $\mu_{\text{Het}} > \mu_{\text{Hopf}}$, inequality (7.71) implies that the tori are attracting.

In Figure 8.1(a) we show a bifurcation diagram for (8.40) which schematically illustrates this behavior and also shows the branch of "mixed mode" equilibria $\{a_1, a_2 \neq 0\}$ which appear at $L = L_1 = 2\pi$ and the branch of traveling waves which bifurcates from these mixed modes. The traveling wave bifurcation appears as a pitchfork for the reduced system corresponding to (8.40) in the form (7.53) and occurs at $\mu = -0.050925\ldots$, but we do not describe this analysis here.

The bifurcation diagram for the "full" system (actually a Galerkin projection including 16 complex Fourier modes) as computed by Kevrekidis *et al.* ([188] Figures 2.1 and 3.2), is similar to Figure 8.1(a) in all but one important respect. For the full system, the bifurcation values at which the branch of modulated traveling waves appears occur in the opposite order ($\mu_{\text{Het}} < \mu_{\text{Hopf}}$) and so the modulated traveling waves are unstable. Moreover, the latter coexist with asymptotically stable traveling waves and the attracting heteroclinic cycles in the narrow range $\mu_{\text{Het}} < \mu < \mu_{\text{Hopf}}$ (it is a *very* narrow range: the branch is almost vertical, as the values tabulated below indicate). Evidently the cubic truncation is

Table 8.1 *Bifurcation values for the Kuramoto–Sivashinsky equation.*

Bifurcation point (L)	Full system	Reduced $\mathcal{O}(3)$ (8.40)	Reduced $\mathcal{O}(4)$ (8.38)
Bifurcation of a_2 mode	4π (12.5664 ...)	4π	4π
Hopf bifurcation	13.1403 ...	12.9162 ...	13.0824 ...
Heteroclinic bifurcation	12.8767 ...	12.9877 ...	12.9833 ...

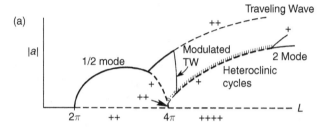

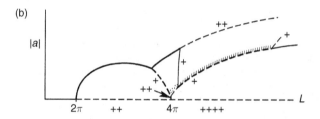

Figure 8.1 The bifurcation diagrams for the reduced systems: (a) (8.40) truncated at third order; (b) (8.38), including fourth order terms. Asymptotically stable branches are shown bold, heteroclinic cycles hatched, unstable cycles dotted. + indicates the number of positive (unstable) eigenvalues on each branch. From Armbruster *et al.* [10]. Copyright ©1989 Society for Industrial and Applied Mathematics. Reprinted with permission. All rights reserved.

not sufficiently accurate and one must at least include fourth order terms in the reduced system, as in (8.38). Armbruster *et al.* [10] did this and, with the help of computer algebra, carried out calculations analogous to those above and in Section 7.4 to obtain the improved estimates:

$$\mu_{\text{Het}} = 0.015799\ldots \quad \text{and} \quad \mu_{\text{Hopf}} = 0.019335\ldots \tag{8.45}$$

Now $\mu_{\text{Hopf}} > \mu_{\text{Het}}$, as in the full system, and the resulting bifurcation diagram is shown in Figure 8.1(b). We also summarize the results in Table 8.1 in terms of the original bifurcation parameter L. The bifurcation values derived for the fourth order reduced system (8.38) lie within 1% of those of the full system.

(a)

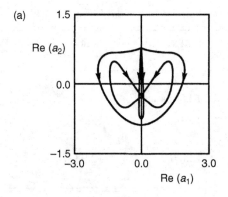

(b)

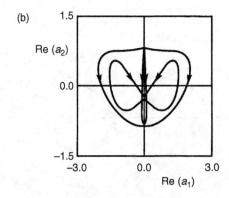

Figure 8.2 Numerical simulations of (a) the four mode Galerkin projection (8.15), and (b) the fourth order reduced system (8.38), both projected onto the $(\text{Re}(a_1), \text{Re}(a_2))$-plane. Parameter value $L = 13.0291$. Note coexistence of attracting heteroclinic cycle and traveling waves in both cases. From Armbruster *et al.* [10]. Copyright ©1989 Society for Industrial and Applied Mathematics. Reprinted with permission. All rights reserved.

It is perhaps even more striking that simulations of the fourth order reduced system (8.38) and of the four complex mode truncation (8.15) of the full system agree so well, as shown in Figure 8.2. (Truncations including 8, 16, or more modes, as in [188], give almost identical behavior.) The similarity, both qualitative and quantitative, is remarkable.

We conclude this section by showing reconstructions of the solution $u(x, t)$ of (8.1) derived from the fourth order reduced system via

$$u(x,t) = \sum_{k=-4}^{4} a_k(t)\phi_k(t), \quad a_{-k}(t) = a_k(t), \tag{8.46}$$

with $a_1(t)$ and $a_2(t)$ obtained by numerical integration of (8.38) and $a_3(t)$ and $a_4(t)$ from the center-unstable manifold approximation (8.39). Figure 8.3 shows examples of traveling waves and heteroclinic cycles for $L = 13.0291$ (as in Figure 8.2). We show the value of $u(x, t)$ over a portion of the (x, t)-plane in a three-dimensional representation. In the case of solutions approaching a heteroclinic cycle we see relatively long periods in which $u(x, t)$ remains close to the pure second Fourier mode, punctuated by brief events, corresponding to traversal of the heteroclinic connections in the phase space, in which the first

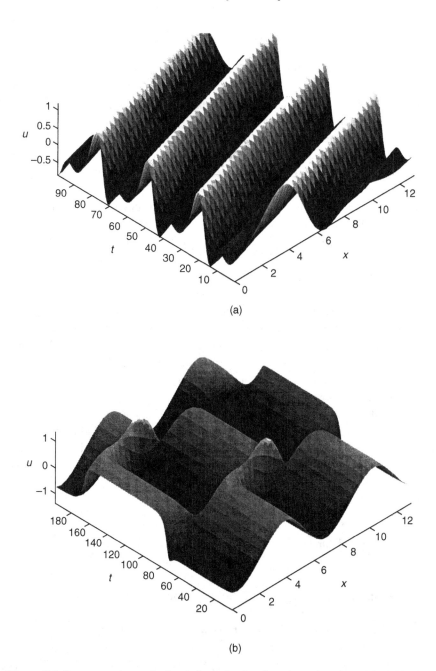

Figure 8.3 Reconstructions of $u(x, t)$ from the fourth order reduced model (8.38) for $L = 13.0291$. (a) A traveling wave, (b) a heteroclinic cycle. The time scale is that of the original PDE (8.1).

and other modes are active. After each of these events the solution again settles near the second Fourier mode, but with a spatial shift of $L/4$ with respect to its previous state, corresponding to a phase shift of π. The perspective view reproduced in Figure 5.1b of [188] also shows the phase shift clearly. Similar behavior reappears in various guises in the next section and in Chapters 10 and 11.

8.4 Spatio-temporal chaos

The analysis sketched above almost justifies our use of a low- (four-) dimensional model to represent the dynamics of the Kuramoto–Sivashinsky equation in the parameter range $L \approx 4\pi$. (Rigorous justification would be complete if we could prove the existence of a four-dimensional global inertial manifold, rather than merely a local center-unstable manifold.) Moreover, we believe that the key features of this model of the "short system" – heteroclinic cycles and (modulated) traveling waves – are also characteristic of solutions of the Kuramoto–Sivashinsky equation in much longer spatial domains. In this final section we indicate why we believe this to be so.

In the last section of Chapter 4 we pointed out that derivation of low-dimensional models of turbulent flows typically demands localization in physical as well as wavenumber space. One has to extract small subdomains of a much larger flow region as well as truncating wavenumbers in the inertial and dissipative ranges. Somewhat glibly, at that point we remarked on the usual strategy of selecting suitable length scales and applying periodic boundary conditions in statistically homogeneous directions. We indicate how this is done for the boundary layer model in Chapter 10, but the Kuramoto–Sivashinsky example provides a simpler test case. In the following we argue that short, periodized systems can indeed reproduce all the key ingredients of coherent structure dynamics, at least for this system. We caution the reader that this section is more speculative than the rest of this chapter, but we believe it is also more interesting and perhaps ultimately more relevant for modeling of turbulent flows.

Figure 8.4 shows a spatio-temporal representation of a solution on the interval $L = 400 \approx 64 \times 2\pi = L_{64}$. To obtain this, a 1024-dimensional system including 512 (complex) Fourier modes was integrated using a pseudospectral code, as described in Berkooz *et al.* [42]. Rather than solving Equation (8.1) directly, for which, as (8.14) indicates, the spatial average

$$|a_0(t)| = \int_0^L |u(x,t)| \, dx$$

typically becomes unbounded, the equation

$$v_t + v_{xx} + v_{xxxx} + v v_x = 0 \tag{8.47}$$

for the spatial derivative $v = u_x$ was used for this simulation. Now we have

$$\frac{d}{dt} \int_0^L v(x,t) \, dx = \int_0^L v_t(x,t) \, dx = \int_0^L (-v_{xx} - v_{xxxx} - v v_x) \, dx$$

$$= -v_x \Big|_0^L - v_{xxx} \Big|_0^L - \frac{v^2}{2} \Big|_0^L \equiv 0, \tag{8.48}$$

due to the periodic boundary conditions, so that the mean remains constant. It is set to zero for the computations described here. In Figure 8.4 the pointwise values of v are indicated by a gray scale (black = positive, white = negative).

In this figure we see a predominant wavenumber $k \approx 45/64 \approx 1/\sqrt{2}$, indicated by the average spacing between the dark and light stripes corresponding to wave crests and troughs. The Fourier (power) spectrum $S(k)$, obtained by averaging the spatial transform

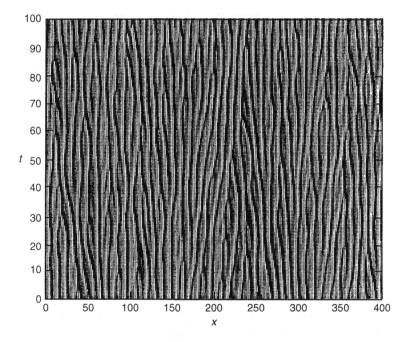

Figure 8.4 A numerical simulation of Equation (8.47) with periodic boundary conditions on the interval of length $L = 400$. From Berkooz *et al.* [42].

$$\hat{u}(k, t) = \int_0^L u(x, t) e^{-ikx} dx, \tag{8.49}$$

with respect to time:

$$S(k) = \lim_{T \to \infty} \frac{1}{T} \int_0^T |\hat{u}(k, t)|^2 dt \tag{8.50}$$

confirms this. A plot of this spatial power spectrum appears in Figure 8.5; similar spectra, computed for intervals of other lengths ($L = 128, 512$) appear in [370]. The spectrum peaks at $k = 1/\sqrt{2}$ and falls off rapidly after $k \approx 1.25$. It is therefore broadly similar to the spectrum $k^2(1 - k^2)$ of the linear part of the equation, as in (8.3).

In Figure 8.5 one can see three clear ranges of behavior: a relatively flat low frequency part ($0 < k < 0.4$), the peak region ($0.4 < k < 1.25$), and the rapid decay ($k > 1.25$), where dissipation dominates. (With imagination, or suitable bias, one can even see an "inertial range" ($1/\sqrt{2} < k < 1.25$) with power-law decay, preceding the true dissipative range in which the decay is exponential.) It is natural to identify the dominant wavenumber range around $k = 1/\sqrt{2}$ with the coherent structures of the problem which account for the bulk of the energy. Here the structures are simply waves and it is their selection of a preferred wavenumber that lends them coherence. Examining Figure 8.4 more closely, we can see the mechanism by which these structures are created, destroyed, and transformed. One sees localized annihilation events, in which neighboring wave crests or troughs merge and $v(x, t)$ loses a pair of adjacent zeros, and creation events in which the reverse occurs. It

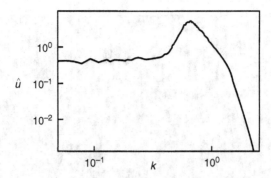

Figure 8.5 Log–log plot of the power spectrum $S(k)$, obtained from a sample of 6×10^6 time values of the solution $u(x, t)$ computed for $L = 400$. From Dankowicz *et al.* [89], following [42].

is as if, while neighboring waves jostle for space, their wavelengths growing or shrinking in the process, the system continually readjusts to maintain the "correct" average wavelength.

Comparing these local events with the behavior of $u(x, t)$ for the short system $L \approx 4\pi$, we see several common features. In particular there are relatively long temporal ranges in which the local pattern remains approximately periodic, ranges initiated and terminated by creation or annihilation events. Sometimes the waves travel to the left or right during these periods. Although we do not see the regularity of Figure 8.3(b) – annihilation *always* immediately followed by creation, with the associated lateral translation – this observation does suggest that the low-dimensional model obtained for the short system may help us to understand at least some aspects of spatially localized behavior in the long system. The reduced system studied in Section 8.3 evidently contains many of the local ingredients – spatially periodic equilibria, (modulated) traveling waves and heteroclinic cycles – which we see sprinkled apparently randomly over space-time for the long system.

At this point it is helpful to recall in more detail the strategy we have proposed for modeling a complex spatio-temporally turbulent field, namely to extract empirical eigenfunctions via proper orthogonal decomposition from an ensemble of "experimental" data and use them to produce a low-dimensional model by Galerkin projection of the governing equations and suitable modeling to account for neglected terms. In the present case, as we have remarked, the empirical eigenfunctions are Fourier modes. The power spectrum of Figure 8.5 reveals the ordering of these eigenfunctions in terms of energy and so effectively specifies the empirical eigenvalues $\lambda(k)$ in terms of wavenumber.

In choosing a finite-dimensional subspace on which to project the governing equation, we must first pick a suitable spatial (sub-) domain. Here we could, of course, take the full interval $L = 400$, in which case we must include around 100 (complex) Fourier modes to cover the energy bearing range and the beginning of the dissipative range. (The 512 mode simulation includes wavenumbers up to $k = \frac{2\pi}{400} \times 512 \approx 8$ and so contains essentially *all* dissipative scales. In the language of computational fluid dynamics, it is a direct numerical simulation.) But to obtain a model of reasonable dimension we must select a short subinterval of length, say, $L_s = L/N \ll L$. This is entirely analogous to the choice of the spanwise (box) dimension for the boundary layer example discussed in Chapter 10.

When constructing such a spatially localized model, it is clear from the discussion above that we must at the very least include two structures – two periods of the dominant wave – in order to accommodate the creation and annihilation events. Since there are about $45 \approx \frac{400}{2\pi} \times \frac{1}{\sqrt{2}}$ periods in the full interval, this would imply a minimum subinterval of $2 \times \frac{400}{45} \approx 18$ or, say, $L_s = 400/22$. Alternatively, we can observe that selecting L_s is equivalent to selecting the lowest wavenumber $k_{min} = (2\pi/L_s)$ that is to be present in the model. All other wavenumbers are integer multiples of k_{min} and so, to obtain non-trivial interactions, for which we need at least two "active" wavenumbers, we must take $2k_{min} < 1$ or $L_s > L_2 = 4\pi$. One can view the choice of L_s as the issue of placing wavenumbers of selected Fourier modes intelligently in the "experimental" spectrum. Figure 8.6 indicates some possibilities.

In deciding how many wavenumbers to include in the dissipative range, it is important to recognize a structural feature of the equations deriving from the quadratic interaction term. We first note that, from the general form of the projected Equations (8.13), any subspace spanned by the lth mode and multiples of it is invariant (consider that spanned by a_2 and a_4 in Equations (8.15) for example). Now suppose that the lth mode is linearly unstable ($l^2[1 - (2\pi l/L)^2] > 0$) and consider the invariant subspace spanned by the lth, $2l$th, ... modes. If we do not include the $2l$th mode, then there is no direct quadratic interaction term to shift energy from the unstable lth mode up to the higher wavenumbers where it can be dissipated. A simple mathematical consequence of this is that, when we seek a fixed point by solving the simultaneous equations

$$\left. \begin{array}{l} l^2 \left[1 - (2\pi l/L)^2 \right] a_l - 2l^2 a_l^* a_{2l} = 0, \\ 4l^2 \left[1 - (4\pi l/L)^2 \right] a_{2l} + \frac{1}{2} l^2 a_l a_l = 0, \end{array} \right\} \tag{8.51}$$

eliminating a_{2l}, we find

$$|a_l| = 2\sqrt{\left[1 - (2\pi l/L)^2 \right] \left[(4\pi l/L)^2 - 1 \right]}. \tag{8.52}$$

Bounded fixed points can similarly be found if a_{3l} and higher modes are included, but if a_{2l} is *not* included then the equation on the one- (complex-) dimensional a_l-subspace is purely linear and is not balanced by any other equation, so that a_l grows exponentially. Thus we must include wavenumbers at least up to twice that of the highest unstable mode $l \approx L_s/2\pi$. Failing this, a center-unstable or inertial manifold approximation similar to that of Section 8.3 can be used.

Anticipating the material of Chapters 10 and 11, it is useful at this point to recall the experimental observations on the wall region of the turbulent boundary layer and in particular the flow visualizations of Kline *et al.* [195] discussed in Section 2.5. The similarity of Figures 2.14 and 8.4 is more than coincidental, for the energy distribution in the boundary layer with respect to spanwise wavenumber, revealed by the empirical eigenvalues $\lambda_n(k_1, k_3)$ and shown below in Figure 10.2(a), is also qualitatively similar to the Kuramoto–Sivashinsky spectrum of Figure 8.5. In both cases a clear peak signifies the presence of a dominant wavenumber *on average*, but it is important to recognize that

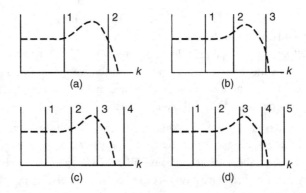

Figure 8.6 Four choices of L_s. (a) $L_s = 400/30 = 13.33$, (b) $L_s = 400/22 = 18.18$, (c) $L_s = 400/20 = 20$, (d) $L_s = 400/16 = 25$.

this spectral peak emerges from the dynamical interactions of a range of wavenumbers. The coherent structures are not static objects convecting "rigidly," rather they are localized spatio-temporal structures whose dynamical behavior is obscured by the integrations in space and time necessary to produce the spectrum. We must therefore include enough active wavenumbers in the local model to allow it to recreate the dynamics which, when averaged, lead to that spectrum.

But how many is enough? In selecting L_s we have no rigorous results such as the center manifold theorem to which we can appeal; nor does there appear to be much experience available to draw on. Here we examine simulations of systems with several choices of L_s comparable to those indicated in Figure 8.6. For these computations we integrated a Galerkin projection containing ten complex Fourier modes: sufficient to cover the dissipative range adequately for all $L_s < 10\pi$, as discussed above.

The case of two active wavenumbers with $L_s > 4\pi$ but $L_s - 4\pi$ small has already been covered in Section 8.3. Further increasing L_s, we obtain the schematic (and incomplete) bifurcation diagram of Figure 8.7, adapted from Kevrekidis *et al.* [188]. The following assertions can be checked by reference to that paper or by local bifurcation analyzes similar to those of Sections 8.2 and 8.3. The branch of "mode two" saddle points persists until it coalesces at $L = 25.3\ldots$ with the branch of "mode four" equilibria which emerges from the trivial solution at $L_s = 8\pi$. For L_s near 4π, the equilibria in these circles are saddle points connected by (attracting) heteroclinic cycles, as studied in Section 8.3. They stabilize to form an attracting circle of equilibria at $L_s = 14.92\ldots$ and then again destabilize at $L_s = 17.31\ldots$, in a Hopf bifurcation in which a complex conjugate pair of eigenvalues of the linearization crosses into the right-hand half plane. Stable periodic solutions (standing waves) are found near these equilibria from $L_s = 17.31$ to around 18.5. In the approximate range $18.5 < L_s < 20.8$, diametrically opposite saddle points on this circle are again connected by heteroclinic cycles, but the departures from and approaches to the equilibria are now oscillatory, due to the presence of two complex conjugate pairs of eigenvalues, with negative and positive real parts respectively, in the linearization. For $L_s > 20.8$, irregular solutions exhibiting ghosts of heteroclinic jumps appear. Two typical

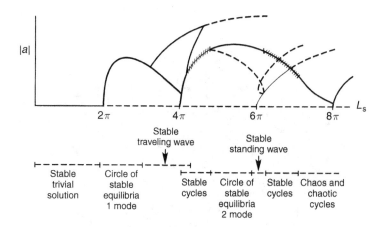

Figure 8.7 A partial bifurcation diagram for the Kuramoto–Sivashinsky equation on lengths $0 < L_s < 26$. After Kevrekidis *et al.* [188]. Copyright ©1990 Society for Industrial and Applied Mathematics. Reprinted with permission. All rights reserved.

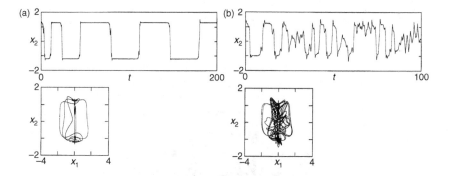

Figure 8.8 The real part of the second Fourier mode as a function of time, and a projection onto the subspace spanned by the real parts of the first and second modes, for the lengths (a) $L_s = 20$, and (b) $L_s = 21$.

examples of numerical simulations are shown in Figure 8.8, along with reconstructions of the solutions $v(x, t)$ from the Fourier modes in Figure 8.9.

These reconstructions, when plotted on the same length and time scales as the simulations of the "full" system of length $L = 400$ in Figure 8.4, exhibit strikingly similar features, both quantitatively and qualitatively. Especially in the case of mildly chaotic motions, such as those in Figures 8.8(b) and 8.9(b), we see irregular annihilation and creation events. While one has to choose the "right" range of subdomain lengths L_s to obtain such results, they do indicate that models containing only a few modes are capable of reproducing local aspects of systems with many interacting modes. We note that Berkooz *et al.* [42, 101] have made similar studies using projections of the Kuramoto–Sivashinsky equation onto wavelet bases, which are in many respects better adapted to produce local models, since their supports are already spatially localized. However, wavelet projections

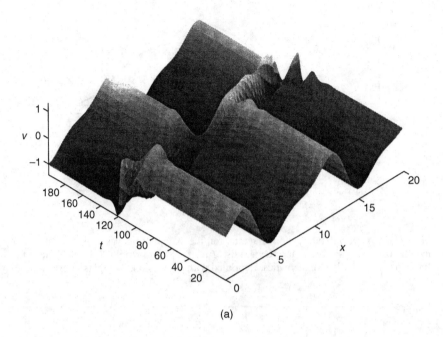

(a)

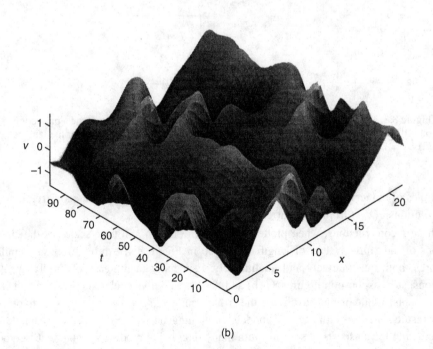

(b)

Figure 8.9 Reconstructions of $v(x, t)$ from the tenth order Galerkin projection for (a) $L_s = 20$, and (b) $L_s = 21$.

do not retain the full $O(2)$-equivariance of the original problem, which as we have seen is important in determining its dynamical behavior, see Myers *et al.* [254].

A more systematic comparison of the dynamics of short subsystems with the "true" behavior can be made as follows. One first solves the Kuramoto–Sivashinsky equation on the full domain, $L = 400$ in this case, obtaining a space-time field $v_1(x, t)$ as indicated in Figure 8.4. One then selects a particular subdomain $S = [\bar{x} - L_s/2, \bar{x} + L_s/2]$ of length L_s and, taking $v_1(x, t)$ restricted to S and periodically extending it, Fourier transforms the resulting field $v_2(x, t)$ to obtain the discrete spatial Fourier coefficients $\hat{v}_2(k_j, t)$ as functions of time. The projected system for (8.47) analogous to (8.13) is then solved using the Fourier coefficients $\hat{v}(k_j, 0)$ as initial data and a second space-time field $v_3(x, t)$ generated for direct comparison with $v_1(x, t)$ on S. The individual Fourier coefficients $\hat{v}_3(k_j, t)$ of the short subsystem can also be compared with the localized coefficients $\hat{v}_2(k_j, t)$ from the full one. As in the computations shown in Figures 8.8 and 8.9, ten Fourier modes were included in the short models (although this is certainly overconservative for the smaller values of L_s). We also note that the local mean $\int_{\bar{x}-L_s/2}^{\bar{x}+L_s/2} v_1(x, 0)dx$ is not generally zero, only the average over the full interval $[0, L]$ remains zero, as in (8.48), so that the mean must be included among the Fourier components of the short model. However, the mean was not allowed to vary during integrations of the latter.

Figure 8.10 illustrates the comparison between $v_1(x, t)$ and $v_3(x, t)$ for three different lengths: $L_s = 14.06$, $L_s = 20.31$, and $L_s = 30.47$, all centered at the same point $\bar{x} = 235$. In (a) we reproduce the relevant piece of the "true" solution and, in (b)–(d), we show $v_3(x, t)$ constructed from the Fourier components of the short models. It is clear that, for appropriately chosen lengths, the solutions of the short systems can track the full solutions reasonably well in the time interval shown, at least in the central part of the subdomain, where the artificially imposed periodic boundary conditions exert less influence. In the shortest case (b), with only two active modes, the coalescence event occurs a little early, and the pattern displays an additional drift to the right absent in the true solution. The longer model (d), with four active modes, does significantly better, although the intermediate length model (c) does less well. Comparisons of the amplitudes of individual Fourier modes are given in Figure 8.11. Further details may be found in [89].

This (cursory) investigation suggests that an extended turbulent field might be modeled as a set of spatially localized processes or subsystems, each of which involves relatively few modes, exhibits a greater or lesser degree of instability and communicates weakly with its neighbors. If we imagine an "instability parameter" analogous to the length L_s in the present example, varying from subsystem to subsystem and time to time, we obtain just the spatio-temporally random sprinkling of events that the extended field exhibits. As we have seen in Chapter 2 and see again in Chapters 11 and 12, this picture is strongly reminiscent of other turbulent systems such as the boundary layer.

Here it is important to realize that, even if the model admits a stochastic component, in, say, parameter variation, the instabilities and attractors of the local models are nonetheless *primarily* deterministic in nature. Thus, our ability to analyze the behavior of these small dynamical systems in some detail offers hope that one can not only better understand the dynamical mechanisms of turbulence generation, but may also be able to develop more effective control strategies.

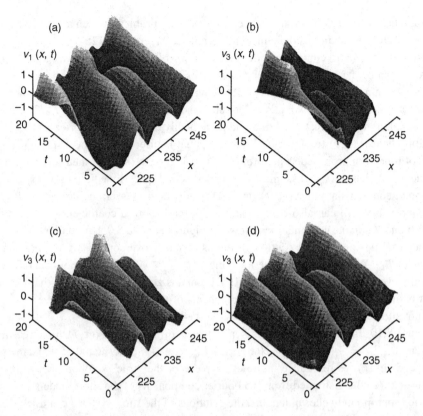

Figure 8.10 Comparison of solutions $v_1(x, t)$ of the full system and $v_3(x, t)$ of tenth order Galerkin projections onto subintervals. (a) The full solution, (b) $L_s = 14.06$, (c) $L_s = 20.31$, (d) $L_s = 30.47$. From Dankowicz *et al.* [89].

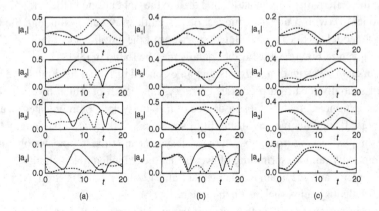

Figure 8.11 Comparison of amplitudes of the first four Fourier modes $\hat{v}_2(k_j, t)$ (solid) and $\hat{v}_3(k_j, t)$ (dashed) from the full and short system solutions shown in Figure 8.10: (a) $L_s = 14.06$, (b) $L_s = 20.31$, (c) $L_s = 30.47$. From Dankowicz *et al.* [89].

Additional work on local models for the Kuramoto–Sivashinsky equation can be found in [389, 390]. Projecting into a spline wavelet basis [90, 278] that allows localization in both spatial location and scale, and integrating the resulting ODEs with successive spatial scales removed, it was found in [389] that the most energetic scales (those around the spectral peak in Fig. 8.5) are essential to the dynamics and that smaller scales are slaved to larger ones, as one would expect. However, the largest spatial scales, which contain less than 5% of the total energy, contribute a slow near-Gaussian background excitation that is also crucial in maintaining spatio-temporal chaos. Nonetheless, by eliminating all coupling between wavelets separated by a given distance l and systematically decreasing l, a characteristic interaction length $l_c \approx 25$ was identified below which coupling appears essential, but above which it has little effect on the most energetic scales.

In [390] these observations were used to construct spatially-localized, periodized models, externally driven at the largest scale included in the domain. Provided that the spatial extent of these systems exceeded l_c (thereby containing at least three or four unstable Fourier modes), both colored noise and temporally-periodic forcing produced spatio-temporal chaos with similar statistics to those of an autonomous system many times larger.

See Sections 12.8 and 13.4 for further comments and references on spatial localization.

9

Randomly perturbed systems

As we have described in Part One, attempts to build low-dimensional models of truly turbulent processes are likely to involve averaging or, more generally, modeling to account for neglected modes that are dynamically active in the sense that their states cannot be expressed as an algebraic function of the modes included in the model. Such models are in turn likely to involve probabilistic elements. Here, "neglected modes" may refer to (high wavenumber) modes in the inertial and dissipative ranges or to mid-range, active modes whose wavenumbers might be linearly unstable. They also may refer to spatial locations that are omitted, in selecting a subdomain of a large or infinite physical spatial extent. The boundary layer model of Chapter 10, for example, contains a forcing term representing a pressure field, unknown a priori, imposed on the outer edge of the wall region. While estimates of this term can be obtained from direct numerical simulations (e.g. [244]), a natural simplification is to replace it with an external random perturbation of suitably small magnitude and appropriate power spectral content. More generally, many processes modeled by nonlinear differential equations involve random effects, in either multiplicative form (coefficient variations) or additive form, and it is therefore worth making a brief foray into the field of stochastic dynamical systems to sample some of the tools available.

In this chapter we give a very selective and cursory description of how one can analyze the effect of additive white noise on a system linearized near an equilibrium point. The theory of stochastic differential equations and dynamical systems involves excursions deep into probability theory, but we try to avoid appeal to results outside the usual repertoire of turbulence theorists familiar with, say Lumley's book *Stochastic Tools in Turbulence* [221].

To introduce the main ideas, we first consider a classical problem in Section 9.1: the motion of a particle in a quadratic potential well, governed by first order dynamics and subject to white noise excitation. This allows us to introduce and find a particular solution of the Fokker–Planck or Kolmogorov equation, which governs the probability density function describing the evolution of an ensemble of solutions. We then go on to study the analogous unstable process and, in Section 9.2, to combine the two explicit solutions to obtain estimates for the distribution of "circuit times" for solutions attracted to a heteroclinic cycle weakly perturbed by white noise. We find that these distributions are asymmetrical, with pronounced exponential tails, and we show how they scale with the strongest unstable eigenvalue of the saddle points in the cycle and the root-mean-square (r.m.s.) level of the white noise. We call on these results in describing the behavior of a class of boundary layer models in Chapter 10. In Section 9.3 we show how the (temporal) power spectra of such perturbed heteroclinic and homoclinic attractors can be estimated, providing a simple example of coherent structures in the time domain. Section 9.4 contains brief comments on the effects of other, deterministic symmetry breaking perturbations on the behavior of heteroclinic attractors.

Much of the material of this chapter is drawn from a series of papers by Stone and Holmes [162, 355–357]. Also see Coller *et al.* [79–81]. For background on probability theory and stochastic differential equations, see [57, 86, 108] and [14], respectively.

9.1 An Ornstein–Uhlenbeck process

In studying stochastic ODEs one is not interested in individual solutions with specific initial conditions and realizations of the random terms. Rather one seeks to describe the range of possible behaviors for all initial data and random processes in some appropriate class. In this respect the emphasis is similar to that in deterministic dynamical systems, but now one has to face the additional complication of dealing with an ensemble of random processes. The notion of a phase portrait no longer applies; each process yields a different set of solutions. One therefore seeks a probabilistic description, the most complete characterization being in the form of a conditional probability density function $p(x, t|x_0)$ – a non-negative function defined over $(-\infty, +\infty) \times [0, +\infty)$ and conditioned on a given distribution of initial conditions x_0. We illustrate this with the simplest possible example, the analysis of which is used in Section 9.2.

Consider the linear scalar ODE modeling a particle in a quadratic potential well subject to first order dynamics and random excitation:

$$\dot{x} = \lambda x + \epsilon \eta(t), \tag{9.1}$$

where $\eta(t)$ (white noise) is the formal derivative of a zero mean Wiener process $W(t)$. Thus, if angle brackets $\langle \eta \rangle$ denote the ensemble average over realizations of η, which we interpret here as a time average, we have

$$\langle \eta(t) \rangle = 0, \quad \text{and} \quad \langle \eta(t)\eta(t + \tau) \rangle = \delta(\tau), \tag{9.2}$$

where ϵ is the r.m.s. noise level. Actually, since W is nowhere differentiable, η is a distribution rather than a function, so (9.1) only makes sense in integrated form, where the second integral is of Itô type [14]:

$$x(t) = x(0) + \lambda \int x dt + \epsilon \int dW, \tag{9.3}$$

with $\langle dW \rangle = 0$ and $\langle dW^2 \rangle = dt$. However, it suffices here to work with the formal expression of (9.1).

Initially we assume that the parameter λ in (9.1) is negative, so that, in the absence of noise, the system has a stable sink at $x = 0$. Equation (9.1) is then an example of a stable Ornstein–Uhlenbeck process [57].

Before starting our analysis we make some general remarks about Equation (9.1). The right-hand side contains two competing effects. For $\epsilon = 0$, solutions $x(t) = x_0 e^{\lambda t}$ all decay towards the stable sink and, whatever the ensemble of initial conditions, we expect the conditional probability density function $p(x, t | x_0)$ to approach a delta function centered on $x = 0$ as $t \to \infty$. In contrast, if $\lambda = 0$ there is no deterministic "drift" term and, for $\epsilon \neq 0$, the particle is buffeted only by the random force. This is classical one-dimensional Brownian motion. If the particle is released from the origin ($x_0 = 0$), the expectation of the distance traveled, $\langle |x(t)| \rangle$, increases like $\sqrt{t}$. In this case $p(x, t | x_0)$ approaches zero at any point $|x| < \infty$, since with probability one the particle displacement eventually exceeds and remains outside any bounded set as t increases.

To obtain an explicit solution, we turn to the partial differential equation that governs the evolution of $p(x, t | x_0)$, variously called the forward Kolmogorov or Fokker–Planck equation:

$$\frac{\partial p}{\partial t} = -\frac{\partial}{\partial x}(\lambda x p) + \frac{\epsilon^2}{2}\frac{\partial^2 p}{\partial x^2}. \tag{9.4}$$

A derivation of this equation may be found in [14]. The first term on the right-hand side represents the drift due to the deterministic part of the vector field, the second is the diffusion term representing the effect of the random excitation. For $\lambda = 0$, (9.4) reduces to the heat equation and any distribution of initial conditions diffuses, "thinning out" as $t \to \infty$. As one might expect, with both terms present a balance is achieved for $\lambda < 0$ and an equilibrium distribution emerges. Taking the special ensemble of identical initial data $x(0) \equiv x_0$, so that the delta function initial condition

$$p(x, 0 | x_0) = p(x(0)) = \delta(x_0) \tag{9.5}$$

applies to (9.4), the reader can check that the solution is given by

$$p(x, t | x_0) = \mathcal{N}(x_0 e^{\lambda t}, \frac{\epsilon^2}{2\lambda}(e^{2\lambda t} - 1)). \tag{9.6}$$

Here

$$\mathcal{N}(\mu, \sigma^2) = \frac{1}{\sqrt{2\pi\sigma^2}} e^{-(x-\mu)^2/(2\sigma^2)} \tag{9.7}$$

denotes the Gaussian or normal distribution with mean μ and variance σ^2. Thus the distribution of solutions broadens and its centroid drifts towards the origin so that, as $t \to \infty$, it approaches the equilibrium distribution with zero mean and variance proportional to ϵ^2 divided by λ:

$$p(x, \infty | x_0) = \mathcal{N}(0, -\frac{\epsilon^2}{2\lambda}). \tag{9.8}$$

(Recall that we assumed $\lambda < 0$.) Note that, when $\lambda \to 0^+$, (9.6) reduces to $\mathcal{N}(x_0, \epsilon^2 t)$ and we recover the linear growth in variance characteristic of Brownian motion.

In deriving and solving the Kolmogorov Equation (9.4), we did not actually restrict the sign of λ to be negative. The solution (9.6) works equally well for $\lambda > 0$ – the unstable Ornstein–Uhlenbeck process. In that case, however, as $t \to \infty$, the variance *increases* at an exponential rate and, unless $x_0 = 0$, the mean also drifts to infinity exponentially fast. This is not surprising, for now the drift and diffusion terms reinforce, both tending to drive

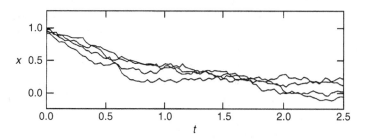

Figure 9.1 A sample of solutions to (9.1) with $\lambda = -1$, $\epsilon^2 = 0.025$, all started at $x(0) = 1.0$.

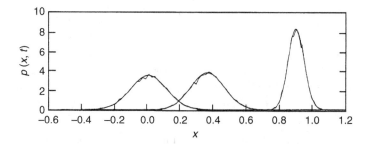

Figure 9.2 Probability density functions for the solutions of (9.1) at times t = 0.1, 1.0, 5.0, with parameters as in Figure 9.1. Jagged lines: computation from 10^4 samples; smooth curve, the prediction of Equation (9.6).

solutions away from the origin. Here there is no useful limiting distribution but, as we shall see in the next section, the explicit information on how $p(x, t|x_0)$ spreads in the unstable case is just what we need to estimate typical passage times for solutions grazing a saddle point during their recurrent passages near a randomly perturbed heteroclinic cycle.

Figures 9.1 and 9.2 illustrate the behavior of the stable process, showing members of an ensemble of solutions all started at the same point $x(0) = 1$ and the associated density function $p(x, t|1)$ at several times during the evolution. The simulations were carried out using a fourth order fixed step Runge–Kutta method and a numerical random number generator rescaled to produce a Gaussian distribution with zero mean and unit variance.

Finally, it is useful to note that the particular solution (9.6) shows that the probability density remains Gaussian for all t: only the mean and variance evolve. Consider an unstable process and suppose that the initial distribution is $\mathcal{N}(0, \sigma_0^2)$ for some non-zero σ_0 instead of the limiting delta function. Then the subsequent evolution is identical to that of (9.6) translated in time by the amount

$$t' = \frac{1}{2\lambda} \ln \left(\frac{2\sigma_0^2 \lambda}{\epsilon^2} + 1 \right) \tag{9.9}$$

required for the delta function initial distribution to spread to $\mathcal{N}(0, \sigma_0^2)$. Using this in (9.6), we obtain

$$p(x, t|\mathcal{N}(0, \sigma_0^2)) = \mathcal{N} \left(0, \sigma_0^2 e^{2\lambda t} + \frac{\epsilon^2}{2\lambda} (e^{2\lambda t} - 1) \right). \tag{9.10}$$

9.2 Noisy heteroclinic cycles

We now use the results of the previous section to estimate the distribution of recurrence times for an attracting heteroclinic cycle subject to additive random perturbations. For simplicity we focus on a two-dimensional model with a planar homoclinic loop or pair of such loops. For more detailed arguments extending the analysis to higher-dimensional cases, see Stone and Holmes [356].

We consider a two-dimensional system of the form

$$\dot{\mathbf{x}} = \mathbf{f}(\mathbf{x}) + \epsilon \boldsymbol{\eta}(t), \tag{9.11}$$

for which, when $\epsilon = 0$, $\mathbf{x} = \mathbf{0}$ is a saddle point with eigenvalues λ_s and λ_u satisfying $-\lambda_s > \lambda_u > 0$, and where $\boldsymbol{\eta}$ is a vector-valued white noise process with suitably uncorrelated components further specified below. For $\epsilon = 0$ we assume that the system has a homoclinic loop as in Figure 7.4(a). In view of the condition on the eigenvalues, this loop is attracting. (Actually, to guarantee that *all* orbits passing near the saddle return to its neighborhood, we have to assume the existence of a double homoclinic loop, as in Figure 9.3 below. This is automatic, due to the symmetry, in our ultimate application to $O(2)$-equivariant heteroclinic attractors.)

Recalling the analysis with which we established stability or instability of a planar homoclinic loop in Section 7.3, we again approximate the return map by splitting it into two components P_0 and P_1. Here P_0 is the local map from a cross section Σ_0 straddling the local stable manifold to a second section Σ_1 straddling the local unstable manifold; its behavior is determined by solutions in the neighborhood U_δ of the system linearized at the saddle:

$$\left. \begin{aligned} \dot{x}_1 &= \lambda_s x_1 + \epsilon \eta_1(t), \\ \dot{x}_2 &= \lambda_u x_2 + \epsilon \eta_2(t). \end{aligned} \right\} \tag{9.12}$$

The size $\mathcal{O}(\delta)$ of U_δ is fixed independent of ϵ.

The global return map, P_1, follows solutions as they circulate near the homoclinic loop for a finite time to return to Σ_0. As argued in Section 7.3, it therefore imposes a bounded distortion. Outside U_δ the deterministic vector field $\mathbf{f}(\mathbf{x})$ near the homoclinic orbit is bounded below, and so for sufficiently small ϵ the deterministic component is dominant and it is reasonable to ignore the random terms on this "leg" of the circuit. Thus, if a given distribution of solutions emerges from U_δ on Σ_1 it gives rise to a similar distribution, perhaps modestly broadened, on Σ_0, and all solutions experience the same time of flight, at leading order, from Σ_1 to Σ_0. In contrast, since $\mathbf{f}(\mathbf{x}) = \mathbf{0}$ at $\mathbf{x} = \mathbf{0}$, the effect of the random component is crucial to the behavior of orbits grazing the saddle as they pass through U_δ, no matter how small ϵ.

To estimate this local effect we shall analyze the linearized stochastic ODEs (9.12), but first we consider what happens to the means $\mu_j = \langle x_j \rangle$, which evolve according to the averaged equations obtained from (9.12):

$$\left. \begin{aligned} \dot{\mu}_1 &= \lambda_s \mu_1, \\ \dot{\mu}_2 &= \lambda_u \mu_2. \end{aligned} \right\} \tag{9.13}$$

Repeating the analysis of Section 7.3, including the global return map, we conclude that $\mu_j \to 0$ as $t \to \infty$: with additive random perturbations solutions still converge to the homoclinic attractor in mean. After sufficiently many circuits, then, we may assume that the outgoing distribution on Σ_1 has equilibrated to that of the stable, zero mean, Ornstein–Uhlenbeck process given by (9.8). As we have noted, this is the same as the incoming distribution on Σ_0 (up to a bounded distortion) and so our problem is reduced to that of predicting escape times from U_δ, in the unstable x_2-direction, for a zero mean Gaussian distribution of solutions with variance $\sigma_0^2 = -(\epsilon^2/2\lambda_s)$, entering on Σ_0.

If we assume that the components η_j of the random excitation in (9.12) are uncorrelated, the unstable process uncouples and, in view of (9.9) and (9.10), its evolution is equivalent to that of a delta function started at time $-t'$, or

$$p(x_2, s|0) = \mathcal{N}\left(0, \frac{\epsilon^2}{2\lambda_u}(e^{2\lambda_u s} - 1)\right); \quad s = t + \frac{1}{2\lambda_u}\ln\left(1 - \frac{\lambda_u}{\lambda_s}\right). \tag{9.14}$$

Now, the probability that a solution lies outside the interval $-\delta \le x_2 \le +\delta$ at time s is

$$P_E(s) = 1 - \int_{-\delta}^{\delta} p(x_2, s|0)dx_2, \tag{9.15}$$

where the subscript denotes "escape." Letting

$$\xi = \left[\frac{\lambda_u}{\epsilon^2(e^{2\lambda_u s} - 1)}\right]^{1/2} x_2, \quad \Delta(s) = \left[\frac{\lambda_u}{\epsilon^2(e^{2\lambda_u s} - 1)}\right]^{1/2}\delta, \tag{9.16}$$

and using (9.14), the integral in (9.15) can be expressed as

$$P_E(s) = 1 - 2\int_0^{\Delta(s)} \frac{1}{\sqrt{\pi}}e^{-\xi^2}d\xi$$

$$= 1 - \operatorname{erf}\Delta(s), \tag{9.17}$$

where erf denotes the error function. Note that, as s increases from 0, $\Delta(s)$ decreases monotonically from $+\infty$ and approaches 0, so that $P_E(s)$ increases monotonically from 0 at $s = 0$ and is asymptotic to 1 as $s \to \infty$. This is what we expect: almost all solutions have escaped from U_δ after sufficient time has elapsed.

The probability density of escape times from U_δ is now given by

$$\mathcal{P}(s) = \frac{d}{ds}P_E(s) = -\frac{d}{ds}\operatorname{erf}\Delta(s)$$

$$= -\frac{2}{\sqrt{\pi}}e^{-\Delta^2(s)}\Delta'(s)$$

or

$$\mathcal{P}(s) = \frac{2\lambda_u\Delta(s)e^{-\Delta^2(s)}}{\sqrt{\pi}(1 - e^{-2\lambda_u s})}. \tag{9.18}$$

We observe that

$$\int_0^\infty \mathcal{P}(s)ds = 1 - \operatorname{erf}\Delta(\infty) = 1,$$

as is required for a probability density. Also, $\mathcal{P}(s)$ is strongly asymmetrical and has a pronounced exponential tail. In fact as $s \to \infty$ we have

$$\mathcal{P}(s) \sim e^{-\lambda_u s}. \tag{9.19}$$

The mean passage time can now easily be estimated from (9.18) in terms of the relevant parameters. It is given by the centroid of the distribution, that is the point below and above which the areas under $\mathcal{P}(s)$ are both equal to one half:

$$\int_0^S \mathcal{P}(s)ds = P_{\mathrm{E}}(S) = 1/2$$

or

$$\mathrm{erf}\,\Delta(S) = 1/2. \tag{9.20}$$

Letting α (≈ 0.477) denote $\mathrm{erf}^{-1}(1/2)$, we have

$$\Delta(S) = \frac{\delta}{\epsilon}\sqrt{\frac{\lambda_u}{e^{2\lambda_u S} - 1}} = \alpha,$$

or

$$S = \frac{1}{2\lambda_u}\ln\left(\frac{\delta^2 \lambda_u}{\epsilon^2 \alpha^2} + 1\right). \tag{9.21}$$

In the limit of weak random excitation ($\epsilon \ll \delta$) the leading term in the logarithm dominates, giving

$$\begin{aligned}
S &= \frac{1}{\lambda_u}\left[\ln\left(\frac{\delta\sqrt{\lambda_u}}{\epsilon\alpha}\right) + \mathcal{O}(\epsilon^2)\right] \\
&= \frac{1}{\lambda_u}\left[\ln\left(\frac{\delta}{\epsilon}\right) + \ln\left(\frac{\sqrt{\lambda_u}}{\alpha}\right) + \mathcal{O}(\epsilon^2)\right] \\
&= \frac{1}{\lambda_u}\left[\ln\left(\frac{\delta}{\epsilon}\right) + \mathcal{O}(1)\right].
\end{aligned} \tag{9.22}$$

Finally, in terms of the original time coordinate, we have the mean passage time

$$\tau = S - t' = \frac{1}{\lambda_u}\ln\left(\frac{\delta}{\epsilon}\right) - \frac{1}{2\lambda_u}\ln\left(1 - \frac{\lambda_u}{\lambda_s}\right) + \mathcal{O}(1),$$

or

$$\tau = \frac{1}{\lambda_u}\left[\ln\left(\frac{\delta}{\epsilon}\right) + \mathcal{O}(1)\right]. \tag{9.23}$$

Equations (9.18) and (9.23) provide estimates for the distribution and mean passage time for solutions in U_δ, the neighborhood of the saddle point. A further $\mathcal{O}(1)$ correction is necessary for the global time of flight from Σ_1 back to Σ_0, which depends on the particular problem at hand. Thus our final estimate for the overall mean circuit time for solutions in the neighborhood of a perturbed heteroclinic attractor containing a single saddle point takes the form

$$\tau_{\mathrm{total}} = K_1 + \frac{1}{\lambda_u}\left[\ln\left(\frac{\delta}{\epsilon}\right) + K_0\right], \tag{9.24}$$

where the "constants" K_1 and K_0 depend upon the global structure of the deterministic vector field near the homoclinic loop(s) and the choice of the neighborhood U_δ, but are independent of the r.m.s. noise level ϵ.

Note that the dominant term in (9.23) and (9.24) depends only on the unstable eigenvalue λ_u and the ratio δ/ϵ. In higher-dimensional situations the drift term is again dominated by the largest unstable eigenvalue and so a similar estimate applies with $\lambda_u = \max[\operatorname{Re}(\lambda)]$.

The theory developed above involves several approximations, notably that of linearization and neglect of the effects of random perturbations along the "global" leg of the homoclinic orbit outside the neighborhood U_δ of the saddle point, the assumption that the outgoing distribution has equilibrated to the asymptotic limit of a stable Ornstein–Uhlenbeck process, and, in the case of mean passage times, neglect of all but the leading term of order $\ln(1/\epsilon)$. The latter two approximations were tested by Stone and Holmes [356] and shown to be acceptable by comparison with numerical simulations of the linear process (9.12). Here we shall only illustrate the theory with a numerical simulation of the perturbed Duffing equation, also taken from Stone and Holmes, which also effectively tests the approximation incurred by linearization. We consider the system

$$\left.\begin{aligned} \dot{x}_1 &= x_2 + \epsilon\eta_1(t), \\ \dot{x}_2 &= x_1 - x_1^3 - \gamma x_2 + \beta x_1^2 x_2 + \epsilon\eta_2(t). \end{aligned}\right\} \tag{9.25}$$

For $\epsilon = 0$ this system has a saddle point at the origin and values of the parameters β and γ can be chosen so that it has an attracting figure-of-eight homoclinic loop as shown in Figure 9.3 (cf. Figure 6.13 and the related discussion in Section 6.4). In fact one can estimate the relation $\beta = 5\gamma/4 + \mathcal{O}(\gamma^2)$ by the perturbation method of Melnikov as described in Guckenheimer and Holmes ([144] Sections 4.5 and 7.3). Also, provided $\gamma > 0$, the eigenvalues at the saddle:

$$\lambda_{u,s} = \frac{1}{2}(-\gamma \pm \sqrt{\gamma^2 + 4})$$

satisfy $-\lambda_s > \lambda_u$ and so careful selection of β, γ near the line $\beta = 5\gamma/4$ by numerical search yields a planar system having a pair of attracting homoclinic cycles. Two cases were considered: $\beta = 0.498$, $\gamma = 0.4$, corresponding to $\lambda_s = -1.22$, $\lambda_u = 0.82$ and $\beta = 0.1$, $\gamma = 0.08$, corresponding to $\lambda_s = -1.04$, $\lambda_u = 0.96$.

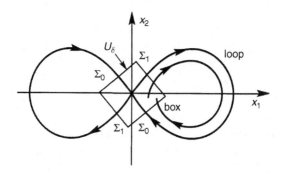

Figure 9.3 A double homoclinic loop for the Duffing Equation (9.25).

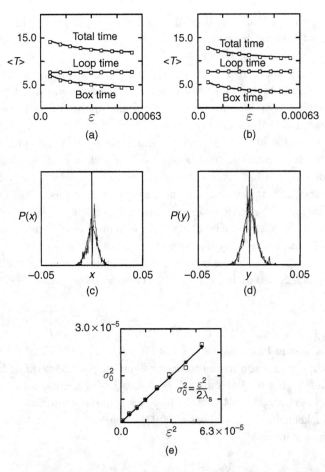

Figure 9.4 Simulations of the Duffing equation (9.25); 1000 trials in each case: (a) $\beta = 0.498$, $\gamma = 0.4$; (b) $\beta = 0.1$, $\gamma = 0.08$. Plots show mean passage times through U_δ and around global part of loops, as a function of ϵ, compared with the estimates (9.23) and (9.26). (c) and (d) show incoming and outgoing distributions on the cross sections Σ_0 and Σ_1, compared with the equilibrium distribution (9.8) of the stable process assumed in our analysis. (e) Shows that the variance of the exit distribution scales linearly with ϵ^2, as expected. From Stone and Holmes [356]. Copyright ©1990 Society for Industrial and Applied Mathematics. Reprinted with permission. All rights reserved.

   In Figure 9.4(a) and (b) we show the mean passage times through a square neighborhood U_δ with sides of length 2δ aligned parallel to the stable and unstable eigenvectors. Somewhat arbitrarily we chose $\delta = 0.1$, in which case one can estimate the time taken to traverse the "global" part of the loop from the exit points $(\pm\delta/\sqrt{2}, \pm\delta/\sqrt{2})$ to the entry points $(\pm\delta/\sqrt{2}, \mp\delta/\sqrt{2})$ from the expression $x_1(t) = \pm\sqrt{2}\mathrm{sech}(t)$ for the unperturbed homoclinic loops. Using the fact that the travel time is twice that required for the solution starting at the furthest point on the loop to reach U_δ, we obtain

$$\frac{\delta}{\sqrt{2}} = \sqrt{2}\,\mathrm{sech}\left(\frac{T_{\text{global}}}{2}\right) \Rightarrow T_{\text{global}} \approx 7.377. \qquad (9.26)$$

This "loop time" is also indicated on Figure 9.4(a) and (b). In (c) and (d) we compare the incoming and outgoing distributions of solutions in the cross sections with the assumed stable equilibrium distribution, and in Figure 9.4(e) we demonstrate that the variance of the latter does scale with the r.m.s. noise level ϵ in the expected way.

Our final example is that of a system possessing $O(2)$-equivariant heteroclinic cycles as studied in the preceding chapters. We add complex-valued independent random processes η_1, η_2 with independent real and imaginary components to the normal form of Equation (7.51) to obtain

$$\left. \begin{aligned} \dot{z}_1 &= z_1^* z_2 + (\mu_1 + e_{11}|z_1|^2 + e_{12}|z_2|^2)z_1 + \epsilon\eta_1(t), \\ \dot{z}_2 &= \pm z_1^2 + (\mu_2 + e_{21}|z_1|^2 + e_{22}|z_2|^2)z_2 + \epsilon\eta_2(t). \end{aligned} \right\} \tag{9.27}$$

We take the parameter values

$$e_{11} = -4, \ e_{12} = -1, \ e_{21} = e_{22} = -2, \ \mu_2 = 0.2,$$

as in Section 7.4, and choose μ_1 to guarantee the existence of attracting heteroclinic cycles for $\epsilon = 0$. Under these assumptions the system has a circle of saddle points $|z_2| = \sqrt{-\mu_2/e_{22}}$ with eigenvalues

$$\lambda_{u,s} = \mu_1 - \frac{\mu_2 e_{12}}{e_{22}} \pm \sqrt{-\frac{\mu_2}{e_{22}}},$$

$$\lambda_i = -2\mu_2,$$

$$\lambda_0 = 0.$$

The analysis of Armbruster *et al.* [9], sketched in Section 7.4, shows that the latter two eigenvalues, with eigenvectors spanning the a_2-plane, are irrelevant to the stability of the cycles, which connect diametrically opposite points on the circle of equilibria. The methods developed in this section may therefore be applied to determine passage time distributions and in deriving the estimates we need only consider the value of the single unstable eigenvalue λ_u. However, since each circuit of the cycle involves passage by two saddle points, the local estimate of (9.23) must be counted twice, leading to a modification of our general result (9.24):

$$T_{O(2)} = 2K_1 + \frac{2}{\lambda_u}\left[\ln\left(\frac{1}{\epsilon}\right) + K_0\right]. \tag{9.28}$$

Note that the constant δ of Equation (9.24), characteristic of the neighborhood size, can be assumed into K_0.

Figure 9.5 shows mean passage times versus r.m.s. noise level ϵ for e_{ij} and μ_2 as given above, and three choices of μ_1 to provide a variation of λ_u from 0.196 to 0.266. The curves represent Equation (9.28) with K_1 chosen as the global passage times to complete the two legs of the cycle and K_0 chosen to fit the local conditions at the saddles. Figure 9.6 shows a probability density of circuit times compared with the prediction of Equation (9.18).

We close this section by remarking that, while our analysis has considered only white noise excitation, colored (low pass filtered) noise and deterministic perturbations, such as periodic forcing, can also give rise to distributions of circuit times of a similar form, with a mean that scales as Equation (9.24) and a distribution approximated by an equation similar to (9.18). Stone and Holmes [357] give examples of both the Duffing equation and the $O(2)$-equivariant system (9.27) subject to such perturbations. In the deterministic

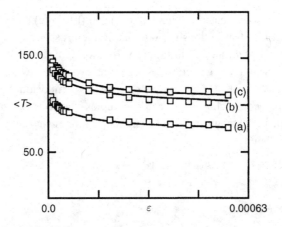

Figure 9.5 Mean circuit times for simulations of Equation (9.27) with the parameters specified in the text and (a) $\mu_1 = 0.05 \Rightarrow \lambda_u = 0.266$; (b) $\mu_1 = -0.01 \Rightarrow \lambda_u = 0.206$; (c) $\mu_1 = -0.02 \Rightarrow \lambda_u = 0.196$; sample sizes: 100 trials each. From Stone and Holmes [356]. Copyright ©1990 Society for Industrial and Applied Mathematics. Reprinted with permission. All rights reserved.

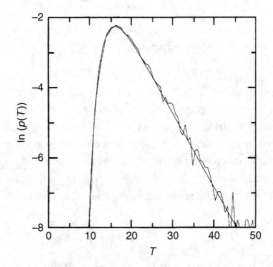

Figure 9.6 Probability density of circuit times for Equation (9.27) with the parameters e_{ij} specified in the text and $\mu_1 = 0.04$, $\mu_2 = 0.10$, $\epsilon = 6.7 \times 10^{-4}$. Jagged line: computation from 16 000 samples; smooth curve, the prediction of Equation (9.18).

case, periodic forcing causes the continuous families of saddle connections to split, leading to transverse intersections of stable and unstable manifolds. As Poincaré [282] suspected and Birkhoff [51] and Smale [343] later proved, such intersections imply the existence of a chaotic invariant set (a Smale horseshoe, see Guckenheimer and Holmes [144] for more details). In the present context, this implies that the perturbed deterministic dynamics near the cycles itself produces a distribution of solutions at the entering cross section to the neighborhood of the saddle point similar to that due to the random process. This distribution now evolves subject only to the drift term of the Kolmogorov equation (9.4), but as the results of [357] show, this still yields a distribution of passage times with an exponential tail. Thus the conclusions of this section appear relevant to a fairly broad range of systems having weakly perturbed homoclinic or heteroclinic attractors.

We may summarize these results as follows. The solutions captured by perturbed heteroclinic (or homoclinic) attractors exhibit well-defined "events" whose temporal signatures are close to those of the unperturbed heteroclinic excursions. These events occur, however, at random times with a distribution determined *jointly* by the r.m.s. level of the random excitation and the dominant unstable eigenvalue. In particular, the mean circuit or passage time scales as the product of the inverse eigenvalue with the logarithm of the r.m.s. level: it exhibits a mixture of deterministic and random effects. The random process has introduced a time scale without significantly affecting the structure of solutions in phase space, much as Busse and Heikes [67] envisaged in their discussion of statistical limit cycles.

As suggested at the end of Chaper 2, this provides an appealing mathematical metaphor for coherent structures in turbulence. If the solutions follow a restricted route in phase space, then the corresponding sequence of temporal events and spatial patterns obtained on reconstructing the fluid velocity field via a modal representation is similarly constrained. Similar cycles of well-defined events occur repeatedly, but effectively at random times. In Chapters 10 and 11 we show that this is more than just a metaphor.

9.3 Power spectra of homoclinic attractors

In this section we briefly describe an idea due to Gol'dshtik and Shtern (see [131]), which was further explored by Brunsden *et al.* [63, 64] and to which we have already referred in Chapter 2. If one supposes that a chaotic temporal signal is formed by superposition of localized deterministic events occurring at randomly distributed times and with random phases, then the power spectral density of the signal is easy to predict. As we have just shown, the case of a perturbed homoclinic or heteroclinic attractor provides an excellent example of such a process.

We describe the analysis for the Duffing example (9.25) in which the temporal event is the unperturbed homoclinic orbit given by

$$\sigma(t) \ (= x_1(t)) = \sqrt{2} \operatorname{sech}(t). \tag{9.29}$$

Letting T_j, $-\infty < j < \infty$, denote the (random) times at which the events occur, and noting that each event may be either positive or negative, corresponding to a trip around either the right or the left lobe of Figure 9.3, we arrive at our representation of a typical solution:

$$x_1(t) = \sum_{j=-\infty}^{\infty} (-1)^{a_j} \sigma(t - T_j), \tag{9.30}$$

where a_j is a random variable that takes the values 0 and 1 with equal probability. (Brunsden *et al.* [63] also consider the case of an unsymmetric attractor, for which the probabilities are biased.)

To compute the power spectrum of $x_1(t)$ we multiply it by a "window" function $g_L(t)$ of width $2L$:

$$g_L(t) = \begin{cases} 1 & \text{if } |t| < L \\ 0 & \text{if } |t| > L \end{cases}$$

and rewrite the (integrable) windowed function as a convolution

$$x_L(t) = \int_{-\infty}^{\infty} a_L(t' - t)\sigma(t')dt', \tag{9.31}$$

where

$$a_L(s) = \sum_{j=-\infty}^{\infty} g_L(s)(-1)^{a_j}\delta(s + T_j) \tag{9.32}$$

is a (finite) random sequence of (signed) delta functions – the shot noise decomposition of Campbell (see [303]). The Fourier transform of $x_L(t)$ is then the product $\hat{a}_L\hat{\sigma}$ of the transforms of $a_L(t)$ and $\sigma(t)$ and the power spectral density may in turn be written as

$$E_{x_1}(f) = \lim_{L\to\infty} \frac{1}{2L}|\hat{x}_L(f)|^2 = \lim_{L\to\infty} \frac{1}{2L}|\hat{a}_L(f)|^2|\hat{\sigma}(f)|^2. \tag{9.33}$$

From (9.32) we have

$$\hat{a}_L(f) = \int_{-\infty}^{\infty} \sum_{j=-\infty}^{\infty} g_L(s)(-1)^{a_j}\delta(s + T_j)e^{-2\pi i f s}\, ds$$

$$= \sum_{j=-J_1}^{J_2} (-1)^{a_j} e^{2\pi i f T_j}, \tag{9.34}$$

where J_1 and J_2 are the largest integers such that the events at T_{-J_1} and T_{J_2} lie within the window. Thus $J_1 + J_2 \approx 2L/\tau$, where τ is the mean time between events. From (9.34) we have

$$|\hat{a}_L(f)|^2 = (J_1 + J_2 + 1) + \sum_{j,k;\ j\neq k} (-1)^{a_j+a_k} e^{2\pi i f(T_j - T_k)}. \tag{9.35}$$

Since both a_j and T_j are random variables, by the central limit theorem the "cross-terms" in the double sum are of $o(L)$ and so, after substitution of (9.35) into (9.33) and taking the limit, we may ignore these terms. We therefore obtain the appealingly simple estimate:

$$E_{x_1}(f) = \frac{1}{\tau}|\hat{\sigma}(f)|^2. \tag{9.36}$$

For the Duffing equation, using the Fourier transform of (9.29), we obtain the specific expression

$$E_{x_1}(f) = \frac{2\pi^2}{\tau}\text{sech}^2(\pi^2 f), \tag{9.37}$$

in which we may use the estimates (9.23) and (9.24) obtained in Section 9.2 for τ. We note that, for large frequencies, the spectrum predicted by (9.37) decays exponentially. More significantly, and along the lines of the comments in the penultimate paragraph of

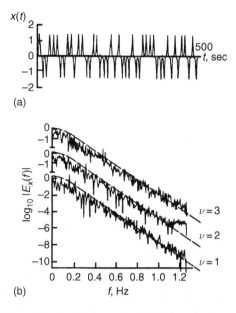

Figure 9.7 (a) A solution of the Duffing equation (9.25) for $\gamma = 0.4$, $\beta = 0.498$ with periodic forcing as described in Brunsden and Holmes [64]; (b) Corresponding power spectra for three different excitation frequencies, compared with the prediction of Equation (9.37).

Section 9.2, we observe that the functional form of the power spectrum (9.36) derives from the deterministic expression (9.29) for the homoclinic orbit, while the random process affects only its level (the more frequent the events, the greater the total power).

Brunsden *et al.* [63, 64] give more details and comparisons of this simple theory with both numerical simulations and experimental data from a vibrating cantilever beam. Figure 9.7 shows an example of these results. In that work the authors were primarily interested in the case of deterministic (periodic) excitation, but, as pointed out at the end of Section 9.2, the chaotic solutions induced by periodic perturbations give rise to essentially the same passage time statistics as do white or colored noise.

9.4 Symmetry breaking

We close this chapter with a brief discussion of the effects of deterministic symmetry-breaking perturbations on systems that possess continuous families of solutions, such as the heteroclinic cycles of the $O(2)$-equivariant ODEs discussed in Section 7.4 and Chapter 8. We start with a planar system that undergoes an $O(2)$-symmetric pitchfork bifurcation. As shown in Section 7.2 (Equation (7.24)), the appropriate normal form, after truncation at cubic terms, is

$$\begin{pmatrix} \dot{x}_1 \\ \dot{x}_2 \end{pmatrix} = [\mu + a_3(x_1^2 + x_2^2)] \begin{pmatrix} x_1 \\ x_2 \end{pmatrix}. \tag{9.38}$$

To this we add a (small) perturbation which breaks the continuous symmetry, replacing it with the discrete dihedral group D_4 – the symmetries of the square (see Equation (7.3)). The easiest way to recognize the new normal form is to pass to the complex representation $z = x_1 + ix_2$ used in Section 7.4 and Chapter 8. With respect to these coordinates, the discrete rotations through $m\pi/2$ of D_4 are given by

$$z \mapsto e^{im\pi/2}z; \quad m = 1, 2, \ldots, \tag{9.39}$$

and the normal form is (to cubic order)

$$\dot{z} = \mu z + a_3|z|^2 z + \epsilon z^{*3}. \tag{9.40}$$

One can easily check that no other cubic monomial of the form $z^k z^{*l}$ with $k, l \geq 0$, $k+l = 3$ is equivariant under (9.39). In polar coordinates $z = re^{i\theta}$, (9.40) becomes

$$\left. \begin{array}{l} \dot{r} = \mu r + a_3 r^3 + \epsilon r^3 \cos 4\theta, \\ \dot{\theta} = -\epsilon r^2 \sin 4\theta. \end{array} \right\} \tag{9.41}$$

For $\epsilon = 0$, Equation (9.41) simply reduces to the uncoupled polar system of Equation (7.25), which has an invariant circle $r = \sqrt{-\mu/a_3}$ filled with fixed points provided $-\mu/a_3 > 0$. When $\epsilon \neq 0$, only a finite number of these fixed points survives. The condition $\sin 4\theta = 0$ from the second component of (9.41) gives $\theta = m\pi/4$, $m = 0, 1, \ldots, 7$ and substituting $\cos \theta = \pm 1$ into the first component yields $r = \sqrt{-\mu/(a_3 \pm \epsilon)}$. Linearizing at these eight fixed points we find that they alternate in stability type as one goes around the origin. For $\mu > 0 > a_3$ (and ϵ sufficiently small), there are alternately sinks on the rays $\theta = 0$, $\pi/2$, π, $3\pi/2$ and saddles on the rays $\theta = \pi/4$, $3\pi/4$, $5\pi/4$, $7\pi/4$. These rays are invariant, since if $\sin 4\theta = 0$ then $\dot{\theta} = 0$ also, see Figure 9.8.

This phase portrait shows that, while the circle of fixed points is destroyed by the perturbation, a ghost of it survives in the form of an invariant closed curve on which the dynamics is relatively slow. (The eigenvalues of the fixed points with eigenvectors tangent to this circle are of $\mathcal{O}(\epsilon)$.) This is a rather typical situation: for $\epsilon = 0$ the individual, non-hyperbolic fixed points are not structurally stable, but the invariant circle as a whole is *normally hyperbolic*. This means that the rate at which solutions approach it is significantly greater than

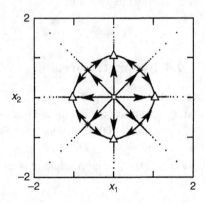

Figure 9.8 The phase portrait of Equation (9.40), with $\mu = 1$, $a_3 = -1$, and $\epsilon = 0.1$.

the rate at which they move within it. Under such conditions, the circle as a whole survives small perturbations, although it does not usually remain a geometric circle and, as in this example, new dynamical phenomena typically appear within it.

A similar situation occurs when weak perturbations having discrete symmetries are applied to the larger $O(2)$-equivariant system (7.51). Campbell and Holmes [69] studied this case, motivated in part by the desire to understand the effects of loss of translation symmetry in the boundary layer due to the addition of riblets, which replace the continuous group $O(2)$ with a discrete group D_k, the order k depending on the period of the riblets relative to the spanwise domain of the model. The representation of D_4 on the two (Fourier) mode phase space is given by

$$(z_1, z_2) \mapsto (e^{im\pi/2}z_1, e^{im\pi}z_2), \quad m = 0, 1, \ldots$$
$$(z_1, z_2) \mapsto (z_1^*, z_2^*),$$

and, as in the simple example above, certain subspaces analogous to the rays $\theta = m\pi/4$ remain invariant under a D_4-equivariant perturbation.

In particular, recalling the discussion in Section 7.4, the real subspace $\{y_1 = y_2 = 0\}$ and the discrete rotation of it given by $\{x_1 = y_2 = 0\}$ are still invariant and the pure mode saddle points in the intersection of these planes, while moving slightly, retain their stability types. The heteroclinic cycles lying in these planes therefore persist, as do two other pairs of cycles in the planes $\{x_1 \pm y_1 = x_2 = 0\}$, but all the rest of the original continuous family of cycles is destroyed. Depending on specific coefficients in the perturbation, the surviving cycles may be stable or unstable in the slow direction along the invariant circle, much as in the example above the stability types of the eight fixed points depend upon the sign of ϵ. This has the effect of "locating" the stable cycles in phase space and so, via a Fourier mode representation, weakly pinning them with respect to translations in physical space. Traveling waves and modulated traveling waves also persist in other parameter ranges, and certain new phenomena appear, including modulated traveling waves which quasiperiodically reverse their propagation direction.

PART THREE

The boundary layer

10

Low-dimensional models

In the preceding nine chapters we have developed our basic tools and techniques. In this chapter and the next we illustrate their use in the derivation and analysis of low-dimensional models of the wall region of a turbulent boundary layer. First, the Navier–Stokes equations are rewritten in a form that highlights the dynamics of the coherent structures (CS) and their interaction with the mean flow. To do this, both the neglected (high) wavenumber modes and the mean flow must be modeled, unlike a large eddy simulation (LES), in which only the neglected high modes are modeled. Second, using physical considerations, we select a family of empirical subspaces upon which to project the equations. Galerkin projection is then carried out. In doing this, we restrict ourselves to a small physical flow domain, and so the response of the (quasi)local mean flow to the coherent structures must also be modeled. This chapter describes each step of the process in some detail, drawing on material presented in Chapters 2, 3, and 4. After deriving the family of low-dimensional models, in the last three sections we discuss in more depth the validity of assumptions used in their derivation. In Chapter 11 we describe use of the dynamical systems ideas presented in Chapters 6 through 9 in the analysis of these models, and interpret their solutions in terms of the dynamical behavior of the fluid flow.

Our presentation is based on a series of papers, beginning with [22] and including [24, 43, 44, 158, 161]. We have selected the boundary layer as our main illustrative example largely because we are most familiar with it. However, we believe that the problems and modeling challenges it presents are characteristic of many other turbulent flows. Most of the difficulties we encounter spring from our wish to develop truly low-dimensional models. In doing this, we must remove and model relatively low wavenumber, dynamically active modes in the inertial range and also cope with spatial localization. Our models therefore exclude much. It is important to recall here that we do not appeal to empirical eigenfunctions to make more efficient and economical computations, but to develop "simple" models that will lead to better understanding of open, turbulent flows. Often, science progresses by a kind of creative ignorance in which one attempts to focus on key phenomena and ignore second order effects. If our critics could not point to approximations and omissions in our models, we would certainly have failed in our main goal.

In addition to their relevance for turbulent boundary layers, this and the following chapter can be read as a case study employing the methods we have introduced.

255

10.1 Equations for coherent structures

In this section we derive evolution equations for the coherent structures from the governing Navier–Stokes equations.

Let Ω be the domain of interest in physical space occupied by fluid obeying the incompressible Navier–Stokes equations. In the wall layer this is a small domain of streamwise length L_1, spanwise width L_3, and wall-normal height X_2: $x_2 = 0$ denotes the wall itself. We subsequently discuss the meaning of "small" and make specific choices for the three length scales. Throughout we use a Cartesian coordinate system $\mathbf{x}$ in which (x_1, x_2, x_3) denote the streamwise, wall-normal, and cross-stream directions respectively. As described in Chapter 4, the Navier–Stokes equations are defined on a suitable subspace of $L^2(\Omega)$ (the space of square integrable, vector-valued functions), consisting of those divergence-free velocity fields which also satisfy appropriate boundary conditions on $\partial\Omega$. We return to boundary conditions in more detail after presenting the equations, but for the moment note that we assume periodicity in the streamwise and spanwise directions. We refer to these as homogeneous, since both the equations, and averages of typical ensembles of experimental or numerical realizations of fully developed turbulent boundary layer flows display translation invariance in these directions (Section 3.3.3). This motivates our choice of periodic boundary conditions. In addition, the governing equations are invariant under spanwise reflections. These symmetries have important implications for our low-dimensional models.

We employ two types of averaging operator: the time average, denoted in this chapter by an overbar, and a partial spatial average over the domain in the x_1- and x_3-directions, denoted by angle brackets $\langle \cdot \rangle$:

$$\bar{f} = \lim_{T \to \infty} \frac{1}{T} \int_0^T f(\mathbf{x}, t)\, dt, \tag{10.1}$$

$$\langle f \rangle = \frac{1}{L_1 L_3} \int_0^{L_1} \int_0^{L_3} f(\mathbf{x}, t)\, dx_3\, dx_1. \tag{10.2}$$

Formally, overbarred quantities depend on x_1, x_2, and x_3 but not on t, while a $\langle \cdot \rangle$ quantity depends on x_2 and t. See Chapters 2 and 3 for further discussions of averaging.

We start with the Navier–Stokes equations without body forces:

$$\frac{\partial v_i}{\partial t} + v_j v_{i,j} = -\frac{1}{\rho} \pi_{,i} + \nu v_{i,jj}, \tag{10.3}$$

$$v_{i,i} = 0, \tag{10.4}$$

where $\mathbf{v}(\mathbf{x}, t) = (v_1(x_1, x_2, x_3, t), v_2(\ldots), v_3(\ldots))$ is the fluid velocity field (recall that bold symbols indicate vectors). Boundary conditions are introduced below. The (constant) density of the fluid is denoted by ρ, its kinematic viscosity by ν, and the pressure by $\pi = \pi(x_1, x_2, x_3, t)$. We perform a Reynolds decomposition: $v_i = \langle v_i \rangle + u_i$, $\pi = \langle \pi \rangle + p$, and denote $\langle v_i \rangle = U_i$ and $\langle \pi \rangle = P$. Here $\mathbf{U} = (U_1, U_2, U_3)$ represents the mean flow and $\mathbf{u} = (u_1, u_2, u_3)$ is the fluctuating part of the velocity field, at this point containing coherent structures as well as fine-scale, incoherent turbulent motions. By definition (10.2), U_1 is independent of x_1 and x_3. We make the assumption that $U_2 = U_3 = 0$; the physical implication is that, on scales much larger than those of the coherent structures, there is no flow in the wall-normal or spanwise directions and homogeneity prevails in the spanwise and streamwise directions. Henceforth we write $U_1(x_2, t) = U$. We additionally assume that U is slowly varying in time compared to the time scale of the coherent structures, so that $|\partial U/\partial t| \ll |\partial u_i/\partial t|$. This is motivated by an assumption of ergodicity in turbulent flows: if the spatial averaging domain $L_1 \times L_3$ were sufficiently large, U should be time independent. We subsequently return, in Sections 10.7–10.9, to examine the validity of these and other assumptions.

We next substitute the Reynolds decomposition into the equations of motion and, ignoring $\partial U/\partial t$, obtain:

$$\frac{\partial u_i}{\partial t} + u_{i,1}U + U_{,2}u_2\delta_{i1} + u_{i,j}u_j = -\frac{1}{\rho}(p_{,i} + P_{,i}) + v(u_{i,jj} + U_{,22}\delta_{i1}). \tag{10.5}$$

Upon taking the $\langle \cdot \rangle$ average of (10.3), and assuming that the derivatives of the averaged quantities in the homogeneous directions are also negligible, we obtain a relationship between the divergence of the Reynolds stress, the mean pressure and mean velocity:

$$\langle u_{i,j}u_j \rangle = -\frac{1}{\rho}P_{,i} + vU_{,22}\delta_{i1}. \tag{10.6}$$

Subtracting Equation (10.6) from (10.5) we then have

$$\frac{\partial u_i}{\partial t} + u_{i,1}U + U_{,2}u_2\delta_{i1} + (u_{i,j}u_j - \langle u_{i,j}u_j \rangle) = -\frac{1}{\rho}p_{,i} + vu_{i,jj},$$

$$u_{i,i} = 0. \tag{10.7}$$

Note that, in view of incompressibility, $u_{i,j}u_j = (u_i u_j)_{,j}$.

The boundary conditions we impose on (10.7) are:

$$\mathbf{u}(x_1, x_2, x_3 + L_3, t) = \mathbf{u}(x_1 + L_1, x_2, x_3, t) = \mathbf{u}(x_1, x_2, x_3, t), \tag{10.8}$$

and

$$\mathbf{u}(x_1, 0, x_3, t) = \mathbf{0} \ ; \quad \mathbf{u}(x_1, X_2, x_3, t) = \mathbf{f}(x_1, x_3, t). \tag{10.9}$$

The dimensions L_1 and L_3 are chosen so that the domain is larger than that of typical observed structures, yet not so large that too many structures inhabit it for a low-dimensional description to apply. The value of X_2 is chosen so that Ω lies in the wall layer (in wall units, $X_2^+ < 100$: in fact $X_2^+ = 40$ was taken in [22]). Specific choices of L_1, X_2, and L_3 are considered in Section 10.6. The (as yet undetermined) function $\mathbf{f}(x_1, x_3, t)$ represents a velocity boundary condition at the "free surface" $x_2 = X_2$ between the wall region and the outer part of the boundary layer. We discuss it further at the end of Section 10.6.

One of the unique features of the approach of [22] was the introduction of a model that accounts for the effect of coherent structures on the mean flow. The model adopted in that paper expresses the shape of the mean flow as a balance between the effects of pressure and those of the coherent structures:

$$U(x_2, t) = \frac{1}{v}\int_0^{x_2} \langle u_1 u_2 \rangle(x_2', t)\, dx_2' + \frac{u_\tau^2}{v}\left(x_2 - \frac{x_2^2}{2H}\right). \tag{10.10}$$

In this embodiment, u_τ is the friction velocity (Section 2.5), H is half the height of a (rectangular) channel that is supposed to contain the flow, and we assume $x_2 < H$ in Ω, consistent with modeling a thin wall region in a fully developed flow. We discuss the derivation and validity of this model in Sections 10.7 and 10.8.

The next step is to substitute a decomposition of the fluctuating field into resolved and unresolved components, which, roughly speaking, represent the coherent structures and the incoherent (or less coherent) smaller-scale turbulent motions. We write $\mathbf{u} = \mathbf{u}_< + \mathbf{u}_>$ where $\mathbf{u}_<$ represents the resolved components and $\mathbf{u}_>$ represents the unresolved, smaller-scale components. We assume that the operations of projection and taking a derivative commute, and we project Equation (10.7) onto the subspace of the resolved modes:

$$\frac{\partial u_{i<}}{\partial t} + u_{i<,1}U + U_{,2}u_{2<}\delta_{i1}$$

$$+ (u_{i<}u_{j<})_{<,j} + (u_{i>}u_{j<} + u_{i<}u_{j>} + u_{i>}u_{j>})_{<,j}$$

$$- \langle u_{i<}u_{j<}\rangle_{<,j} - \langle u_{i>}u_{j<} + u_{i<}u_{j>} + u_{i>}u_{j>}\rangle_{<,j}$$

$$= -\frac{1}{\rho}p_{<,i} + \nu u_{<i,jj}. \quad (10.11)$$

To close this equation we must specify $\mathbf{u}_>$ in terms of $\mathbf{u}_<$. We use the Heisenberg spectral transfer model, or equivalently the Smagorinsky subgrid scale model ([32], Section 7.5; [93]). This may be described as follows.

Mixed terms of the form $(\mathbf{u}_<\mathbf{u}_>)_<$ (Leonard stresses) are neglected. We justify this for a special class of models in Section 10.9. The remaining tensorial term $(u_{i>}u_{j>})_{<,j} - \langle u_{i>}u_{j>}\rangle_{<,j}$ can be split into a deformational component and a pseudo-pressure, as follows:

$$(u_{i>}u_{j>})_{<,j} - \langle u_{i>}u_{j>}\rangle_{<,j} =$$

$$\underbrace{\left[(u_{i>}u_{j>})_< - \langle u_{i>}u_{j>}\rangle_< - \frac{1}{3}\delta_{ij}\big((u_{k>}u_{k>})_< - \langle u_{k>}u_{k>}\rangle_<\big)\right]_{,j}}_{\text{deformational terms}}$$

$$+ \underbrace{\frac{1}{3}\left[\delta_{ij}\big((u_{k>}u_{k>})_< - \langle u_{k>}u_{k>}\rangle_<\big)\right]_{,j}}_{\text{pseudo-pressure}}. \quad (10.12)$$

The first bracketed term on the right-hand side of the equation is the deformation work done by the resolved scales against the stresses produced by the unresolved scales. It can be replaced by an eddy viscosity model:

$$- 2\alpha_1 \nu_T S_{ij<,j} = -\alpha_1\nu_T(u_{i<,j} + u_{j<,i})_{,j} = -\alpha_1\nu_T u_{i<,jj}, \quad (10.13)$$

where $S_{ij<}$ is the strain-rate tensor of the resolved field. In our case we can compute an estimate for ν_T from the eigenvalues and eigenvectors of the correlation tensor as outlined in Section 10.9. In addition, α_1 is included as an (order one) adjustable parameter.

The second bracketed term represents the i-gradient of the departure of kinetic energy of the unresolved scales from its mean value. The kinetic energy anomaly itself is assumed proportional to the excess in the rate of production of the kinetic energy of the unresolved scales, using the parameterization of the deformation work:

$$\frac{1}{3}\alpha_2 \, l_>^2 [(u_{i<,j} + u_{j<,i})(u_{i<,j} + u_{j<,i}) - \langle(u_{i<,j} + u_{j<,i})(u_{i<,j} + u_{j<,i})\rangle]. \quad (10.14)$$

Here $l_>$ is a length scale characteristic of the unresolved modes.

In this modeling process we have introduced the two "free" parameters α_1 and α_2, which should be both of order one, but which can be adjusted, within bounds, to obtain the correct energy flow to the unresolved modes.

Replacing the $\mathbf{u}_>$ terms in Equation (10.11) by the expressions (10.13) and (10.14), the dynamical equations for that part of the fluctuating velocity field representing the coherent structures assume the form:

$$\frac{\partial u_i}{\partial t} + u_{i,1}U + U_{,2}u_2\delta_{i1} + (u_{i,j}u_j) - \langle u_{i,j}u_j\rangle - \alpha_1 \nu_T u_{i,jj}$$
$$+ \frac{1}{3}\alpha_2 l_>^2 [(u_{k,l} + u_{l,k})(u_{k,l} + u_{l,k}) - \langle(u_{k,l} + u_{l,k})(u_{k,l} + u_{l,k})\rangle]_{,i}$$
$$= -\frac{1}{\rho}p_{,i} + \nu u_{i,jj}. \quad (10.15)$$

Here and henceforth we drop the subscript $<$. In this equation, α_1 and α_2 may be thought of as bifurcation parameters, while ν and ρ are (fixed) material properties and ν_T and $l_>$ are model parameters implicitly dependent on the domain size and wavenumber cut-off.

The final step is to normalize the equations. We normalize with wall units, using the friction velocity u_τ and ν, so that our units of length, time, and velocity are ν/u_τ, ν/u_τ^2, and u_τ (see Section 2.5). Our unit of pressure is ρu_τ^2, and pressure so normalized is designated by p^τ. Otherwise the normalized variables and parameters are designated by the same symbols as the unnormalized ones. The advantage of working in wall variables is that all wall bounded flows with the same pressure gradient look the same. The Reynolds number based on wall units is unity, so that viscosity does not appear in the normalized equations. Substitution of (10.10) into (10.15) yields the set of equations to be studied:

$$\frac{\partial u_i}{\partial t} + u_{i,1}\left[\int_0^{x_2} \langle u_1 u_2\rangle \, dx_2' + \left(x_2 - \frac{x_2^2}{2H}\right)\right]$$
$$+ u_2\delta_{i1}\left[\langle u_1 u_2\rangle + \left(1 - \frac{x_2}{H}\right)\right] + [(u_{i,j}u_j) - \langle u_{i,j}u_j\rangle]$$
$$- \alpha_1 \nu_T u_{i,jj} + \frac{1}{3}\alpha_2 l_>^2 [(u_{k,l} + u_{l,k})(u_{k,l} + u_{l,k})$$
$$- \langle(u_{k,l} + u_{l,k})(u_{k,l} + u_{l,k})\rangle]_{,i}$$
$$= -p_{,i}^\tau + u_{i,jj}. \quad (10.16)$$

10.2 The eigenfunction expansion

We expand each component of the velocity field as follows:

$$u_i(x_1, x_2, x_3, t) = \frac{1}{\sqrt{L_1 L_3}} \sum_{n=1}^{\infty} \sum_{\substack{k_1=-\infty \\ k_3=-\infty}}^{\infty} a_{k_1 k_3}^{(n)}(t) e^{2\pi i(\frac{k_1 x_1}{L_1} + \frac{k_3 x_3}{L_3})} \phi_{i k_1 k_3}^{(n)}(x_2) \quad (10.17)$$

for $i = 1, 2, 3$. Here $\phi_{ik_1k_3}^{(n)}$ denotes the ith component of the nth empirical eigenfunction $\varphi_{k_1k_3}^{(n)}$ for the wavenumber pair k_1, k_3, see Section 3.3, Equation (3.34). Division by $\sqrt{L_1L_3}$ yields an orthonormal set of basis elements. To distinguish the particular basis used in the wall layer work from the general (vector-valued) bases φ_n of Chapters 3 and 4, we employ the symbol $\varphi_{k_1k_3}^{(n)}$ in this chapter.

The triple integral defining projection onto each basis function (10.17) is Fourier transformation in the x_1- and x_3-directions, followed by integration over x_2. The Fourier transform of each velocity component u_i is

$$\hat{u}_i(x_2, t; k_1, k_3) = \int_0^{L_1} \int_0^{L_3} u_i(x_1, x_2, x_3, t)$$
$$\times \left[\frac{1}{\sqrt{L_1L_3}} e^{-2\pi i(\frac{k_1x_1}{L_1} + \frac{k_3x_3}{L_3})} \right] dx_3\, dx_1, \tag{10.18}$$

and, substituting from (10.17), the Fourier transformed velocity field is therefore

$$\hat{u}_i(x_2, t; k_1, k_3) = \sum_{m=1}^{\infty} a_{k_1k_3}^{(m)}(t)\phi_{ik_1k_3}^{(m)}(x_2). \tag{10.19}$$

To complete projection of the transformed velocity field $\hat{\mathbf{u}}$ onto a given basis element φ, we take the inner product of $\hat{\mathbf{u}}$ and φ and integrate from 0 to X_2:

$$(\hat{\mathbf{u}}, \varphi) = \int_0^{X_2} \hat{\mathbf{u}} \cdot \varphi^* \, dx_2. \tag{10.20}$$

Recall that $(\cdot)^*$ denotes the complex conjugate. Moreover, the empirical eigenfunctions are orthonormal with respect to n for each wavenumber pair. Thus, in computing (10.20), one has

$$(\varphi_{k_1k_3}^{(m)}, \varphi_{k_1k_3}^{(n)*}) = \int_0^{X_2} \phi_{ik_1k_3}^{(m)}(x_2)\phi_{ik_1k_3}^{(n)*}(x_2)\, dx_2 = \delta_{mn},$$

with summation from 1 to 3 implied over the repeated subscript i, but *not* over the wavenumbers k_j.

Equating (10.18) and (10.19) and taking the inner product with the (k_1, k_3, n)th eigenfunction gives:

$$a_{k_1k_3}^{(n)}(t) = \int_0^{X_2} \left[\frac{1}{\sqrt{L_1L_3}} \int_0^{L_1} \int_0^{L_3} \mathbf{u}\, e^{-2\pi i(\frac{k_1x_1}{L_1} + \frac{k_3x_3}{L_3})}\, dx_1\, dx_3 \right] \cdot \varphi_{k_1k_3}^{(n)*}\, dx_2$$
$$= \frac{1}{\sqrt{L_1L_3}} \int_0^{L_1} \int_0^{L_3} \int_0^{X_2} (u_1\phi_{1k_1k_3}^{(n)*} + u_2\phi_{2k_1k_3}^{(n)*} + u_3\phi_{3k_1k_3}^{(n)*})$$
$$\times e^{-2\pi i(\frac{k_1x_1}{L_1} + \frac{k_3x_3}{L_3})}\, dx_2\, dx_3\, dx_1. \tag{10.21}$$

10.3 Symmetries

There are a number of symmetries among the modal coefficients $a_{k_1k_3}^{(n)}$ and the eigenfunction components $\varphi_{k_1k_3}^{(n)} = (\phi_{1k_1k_3}^{(n)}, \phi_{2k_1k_3}^{(n)}, \phi_{3k_1k_3}^{(n)})$ that arise because the velocity field is real and the flow from which the eigenfunctions are derived has certain symmetries. The original study of [22] used eigenfunctions derived from the database of Herzog [155], who conducted experiments in a circular pipe. However, the rectangular channel flows of Moin *et al.* [190, 244]) also exhibit the same symmetries in the wall region.

In particular, the original two point correlation tensor of [155] enjoys the following features:

- The entries of the tensor are real.
- The tensor is reflection symmetric about a mid-plane in the spanwise direction.
- The field that generated the tensor is divergence-free.

In terms of the eigenfunctions, these properties imply that:

$$
\left.
\begin{aligned}
\varphi^{(n)}_{-k_1 k_3} &= (\phi^{(n)*}_{1 k_1 k_3}, \phi^{(n)*}_{2 k_1 k_3}, -\phi^{(n)*}_{3 k_1 k_3}), \\
\varphi^{(n)}_{k_1 -k_3} &= (\phi^{(n)}_{1 k_1 k_3}, \phi^{(n)}_{2 k_1 k_3}, -\phi^{(n)}_{3 k_1 k_3}), \\
\varphi^{(n)}_{-k_1 -k_3} &= (\phi^{(n)*}_{1 k_1 k_3}, \phi^{(n)*}_{2 k_1 k_3}, \phi^{(n)*}_{3 k_1 k_3}).
\end{aligned}
\right\}
\tag{10.22}
$$

From this symmetry of the eigenfunctions and from the form of the modal decomposition (10.17), we obtain the conjugate symmetry of the modal amplitudes, required for reality of the velocity field:

$$
a^{(n)}_{k_1 k_3} = a^{(n)*}_{-k_1 -k_3}.
\tag{10.23}
$$

This implies that, after projection, it is only necessary to solve approximately half of the resulting complex ODEs. For example, if one truncates the integer wavenumbers k_1 and k_2 at K_1 and K_2 and the eigenfunction order at $n = N$, there are $N(2K_1 + 1)(2K_3 + 1)$ modal coefficients in the expansion (10.17). Use of (10.23) reduces the number of ODEs one needs to solve to $N[(2K_1 + 1)K_3 + K_1]$. In calculating this number, we discount the modes $a^{(n)}_{00}$, corresponding to the mean flow U, which has already been accounted for in our model (Equation (10.10)).

In addition to the symmetries of the empirical eigenfunctions themselves, and that among the modal coefficients due to reality of the velocity field, both the original Navier–Stokes equations (10.3) themselves and the reduced equations (10.16) for the coherent structures display symmetries. In the language of Chapter 7, they are equivariant under translations and reflections in the spanwise direction, and under translations in the streamwise direction:

$$
x_3 \mapsto x_3 + \xi_3, \quad x_3 \mapsto -x_3,
\tag{10.24}
$$

and

$$
x_1 \mapsto x_1 + \xi_1.
\tag{10.25}
$$

In writing (10.24) and (10.25), we also rely on the use of periodic boundary conditions in these directions, without which, of course, these symmetries may not hold. In view of (10.17), these two actions respectively become, in terms of the modal coefficients:

$$
a_{k_1 k_3}^{(n)} \mapsto e^{2\pi i(\frac{k_3}{L_3}\xi_3)} a_{k_1 k_3}^{(n)}, \quad a_{k_1 k_3}^{(n)} \mapsto a_{k_1 -k_3}^{(n)*}, \tag{10.26}
$$

and

$$
a_{k_1 k_3}^{(n)} \mapsto e^{2\pi i(\frac{k_1}{L_1}\xi_1)} a_{k_1 k_3}^{(n)}. \tag{10.27}
$$

The ODEs derived in the next section are therefore equivariant under (10.26) and (10.27). Note that these are the groups $O(2)$, of rotations and reflections, and $SO(2)$, of rotations alone, in Fourier space. See the discussions in Chapters 7 and 8 for more information and introductory examples.

10.4 Galerkin projection

The partial differential equation we are working with, (10.16), may be rewritten as

$$
\begin{aligned}
\frac{\partial u_i}{\partial t} = -u_{i,1} &\left[\int_0^{x_2} \langle u_1 u_2 \rangle \, dx_2' + \left(x_2 - \frac{x_2^2}{2H} \right) \right] \\
&- u_2 \delta_{i1} \left[\langle u_1 u_2 \rangle + \left(1 - \frac{x_2}{H} \right) \right] \\
&+ (1 + \alpha_1 \nu_T) u_{i,jj} - [(u_{i,j} u_j) - \langle u_{i,j} u_j \rangle] \\
&- \frac{1}{3} \alpha_2 l_>^2 [(u_{k,l} + u_{l,k})(u_{k,l} + u_{l,k}) \\
&\qquad - \langle (u_{k,l} + u_{l,k})(u_{k,l} + u_{l,k}) \rangle]_{,i} - p_{,i}^\tau, \tag{10.28}
\end{aligned}
$$

where we have reordered the terms to match more closely the form of the final system of ordinary differential equations, given in (10.46) below. As noted in Section 10.2, each term of the PDE is converted to the corresponding term of that final dynamical system in two steps. Fourier transform is first applied, followed by projection onto the basis function $\varphi_{k_1 k_3}^{(n)}(x_2)$. The resulting terms appear on the right-hand side of the (k_1, k_2, n)th ODE. For example, the time-derivative term is transformed as follows. The Fourier transform is

$$
\frac{\partial}{\partial t} u_i(x_1, x_2, x_3, t) \xrightarrow{\mathcal{F}_{k_1 k_3}} \frac{\partial}{\partial t} \hat{u}_i(x_2, t; k_1, k_3).
$$

The final projection is, via (10.19) and (10.20):

$$
\frac{\partial}{\partial t} \hat{u}_i(x_2, t; k_1, k_3) \xrightarrow{\int_0^{X_2} (\cdot)\phi^* dx_2}
$$

$$
\int_0^{X_2} \frac{\partial}{\partial t} \left(\sum_{m=1}^\infty a_{k_1 k_3}^{(m)}(t) \phi_{i k_1 k_3}^{(m)} \right) \cdot \phi_{i k_1 k_3}^{(n)*} \, dx_2 = \frac{d}{dt} a_{k_1 k_3}^{(n)}(t).
$$

At several points we need to compute the derivatives of the velocity field. Using the orthogonal expansion (10.17), we can write:

$$u_{i,j}(x_1, x_2, x_3, t) = \frac{1}{\sqrt{L_1 L_3}} \sum_{n=1}^{\infty} \sum_{\substack{k_1=-\infty \\ k_3=-\infty}}^{\infty} a_{k_1 k_3}^{(n)}(t) e^{2\pi i (\frac{k_1 x_1}{L_1} + \frac{k_3 x_3}{L_3})} \Omega_j \phi_{ik_1 k_3}^{(n)}(x_2), \quad (10.29)$$

where, to simplify the formula, we denote the spatial derivative by the operator Ω_j, defined as follows:

$$\Omega_j \phi_{ik_1 k_3}^{(n)} = \begin{cases} \dfrac{2\pi i k_1}{L_1} \phi_{ik_1 k_3}^{(n)} & \text{if } j = 1, \\[2mm] \dfrac{d}{dx_2} \phi_{ik_1 k_3}^{(n)} & \text{if } j = 2, \\[2mm] \dfrac{2\pi i k_3}{L_3} \phi_{ik_1 k_3}^{(n)} & \text{if } j = 3. \end{cases} \quad (10.30)$$

The Reynolds stress $\langle u_1 u_2 \rangle$ appears in the model for the mean velocity $U(x_2)$. Multiplying out the orthogonal expansions for u_1 and u_2 we have:

$$\langle u_1 u_2 \rangle = \frac{1}{L_1 L_3} \int_0^{L_1} \int_0^{L_3} \left[\frac{1}{\sqrt{L_1 L_3}} \sum_{\substack{k_1, k_3 \\ n}} a_{k_1 k_3}^{(n)} e^{2\pi i (\frac{k_1}{L_1} x_1 + \frac{k_3}{L_3} x_3)} \phi_{1 k_1 k_3}^{(n)} \right.$$

$$\left. \times \frac{1}{\sqrt{L_1 L_3}} \sum_{\substack{k_1', k_3' \\ n'}} a_{k_1' k_3'}^{(n')} e^{2\pi i (\frac{k_1'}{L_1} x_1 + \frac{k_3'}{L_3} x_3)} \phi_{2 k_1' k_3'}^{(n')} \right] dx_3 \, dx_1$$

$$= \frac{1}{L_1 L_3} \sum_{\substack{k_1, k_3 \\ n}} \sum_{\substack{k_1', k_3' \\ n'}} a_{k_1 k_3}^{(n)} a_{k_1' k_3'}^{(n')} \phi_{1 k_1 k_3}^{(n)} \phi_{2 k_1' k_3'}^{(n')}$$

$$\times \frac{1}{L_1 L_3} \left[\int_0^{L_1} \int_0^{L_3} e^{2\pi i (\frac{k_1 + k_1'}{L_1} x_1 + \frac{k_3 + k_3'}{L_3} x_3)} \right] dx_3 \, dx_1$$

$$= \frac{1}{L_1 L_3} \sum_{\substack{n=1 \\ n'=1}}^{\infty} \sum_{\substack{k_1=-\infty \\ k_3=-\infty}}^{\infty} a_{k_1 k_3}^{(n)} a_{-k_1 -k_3}^{(n')} \phi_{1 k_1 k_3}^{(n)} \phi_{2 -k_1 -k_3}^{(n')}. \quad (10.31)$$

Using the symmetries of (10.22) and (10.23), we may further simplify this to read:

$$\langle u_1 u_2 \rangle = \frac{1}{L_1 L_3} \sum_{\substack{n=1 \\ n'=1}}^{\infty} \sum_{\substack{k_1=-\infty \\ k_3=-\infty}}^{\infty} a_{k_1 k_3}^{(n)} a_{k_1 k_3}^{(n')*} \phi_{1 k_1 k_3}^{(n)} \phi_{2 k_1 k_3}^{(n')*}. \quad (10.32)$$

Applying first Fourier transformation, via (10.18), and then Galerkin projection (10.20) to the various terms of the right-hand side of (10.28), we successively obtain:

$$u_{i,1} \int_0^{x_2} \langle u_1 u_2 \rangle \, dx_2' \xrightarrow{\mathcal{F}_{k_1 k_3}} \left(\sum_{l=1}^{\infty} a_{k_1 k_3}^{(l)} \frac{2\pi i k_1}{L_1} \phi_{i k_1 k_3}^{(l)} \right)$$

$$\times \left(\frac{1}{L_1 L_3} \int_0^{x_2} \sum_{\substack{k_1', k_3' = -\infty \\ m, q = 1}}^{\infty} a_{k_1' k_3'}^{(m)} a_{k_1' k_3'}^{(q)*} \phi_{1 k_1' k_3'}^{(m)} (x_2') \phi_{2 k_1' k_3'}^{(q)*} (x_2') \, dx_2' \right)$$

$$\xrightarrow{\int_0^{X_2} (\cdot) \phi^* dx_2} \frac{2\pi i k_1}{L_1} \frac{1}{L_1 L_3} \sum_{\substack{k_1', k_3' = -\infty \\ l, m, q = 1}}^{\infty} a_{k_1 k_3}^{(l)} a_{k_1' k_3'}^{(m)} a_{k_1' k_3'}^{(q)*}$$

$$\times \left(\int_0^{X_2} \phi_{i k_1 k_3}^{(l)} \phi_{i k_1 k_3}^{(n)*} \int_0^{x_2} \phi_{1 k_1' k_3'}^{(m)} (x_2') \phi_{2 k_1' k_3'}^{(q)*} (x_2') \, dx_2' \, dx_2 \right), \qquad (10.33)$$

$$u_{i,1} \left(x_2 - \frac{x_2^2}{2H} \right) \xrightarrow{\mathcal{F}_{k_1 k_3}}$$

$$\left(\sum_{l=1}^{\infty} a_{k_1 k_3}^{(l)} \frac{2\pi i k_1}{L_1} \phi_{i k_1 k_3}^{(l)} \right) \left(x_2 - \frac{x_2^2}{2H} \right) \xrightarrow{\int_0^{X_2} (\cdot) \phi^* dx_2}$$

$$\frac{2\pi i k_1}{L_1} \sum_{l=1}^{\infty} a_{k_1 k_3}^{(l)} \int_0^{X_2} \left(x_2 - \frac{x_2^2}{2H} \right) \phi_{i k_1 k_3}^{(l)} \phi_{i k_1 k_3}^{(n)*} \, dx_2, \qquad (10.34)$$

$$u_2 \langle u_1 u_2 \rangle \delta_{i1} \xrightarrow{\mathcal{F}_{k_1 k_3}}$$

$$\frac{1}{L_1 L_3} \sum_{l=1}^{\infty} a_{k_1 k_3}^{(l)} \phi_{2 k_1 k_3}^{(l)} \sum_{\substack{k_1', k_3' = -\infty \\ m, q = 1}}^{\infty} \left(a_{k_1' k_3'}^{(m)} a_{k_1' k_3'}^{(q)*} \phi_{1 k_1' k_3'}^{(m)} \phi_{2 k_1' k_3'}^{(q)*} \right) \delta_{i1}$$

$$\xrightarrow{\int_0^{X_2} (\cdot) \phi^* dx_2} \frac{1}{L_1 L_3} \sum_{\substack{k_1', k_3' = -\infty \\ l, m, q = 1}}^{\infty} a_{k_1 k_3}^{(l)} a_{k_1' k_3'}^{(m)} a_{k_1' k_3'}^{(q)*}$$

$$\times \int_0^{X_2} \phi_{2 k_1 k_3}^{(l)} \phi_{1 k_1 k_3}^{(n)*} \phi_{1 k_1' k_3'}^{(m)} \phi_{2 k_1' k_3'}^{(q)*} \, dx_2, \qquad (10.35)$$

$$u_2 \left(1 - \frac{x_2}{H} \right) \delta_{i1} \xrightarrow{\mathcal{F}_{k_1 k_3}} \sum_{l=1}^{\infty} a_{k_1 k_3}^{(l)} \phi_{2 k_1 k_3}^{(l)} \left(1 - \frac{x_2}{H} \right) \delta_{i1}$$

$$\xrightarrow{\int_0^{X_2} (\cdot) \phi^* dx_2} \sum_{l=1}^{\infty} a_{k_1 k_3}^{(l)} \int_0^{X_2} \left(1 - \frac{x_2}{H} \right) \phi_{2 k_1 k_3}^{(l)} \phi_{1 k_1 k_3}^{(n)*} \, dx_2, \qquad (10.36)$$

$$u_{i,jj} = \frac{1}{\sqrt{L_1 L_3}} \sum_{\substack{k_1', k_3' = -\infty \\ m=1}}^{\infty} a_{k_1' k_3'}^{(m)} \left[e^{\left(\frac{2\pi i k_1'}{L_1} x_1 + \frac{2\pi i k_3'}{L_3} x_3\right)} \phi_{i k_1 k_3}^{(m)} \right]_{,jj}$$

$$\xrightarrow{\mathcal{F}_{k_1 k_3}} \sum_m a_{k_1 k_3}^{(m)} \left\{ \phi_{i k_1 k_3}^{(m)} \left[\left(\frac{2\pi i k_1}{L_1}\right)^2 + \left(\frac{2\pi i k_3}{L_3}\right)^2 \right] + \frac{d^2}{dx_2^2} \phi_{i k_1 k_3}^{(m)} \right\}$$

$$\xrightarrow{\int_0^{X_2} (\cdot) \phi^* \, dx_2} \sum_m a_{k_1 k_3}^{(m)} \left\{ \delta_{mn} \left[\left(\frac{2\pi i k_1}{L_1}\right)^2 + \left(\frac{2\pi i k_3}{L_3}\right)^2 \right] \right.$$

$$\left. + \int_0^{X_2} \left(\frac{d^2}{dx_2^2} \phi_{i k_1 k_3}^{(m)}\right) \phi_{i k_1 k_3}^{(n)*} \, dx_2 \right\}, \quad (10.37)$$

$$u_{i,j} u_j \xrightarrow{\mathcal{F}_{k_1 k_3}} \frac{1}{\sqrt{L_1 L_3}} \sum_{\substack{k_1', k_3' = -\infty \\ m,q=1}}^{\infty} a_{k_1' k_3'}^{(m)} a_{k_1 - k_1' k_3 - k_3'}^{(q)}$$

$$\times \left[\frac{2\pi i (k_1 - k_1')}{L_1} \phi_{1 k_1' k_3'}^{(m)} \phi_{i k_1 - k_1' k_3 - k_3'}^{(q)} + \phi_{2 k_1' k_3'}^{(m)} \frac{d}{dx_2} \phi_{i k_1 - k_1' k_3 - k_3'}^{(q)} \right.$$

$$\left. + \frac{2\pi i (k_3 - k_3')}{L_3} \phi_{3 k_1' k_3'}^{(m)} \phi_{i k_1 - k_1' k_3 - k_3'}^{(q)} \right]$$

$$\xrightarrow{\int_0^{X_2} (\cdot) \phi_i^* \, dx_2} \frac{1}{\sqrt{L_1 L_3}} \sum_{\substack{k_1', k_3' = -\infty \\ m,q=1}}^{\infty} a_{k_1' k_3'}^{(m)} a_{k_1 - k_1' k_3 - k_3'}^{(q)}$$

$$\times \int_0^{X_2} \left(\phi_{l k_1' k_3'}^{(m)} \Omega_l \phi_{i k_1 - k_1' k_3 - k_3'}^{(q)} \phi_{i k_1 k_3}^{(n)*} \right) dx_2 \equiv \Upsilon, \quad (10.38)$$

where Ω_j is defined analogously to Ω_j of (10.30):

$$\Omega_j \phi_{i k_1 - k_1' k_3 - k_3'}^{(n)} = \begin{cases} \dfrac{2\pi i (k_1 - k_1')}{L_1} \phi_{i k_1 - k_1' k_3 - k_3'}^{(n)} & \text{if } j = 1, \\[2mm] \dfrac{d}{dx_2} \phi_{i k_1 - k_1' k_3 - k_3'}^{(n)} & \text{if } j = 2, \\[2mm] \dfrac{2\pi i (k_3 - k_3')}{L_3} \phi_{i k_1 - k_1' k_3 - k_3'}^{(n)} & \text{if } j = 3. \end{cases} \quad (10.39)$$

We observe that the Fourier transform of the spatial average $\langle u_{i,j} u_j \rangle$ is Υ for $k_1 = k_3 = 0$ and zero otherwise. Thus we have

$$u_{i,j} u_j - \langle u_{i,j} u_j \rangle = \Upsilon (1 - \delta_{\substack{k_1=0 \\ k_3=0}}). \quad (10.40)$$

We have now computed the projections of all terms on the right-hand side of (10.28) except the pressure term and the model for the pseudo-pressure term. As in Chapter 4, to project these terms onto divergence-free basis functions, we appeal to the divergence theorem, which supplies the following (Green's) identity for the inner product of the gradient of a scalar f and a divergence-free vector $\mathbf{g}$:

$$\int_\Omega \nabla f \cdot \mathbf{g} \, dx = \int_\Omega f_{,i} g_i \, dx = \int_\Omega (fg_i)_{,i} \, dx = \int_{\partial\Omega} \mathbf{n} \cdot (f\mathbf{g}) \, dS. \qquad (10.41)$$

Here $\partial\Omega$ denotes the boundary surface of the domain, $\mathbf{n}$ its unit normal, and dS the area element. Thus, in our particular case, after Fourier transformation and projection, we have:

$$\int_0^{L_1} \int_0^{L_3} \int_0^{X_2} \frac{1}{\sqrt{L_1 L_3}} f_{,i} e^{-2\pi i(\frac{k_1}{L_1} x_1 + \frac{k_3}{L_3} x_3)} \phi_{ik_1k_3}^{(n)*}(x_2) \, dx_2 \, dx_3 \, dx_1$$

$$= \int_0^{L_1} \int_0^{L_3} (0, 1, 0) \cdot \frac{1}{\sqrt{L_1 L_3}} f \, e^{-2\pi i(\frac{k_1 x_1}{L_1} + \frac{k_3 x_3}{L_3})}$$

$$\times (\phi_{1k_1k_3}^{(n)*}, \phi_{2k_1k_3}^{(n)*}, \phi_{3k_1k_3}^{(n)*}) \, dx_3 \, dx_1 \Big|_0^{X_2}$$

$$= \hat{f}_{k_1k_3}(X_2) \phi_{2k_1k_3}^{(n)*}(X_2). \qquad (10.42)$$

Note that the remaining surface integral terms of (10.41), on $x_1 = 0$, L_1 and on $x_3 = 0$, L_3, cancel identically due to periodicity of the basis functions in these directions, see Section 4.3. The part of (10.42) on $x_2 = 0$ also vanishes, since the empirical eigenfunctions satisfy the no-slip condition at the wall.

The pressure term in (10.28) therefore transforms as

$$p_{,i}^\tau \to \hat{p}_{k_1k_3}^\tau(X_2) \phi_{2k_1k_3}^{(n)*}(X_2). \qquad (10.43)$$

As noted in Section 4.4, the eigenfunctions do not vanish on the upper edge of the wall region, and so this boundary pressure term survives. It represents communication between the wall region and the outer part of the boundary layer.

For the pseudo-pressure term,

$$\frac{1}{3} \alpha_2 l_>^2 \left[(u_{k,l} + u_{l,k})(u_{k,l} + u_{l,k}) - \langle (u_{k,l} + u_{l,k})(u_{k,l} + u_{l,k}) \rangle \right]_{,i},$$

we first compute

$$u_{k,l} + u_{l,k} = \frac{1}{\sqrt{L_1 L_3}} \sum_{\substack{k_1', k_3' = -\infty \\ m=1}}^{\infty} a_{k_1'k_3'}^{(m)} e^{2\pi i(\frac{k_1' x_1}{L_1} + \frac{k_3' x_3}{L_3})} (\Omega_l \phi_{kk_1'k_3'}^{(m)} + \Omega_k \phi_{lk_1'k_3'}^{(m)}),$$

so that, using the notations (10.30) and (10.39), the Fourier transform of the product may be written:

$$(u_{k,l} + u_{l,k})(u_{k,l} + u_{l,k}) \xrightarrow{\mathcal{F}_{k_1k_3}} \frac{1}{\sqrt{L_1 L_3}} \sum_{\substack{k_1', k_3' = -\infty \\ m,q=1}}^{\infty} a_{k_1'k_3'}^{(m)} a_{k_1-k_1'k_3-k_3'}^{(q)}$$

$$\times (\Omega_l \phi_{kk_1'k_3'}^{(m)} + \Omega_k \phi_{lk_1'k_3'}^{(m)})(\Omega_l \phi_{kk_1-k_1'k_3-k_3'}^{(q)} + \Omega_k \phi_{lk_1-k_1'k_3-k_3'}^{(q)})$$

$$\equiv \Psi(x_2).$$

The first (non-averaged) part of the pseudo-pressure term therefore transforms to:

$$[(u_{k,l} + u_{l,k})(u_{k,l} + u_{l,k})]_{,i} \rightarrow \frac{1}{\sqrt{L_1 L_3}} \sum_{\substack{k_1', k_3' = -\infty \\ m,q=1}}^{\infty} a_{k_1' k_3'}^{(m)} a_{k_1 - k_1' k_3 - k_3'}^{(q)}$$

$$\times [\Omega_l \phi_{k k_1' k_3'}^{(m)} (X_2) + \Omega_k \phi_{l k_1' k_3'}^{(m)} (X_2)]$$

$$\times [\Omega_l \phi_{k k_1 - k_1' k_3 - k_3'}^{(q)} (X_2) + \Omega_k \phi_{l k_1 - k_1' k_3 - k_3'}^{(q)} (X_2)] \phi_{2 k_1 k_3}^{(n)*} (X_2), \tag{10.44}$$

and as above in (10.40), subtracting the transforms of the spatially averaged terms corresponds to multiplying by $(1 - \delta_{\substack{k_1=0 \\ k_3=0}})$. We also note that, since summation is implied on the repeated indices k and l, the four terms resulting from expansion of the expressions in parentheses in fact yield only two products, so that one finally obtains:

$$\frac{2}{\sqrt{L_1 L_3}} \left(1 - \delta_{\substack{k_1=0 \\ k_3=0}}\right) \sum_{\substack{k_1', k_3' = -\infty \\ m,q=1}}^{\infty} a_{k_1' k_3'}^{(m)} a_{k_1 - k_1' k_3 - k_3'}^{(q)}$$

$$\times [\Omega_l \phi_{k k_1' k_3'}^{(m)} (X_2) \Omega_l \phi_{k k_1 - k_1' k_3 - k_3'}^{(q)} (X_2)$$

$$+ \Omega_l \phi_{k k_1' k_3'}^{(m)} (X_2) \Omega_k \phi_{l k_1 - k_1' k_3 - k_3'}^{(q)} (X_2)] \phi_{2 k_1 k_3}^{(n)*} (X_2). \tag{10.45}$$

This may also be seen at the outset by noting that

$$u_{k,l} u_{k,l} + u_{k,l} u_{l,k} + u_{l,k} u_{k,l} + u_{l,k} u_{l,k} = 2(u_{l,k} u_{l,k} + u_{l,k} u_{k,l}).$$

Finally, using the terms (10.33)–(10.45), we can now write the projected ODEs as follows:

$$\dot{\mathbf{a}} = [\mathbf{A}_1 + (1 + \alpha_1 \nu_T) \mathbf{A}_2] \mathbf{a}$$

$$+ [\mathbf{Q}_1(\mathbf{a}, \mathbf{a}) + \tfrac{2}{3} \alpha_2 l_>^2 \mathbf{Q}_2(\mathbf{a}, \mathbf{a})] + \mathbf{C}(\mathbf{a}, \mathbf{a}, \mathbf{a}) + \mathbf{d}(t), \tag{10.46}$$

where $\mathbf{a}$ denotes the vector of modal components $a_{k_1, k_3}^{(n)}(t)$. The linear, quadratic, and cubic terms in this (set of) complex equations are perhaps most easily specified in component form, the (k_1, k_3, n)th row of each matrix product in turn being:

$\mathbf{A}_1 \mathbf{a}$:

$$-\sum_{l=1}^{\infty} a_{k_1 k_3}^{(l)} \left[\frac{2\pi i k_1}{L_1} \int_0^{X_2} \left(x_2 - \frac{x_2^2}{2H} \right) \phi_{i k_1 k_3}^{(l)} \phi_{i k_1 k_3}^{(n)*} \, dx_2 \right.$$

$$\left. + \int_0^{X_2} \left(1 - \frac{x_2}{H} \right) \phi_{2 k_1 k_3}^{(l)} \phi_{1 k_1 k_3}^{(n)*} \, dx_2 \right], \tag{10.47}$$

A_2a :

$$
-\sum_{l=1}^{\infty} a_{k_1 k_3}^{(l)} \left\{ \left[\left(\frac{2\pi k_1}{L_1} \right)^2 + \left(\frac{2\pi k_3}{L_3} \right)^2 \right] \delta_{ln} \right.
$$
$$
\left. - \int_0^{X_2} \frac{d^2}{dx_2^2} \phi_{ik_1 k_3}^{(l)} \, \phi_{ik_1 k_3}^{(n)*} \, dx_2 \right\}, \tag{10.48}
$$

$Q_1(\mathbf{a}, \mathbf{a})$:

$$
-\frac{1}{\sqrt{L_1 L_3}} \left(1 - \delta_{\substack{k_1=0 \\ k_3=0}} \right) \sum_{\substack{k_1', k_3' = -\infty \\ m,q=1}}^{\infty} a_{k_1' k_3'}^{(m)} a_{k_1 - k_1' k_3 - k_3'}^{(q)}
$$
$$
\times \int_0^{X_2} \phi_{l k_1' k_3'}^{(m)} \, \Omega_l \phi_{k_1 - k_1' k_3 - k_3'}^{(q)} \, \phi_{ik_1 k_3}^{(n)*} \, dx_2, \tag{10.49}
$$

$Q_2(\mathbf{a}, \mathbf{a})$:

$$
-\frac{1}{\sqrt{L_1 L_3}} \left(1 - \delta_{\substack{k_1=0 \\ k_3=0}} \right) \sum_{\substack{k_1', k_3' = -\infty \\ m,q=1}}^{\infty} a_{k_1' k_3'}^{(m)} a_{k_1 - k_1' k_3 - k_3'}^{(q)}
$$
$$
\times [\Omega_l \phi_{k_1' k_3'}^{(m)} (X_2) \Omega_l \phi_{k_1 - k_1' k_3 - k_3'}^{(q)} (X_2)
$$
$$
+ \Omega_l \phi_{k_1' k_3'}^{(m)} (X_2) \Omega_k \phi_{l k_1 - k_1' k_3 - k_3'}^{(q)} (X_2)] \phi_{2 k_1 k_3}^{(n)*} (X_2), \tag{10.50}
$$

$C(\mathbf{a}, \mathbf{a}, \mathbf{a})$:

$$
-\frac{1}{L_1 L_3} \sum_{\substack{k_1', k_3' = -\infty \\ l=1, m=1, q=1}}^{\infty} a_{k_1 k_3}^{(l)} a_{k_1' k_3'}^{(m)} a_{k_1' k_3'}^{(q)*}
$$
$$
\times \int_0^{X_2} \left[\frac{2\pi i k_1}{L_1} \phi_{ik_1 k_3}^{(l)} \, \phi_{ik_1 k_3}^{(n)*} \left(\int_0^{x_2} \phi_{1 k_1' k_3'}^{(m)} \, \phi_{2 k_1' k_3'}^{(q)*} \, dx_2' \right) \right.
$$
$$
\left. + \phi_{2 k_1 k_3}^{(l)} \, \phi_{1 k_1 k_3}^{(n)*} \, \phi_{1 k_1' k_3'}^{(m)} \, \phi_{2 k_1' k_3'}^{(q)*} \right] dx_2 \tag{10.51}
$$

$\mathbf{d}(t)$:

$$
- \hat{p}_{k_1 k_3}^{\tau} (X_2, t) \phi_{2 k_1 k_3}^{(n)*} (X_2). \tag{10.52}
$$

Recall that the differential operator Ω_l was defined in (10.30) and (10.39), and that the definition depends upon the Fourier wavenumbers in the basis function upon which it operates:

$$
\Omega_j \phi_{ik_1 k_3}^{(n)} = \begin{cases} \dfrac{2\pi i k_1}{L_1} \phi_{ik_1 k_3}^{(n)}, & \text{if } j = 1, \\[2mm] \dfrac{d}{dx_2} \phi_{ik_1 k_3}^{(n)}, & \text{if } j = 2, \\[2mm] \dfrac{2\pi i k_3}{L_3} \phi_{ik_1 k_3}^{(n)}, & \text{if } j = 3. \end{cases}
$$

These ODEs contain linear, quadratic, and cubic terms as well as "external" forcing terms. The first two types derive directly from the Navier-Stokes equation and the models for the constant (pressure driven) part of the mean flow U and the losses to neglected modes. They therefore contain both stabilizing (viscous) terms $\mathbf{A}_2$ and energy production terms $\mathbf{A}_1$, as well as the quadratic interaction inherited from the convective derivative in $\mathbf{Q}_1$. The cubic term is due to the Reynolds stress terms in the mean flow model, which provide a kind of negative feedback, as we see below. Finally, the forcing term $\mathbf{d}(t)$ represents communication between the wall region and the outer flow, via the pressure field on the upper boundary. The second quadratic (pseudo-pressure) term $\mathbf{Q}_2$ also arises due to this "free surface." Both these terms would vanish identically if the domain Ω spanned the whole channel, with the basis elements $\phi_{2k_1k_3}^{(n)*}$ satisfying a no-slip condition on the upper boundary. Such a case is described in Section 12.5, also see [347, 348].

Boundary conditions, or the lack of them, can give rise to pressure terms in other situations. In [268] the authors consider an incompressible shear flow, an open flow for which, as above, the pressure term does not vanish. They replace the full velocity terms in the Poisson equation (4.67) by their low-dimensional representations and thereby transform the projection of the pressure into a second set of quadratic, state-dependent terms in the projected ODEs. The coefficients of these terms are inner products of the empirical basis functions and ∇p that may be computed recursively from the "partial pressures" associated with each empirical eigenfunction, or approximated from (numerical) pressure field data. Generalizing the work of [301, 302], modal energy balances are used to analyze the effects of the pressure terms, and the method is applied to a 2-dimensional laminar shear layer and to a 3-dimensional transitional shear layer.

All the coefficients in (10.46) contain integrals of products of the empirical eigenfunctions and their derivatives. In this, as in any other case derived from experimental or numerical data, these functions are known only in tabular numerical form (or perhaps in terms of polynomial spline or other functions with numerically fitted coefficients). There may be a significant margin of error in the coefficients, especially those involving derivatives. Here the adjustable parameters α_j can play a useful rôle, permitting one to explore a range around the experimental values. The notion of structural stability, introduced in Section 6.3, is, however, even more relevant; for if we find interesting dynamical behavior in a particular case of (10.46), then we certainly want to be sure that it will not vanish under small changes in the coefficients. If the system is structurally stable, then this robustness is guaranteed.

10.5 Geometrical structure of the model

We have already remarked, in Section 10.3, on some basic symmetries that the empirical basis functions and the projected equations inherit from the governing PDEs and the boundary conditions. In this section we briefly describe some other structural features that the model equations (10.46) enjoy in their phase space. We first consider a minimal truncation with only one family of eigenfunctions ($N = 1$) and a single streamwise Fourier mode ($K_1 = 0$), as taken in [22]. That this is reasonable, at least for preliminary studies, is argued in the next section. The resulting velocity fields are "two-and-a-half"-dimensional, being independent of the streamwise coordinate x_1:

$$\mathbf{u}(\mathbf{x}, t) = \sum_{k_3=-K_3}^{K_3} a_{0,k_3}^{(1)}(t) e^{2\pi i \frac{k_3 x_3}{L_3}} \boldsymbol{\varphi}_{0,k_3}^{(1)}(x_2), \qquad (10.53)$$

and the ODEs simplify considerably, taking the form:

$$\dot{a}_k = \left[A_k^1 + (1 + \alpha_1 \nu_T)A_k^2\right]a_k$$

$$+ \left(\sum_{k'} Q_{k\ k-k'}^1 a_{k'} a_{k-k'} + \frac{2}{3}\alpha_2 l_>^2 \sum_{k'} Q_{k\ k-k'}^2 a_{k'} a_{k-k'}\right)$$

$$+ \left(\sum_{k'} C_{kk'}|a_{k'}|^2\right)a_k + d_k(t), \quad k = -K_3, \ldots K_3. \qquad (10.54)$$

Here, for brevity, we have replaced the double subscript $0, k_3$ by k ($= k_3$) and omitted the superscript $^{(1)}$. Note that the matrices $\mathbf{A}_1$ and $\mathbf{A}_2$ become diagonal and the cubic term also takes a simple form. In fact, as the reader can check by comparison with equations (8.13) and (8.39), for $N = 1$, $K_1 = 0$, the model has a similar form to Fourier projections of the one-space-dimensional Kuramoto–Sivashinsky equation studied in Chapter 8. In particular, with the (boundary) pressure terms d_k set to zero, (10.54) is equivariant under the group $O(2)$ of rotations and reflections in Fourier space, corresponding to spanwise translation by ξ_3 and reflection:

$$\left.\begin{array}{l} T_\xi : a_k \mapsto e^{\frac{2\pi i k}{L_3}\xi_3} a_k, \\[2mm] R_f : a_k \mapsto a_k^*. \end{array}\right\} \qquad (10.55)$$

This is just the restriction of (10.26) to the subspace $k_1 = 0$. Also see (8.16). The symmetry (10.27) under streamwise translations does not appear explicitly in these ODEs, for all non-zero streamwise modes have been excluded.

As we have noted (Equation (10.23)), reality of the velocity field implies that we may use the relation

$$a_{-k}(t) = a_k^*(t), \qquad (10.56)$$

and so we need solve only a set of K_3 complex ODEs. (The mode $k = 0$ represents modification to the mean velocity U, which is already included in the model (10.10).) We discuss choices for K_3 in the next section. In Chapter 11, we denote the real and imaginary components of each modal coefficient as $a_k = x_k + iy_k$, or, in terms of amplitude and phase, by $a_k = r_k e^{i\theta_k}$.

Within the K_3-dimensional complex, or $2K_3$-dimensional real, phase space of (10.54), there are several invariant subspaces. First, due to the symmetries (10.55), the purely real subspace $\{(x_k, y_k)|\ y_k = 0\}$ is invariant, as is any rotation of it under T_ξ. Second, due to the structure of the quadratic and cubic terms of the ODEs, inherited from that of the Navier–Stokes equations themselves, the subspace spanned by even Fourier modes ($k = 2, 4, \ldots$) is also invariant. (Here, as above, we set d_k to zero: this term provides an important

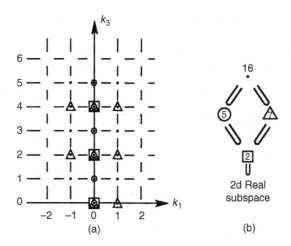

Figure 10.1 (a) Hierarchies of nested systems for the $N = 1$ case: $\square$: $K_1 = 0$, $K_3 = 5$, even modes; $\circ$: $K_1 = 0$, $K_3 = 5$, all modes; $\triangle$: $K_1 = 1$, $K_3 = 5$, even modes; $\bullet$: $K_1 = 1$, $K_3 = 5$, all modes. (b) Inclusion of subsystems in larger systems and complex dimension. From Holmes [158].

symmetry-breaking effect, described in Section 11.4.) The same symmetries are shared by the Fourier projections (8.13) of the one-space-dimensional Kuramoto–Sivashinsky equation. These lower-dimensional subspaces allow us to "parse" the system and thus enable us to reach at least a partial understanding of its bifurcational and dynamical behavior.

Similarly, an understanding of the $K_1 = 0$ case is important not only for its own sake, but also because systems of this type appear as invariant subsystems for the $N = 1$, $K_1 \neq 0$ truncations, as the reader can see by reference to Equations (10.47–10.51) (see Holmes [158]). Figure 10.1 illustrates this nesting property schematically for the case of $N = 1$, $K_1 = 1$, $K_3 = 5$. The absolutely minimal non-trivial system containing only the two modes $N = K_1 = 0$, $K_3 = 2$, has precisely the structure of the normal form of the codimension two $O(2)$-equivariant ODE, Equation (7.50), analyzed in Section 7.4, as does the invariant subspace $k = 2, 4$ of the system $K_1 = 0$, $K_3 = 5$. In Chapter 11 we build up our understanding of more realistic truncations, starting with this simplest case.

When more than one family of eigenfunctions is included ($N \geq 2$), the linear and cubic terms (10.47), (10.48), and (10.51) are no longer diagonal, and coupling between the modal coefficients occurs even at the linear level. Little appears to be known about the geometrical structure of such systems, whose dimension is, in any case, quite high. Aubry and Sanghi [25, 323] have carried out preliminary studies of systems with up to $N = 2$, $K_1 = 1$, $K_3 = 5$ modes: 32 complex or 64 real dimensions. We briefly summarize some of their findings in Chapter 11.

We now turn to the problem of specific choices of scales L_j and orders of truncation N, K_j.

10.6 Choosing subspaces and domains

The domain Ω in which we work has dimensions L_1 in the streamwise direction, L_3 in the cross-stream direction, and X_2 normal to the wall. In the initial study of [22], the authors used the values $L_1 = L_3 = 333$, $X_2 = 40$ in wall variables. (These are, properly speaking, values of $L_1^+ = L_1 u_\tau / \nu$, and so forth, but we have agreed to work in variables normalized in this way, and to suppress the $(\cdot)^+$ notation.) We must now justify these choices.

The coherent structures in the wall region of the turbulent boundary layer are observed to have a cross-stream wavelength (corresponding to an eddy pair) in the neighborhood of 100 wall units, based both on the measurements of Kline *et al.* [195] and on the coherent structures extracted from the direct numerical simulations of Moin and Moser [244]. Our choice of L_3 determines, at least in part, the complexity of the cross-stream interactions that we can reproduce. If we restrict our domain to contain only a single eddy pair, we exclude cooperative behavior involving two or more such pairs. We can determine how important this might be by examining the eigenvalue spectrum as a function of spanwise wavenumber (see Figure 10.2(a)). We find an off-axis peak corresponding to spanwise wavelengths in the neighborhood of 165 wall units; this corresponds to the wavelength for recurrence of eddy pairs. The eigenvalue spectrum is relatively simple in form, falling away monotonically from the peak. To resolve the detail in the crudest possible way, we need a minimum of one point near the peak and one point between the peak and the axis; this roughly determines the spacing of the points, and hence the spanwise extent of the domain. We then add Fourier modes at the same spacing to cover the higher wavenumber part of the eigenvalue spectrum. Five non-zero modes ($K_3 = 5$) provides a reasonable coverage, and this was the choice adopted in [22], after (unpublished) studies of truncations with K_3 between 4 and 8. There seemed no great advantage in including modes beyond $K_3 = 5$. See [19, 354] for further information.

These considerations give a lowest wavenumber corresponding to a cross-stream distance of $2 \times 165 \approx 333$, which permits the interaction of two eddy pairs. In choosing this width, we deliberately exclude interactions (e.g. clustering) involving three or more pairs. We hope that such interactions are less probable, just as simultaneous collisions of three or more molecules in a gas are much less probable than pairwise collisions. We are in any case interested in developing *simple* models, rather than capturing higher order effects. In this connection the reader should refer to the discussion of localized models of spatio-temporally chaotic fields in Section 8.4.

The choice of L_1 is less clear, but it also matters less. The eigenvalue spectrum as a function of streamwise wavenumber peaks at the origin (Figure 10.2(b)). This implies that there is no characteristic recurrence period for coherent structures in the streamwise direction. It is not possible to tell from visual examination of the spectrum what the typical size of the coherent structures in the streamwise direction might be. A streamwise length scale can be associated with the spectrum by various techniques. Let $S(\kappa)$ be the eigenvalue spectrum versus streamwise wavenumber. In meteorology, for example, one plots $\kappa S(\kappa)$ versus $\ln(\kappa)$. This is area/energy conserving, and exhibits an off-axis peak. The location of this peak provides a length scale (see, for example, Lumley and Panofsky [227]). We can also proceed more directly, and consult visual observation of coherent structures in

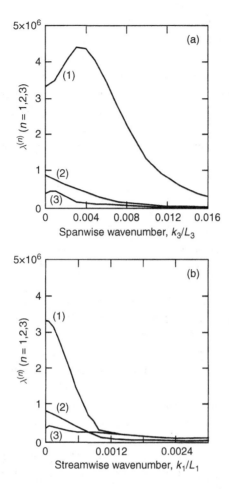

Figure 10.2 Empirical eigenvalue spectra for the near-wall region from the data of [155]: (a) $\lambda^{(n)}_{0k_3}$ plotted versus spanwise wavenumber k_3/L_3, (b) $\lambda^{(n)}_{k_10}$ plotted versus streamwise wavenumber k_1/L_1.

the boundary layer, which are observed to have a mean streamwise length of about 1000 wall units. If we include streamwise variation ($K_3 \geq 1$), we certainly want our domain to cover at least one whole coherent structure in that direction, and hence need a domain at least 1000 wall units long, preferably 2000. However, following [22], we intend at first to consider only coherent structures having no streamwise variation, and consequently the question is moot. With no variation in the streamwise direction, it does not matter how long the domain is in that direction (Equation (10.53)).

In the direction normal to the wall we have chosen the distance $X_2 (= X_2^+) = 40$ in wall units. The quantity X_2^+ is actually a Reynolds number based on distance from the wall. This Reynolds number governs the turbulence in the vicinity of the wall (see Section 2.5). Further from the wall, the appropriate Reynolds number is that based on the thickness of the boundary layer, or, for a channel flow, the half-width H. Since the governing Reynolds number in the vicinity of the wall is small, the turbulent flow there is relatively simple; few degrees of freedom are excited, and the dimension necessary in a model is small, as discussed in Section 4.4. As the size of the domain in the direction normal to the wall is increased, the flow becomes more and more complicated, the number of excited degrees of freedom increases, and with it, the requisite dimension of models (see Figure 4.1). We wish to stay close enough to the wall to permit a low-dimensional model. At the same time, the coherent structures near the wall extend into the flow a considerable distance, and we wish to resolve as much of them as possible. We thus have conflicting requirements. The choice $X_2 = 40$ is sufficient to resolve more than half of the coherent structures, but does not yet require excessive complexity. The border between the viscously dominated region near the wall, and the logarithmic region, is usually placed at $x_2 = 12.5$, in the middle of the buffer zone. Hence our choice of upper boundary at $X_2 = 40$ includes a substantial part of the lower edge of the logarithmic layer.

Our phase space is the span of the empirical eigenfunctions: a subspace of square integrable vector-valued functions that satisfy the no-slip boundary condition on the wall and are periodic in the spanwise and streamwise directions. The empirical basis functions are additionally divergence-free, because we take the individual velocity field realizations in the boundary layer or channel flow to be incompressible and the basis inherits this property from the averaged autocorrelation tensor; see Section 3.3. In experimental work, such as that of [155], incompressibility is sometimes invoked to compute the third velocity component after only two have been measured. How good is the theoretical assumption of incompressibility?

It is straightforward to show (using a little thermodynamics) that $u_{i,i} \approx m^2$, where $u_{i,i}$ is non-dimensionalized by local scales of length and velocity, and m is the Mach number based on the turbulent fluctuating velocity, u/c. Here c is the isentropic speed of sound, given in a gas by $\sqrt{\gamma RT}$, and $\gamma = c_p/c_v$ is the ratio of the specific heats at constant pressure and at constant volume. Generally, in a channel flow, the ratio $M/m \approx 30$, where $M = U/c$ is the mean flow Mach number. The specification of this latter Mach number is part of the statement of the problem. We deliberately pick a value of $M \ll 1$ in order to avoid the complications of compressibility. Also see the remarks in Section 2.1.

All realizations of the velocity field also satisfy the no-slip condition on the boundary of the channel or boundary layer. That is, $u_i = 0$; $i = 1, 2, 3$. A rationale for this at the molecular level was sketched in Section 2.1. Hence, all members of our subspace must have zero velocity at the wall $x_2 = 0$. Since $u_1 = 0$ at all values of x_1, we must have $u_{1,1} = 0$, and similarly, since $u_3 = 0$ at all values of x_3, we must have $u_{3,3} = 0$. Use of the continuity equation then gives us $u_{2,2} = 0$. (One or another of these is often used to supplement, or substitute for, the conditions $u_i = 0$.)

The conditions at $X_2 = 40$, the upper edge of the wall region, are more problematic. Far from the wall (essentially beyond the edge of the boundary layer), we could specify that all empirical basis functions decrease exponentially, or $u_{i,2} = -cu_i$ for some c. This seems reasonable because turbulent fluctuations decay as one moves into the free stream. However, $X_2 = 40$ lies deep within the boundary layer, and the behavior of the velocity field there cannot realistically be restricted in this way. Fortunately, in considering the projected ODEs (10.46), we do not need to specify the boundary conditions on individual realizations of the velocity field at $X_2 = 40$. Since we express each realization in terms of the empirical eigenfunctions, and the behavior of these eigenfunctions at the outer edge of the domain is consistent with that of the realizations that contributed to the averaged autocorrelation kernel, we can be sure that the (average) behavior of the reconstituted realizations is also appropriate at the outer edge of the domain. (There is an unresolved technical point here: the Karhunen–Loève theory of Chapter 3 guarantees convergence of empirical representations only in the L^2 sense, and not at individual points of the domain or on boundaries, but if the fields of interest are smooth, as they are "physically," this point is moot.)

More specifically, as we have shown in Section 10.4 (cf. Section 4.4), the undetermined vectorial boundary condition $\mathbf{u}(\mathbf{x}, t) = \mathbf{f}(x_1, x_3, t)$ of (10.9), which must be supplied to make (10.7) a well-posed PDE, is replaced, in the projected ODEs, by the scalar pressure terms of (10.52). If the Fourier components $\hat{p}_{k_1 k_3}^{\tau}(X_2, t)$ of the pressure at the outer edge can be specified, then they and the components $\phi_{2k_1 k_3}^{(n)*}(X_2)$ of the (vector-valued) basis functions implicitly determine the missing boundary condition. (Also see Section 13.4, which sketches an argument under which the specification of boundary pressure terms is sufficient for well posedness of the PDE initial-boundary value problem in the domain Ω.) Now we cannot determine the pressure field at $X_2 = 40$ a priori, but we can estimate its magnitude and we can also determine typical fields by appeal to direct numerical simulations. Estimates obtained from the latter reveal that $|\hat{p}_{k_1 k_3}^{\tau}|$ is two orders of magnitude smaller than typical modal amplitudes and so the authors of [22] neglected this term in their initial computations. Subsequently, as we have indicated in Chapter 9, Stone and Holmes [162, 355, 356] demonstrated its potential importance in setting a characteristic time scale for the dynamics.

We return to this issue in Chapters 11 and 13. We also describe, in Chapter 12, an idea due to Zhou and Sirovich [394] which obviates the need for "forcing" terms such as (10.52), at the expense of including a weighted influence of the outer flow and loss of universal scaling. Neither of these approaches is without its problems, which seem unavoidable if one is to restrict one's attention to the near-wall region in the hope of obtaining a model of reasonably low dimension.

In the remainder of this chapter we revisit some of the issues involved in the models for mean flow and losses to neglected modes, as well as considering in more detail the constraints imposed by use of empirical eigenfunctions in low-dimensional truncations.

10.7 The energy budget

It is instructive to examine the energy budget of the low-dimensional models. We consider both creation and dissipation of kinetic energy. The material here and in the next two sections follows [37] and [44].

As is well known, the production of turbulent kinetic energy is due to the term $-\langle u_1 u_2 \rangle U_{1,2}$ in the equation for the mean turbulent kinetic energy, see [368] and Chapter 2, Equations (2.32–2.34). The study of production may be divided into two parts: consideration of $\langle u_1 u_2 \rangle$ and of $U_{1,2}$. The latter is clearly positive (mean velocity increases away from the wall), and thus, for production, we expect the former to be negative. We discuss these terms separately.

10.7.1 The ratio $\langle u_1 u_2 \rangle / \langle u_i u_i \rangle$

We first express $\langle u_1 u_2 \rangle / \langle u_i u_i \rangle$ in terms of the empirical eigenfunctions. We know that, in the wall region, this term should contribute to production of kinetic energy, at least for the low Fourier wavenumbers included in our truncations. Specifically we observe that for models with a single family of eigenfunctions ($N = 1$) we have the representation

$$u_i(\mathbf{x}, t) = \frac{1}{\sqrt{L_1 L_3}} \sum_{k_1, k_3} a_{k_1 k_3}^{(1)}(t) e^{2\pi i (\frac{k_1 x_1}{L_1} + \frac{k_3 x_3}{L_3})} \phi_{i k_1 k_3}^{(1)}(x_2). \tag{10.57}$$

As shown above (Equation (10.32)), we can express the Reynolds stress term as:

$$\langle u_1 u_2 \rangle (x_2, t) = \frac{1}{L_1 L_3} \sum_{k_1, k_3} a_{k_1 k_3}^{(1)}(t) a_{k_1 k_3}^{(1)*}(t) \phi_{1 k_1 k_3}^{(1)}(x_2) \phi_{2 k_1 k_3}^{(1)*}(x_2). \tag{10.58}$$

Thus the ratio of $\langle u_1 u_2 \rangle / \langle u_i u_i \rangle$ becomes:

$$
\begin{aligned}
\frac{\langle u_1 u_2 \rangle}{\langle u_i u_i \rangle} &= \frac{\sum_{k_1, k_3} a_{k_1 k_3}^{(1)} a_{k_1 k_3}^{(1)*} \phi_{1 k_1 k_3}^{(1)} \phi_{2 k_1 k_3}^{(1)*}}{\sum_{k_1, k_3} a_{k_1 k_3}^{(1)} a_{k_1 k_3}^{(1)*} \phi_{j k_1 k_3}^{(1)} \phi_{j k_1 k_3}^{(1)*}} \\
&= \frac{\sum_{k_1, k_3} |a_{k_1 k_3}^{(1)}|^2 \phi_{1 k_1 k_3}^{(1)} \phi_{2 k_1 k_3}^{(1)*}}{\sum_{k_1, k_3} |a_{k_1 k_3}^{(1)}|^2 |\varphi_{k_1 k_3}^{(1)}|^2} \\
&= \frac{2 \sum_{k_1, k_3 > 0} |a_{k_1 k_3}^{(1)}|^2 \operatorname{Re}(\phi_{1 k_1 k_3}^{(1)} \phi_{2 k_1 k_3}^{(1)*})}{2 \sum_{k_1, k_3 > 0} |a_{k_1 k_3}^{(1)}|^2 |\varphi_{k_1 k_3}^{(1)}|^2},
\end{aligned} \tag{10.59}
$$

where $|\varphi_{k_1 k_3}^{(1)}|^2 = \phi_{j k_1 k_3}^{(1)}(x_2) \phi_{j k_1 k_3}^{(1)*}(x_2)$.

The following elementary algebraic observation will lead to the desired bounds. If $a_i > 0$, $y_i > 0$ and the x_i are real, then we have

$$\min_{j,l} \frac{x_j}{y_l} \le \frac{\sum_i a_i x_i}{\sum_i a_i y_i} \le \max_{j,l} \frac{x_j}{y_l}. \tag{10.60}$$

Applying the inequality (10.60) to the ratio (10.59), we obtain:

$$\min_{k_1, k_3, k_1', k_3'} \frac{\operatorname{Re}(\phi_{1 k_1' k_3'}^{(1)} \phi_{2 k_1' k_3'}^{(1)*})}{|\varphi_{k_1 k_3}^{(1)}|^2} \le \frac{\langle u_1 u_2 \rangle}{\langle u_i u_i \rangle} \le \max_{k_1, k_3, k_1', k_3'} \frac{\operatorname{Re}(\phi_{1 k_1' k_3'}^{(1)} \phi_{2 k_1' k_3'}^{(1)*})}{|\varphi_{k_1 k_3}^{(1)}|^2}. \tag{10.61}$$

In the case of the models studied in [22], we may use this observation to show that $\langle u_1 u_2 \rangle / \langle u_i u_i \rangle (x_2, t)$ is contained in an experimentally determined interval, bounded away from zero, as in [44]. In Figure 10.3, we show the empirical eigenfunction components $\phi_{j 0, k_3}^{(1)}$. We note that the streamwise and normal components ϕ_1 and ϕ_2 do not change sign through the wall layer, and that their product is strictly negative, as expected. Moreover, the bounds supplied by (10.61), extremized over the modes included in each finite-dimensional

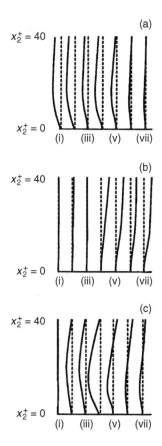

Figure 10.3 Components of the first family of empirical eigenfunctions as a function of x_2^+, weighted by the square root of the corresponding eigenvalues, for different spanwise wavenumbers and zero streamwise wavenumber. (a), (b), and (c) show ϕ_1, ϕ_2, ϕ_3 respectively. From the experimental data of Herzog [155]. Note that all products $\phi_{10,k_3}\phi_{20,k_3}$ are strictly negative for $k_3 \neq 0$.

model, are experimentally determined through the eigenfunctions. Thus we may expect them to be close to the experimental values of $\langle u_1 u_2\rangle/\langle u_i u_i\rangle$. This is a consequence of the fact that the first family of eigenfunctions, when combined with amplitudes equal to the square roots of the empirical eigenvalues, reproduces the individual averages $\langle u_1 u_2\rangle$ and $\langle u_i u_i\rangle$ quite well. Thus, projection onto empirical eigenfunctions effectively constrains the energy production term $-\langle u_1 u_2\rangle$ to an appropriate range.

In a similar vein, we observe that projection into the $k_1 = 0$ streamwise invariant subspace spanned by the empirical eigenfunctions effectively averages the streamwise variations that are ultimately responsible for extraction of energy from the mean flow $U(x_2, t)$. As is well known (and is described in detail in [43] for low-dimensional models), coupling between streamwise and cross-stream fluctuations is required for sustained extraction of energy from the mean shear U. As one can see by direct examination of the coherent structure equations (10.16) (or indeed of (10.7)), if the u_i are independent of x_1, then the first (linear) energy production term in (10.16) is absent and the second contributes only to the evolution equation for the streamwise component u_1. Thus, if the individual velocity components are allowed to vary independently, it is clear that the cross-stream fluctuations u_2, u_3 will decay due to the viscous terms. Then, as soon as $|u_2|$ falls below some critical level, the remaining linear production term

$$u_2 \delta_{i1} \left(1 - \frac{x_2}{H}\right) \tag{10.62}$$

is overwhelmed by viscous losses and all turbulent fluctuations die out. In commenting on the work of [22], Moffatt [240] noted this fact and questioned the ability of low-dimensional models possessing no non-trivial streamwise modes to produce sustained "turbulence."

In [159], and subsequently in greater detail in [43], the bounds described above were used to answer this question by showing how projections containing only a single family of eigenfunctions (and low Fourier modes) effectively couple streamwise and cross-stream velocity components. Specifically, the linear part of the term (10.62) projects to terms of the form (10.36), which contain integrals over $0 \leq x_2 \leq X_2$ of the products $\phi_{2k_1k_3}^{(l)} \phi_{1k_1k_3}^{(n)*}$. The fact that, at least for the five mode model of [22], all these products are strictly negative in the interior of this range, leads to positive coefficients A_k^1 in (10.54) and so explains the sustained dynamics. In brief, the coupling between streamwise and cross-stream modes due to the vector-valued basis functions employed in the projection, results in the effective production of energy. *Empirical bases, which capture averaged effects, retain aspects of the coupling between velocity components resulting from streamwise variations, even when the PDE is projected on a subspace that ostensibly excludes such effects.* This is a case in which the property of the POD to focus attention on dynamically relevant subspaces proves very useful.

We also remark here that the signs of coefficients of the Reynolds stress terms are likewise negative, leading in turn to strict negativity of all cubic coefficients (such as $C_{kk'}$ in (10.54)). This is important in proving the existence of attracting sets in the next chapter.

In [43] the dynamics of a "decoupled" system, with independent streamwise and cross-stream fluctuations, were also studied in order to establish whether the dynamics of the models studied in [22] are solely due to the coupling introduced through the specific choice of the basis, or whether they are intrinsic to the model. The results show that (decaying) intermittent dynamics are indeed intrinsic. The coupling, which effectively introduces sustained forcing as described above, simply makes these dynamics recurrent.

10.7.2 The mean velocity profile

The model used for the effect of the mean velocity profile on the dynamics of the coherent structures is a crucial element in the low-dimensional models. We now discuss and justify its derivation.

We start with a qualitative analysis and follow with a more careful analysis of feedback experienced by coherent structures through the mean velocity profile in shear flows. The main requirement for the model is that it be "adaptive," i.e. the shear $U_{1,2}$ which the coherent structures experience through their life cycle should depend on where in that cycle

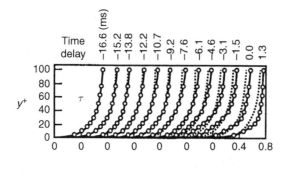

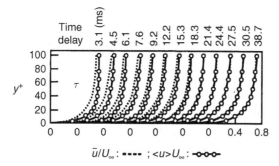

$\bar{u}/U_\infty:$ ━━━━ ; $<u>U_\infty:$ ━○━○━

Figure 10.4 Conditionally averaged mean velocity profiles with positive and negative time delay relative to the point of detection. Note how the mean velocity gradient in the near-wall region is reduced during a burst ($t = -7.6 \rightarrow 1.3$) and increased during a sweep ($t = 3.1 \rightarrow 24.4$). From Blackwelder and Kaplan [52].

they are. This need is born out by physical observation of conditionally sampled mean velocity profiles as shown in Figure 10.4, reproduced from experiments of Blackwelder and Kaplan [52]. It is also supported by the results of *failing* to use an adaptive model. When the long-time average (constant) mean velocity profile is used, the energy of the structures is generally found to grow without bound (Aubry, Holmes, and Lumley, unpublished). A second requirement is that the model be "local" in the sense that the evolution of a coherent structure in physical space should affect the mean velocity profile only in its neighborhood. As we now show, the model adopted by [22], and introduced in Section 10.1 (Equation (10.10)), meets both requirements.

We first observe that in a reflection-symmetric, statistically stationary, homogeneous channel flow the following expression holds:

$$U_1(x_2) = \frac{1}{\nu} \int_0^{x_2} \overline{u_1 u_2} \, dx_2' + \frac{u_\tau^2}{\nu} \left(x_2 - \frac{x_2^2}{2H} \right) \tag{10.63}$$

(see Tennekes and Lumley [368]). Here u_τ is the friction velocity defined by

$$u_\tau^2 = \nu \left. \frac{\partial U}{\partial x_2} \right|_{x_2=0}, \tag{10.64}$$

the overbar denotes the time average, and H is the channel half-width. As noted earlier (see Equations (10.1) and (10.2)), time averages formally depend on all spatial coordinates, but spatial homogeneity implies that effectively (10.63) is independent of x_1 and x_3. The wall region of a boundary layer is similar to the wall region of a channel with the same pressure gradient, represented by $x_2^2/2H$.

In the context of low-dimensional models in a "small" domain, [22] adopted the expression

$$U_1(x_2, t) = \frac{1}{\nu} \int_0^{x_2} \langle u_1 u_2 \rangle \, dx_2' + \frac{u_\tau^2}{\nu} \left(x_2 - \frac{x_2^2}{2H} \right), \tag{10.65}$$

where $\langle \cdot \rangle$ is the local spatial average defined in (10.2). The change in the averaging operator is of no great consequence for the derivation, but it permits the mean velocity U_1 to be time dependent. However, this time dependence implies that neglect of the derivative $\partial U_1/\partial t$ in the derivation of (10.65) must be justified by asymptotic arguments, as we have already suggested in Section 10.1. Supposing that this can be done, in (10.65) we have gained a model that meets our requirements: it is both adaptive and spatially localized within the layer. The adaptiveness is evident from the studies of [22], where its consequence is the cubic damping terms (10.51) which result in an absorbing ball and attractor in phase space, as we show in Section 11.1.

One can provide the following heuristic argument for (10.65), which is further justified in the following section. If there is a difference in time scales between $U_1(x_2, t)$ as defined by $\langle \cdot \rangle$ and the instantaneous field $\mathbf{u}(\mathbf{x}, t)$, then on time scales relevant for the dynamical model, $U_1(x_2, t)$ is slowly varying and its time derivative may be neglected. (While the fluctuations u_i are rapid, spatial averaging and integration of the Reynolds stress term in (10.65) mitigate this.) Since the symmetry of the channel still approximately holds, one can reproduce the argument of [368], merely exchanging the averaging operations. Figure 10.5, derived in [36] (cf. [44]) from a minimal flow unit computation of [175], illustrates evidence for the existence of such well-separated time scales.

In [218] Luchtenburg *et al.* describe a generalized mean field model for the flow around a three-element high-lift airfoil with a flap that can be periodically actuated, thereby entraining shed vortices to the activation frequency and increasing lift. As in the above model (Equations (10.63) and (10.65)), coupling between secondary structures and the (time-varying) mean flow is communicated by the Reynolds stresses. The dominant terms in the velocity field are represented by a slowly-varying mean flow and a pair of harmonics at the natural and actuated shedding frequencies. Neglecting quadratic interactions between the two frequencies, assuming invariance with respect to time shifts, and linearizing the Reynolds (averaged) equation around a steady solution, they derive a low-dimensional model that uses shift modes (Section 4.4) to allow a linear combination of the unactuated and periodically-actuated flows. The model's behavior, expressed via phase portraits, modal amplitudes and lift coefficients, compares well with that of the unsteady Reynolds-averaged Navier–Stokes simulations.

The next section gives a more detailed argument for the boundary layer case, which takes into account the relevant parameters of domain size and distance from the wall.

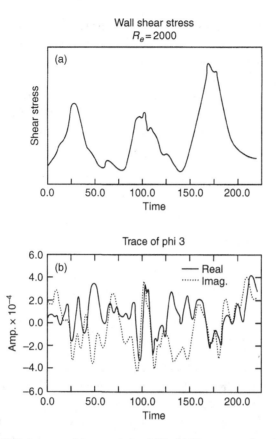

Figure 10.5 (a) Wall shear rate (characteristic of U). (b) Time trace of projection onto an individual eigenfunction (characteristic of u). Note the significant difference in time scales. From Berkooz *et al.* [44].

10.8 Nonlinear feedback

It is well known that as coherent structures grow they affect the local mean velocity profile, which is their ultimate source of energy. As a basis for a nonlinear stability analysis, Stuart [360] assumes the existence of an equilibrium state in which the mean velocity profile does not change and the energy extracted by the coherent structures is equal to the energy the coherent structures lose to stochastic motions. In this work the expression (10.65) holds, but, because of the equilibrium assumption, U_1 is time independent, see also [283].

In contrast to nonlinear stability theory, we are interested here in a dynamical model and not an energy equilibrium, so we must include time dependence. Let us also consider the spatial average. Since we are dealing in general with periodic boundary conditions, and hence a periodic array of coherent structures, a spatial average over an infinite domain would be satisfactory for our purposes. The mean velocity $U_1(x_2, t)$ would still be time dependent, and would provide feedback. However, we are interested ultimately in isolated coherent structures. Hence, we must ask: what are the smallest dimensions L_1 and L_3 for a domain in which (10.65) provides a good *approximation* for the ranges of x_2 we are interested in?

The first stages of the derivation below follow that of (10.63) due to Tennekes and Lumley [368]. Also see [44], where much of the following material appears. In this section we denote by $\tilde{u}_i$ the instantaneous field and by $U_i = \langle \tilde{u}_i \rangle$ the spatial average, and we let $\tilde{u}_i = U_i + u_i$. Until L_1 and L_3 are specified, there is no reason to believe that the average cross-flow terms U_2 and U_3 are zero. All terms are a priori time dependent. Recall that $\langle \cdot \rangle = (1/L_1 L_3) \int_0^{L_1} \int_0^{L_3} (\cdot) \, dx_3 \, dx_1$. We start with the Navier–Stokes equations:

$$\tilde{u}_{i,t} + \tilde{u}_{i,j}\tilde{u}_j = -\frac{1}{\rho}\tilde{p}_{,i} + \nu\tilde{u}_{i,jj}, \tag{10.66}$$

and take the $\langle \cdot \rangle$ average of (10.66) and appeal to periodicity in x_1 and x_3 to obtain

$$U_{i,t} + U_{i,j}U_j + \langle u_{i,j}u_j \rangle = -\frac{1}{\rho}P_{,i} + \nu U_{i,jj}. \tag{10.67}$$

In fact, rather than the "fixed" average of (10.2), we are interested in what is called a moving average, or smoothing operator:

$$\langle \cdot \rangle_x = (1/L_1 L_3) \int_{x_1 - L_1/2}^{x_1 + L_1/2} \int_{x_3 - L_3/2}^{x_3 + L_3/2} (\cdot) \, dx_3 \, dx_1,$$

so that the terms in (10.67) would actually depend (weakly) on x_1 and x_3 as well as (x_2, t). However, for large L_1, L_3, we expect spatial derivatives such as $U_{j,1}$ and $U_{j,11}$ to be small.

We now derive (10.65) from (10.67), using assumptions on the orders of magnitude of various terms. We later return to these assumptions, for an analysis of their validity tells us when (10.65) is valid. Henceforth in this section, ε denotes a suitable small parameter.

Assuming that $U_{2,t}$ and $U_{2,j}U_j$ are of order ε in comparison to $\langle u_2 u_2 \rangle_{,2}$ and using incompressibility $u_{j,j} = 0$, the second (wall-normal) component of the left-hand side of (10.67) may be reduced to

$$\approx \frac{1}{L_1 L_3}\left[\int_0^{L_3}(u_2 u_1)|_{x_1=0}^{L_1}\,dx_3 + \int_0^{L_1}(u_2 u_3)|_{x_3=0}^{L_3}\,dx_1\right] + \langle u_2 u_2 \rangle_{,2}.$$

The term in square brackets vanishes identically for periodic boundary conditions in the streamwise and spanwise directions. More generally, we expect it to be small for dimensions L_1, L_3 larger than the correlation length in the wall layer, since the Reynolds stress terms remain of similar size at $x_1 = 0$, $x_1 = L_1$ and $x_3 = 0$, $x_3 = L_3$, while both are divided by $L_1 L_3$ and each is only integrated with respect to one of the variables x_1 or x_3.

Additionally assuming that $\nu U_{2,jj}$ is of order ε, this yields the well-known estimate [368]:

$$\langle u_2 u_2 \rangle_{,2} \approx -\frac{1}{\rho}P_{,2}, \tag{10.68}$$

which holds in any almost-parallel turbulent flow. Assuming that $U_{1,t}$, $U_{1,j}U_j$ and $U_{1,11}$, $U_{1,33}$ are of order ε compared to $\langle u_1 u_2 \rangle_{,2}$, and using similar arguments, the first component of (10.67) likewise yields:

$$\langle u_1 u_2 \rangle_{,2} \approx -\frac{1}{\rho}P_{,1} + \nu U_{1,22}. \tag{10.69}$$

Now both sides of (10.68) depend spatially on x_2 alone (to order ε) and integrating with respect to that variable, we obtain an integration "constant" $P_0(x_1, x_3, t)$ that in general depends on x_1 and x_3 (and, of course, t):

$$\langle u_2 u_2 \rangle \approx -\frac{1}{\rho} P + \frac{1}{\rho} P_0, \tag{10.70}$$

and P_0 can be identified as the pressure on the wall. Note that the left-hand side still depends only on x_2, so that any dependence on x_1 and x_3 present in P_0 must also be present in P in order to cancel. We note that the mean flow is in the x_1-direction, and hence there can be a driving pressure gradient on the wall only in that direction; further, since U_1 does not change in the x_1-direction, the wall pressure gradient in this direction must be constant. Hence, P_0 is linear in x_1, and independent of x_3. Differentiating (10.70) with respect to x_1, we therefore obtain:

$$-\frac{1}{\rho} P_{,1} \approx -\frac{1}{\rho} \frac{\partial P_0}{\partial x_1} \stackrel{\text{def}}{=} f(t), \tag{10.71}$$

and substitution of this last expression into (10.69) yields:

$$\langle u_1 u_2 \rangle_{,2} \approx f(t) + \nu U_{1,22}. \tag{10.72}$$

Note that, although P itself is a function of x_2 (see Equation (10.68)), $P_{,1}$ is not. Also, if any further proof were needed that f depends on t alone, the fact that the first and third terms in (10.72) are functions only of x_2 and t verifies it. Integrating with respect to x_2 we obtain:

$$\langle u_1 u_2 \rangle \approx x_2 f(t) + \nu U_{1,2}(x_2, t) - \nu U_{1,2}(0, t). \tag{10.73}$$

Integrating once again and using the wall boundary condition $U_1(0) = 0$, this leads to:

$$U_1(x_2, t) \approx \frac{1}{\nu} \int_0^{x_2} \langle u_1 u_2 \rangle (x_2', t) \, dx_2' - \frac{x_2^2}{2\nu} f(t) + x_2 U_{1,2}(0, t). \tag{10.74}$$

Equation (10.74) does not provide a closed equation for the purposes of simulation; rather, it is an integral equation for U_1. The simplest method for closure is to replace the second and third (slow time-dependent) terms on the right-hand side of (10.74) by their time-averaged values. As we shortly see, this is what was done in [22]. For a general flow these quantities are accessible by single point measurements of velocity and wall shear. However in a channel configuration we have more information and can proceed analytically.

Specifically, using the symmetry of the channel and assuming that $\langle \cdot \rangle$ quantities are symmetric to order ε, we get from (10.73):

$$Hf(t) \approx \nu U_{1,2}(0, t), \tag{10.75}$$

where H is the channel half-width and we have used

$$U_{1,2}(H, t) = \langle u_1 u_2 \rangle (H, t) = 0$$

(symmetry about the centerline). Actually, Equation (10.75) is a much more general statement, saying that the pressure drop over a section of channel is equal to the stress integrated over the walls; that is, that the external force pushing the fluid through the channel is transmitted by the fluid to the walls. Hence, a similar statement can be made in any channel, symmetric or not. This leads to the expression

$$U_1(x_2, t) \approx \frac{1}{\nu} \int_0^{x_2} \langle u_1 u_2 \rangle (x_2', t) \, dx_2' + U_{1,2}(0, t) \left(x_2 - \frac{x_2^2}{2H} \right), \tag{10.76}$$

which is equivalent to our assumption (10.65), once the time-average value of $U_{1,2}(0, t)$ from Equation (10.64) is used.

The above derivation reveals that the model used by [22] may be viewed as an approximation of (10.74) in the sense that the "slow" time-dependent terms on the right-hand side are replaced by representative average values, based on the experimental value for the friction velocity.

The question that remains to be addressed is: what are the smallest dimensions L_1 and L_3, and the largest values of x_2, for which the following assumptions on the orders of magnitude are reasonable? These are the orders that must be adhered to if our neglect of various terms above is to be sustained:

$$\left. \begin{array}{c} U_{2,t}, \; \nu U_{2,jj}, \; U_{2,j} U_j \sim O(\varepsilon) \langle u_2 u_2 \rangle_{,2} , \\[2mm] U_{1,t}, \; U_{1,j} U_j \sim O(\varepsilon) \langle u_1 u_2 \rangle_{,2} . \end{array} \right\} \tag{10.77}$$

It is easy to show, using classical techniques developed for the convergence of averages (see Tennekes and Lumley [368]), that variations in U_1 will be of order $u_\tau \mathcal{L}/\sqrt{L_1 L_2}$. Here $\mathcal{L}$ is the correlation length of the fluctuations. Derivatives will be $\mathcal{O}(1/L_1)$ in the 1-direction and $\mathcal{O}(1/L_3)$ in the 3-direction, while in the 2-direction they will be $\mathcal{O}(1/x_2)$ close to the wall and $\mathcal{O}(1/H)$ further from the wall. Hence, $U_{1,1}$ will be of order $u_\tau \mathcal{L}/L_1 \sqrt{L_1 L_3}$. Making use of the continuity equation, we can determine that U_2 and U_3 are of order $u_\tau x_2 \mathcal{L}/L_1 \sqrt{L_1 L_3}$ to a maximum of $u_\tau H \mathcal{L}/L_1 \sqrt{L_1 L_3}$. Proceeding in this way, we can show that all the terms in (10.77) can be made small by suitable choice of L_1 and L_3 relative to $\mathcal{L}$, with the exception of $U_{1,t}$. Finally we show that, under suitable conditions $U_{1,t} = O(\varepsilon) \cdot \langle u_1 u_2 \rangle_{,2}$.

We can estimate the local (unaveraged) value of $U_{1,t} \approx u_\tau / T_B$, where T_B is the mean time between the bursts that cause the mean shear profile to locally collapse (Figure 10.4). Experimentally, we observe the relation $T_B u_\tau^2 / \nu = 2.6 \times 10^2$ (see Lumley and Kubo [226]). In the wall region of the boundary layer, the magnitude of $\langle u_1 u_2 \rangle_{,2} \approx u_\tau^2 / H$. Hence, the local value of the ratio is given by

$$\frac{U_{1,t}}{\langle u_1 u_2 \rangle_{,2}} \approx \frac{H^+}{2.6 \times 10^2}, \tag{10.78}$$

where we use the wall unit normalization: $H = \nu H^+ / u_\tau$.

Finally we must consider the spatial averaging process. Let $\mathcal{L}$ be the cross-stream wavelength of the coherent structures in the wall region. Then, the amplitude of $U_{1,t}$ is reduced by a factor of approximately $\mathcal{L}/\pi L_3$. Thus we have for the averaged value:

$$\frac{U_{1,t}}{\langle u_1 u_2 \rangle_{,2}} \approx \frac{H^+ \mathcal{L}}{2.6 \times 10^2 \pi L_3}. \tag{10.79}$$

In our flow $H^+ \approx 10^2$, while $\mathcal{L} \approx 10^2$ and $L_3 \approx 3 \times 10^2$. Hence, the factor in (10.79) is approximately 4×10^{-2}. This is satisfactorily small.

10.9 Interaction with unresolved modes

The complement to the energy production mechanisms considered in Sections 10.7 and 10.8 is dissipation. Energy is dissipated by small-scale motions which cannot be represented in highly truncated models. Here we consider the projected Navier–Stokes equations and once more restrict ourselves to the case in which P denotes projection onto a subspace R of modes lacking streamwise variation ($k_1 = 0$) (but still containing arbitrarily many spanwise and wall-normal modes). The material in this section first appeared in [161] and [44].

We show first that terms of the type $P(u_{i>,j}u_{j<})$ (the Leonard stresses) in Equation (10.11) vanish identically due to the specific choice of the projection. Recall that these terms were neglected in our Heisenberg model for losses to unresolved modes. The following simple observation justifies that decision, at least for this special case.

We consider only one such term; the others may be dealt with similarly. Observe that, since the resolved (projected) field $u_<$ is independent of x_1, we may write

$$P(u_{i>,j}u_{j<}) = \frac{1}{L_1} \int_0^{L_1} u_{i>,j}(x_1, x_2, x_3, t)u_{j<}(x_2, x_3, t)\,dx_1,$$

$$= u_{j<}(x_2, x_3, t)\frac{1}{L_1} \int_0^{L_1} u_{i>,j}(x_1, x_2, x_3, t)\,dx_1. \quad (10.80)$$

However, since $u_{i>}$ contains only non-zero Fourier modes in the k_1- (x_1-) direction, the integral is clearly zero. In the same way, the other Leonard stresses may be shown to be zero.

The interaction with the unresolved modes comes through the term $P(u_{i>,j}u_{j>} - \langle u_{i>,j}u_{j>}\rangle)$. In the models of [22], as described in Section 10.1 (Equations (10.12–10.14)), the following relations are used:

$$P(u_{i>}u_{j>} - \langle u_{i>}u_{j>}\rangle) - \frac{1}{3}\delta_{ij}P(u_{k>}u_{k>} - \langle u_{k>}u_{k>}\rangle)$$

$$= -\alpha_1 \nu_T(u_{i<,j} + u_{j<,i}). \quad (10.81)$$

This is a Heisenberg spectral model or Smagorinsky subgrid scale type model in which ν_T is an experimentally determined viscosity parameter, the value of which is discussed later. Note that both sides of (10.81) are zero trace tensors. In this type of model, the unresolved modes act as additional dissipation on the resolved modes, see also [113]. To get an expression for the interaction term $P(u_{i>}u_{j>} - \langle u_{i>}u_{j>}\rangle)$, we make the further closure assumption that:

$$P(u_{k>}u_{k>} - \langle u_{k>}u_{k>}\rangle) =$$

$$\alpha_2 l_>^2 \left[(u_{i<,j} + u_{j<,i})(u_{i<,j} + u_{j<,i}) \right.$$

$$\left. - \langle (u_{i<,j} + u_{j<,i})(u_{i<,j} + u_{j<,i})\rangle \right]. \quad (10.82)$$

This supposes that the kinetic energy of the unresolved modes is proportional to the rate of loss of energy by the resolved modes, which is proportional to their rate of deformation. Summing up, the model equation then takes the following form:

$$u_{i<,t} + u_{i<,1}U_1 + U_{1,2}u_{2<}\delta_{i1} + u_{i<,j}u_{j<} - \langle u_{i<,j}u_{j<}\rangle$$

$$= -\frac{1}{\rho}p_{<,i} + (v + \alpha_1 v_T)u_{i<,jj}$$

$$- \frac{2}{3}\alpha_2 l_>^2 [(u_{k<,l} + u_{l<,k})(u_{k<,l} + u_{l<,k})$$

$$- \langle(u_{k<,l} + u_{l<,k})(u_{k<,l} + u_{l<,k})\rangle]_{,i}. \tag{10.83}$$

As observed in [22], the term resulting from the energy of the unresolved modes may be combined with the pressure term. Upon integration over the full channel it vanishes. A free boundary results in a small quadratic term (as Equation (10.50)). Because of its small magnitude (see [22], Appendix C) we may ignore it in the energy balance calculations.

We make two main observations regarding the energy budget:

(1) The interaction term $P(u_{i>,j}u_{j>} - \langle u_{i>,j}u_{j>}\rangle)$ either shuffles energy within R or transfers energy from R to the complementary space $R^\perp$ of neglected modes. There is no energy transfer from $R^\perp$ to R on average.

(2) The Heisenberg model can provide the correct amount of dissipation with an order 1 correction, α_1, to the experimental eddy viscosity v_T.

In the remainder of this section we justify these claims.

We start with the average kinetic energy equation for the resolved modes obtained from (10.83) by multiplying by $u_{i<}$, taking the $\langle\cdot\rangle$-average and rearranging terms:

$$\frac{1}{2}\langle u_{i<}u_{i<}\rangle_{,t} + \langle u_{i<}u_{i<,1}\rangle U_1 + \langle u_{1<}u_{2<}\rangle U_{1,2} + \langle u_{i<,j}u_{j<}u_{i<}\rangle$$

$$+ \langle(P(u_{i>,j}u_{j>} - \langle u_{i>,j}u_{j>}\rangle))u_{i<}\rangle$$

$$= -\frac{1}{\rho}\langle p_{<,i}u_{i<}\rangle + v\langle u_{i<,jj}u_{i<}\rangle. \tag{10.84}$$

Using incompressibility and the facts that $\langle u_{i<}u_{i<,jj}\rangle = \langle u_{i<}u_{i<,j}\rangle_{,j} - \langle u_{i<,j}u_{i<,j}\rangle$ and $u_{i<,jj}u_{i<} = \frac{1}{2}(u_{i<}u_{i<})_{,jj} - 2s_{ij<}s_{ij<} + (u_{i<}u_{j<})_{,ij}$, where $s_{ij<} = \frac{1}{2}(u_{i<,j} + u_{j<,i})$, and denoting by D/Dt the mean material derivative, we obtain:

$$\frac{1}{2}\frac{D\langle u_{i<}u_{i<}\rangle}{Dt} = -\langle u_{1<}u_{2<}\rangle U_{1,2}$$

$$+ \left[-\frac{1}{\rho}\langle p_{<,i}u_{i<}\rangle + \frac{v}{2}(u_{j<}u_{j<})_{,i} + v(u_{i<}u_{j<})_{,j} + \frac{1}{2}\langle u_{j<}u_{j<}u_{i<}\rangle\right]_{,i}$$

$$+ \langle(P(u_{i>,j}u_{j>} - \langle u_{i>,j}u_{j>}\rangle))u_{i<}\rangle + 2v\langle s_{ij<}s_{ij<}\rangle. \tag{10.85}$$

To show (1), we have to study the term

$$\langle(P(u_{i>,j}u_{j>} - \langle u_{i>,j}u_{j>}\rangle))u_{i<}\rangle. \tag{10.86}$$

By the commutation of $P(\cdot)$ and $\langle\cdot\rangle$ and the fact that $\langle u_{i<}\rangle = 0$, (10.86) is equal to $\langle P(u_{i>,j}u_{j>})u_{i<}\rangle$. But we have (since $u_{i<}$ is independent of x_1):

$$\left\langle\left(\frac{1}{L_1}\int_0^{L_1}u_{j>}u_{i>,j}dx_1\right)u_{i<}\right\rangle = \left\langle\left(\frac{1}{L_1}\int_0^{L_1}(u_{i>}u_{j>})_{,j}dx_1\right)u_{i<}\right\rangle =$$

$$\left\langle\frac{1}{L_1}u_{i>}u_{1>}\Big|_0^{L_1}u_{i<} + \frac{1}{L_1}\int_0^{L_1}(u_{i>}u_{2>})_{,2}u_{i<}dx_1\right.$$

$$\left. + \frac{1}{L_1}\int_0^{L_1}(u_{i>}u_{3>})_{,3}u_{i<}dx_1\right\rangle$$

$$= \left\langle\left(\frac{1}{L_1}\int_0^{L_1}u_{i>}u_{2>}u_{i<}dx_1\right)_{,2} + \left(\frac{1}{L_1}\int_0^{L_1}u_{i>}u_{3>}u_{i<}dx_1\right)_{,3}\right.$$

$$\left. -\frac{1}{L_1}\int_0^{L_1}u_{i>}u_{2>}dx_1u_{i<,2} - \frac{1}{L_1}\int_0^{L_1}u_{i>}u_{3>}dx_1u_{i<,3}\right\rangle$$

$$= \langle P(u_{i>}u_{j>}u_{i<})_{,j}\rangle - \langle P(u_{i>}u_{j>})u_{i<,j}\rangle$$

$$= \langle u_{i>}u_{j>}u_{i<}\rangle_{,j} - \langle P(u_{i>}u_{j>})u_{i<,j}\rangle. \tag{10.87}$$

Here $\langle u_{i>}u_{j>}u_{i<}\rangle_{,j}$ is an advection term which is responsible for shifting energy around in R. It vanishes when integrated over a full channel. The term $\langle P(u_{i>}u_{j>})u_{i<,j}\rangle$ represents energy transfer from R to $R^\perp$. It has the form of deformation work done against the unresolved stresses [368]. This shows claim (1).

To verify the second claim, we must compare the actual energy loss given by $\langle P(u_{i>}u_{j>})u_{i<,j}\rangle$ with the result of the model for energy loss which appears as $-2\alpha_1\nu_T\langle s_{ij<}s_{ij<}\rangle$ in the kinetic energy equation for the resolved modes. In the following, $\sim$ denotes equal up to multiplication by a number of $\mathcal{O}(1)$. We use $\sim$ as suggested by experimental observations on single point correlations in turbulence. Using the fact that the small scales of the resolved modes are of the order of the large scales of the unresolved modes, we obtain

$$\frac{\langle P(u_{i>}u_{j>})u_{i<,j}\rangle}{\langle s_{ij<}s_{ij<}\rangle} \sim \frac{\langle u_{i<,j}u_{i<,j}\rangle^{1/2}\langle P(u_{i>}u_{j>})P(u_{i>}u_{j>})\rangle^{1/2}}{\langle(u_{i<,j}+u_{j<,i})(u_{i<,j}+u_{j<,i})\rangle}. \tag{10.88}$$

Using the correlation of the unresolved modes with themselves and the commutation of P and $\langle\cdot\rangle$, we obtain:

$$\sim \frac{\langle u_{i<,j}u_{i<,j}\rangle^{1/2}\langle u_{i>}u_{i>}\rangle}{\langle u_{i<,j}u_{i<,j}\rangle} = \frac{\langle u_{i>}u_{i>}\rangle}{\langle u_{i<,j}u_{i<,j}\rangle^{1/2}} = \nu_T(x_2). \tag{10.89}$$

This is an x_2-dependent eddy viscosity, which may be computed explicitly from the two point velocity correlation tensor when $\langle\cdot\rangle_E$ (the ensemble average) is substituted for $\langle\cdot\rangle$. An appropriate value for a spatially independent eddy viscosity is therefore obtained by averaging over the wall-normal direction:

$$\frac{1}{X_2}\int_0^{X_2}\frac{\langle u_{i>}u_{i>}\rangle}{\langle u_{i<,j}u_{i<,j}\rangle^{1/2}}(x_2)\,dx_2. \tag{10.90}$$

This expression for v_T is similar to the one adopted by [22], which was:

$$\frac{\int_0^{X_2} \langle u_{i>} u_{i>} \rangle \, dx_2}{\left(X_2 \int_0^{X_2} \langle u_{i>,j} u_{i>,j} \rangle \, dx_2 \right)^{1/2}}. \tag{10.91}$$

The difference between (10.90) and the expression (10.91) is practically nullified by the way Aubry *et al.* evaluated the expression for v_T. (They took $\langle u_{i>,j} u_{i>,j} \rangle^{1/2}$ in the denominator, but estimated it using only the lowest wavenumber neglected modes and the first eigenfunction, also see [161].) This establishes claim (2), since we have shown that the experimentally derived eddy viscosity provides the right amount of dissipation within an order 1 constant.

Our first point has an interesting physical interpretation: if one accepts the picture of energy transfer by vortex stretching, an essentially three-dimensional process, then the resultant of any such stretching on elements of R by elements of $R^\perp$, should lie outside R.

11

Behavior of the models

In this chapter we describe the qualitative structure, in phase space, of some of the low-dimensional models derived in the preceding chapter. We also discuss the physical implications of our findings. Drawing on the material introduced in Chapters 6–9, we solve for some of the simpler fixed points (steady, time-independent flows and traveling waves) and discuss their stability and bifurcations under variation of the loss parameters α_j introduced in Section 10.1. We focus on the five mode model ($N = 1$, $K_1 = 0$, $K_3 = 5$) introduced in the original paper of Aubry *et al.* [22], and referred to there as the "six mode model," the $k_3 = 0$ mode being implicitly included in the model of the slowly varying mean flow. The full range of dynamical behavior of even such a draconian truncation as this is bewilderingly complex and still incompletely understood, but we are able to give a fairly complete account of a particular family of solutions – attracting heteroclinic cycles – which appear especially relevant to understanding the burst/sweep cycle which was described in Section 2.5.

In Sections 11.1 and 11.2 we use the nesting properties of invariant subspaces, noted in Section 10.5, to solve a reduced system, containing only two (even) complex modes, for fixed points. We exhibit the bifurcation diagram and discuss the stability of a particular branch of fixed points corresponding to streamwise vortices of the appropriate spanwise wavenumber. Due to the spanwise translation invariance (Section 10.3), circles of such equilibria occur in phase space. We then show that open ranges of the loss parameter exist for which diametrically opposite points on these circles are connected by heteroclinic orbits involving the odd modes, and moreover that the resulting heteroclinic cycles are asymptotically stable (attracting) in parts of these ranges. In Section 11.3 we show that the velocity field, reconstructed during each of these heteroclinic events, captures key features of the burst/sweep cycle. Section 11.4 addresses the effect of the pressure term and in Section 11.5 we briefly consider the behavior of other, higher-dimensional models including non-trivial streamwise modes ($K_1 = 1$, 2) and a second family of eigenfunctions ($N = 2$). Here we draw on papers of Aubry and Sanghi [25, 323]. Section 11.6 contains a summary.

We hope that this chapter will serve to demonstrate that remarkably low-dimensional models can capture, and help explain, major aspects of turbulence generation. We also hope to show how qualitative analyses of these models (which, from a dynamical systems standpoint, are not at all small) can profit from the presence of symmetries and the associated invariant subspaces.

11.1 Backbones for the models

In this and the following three sections we describe the behavior of the ($N = 1$, $K_1 = 0$, $K_3 = 5$) five mode model of [22]. We adopt the convention of Section 10.5, letting $a_k(t)$ denote the modal coefficient that should properly be called $a_{0k_3}^{(1)}(t)$. The form of the equations was given in (10.54). Values of the various linear, quadratic, and cubic coefficients can be found in [22], Appendix C; for convenience, we have reproduced them in the Appendix to this chapter (Section 11.7). We note that, in [22], the modal coefficients a_k were normalized by the empirical eigenvalues, so that, when their dynamical behavior is similar to that experimentally observed, their amplitudes should all be approximately equal and of order one. This was done to avoid numerical overflow or underflow and is important in making physical deductions from the observations below. Moreover, the right-hand sides of all the ODEs were multiplied by the factor $\sqrt{L_1 L_3}$ (=333 for the choice of length scales employed). Unfortunately the corresponding factor was omitted from the left-hand sides, leading to an erroneous compression of time scales by the same factor. By coincidence, this brought the cycle times discussed in Section 11.4 into the experimental range and in [22] a heteroclinic cycle was therefore identified *directly* with an experimentally observed burst. The time scale error was remedied, without much discussion of its consequences, in [323]. We return to it in Sections 11.4 and 11.6.

We do not include details of the computation of the Heisenberg (eddy) viscosity coefficient ν_T or of the length scale $l_>$ here. It depends on the physical situation, the particular transport model employed, and the wavenumber cut-off. Here we are more concerned with illustrating the method of analysis rather than details of the specific application. See [22] and [323] for more information on the boundary layer. It is sufficient to note that a value of $\nu_T = 6.28$ was adopted in the computations described here.

In Section 10.5 we noted that the four-(real)-dimensional subspace of even Fourier modes, spanned by $\boldsymbol{\varphi}_{0,2}^{(1)}$ and $\boldsymbol{\varphi}_{0,4}^{(1)}$ and henceforth called the (a_2, a_4)-subspace, is invariant within the larger ten-dimensional phase space of the model. Restricted to this subspace, we have a system of the form:

$$
\left.
\begin{aligned}
\dot{a}_2 &= [A_2^1 + (1 + \alpha_1 \nu_T)A_2^2]a_2 + (Q_{4-2}^1 + \tfrac{2}{3}\alpha_2 l_>^2 Q_{4-2}^2)a_2^* a_4 \\
&\quad + (C_{22}|a_2|^2 + C_{24}|a_4|^2)a_2 + d_2(t), \\
\dot{a}_4 &= [A_4^1 + (1 + \alpha_1 \nu_T)A_4^2]a_4 + (Q_{22}^1 + \tfrac{2}{3}\alpha_2 l_>^2 Q_{22}^2)a_2^2 \\
&\quad + (C_{42}|a_2|^2 + C_{44}|a_4|^2)a_4 + d_4(t),
\end{aligned}
\right\}
\tag{11.1}
$$

where we have used the relation $a_{-k} = a_k^*$ and grouped and renamed coefficients, so that, for example, Q_{4-2}^j in (11.1) is actually equal to $(Q_{-24}^j + Q_{4-2}^j)$ in (10.54). We first consider Equation (11.1) as a backbone for the larger system, adding the flesh later.

The coefficients satisfy the following inequalities:

$$
A_k^1 > 0 > A_k^2, \quad Q_{4-2}^j > 0 > Q_{22}^j, \quad C_{kk'} < 0, \quad \text{and } |Q_{kl}^1| \gg |Q_{kl}^2|,
$$

so that for practical purposes the latter (pseudo-pressure) terms are negligible. In fact they were not neglected in the computations reported below, but we did take $\alpha_1 = \alpha_2 = \alpha$: a single bifurcation parameter. The additive pressure terms d_k are also small in comparison to typical ($\mathcal{O}(1)$) values of a_k, but as we see in Section 11.4, they may have an important effect. However, until we reach that section, we set $d_k(t) \equiv 0$. Observe that, without these pressure terms, Equation (11.1) is precisely the $O(2)$-equivariant normal form considered in Section 7.4, Equation (7.50). The reader should refer to that section to verify some of the following assertions.

The linearization of (11.1) at the trivial solution $a_k = 0$ is simply the system obtained by discarding quadratic and cubic terms. In terms of the real modal components $a_k = x_k + iy_k$, we have the repeated eigenvalues $[A_2^1 + (1 + \alpha_1 v_T)A_2^2]$ and $[A_4^1 + (1 + \alpha_1 v_T)A_4^2]$, each of multiplicity two. Let $\alpha_{ck} = -(A_k^1 + A_k^2)/v_T A_k^2$ and note that $\alpha_{c2} \approx 2.41 > \alpha_{c4} \approx 1.94$ (see Appendix). Thus, for $\alpha > \alpha_{c2}$ the fixed point $a_k = 0$ is stable, while for $\alpha_{c2} > \alpha > \alpha_{c4}$ it is a saddle point with a two-dimensional unstable manifold, and for $\alpha < \alpha_{c4}$ it is a source (within the four-dimensional (a_2, a_4)-subspace). As for the Kuramoto–Sivashinsky equation considered in Chapter 8, $O(2)$-equivariant pitchfork bifurcations occur at these two critical parameter values as α decreases. This is physically reasonable, for (unrealistically) large values of the loss parameter: the trivial solution is stable and turbulent motions are not sustained. As α decreases, non-trivial solutions emerge.

The bifurcation diagram is readily computed by direct solution of (11.1) for fixed points, as in Chapters 7 and 8. We show it in Figure 11.1, in which we also indicate the stability types of solution *within the* (a_2, a_4)-*subspace*, as well as the branch of traveling waves. The first bifurcation to occur with decreasing α is that of the mixed 2/4 mode at α_{c2}; this is followed by bifurcation of the pure 4 mode at α_{c4}. The precise values of these bifurcation points are not particularly significant, but their relative ordering, established by the numerical values of the coefficients A_k^j, and hence ultimately by the empirical eigenfunctions, *is* important. The mixed mode, which, as in Section 8.2, is (almost) a pure 2 mode at bifurcation, inherits the stability of the trivial solution and remains stable until a pair

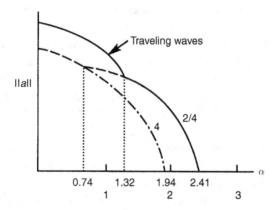

Figure 11.1 Bifurcation diagram for Equation (11.1), showing stability types of solution within the four-dimensional (a_2, a_4)-subspace: ———, sinks; – – –, saddles with one unstable eigenvalue; – · –, saddles with two unstable eigenvalues. From Aubry *et al.* [22].

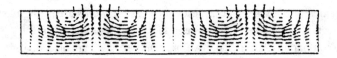

Figure 11.2 Cross-stream (u_2, u_3) components of the velocity field corresponding to the
2/4 branch at a value of $\alpha = 2.0$. From Aubry *et al.* [22].

of traveling waves (left and right going), branches from it at $\alpha \approx 1.32$. (See Section 7.4,
Equations (7.58–7.60).) The mixed mode is characterized, in terms of polar coordinates,
by $|a_2| = r_2 \neq 0$, $|a_4| = r_4 \neq 0$ and phase angle difference $2\theta_2 - \theta_4 = \pi$. It may be
helpful here to refer to the polar coordinate representation of Equation (7.52).

Recalling that the fixed points found thus far are also equilibria for the full ten-
dimensional system, it is appropriate to inquire to what they correspond in the physical
fluid flow. The 2/4 mode starts at α_{c2}, with $|a_2| \gg |a_4|$ and ends by coalescence with
the pure 4 branch at $\alpha \approx 0.74$. Thus, in the middle range, it represents an approximately
equal mixture of modes 2 and 4. Recalling the scaling via empirical eigenvalues, and the
spectrum of Figure 10.2, this implies that the corresponding fluid velocity field contains
about 70% spanwise Fourier mode 2 and 30% mode 4. Thus it displays two pairs of rolls
within the spanwise domain $L_3 = 333$ from the field $\varphi_{0,2}^{(1)}$, modified by the four-pair second harmonic $\varphi_{0,4}^{(1)}$. Figure 11.2 shows a typical reconstruction. Of course, since each of
these equilibria occurs as a member of a whole circle of such, invariant under the rotation
(10.26), the spanwise positions of the vortices are arbitrary: any translation of the whole
pattern is also an equilibrium solution. Now this is precisely what is observed in the wall
region *on average*, and so we are easily persuaded that the 2/4 branch deserves further
study. We now go on to examine its stability in the full ten-(real)-dimensional phase space
of the five mode model.

Without loss of generality, we choose a fixed point lying in the real subspace $\{y_k = 0\}$.
Since $|a_2| = r_2 \overset{\text{def}}{=} \rho_2$, $|a_4| = r_4 \overset{\text{def}}{=} \rho_4$, and $2\theta_2 - \theta_4 = \pi$, there are two such:
$x_2 = \pm\rho_2$, $x_4 = -\rho_4$, all other real modal coefficients being zero. Linearizing the full
system in Cartesian coordinates at the first of these fixed points, we find that the 10×10
Jacobian matrix block diagonalizes into two 2×2 blocks, corresponding to the (x_2, x_4)- and
(y_2, y_4)-subspaces, and two 3×3 blocks, corresponding to the (x_1, x_3, x_5)- and (y_1, y_3, y_5)-
subspaces respectively. This is a direct consequence of the symmetries described in
Sections 10.3 and 10.5: invariance of the real subspace, the (a_2, a_4)-subspace, and, within
that, the (x_2, x_4)-subspace itself. See Equation (31) of [22] for explicit expressions of the
submatrices.

Of the ten eigenvalues of this matrix, one, associated with the (y_2, y_4)-block, is zero,
its eigenvector being tangent to the circle of equilibria. The other eigenvalue of this block
passes through zero as the traveling wave bifurcates off at $\alpha \approx 1.32$, being negative above
that value. Those of the (x_2, x_4)-block remain negative throughout the range of interest. Of
the others, four retain negative real parts and the remaining two change signs in the range
$1.32 < \alpha < 2.41$. Again, for details, see [22], Table 1. For $1.32 < \alpha < 1.61$, the 2/4 mode
equilibria are saddle points with seven-dimensional stable manifolds, two-dimensional

unstable manifolds, and one-dimensional center manifolds; the latter being the circle of equilibria itself. The unstable eigenvalues are a complex conjugate pair with non-zero imaginary parts and the unstable manifold lies in the (x_1, x_3, x_5)-subspace. The invariant subspace spanned by $(y_1, x_2, y_3, x_4, y_5)$ (a rotation of the real subspace) lies within the stable manifold of this fixed point. At the second "real" fixed point, $x_2 = -\rho_2$, $x_4 = -\rho_4$, the eigenvalues are of course the same, but the rôles of the two three-dimensional subspaces are exactly interchanged. This is easily understood if one notes that rotation by phase angles $\theta_k = k\pi/2$ interchanges the two fixed points and maps $(x_1, x_2, x_3, x_4, x_5)$ to $(y_1, -x_2, -y_3, x_4, y_5)$. These observations are key to understanding the global behavior described in the next section.

Finally in this section we sketch a global stability result. The (Liapunov) function

$$S = \sum_{k=1}^{5}(x_k^2 + y_k^2) = \varrho^2 \tag{11.2}$$

defines a family of nine-dimensional (hyper-) spheres S_ϱ, of radius ϱ, in phase space, each enclosing a solid ball about the origin. We claim that, for the constant ϱ chosen sufficiently large, S_ϱ bounds a trapping region for the flow in phase space, which therefore in turn contains an attracting set (see Section 6.5). To see this, consider the derivative of the function S of (11.2) along solution curves of (10.54), which may be written schematically as:

$$\frac{dS}{dt} = 2\sum_{k=1}^{5}(x_k \dot{x}_k + y_k \dot{y}_k)$$

$$= (\text{quadratic terms}) + (\text{cubic terms})$$

$$+ 2\sum_{k=1}^{5}(x_k^2 + y_k^2)\sum_{k'=1}^{5} C_{kk'}(x_{k'}^2 + y_{k'}^2). \tag{11.3}$$

Since all the coefficients $C_{kk'}$ are strictly negative (see the Appendix to this chapter and recall Section 10.7.1), for $\sum_{k=1}^{5}|x_k^2 + y_k^2| = \varrho^2$ sufficiently large, the final term in (11.3) is dominant, leading to the conclusion that the vector field is directed everywhere inwards on the sphere S_ϱ. The sphere therefore bounds a trapping region. This is the expression, in phase space, of the stabilizing feedback effect of turbulence on the mean flow, described in Sections 10.7.2 and 10.8.

11.2 Heteroclinic cycles

We have already met and analyzed heteroclinic cycles in the context of the two complex mode $O(2)$-equivariant normal form in Section 7.4 and Chapter 8. However, the cycles we are about to encounter here are subtly different and geometrically richer. They involve excursions out of the (a_2, a_4)-subspace into the complementary subspace spanned by modes 1, 3, and 5. In them, the (a_2, a_4)-subsystem plays the rôle of the pure z_2-subspace in Section 7.4 and the (a_1, a_3, a_5)-subspace, the rôle of the z_1-subspace. This repetition and elaboration of structure in systems of increasing dimension is seen again in Sanghi and Aubry's studies [323] of models with non-zero streamwise wavenumbers ($K_1 = 1$, 2). There the unstable "fixed points" (actually more complicated invariant sets having $|a_{0k_3}^{(1)}| \neq 0$ for all k_3) lie within the invariant subspace $k_1 = 0$ and the heteroclinic excursions take place in the complementary $k_1 \neq 0$ subspace, see Section 11.5, below.

It is probably easiest to describe the cycles in the manner they were first discovered – numerically, as in [22, 354]. Figure 11.3 shows the variation of real and imaginary parts of the modal coefficients $a_k(t)$ for three α values in the range $1.32 < \alpha < 1.61$. We shall describe the characteristic sequence of events observed directly following a heteroclinic excursion. The solution remains near a fixed point $r_2, r_4 \neq 0$, $r_1 = r_3 = r_5 = 0$ for a relatively long period, during which the amplitudes r_1, r_3, r_5 grow exponentially in an oscillatory fashion. Eventually r_2 and r_4 begin to oscillate themselves, and, with the phase difference $2\theta_2 - \theta_4$ remaining at π, θ_2 rapidly changes by a net value π and θ_4 by 2π. After this relatively violent event, r_2 and r_4 return to their original values and r_1, r_3, r_5 simultaneously collapse to zero. The cycle then repeats. For the computations of Figure 11.3, initial conditions near the purely real subspace were chosen and one can clearly see how the unstable modes switch from the (x_1, x_3, x_5)-subspace to the (y_1, y_3, y_5)-subspace and back at each successive excursion. As α is reduced from ≈ 1.61, when the equilibria become saddles, the typical duration of the "steady" (near-equilibrium) phases decreases until at $\alpha \approx 1.32$, there is no clear steady phase at all. Below $\alpha \approx 1.3$, persistently active dynamics is seen (Figure 11.6, below).

In the light of the linear analysis of Section 11.1, the slow growth phase is easily understood as the solution moving away from one of the saddle type equilibria and winding out near its unstable manifold. The growth rate is controlled by the magnitude of the real part of the positive eigenvalue, which increases as α decreases ([22], Table 1). It is more difficult to see why this growth phase should culminate in the relatively sudden excursion and transit to a diametrically opposite point on the circle of equilibria, but if we postulate the existence of a heteroclinic orbit connecting $x_2 = +\rho_2$, $x_4 = -\rho_4$ to $x_2 = -\rho_2$, $x_4 = -\rho_4$, then this behavior also becomes clear. The numerical evidence for such a heteroclinic connection is overwhelming, and is supported by the observation that, *within the invariant* $(y_1, x_2, y_3, x_4, y_5)$-*subspace*, the latter fixed point is a stable sink. Thus, to establish the existence of the connecting orbit, one need only show that the (two-dimensional) unstable manifold of $x_2 = +\rho_2$, $x_4 = -\rho_4$, which also lies in this five-dimensional subspace, contains points near $x_2 = -\rho_2$, $x_4 = -\rho_4$. Local attractivity of the sink will do the rest. While this is easy to prove analytically in two- and even three-dimensional cases (see [9] and Section 7.4, and [68]), it is difficult explicitly to construct positively invariant domains in four and more dimensions, in order to show that the unstable manifold is not "caught" by some other intervening set. A rigorous proof of connecting orbits is still lacking.

Nonetheless, if a connection exists, then it is structurally stable and robust to small perturbations *within the same symmetry class*, for precisely the same reasons that the simpler cycles of Sections 7.3 and 7.4 are structurally stable. It therefore cannot be dismissed as a freak associated with special coefficient values. Moreover, the existence of a connection from $x_2 = +\rho_2$, $x_4 = -\rho_4$ to $x_2 = -\rho_2$, $x_4 = -\rho_4$, together with the $O(2)$-equivariance, automatically yields a connection in the opposite direction, giving a full heteroclinic cycle.

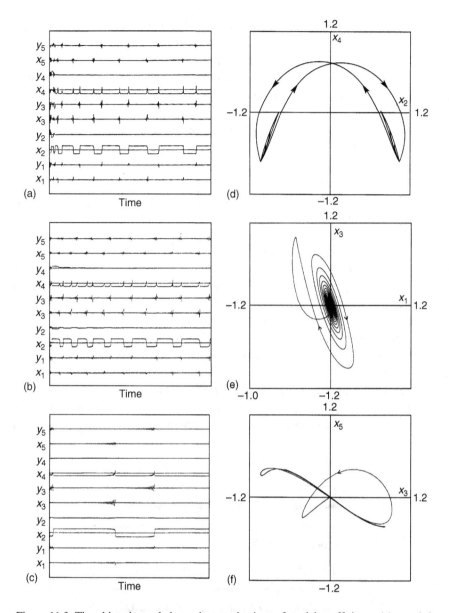

Figure 11.3 Time histories and phase plane projections of modal coefficients: (a) $\alpha = 1.4$, (b) $\alpha = 1.5$, (c, d, e, f) $\alpha = 1.6$. Panel (d) shows a projection onto the (x_2, x_4)-plane of a pair of excursions, showing how the solution leaves a neighborhood of $(+\rho_2, -\rho_4)$ for $(-\rho_2, -\rho_4)$ and returns. Panels (e, f) display the oscillatory (linear) instability. From Aubry *et al.* [22].

In fact, as in Section 7.4, one has such a cycle connecting every diametrically opposite pair of saddles on the circle $|a_2| = \rho_2$, $|a_4| = \rho_4$: a whole circle's worth of structurally stable cycles. Structural stability is crucial, since small errors in the empirical eigenfunctions and hence in the coefficients of the ODEs, are inevitable.

For the range $1.35 < \alpha < 1.61$, the magnitude of the weakest (negative) stable eigenvalue exceeds that of the (positive) real parts of the unstable eigenvalues. Arguments analogous to those of Section 7.3 then show that the heteroclinic cycles are attractive. As described in that section, for solutions near such a cycle, the durations of quiescent periods between heteroclinic excursions should increase without bound, but they are generally limited by numerical precision. See Section 10 of [22] for details specific to the present case; also see Section 7.3 above, and [46]. In Section 11.4 we discuss the rôle of the pressure terms $d_k(t)$ in (10.46), which limit the inter-event durations in a physically relevant fashion.

As α drops below ≈ 1.35, the cycle loses attractivity preparatory to the traveling wave bifurcation at $\alpha \approx 1.32$ (Figure 11.1). A complex interaction of (modulated) traveling waves and heteroclinic cycles then occurs, in which one can see slowly rotating cycles, corresponding to spanwise translating vortices, still with rapid phase changes of π and 2π in θ_2 and θ_4 respectively, but now also showing slow variation in the average magnitudes of real and imaginary components, see Figure 11.4. Campbell and Holmes [68] studied a simpler model retaining only modes 1, 2, and 4 and showed that, after bifurcation to traveling waves in the (a_2, a_4)-subspace, properly no attracting heteroclinic cycles are preserved: only various kinds of modulated traveling wave exist. Further decrease in α leads to increase in the (slow) modulation frequency and the appearance of a less structured and apparently chaotic régime to which we return in the next section. This type of behavior persists down to $\alpha \approx 1.1$, below which rather simple and physically unrealistic traveling and modulated traveling waves occur, see [22] for further details.

As we have noted, for $\alpha > 2.41$ the trivial fixed point $a_k = 0$ is linearly stable within the (a_2, a_4)-subspace. In this range it is linearly stable for the full system also, the other bifurcation points from the zero solution occurring at $\alpha_{c1} \approx 2.32$ (mode 1), $\alpha_{c3} \approx 2.20$

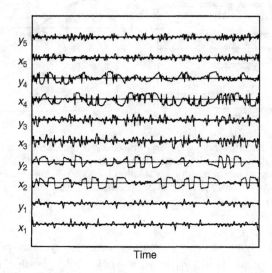

Figure 11.4 Time histories for $\alpha = 1.3$, showing cycles "precessing" in an irregular manner. From Aubry *et al.* [22].

(mode 3), and $\alpha_{c5} \approx 1.62$ (mode 5). Note that these, and the bifurcation values for modes 2 and 4, occur in the sequence $\alpha_{c2} > \alpha_{c1} > \alpha_{c3} > \alpha_{c4} > \alpha_{c5}$, reflecting the relative ordering of energy by Fourier wavenumber revealed by the spectrum of empirical eigenvalues of Figure 10.2(a). This, of course, makes good physical sense.

For $\alpha > 2.41$, both the quadratic and final quartic terms of (11.3) are negative definite, and, since the cubic terms derive from the quadratic, energy conserving terms of the Navier–Stokes equations and the Liapunov function S of (11.2) is the (scaled) kinetic energy, we conclude that the trivial fixed point is in fact *globally* stable in this range.

When α first drops below 2.41, the stability of this zero solution is transferred to the circle of mixed mode 2/4 equilibria which bifurcate from it. In the relatively narrow range $2.0 < \alpha < 2.3$, in which the first and second Fourier modes interact alone and modes 3, 4, and 5 remain essentially passive, one can restrict to a center-unstable manifold much as in Section 8.3 and show that heteroclinic cycles of the same type as those in the Kuramoto–Sivashinsky equation occur. The mixed mode branch restabilizes at $\alpha \approx 2$ and remains stable until the attracting cycles described above appear at $\alpha \approx 1.61$. Thus there are two "windows" of heteroclinic cycles in the range $1.32 < \alpha < 2.41$. This sort of behavior reappears in studies of systems with non-zero streamwise Fourier modes, described in Section 11.5.

11.3 Bursts and sweeps

In Figure 11.2 we showed the cross-stream components of the fluid velocity field corresponding to one of the fixed points in the circle of mixed 2/4 mode equilibria. We remarked that it appears similar to the pattern of streamwise vortices observed in the wall region *on average*. For this figure, a value of α for which the fixed point is stable was selected, but the vortices look much the same in the range $1.32 < \alpha < 1.61$, for which they correspond to unstable saddle points belonging to heteroclinic cycles. The discovery of heteroclinic cycles provides a nice explanation to the puzzle of why such structures, which are evidently unstable in themselves (see Section 2.5), should keep reappearing in the boundary layer. In the phase space of the dynamical system, they belong to an attracting set and so are fated to recur continually. The formation and reformation of streamwise vortices, punctuated by intermittent events, is the expression of the attracting set in the fluid flow itself. We now ask if the specific changes occurring in the physical flow during a heteroclinic excursion match the actual behavior observed in the velocity field of the wall region.

Figure 11.5 shows a series of cross-stream velocity fields at equally spaced intervals during one such heteroclinic excursion. In spite of the extreme truncation, in the reconstructed velocity field one can see evidence of a burst, involving regrouping of vortices, in which the first Fourier mode briefly dominates to give an updraft between a single strong pair, followed by a weaker and more diffuse downdraft or sweep, and reformation of the original pattern of two pairs of rolls, shifted laterally by $\frac{1}{4} \times L_3 \approx 83$ wall units. At this qualitative level, the model evidently captures the gross features of the burst/sweep cycle, including the lateral shift noted in [194], see Section 2.5, above.

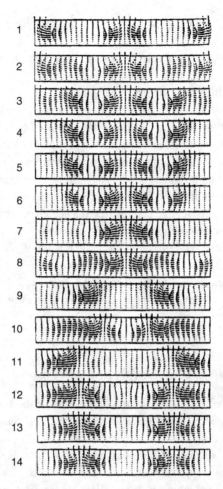

Figure 11.5 Cross-stream (u_2, u_3)-components of the velocity field corresponding to a single heteroclinic excursion at a value of $\alpha = 1.4$. From Aubry *et al.* [22].

The relatively strong attractivity of the cycle keeps solutions close to the union of the real subspace and rotations of it, imposing a strong symmetry on the roll patterns even during the burst and sweep of the excursion itself. But for α values slightly below the range in which attracting cycles occur, as we noted in Section 11.2, we observe "chaotically modulated" cycles. These provide a richer mixture of the component (Fourier mode) vortices, leading to less regular, asymmetric flow fields more characteristic of the real turbulent boundary layer, see [307]. Figure 11.6 shows a typical time series from this range and Figure 11.7 shows the corresponding cross-stream velocity fields, in which evidence of unequal pairs and single rolls can be seen.

In [22] the Reynolds stress

$$-\langle u_1 u_2 \rangle (x_2, t)$$

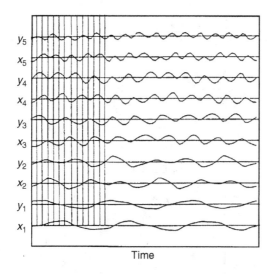

Figure 11.6 A chaotic (or quasiperiodic?) solution for $\alpha = 1.2$. From Aubry *et al.* [22].

was calculated from the reconstructed velocity field by averaging over the flow domain in the spanwise and streamwise directions at various distances x_2 from the wall. For relatively large distances (above $x_2 = 25$), it displays a positive peak during each heteroclinic event, as one expects during bursting. However, the magnitude of the increase is modest (only 20–25% at $x_2 = 35$), and the change is negative over too much of the domain. This is almost certainly due to the drastic truncation. The model retains the basic mechanisms responsible for the instability driving the burst, but insufficient modes to represent the secondary instabilities triggered by that process. For a start, there are no modes with streamwise variations present in this rudimentary model. In addition, Aubry [19] considered the average kinetic energy of each component of the velocity field captured by various truncations and showed that, while the streamwise component u_1 converges fairly fast, the cross-stream components u_2 and u_3 do so considerably more slowly (see Figure 4.1, for another instance of this behavior). This also helps explain why the Reynolds stress increases are too small, see [22] for more details.

11.4 The pressure term

Thus far our model has been essentially isolated from the flow in the outer part of the boundary layer, for $x_2 > 40$, for we have set the boundary terms $d_k(t)$, which derive from the pressure field at $X_2 = 40$, to zero. This is tantamount to assuming that no pressure variations are transmitted though this free boundary, and is clearly unrealistic. We now consider the introduction of terms representative of the pressure field at this location.

Currently it is not experimentally possible to make the time-varying pressure measurements in the body of the flow necessary for estimates of the Fourier components $\hat{p}^{\tau}_{k_1 k_3}(X_2, t)$ (fluctuating pressure can only be measured at the wall). One can, however,

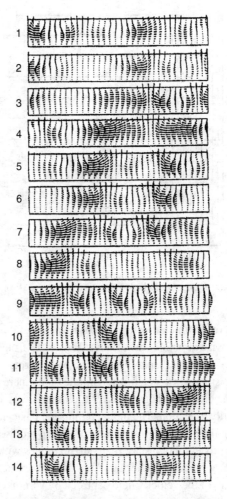

Figure 11.7 Cross-stream (u_2, u_3)-components of the velocity field corresponding to chaotic dynamics at a value of $\alpha = 1.2$. From Aubry *et al.* [22].

use data from numerical simulations of the Navier–Stokes equations, and Aubry *et al.* [22] used such data from the large eddy simulations of Moin for a channel flow at a (center-line) Reynolds number of 13 800 [241]. Projecting the pressure spectrum at X_2 (actually $x_2 = 38$, due to discretization) onto the relevant Fourier components, they computed the real and imaginary parts of the complex functions $d_k(t)$ for $k = 1, \ldots, 5$ and fed these as forcing functions to the ODEs. Figure 11.8 shows the pressure components for the time range available. The maximum amplitude of these signals is two orders of magnitude below typical values of the modal amplitudes themselves, such as those shown in Figures 11.3 and 11.4. They therefore can have little effect on the structure of solutions during heteroclinic excursions (bursts), but during the quiescent periods, when the solution is lingering near a fixed point at which the linear, quadratic, and cubic terms of the ODE collectively vanish, *any* additional perturbation may be expected to have a dramatic effect.

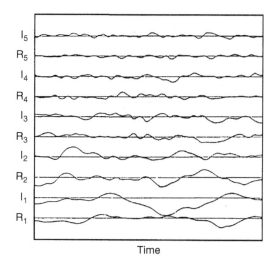

Figure 11.8 Time histories of pressure components, derived from [241]. From Aubry *et al.* [22].

The effect here is to encourage solutions to leave the neighborhood of the saddle points more quickly than they would otherwise. In Sections 9.1 and 9.2, we sketched the analysis due to Stone and Holmes [355,356], which leads to quantitative predictions of the probability distribution of inter-event times for attracting heteroclinic cycles perturbed by additive white noise. In particular, in Section 9.2, Equation (9.28), we found that, for an $O(2)$-symmetric cycle containing a pair of saddle points, the mean inter-event duration scales jointly with the (real part of) the unstable eigenvalue, λ_u, and the r.m.s. noise level, ϵ, according to

$$T_{O(2)} = 2K_1 + \frac{2}{\lambda_u}\left[\ln\left(\frac{1}{\epsilon}\right) + K_0\right]. \tag{11.4}$$

This is the time required for a full circuit of the cycle – two events.

It is clear from Figure 11.8 that the actual pressure signal is far from being a Wiener process with independent, identically distributed components. The various (spanwise) Fourier components are correlated to a rather high degree: in particular, temporal and spatial frequencies are approximately linearly related. For low fluctuation intensities, the turbulence can be regarded as approximately frozen as it passes by (this is known as Taylor's hypothesis); thus the temporal behavior seen is just a reflection of the streamwise spatial behavior. Hence, the fact that the frequency content of the temporal behavior and the wavenumber content of the cross-stream behavior go hand-in-hand is really a statement that the streamwise wavenumber content and the cross-stream wavenumber content are similar, or that the turbulence is crudely isotropic. In spite of these "modal correlations," the simple theory of Chapter 9 nonetheless provides a reasonable guide to the effects of the pressure term, as Figures 11.9 and 11.10 demonstrate. In the second of these figures, the constants K_i of the formula (11.4) were fitted using least squares (T_B versus $1/\lambda_u$), for a single value of ϵ. One can clearly see the effects of both the unstable eigenvalue and the noise amplitude.

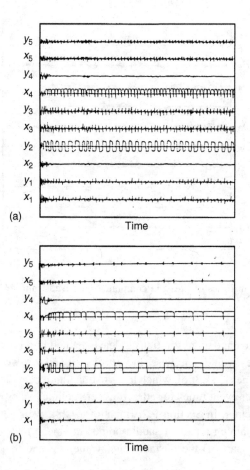

Figure 11.9 Time histories for $\alpha = 1.4$: (a) with the pressure term, (b) without the pressure term. From Aubry *et al.* [22].

The pressure signals were uniformly scaled by multiplication by ϵ to obtain these results; $\epsilon = 1$ corresponds to the actual pressure field of [241]. Figure 11.9 shows that, as expected, the signatures of the heteroclinic events themselves are not much affected by the addition of the pressure signals; it is primarily their timing that is influenced. Attracting cycles in the absence of noise have no characteristic time scale: as shown in Section 7.3 they "die out" in the sense that durations between heteroclinic transits grow without bound (unless limited by numerical effects). External noise perturbations introduce a time scale.

The length of the pressure time series available is too short to yield reliable statistics for the full probability distribution of passage times, as in Section 9.2. However, the authors of [355] examined the five mode model subject to white noise excitations and found remarkably good agreement with the theory. A survey of these results, with reproductions of a number of experimentally measured inter-event distributions from boundary layers, may be found in [162].

In the discussion thus far, we have treated α ($= \alpha_1 = \alpha_2$) as a bifurcation parameter, which may be varied freely to probe the behavior of the model. We must now ask if values

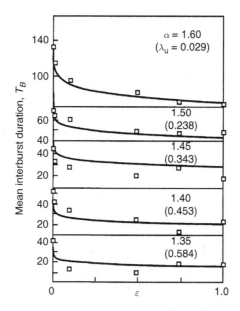

Figure 11.10 Mean interburst durations T_B (compressed by a factor of 333 – see text) as a function of unstable eigenvalue λ_u and pressure signal amplitude ϵ. The solid curve shows the simple theory of Equation (11.4). From Aubry *et al.* [22].

of α may be found for which the inter- and intra-burst timing is physically reasonable. In [22], ignorant of the omitted factor of $\sqrt{L_1 L_3} = 333$, it was noted that for $\alpha \approx 1.5$–1.6, with the pressure signal from [241] present, the mean interburst duration was around 50–75 wall units, while the burst duration itself remained at around 10 wall units, irrespective of the pressure terms. The authors concluded that the overall duration of 60–85 was in reasonable agreement with experimental observations (Section 2.5), which give a mean of about 3×10^2, but with great scatter [226].

Once the time scale is expanded by 333, these figures become unrealistically long. Even for less stable values of α around 1.35, the interburst duration is still one to two orders of magnitude longer than that observed experimentally, and the durations of the events themselves (1000–3000) are also far too long. The interpretation of [22] is clearly in error and in Section 11.6 we provide a new interpretation, involving the convection of structures and events, an effect which was ignored in [22]. In this new analysis the *ratio* between burst duration (K_1 in Equation (11.4)) and inter-event duration $(1/\lambda_u)[\ln(1/\epsilon) + K_0]$ is a crucial factor, rather than the absolute durations themselves.

11.5 More modes and instabilities

In this section we briefly describe the work of Sanghi and Aubry [25, 323], who considered models of the form (10.46) with $N = 1$, $K_1 = 1, 2$, and $K_3 = 5$, and with $N = 2$, $K_1 = 1$, and $K_3 = 5$. As we have remarked, the $N = 1$, $K_1 = 0$, $K_3 = 5$ case of Sections 11.1–11.4 is an invariant subsystem for any K_1, K_3 with $N = 1$ and we can therefore build an analysis of the behavior of the first two of these models on our earlier studies. These particular larger models have real dimensions 32 and 54 respectively (the final one ($N = 2$) has dimension 64), so the existence of invariant subsystems, even if they are only partially understood, is crucial if we are to obtain any real understanding of qualitative behavior.

Here we give only a brief account of the results; the reader should refer to the papers cited for further information. One point to note is that Sanghi and Aubry employed a different eddy viscosity coefficient ν_T than that of [22]; consequently the bifurcation values α_{ck} at which the various modes branch from the trivial solution $a_{k_1 k_3}^{(n)} = 0$ differ from those noted in Sections 11.1–11.2 above. Since the choice of ν_T depends on the wavenumbers of the leading neglected modes, this is inevitable, see Section 10.9. We do not note the specific values here, for we are more concerned with qualitative aspects of the behavior. However, it is important to note that the streamwise extent of the domain, L_1, which is now influential (recall Section 10.6), was chosen to be 666.

A key observation is that, with the inclusion of non-trivial streamwise modes $k_1 \neq 0$, one introduces an additional source of instability for the x_1-independent roll structures described above. In physical terms, the introduction of a fully three-dimensional field, depending on all spatial dimensions, permits the vortex stretching that is excluded from the simple models considered above. (Although, as we noted in Section 10.7.1, the use of empirical eigenfunctions does effectively introduce averaged effects of streamwise variations in the coupling it imposes between their components.) Nonetheless, the addition of streamwise variations does not introduce new instabilities until the loss parameter α is reduced below the ranges in which the "simple" heteroclinic cycles occur, namely those involving only the modes $k_1 = 0$, as described in Section 11.2.

In their studies of the $N = 1$ cases, Sanghi and Aubry [323] find four "windows of intermittency," that is open sets of α values in which heteroclinic cycles of increasing complexity occur, as α is reduced. In the first (window I), corresponding to that for $2.0 < \alpha < 2.3$ in the $K_1 = 0$ model above, cycles essentially involving only the modes $a_{01}^{(1)}$ and $a_{02}^{(1)}$ are seen. In window II, corresponding to that for $1.35 < \alpha < 1.61$, behavior also closely paralleling the $K_1 = 0$ case is seen, all five modes $a_{0k_3}^{(1)}$ being active during the heteroclinic excursions, while the streamwise modes $a_{k_1 k_3}^{(1)}$, $k_1 \neq 0$ are not appreciably excited. Evidently the $k_1 = 0$ subspace is still attractive. However, as α is reduced further, this subspace loses its attractivity. In window III, while the modes $a_{0k_3}^{(1)}$ behave in much the same way as in window II, the growing instability in the odd modes $k_1 = 0$, $k_3 = 1, 3, 5$ which precedes each burst is accompanied by a growth of the modes $k_1 = \pm 1$, $k_3 = 1, 3, 5$, followed shortly afterwards by $k_1 = \pm 1$, $k_3 = 2, 4$. High-frequency oscillations take place in these modes, oscillations which initially grow exponentially and are rapidly quenched after the heteroclinic transit occurs. Figure 11.11 shows typical results from [323] in the case $K_1 = 1$. Similar behavior occurs when a second pair of streamwise modes is included, $K_1 = 2$ (see [323], Figure 6).

In window IV, all the $k_1 = 0$ modes remain persistently active, apparently behaving much as they do in the $K_1 = 0$ model for $1.1 < \alpha < 1.32$. However, intermittent bursts now occur in which all the $k_1 \neq 0$ modes grow, at first exponentially from low amplitude,

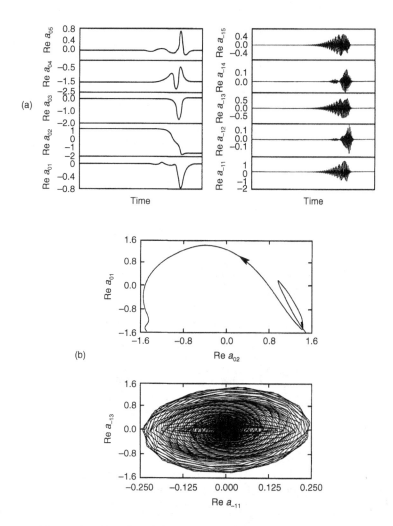

Figure 11.11 (a) Time histories of the real parts of the modes $k_1 = 0$, -1 and $k_3 = 1, \ldots, 5$ for a model with streamwise variations, $K_1 = 1$, in window III. (b) Projection onto two phase planes, showing evidence of heteroclinic cycles. From Sanghi and Aubry [323].

and then in an irregular fashion, and finally collapse. It appears that there are heteroclinic cycles connecting a (possibly chaotic) invariant set in the ($k_1 = 0$)-subspace to itself, see Figure 11.12.

In [323] Sanghi and Aubry also studied the stability of the 2/4 branch of equilibria in the ($k_1 = 0$)-subspace within the larger $k_1 = \pm 1$, ± 2, 32- and 54-dimensional models, much as did Aubry *et al.* [22] for the simpler system as described in Section 11.1. Again the matrices block-diagonalize, due to the presence of invariant (hyper-)planes within the phase space, and one can compute eigenvalues and use them to help interpret the numerical results. This confirms that, in windows I and II, the branch of saddles is indeed stable to perturbations with $k_1 \neq 0$, while in window III, the $k_1 = \pm 1$ non-trivial streamwise modes

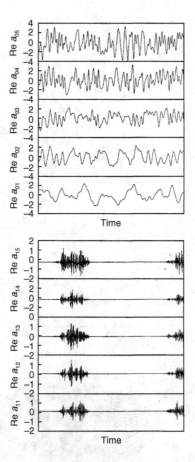

Figure 11.12 Time histories of the real parts of the modes $k_1 = 0$, -1 and $k_3 = 1, \ldots, 5$ for a model with streamwise variations, $K_1 = 1$, in window IV. From Sanghi and Aubry [323].

become unstable. Although the $k_1 = \pm 2$ modes are still stable in this range (when they are included), they nonetheless are excited by the $k_1 = \pm 1$ modes and so one sees activity during the bursts in these modes too. These linear stability studies even explain the lag in response of the ($k_1 = \pm 1$, $k_3 = 2, 4$)-modes, since the unstable eigenvectors lie in the ($k_1 = \pm 1$, $k_3 = 1, 3, 5$)-subspace and the even streamwise modes are excited through nonlinear interactions rather than directly ([323], Section 6).

The inclusion of modes with streamwise variations provides a much more realistic reconstruction of the flow than the $K_1 = 0$ model of [22] discussed earlier. One can see not only the downstream convection of low-speed streaks, but also their development as they move. For example, Figure 11.13, in which the mean (convective) velocity is also included, shows contours of the streamwise component of velocity at a fixed distance from the wall at two times. In terms of modal behavior, the growth of high-frequency oscillations in the $k_1 = \pm 1$, $k_3 = 1, 3, 5$ and $k_1 = \pm 1$, $k_3 = 2, 4$ modes, followed by rapid

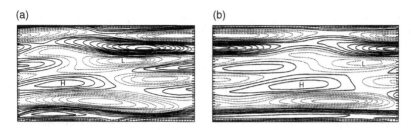

Figure 11.13 Contours of $U_1(x_2, t) + u_1(x_1, x_2, x_3, t)$ at $x_2 = 5.6$ and two successive times, separated by $\Delta t = 9.42$, for a model with streamwise variations, $K_1 = 1$, in window IV. Note downstream convection and distortion of the low speed streaks. H = high speed ($u_1 > 0$) and L = low speed ($u_1 < 0$). From Sanghi and Aubry [323].

quenching, closely mirrors the experimental observations of secondary instability during a burst [194, 195].

The model with two families of eigenfunctions $N = 2$ is significantly more complicated geometrically than those with $N = 1$. As the reader can see from the coefficient formulae in (10.47) and (10.48), with two or more distinct families, the matrices of linear terms in the ODEs (10.46) are no longer diagonal. In particular, the $N = 1$ systems do not occur as invariant subsystems in $N > 1$ models (although the ($k_1 = 0$)-subspace is still invariant within the $k_1 > 0$ models). This makes analysis much more difficult and virtually nothing is known. However, [323] briefly describes some numerical simulations that display windows of intermittency analogous to III and IV for the $N = 1$ models.

As we have noted, Aubry and Sanghi corrected the erroneous time scales of [22] (see Equation (6) of [323]), and they noted that their typical inter-event durations of 4000 were unrealistically long. While they commented on it, they did not analyze the effects of the pressure terms, which they set to zero throughout their studies. However, one can expect that these terms will serve to trigger bursting in windows I–III in much the same way as for the $K_1 = 0$ model. In window IV, where the solution evidently never settles near a fixed point and the right-hand side of the ODEs remains relatively large, the small additional perturbations due to $d_k(t)$ may have less appreciable effects. In any event, as above, small pressure perturbations will not appreciably affect the event durations themselves, which from Figures 3 and 5 of [323] (reproduced as Figures 11.11 and 11.12 above) are approximately 1000 wall units, much as in the corrected study of [22]. Even with streamwise modes present, these time scales are still too long for a direct identification of cycles with experimentally observed bursts.

11.6 A tentative summary

Perhaps the most significant finding to emerge from the low-dimensional models described in this chapter concerns the bursting phenomenon. Even the most rudimentary model, containing only five spanwise modes with no streamwise variations, exhibits structurally stable heteroclinic cycles over significant ranges of loss parameter. The velocity fields reconstructed from the time-varying modal coefficients in these ranges reproduce the main features of the burst/sweep cycle, which begins and ends with quasisteady streamwise vortices. This suggests not only that the energy source for the instability of these vortices, which drives the burst, resides in the wall region, but also that the mechanism promoting their reformation is local to the wall. The reader should recall the discussion at the end of Section 2.5, especially the outline of [151]. In phase space, at least, we have an account of the etiology of the rolls as well as their dynamical interaction. Their behavior is not primarily responsive to the outer flow; rather they dance, collapse, and reform of their own accord. The pressure signal – communication with the outer layer – seems to serve primarily as a trigger for bursting. Thus, this model predicts that events *during* a burst (or sweep) should scale with wall variables, while the times between bursts should scale jointly with inner and outer variables. Specifically, in Equation (11.4), λ_u is a property of the inner layer, derived ultimately from eigenfunctions near the wall, while the r.m.s. level ϵ characterizes the pressure at the outer edge of the wall region, and is hence associated with outer variables.

Before proceeding, however, we must address the issue of time scales. In both [22] and the later studies of [25, 323], as we have noted, the (corrected) heteroclinic event and inter-event durations are unreasonably long, even though the latter may be adjusted somewhat by choice of the loss parameter α and the noise level characteristic of the outer flow. Noting additionally that direct numerical simulations on a "minimal flow unit" of [175], similarly sized in $L_1 \times L_3$, also yield much slower cyclic behavior, we now believe that this effect is due to our spatially localized model's retention of only a single structure in the streamwise direction. Indeed, use of periodic boundary conditions in the streamwise direction implies that the model repeatedly recycles the same coherent structure; effectively, the model domain is convected with the structure. Here the term *structure* denotes a streamwise roll or pair of such rolls.

In a typical experiment, a succession of quasi-independent structures of typical streamwise extent L_1 is swept past a stationary sensor, at a (convection) velocity U characteristic of the wall-normal location of the sensor. The structures are consequently sampled at a rate U/L_1 and the expected inter-burst duration may be obtained by averaging over the probability of m structures passing in succession, none of which is bursting. We identify T_{on} with the transit time K_1 of one leg of the cycle in Equation (11.4) – the duration of a single burst-sweep event in the moving Lagrangian frame – and $\langle T_{off} \rangle = (1/\lambda_u)[\ln(1/\epsilon) + K_0]$ with the expected inter-burst duration. We ignore the (rare) case of a "quiet" structure beginning to burst while passing the sensor. The probability of passage of m quiet structures is then $(1-p)^m$, where $p = T_{on}/(T_{on} + \langle T_{off} \rangle)$; in this case, the quiet time is mL_1/U. It follows that the expected quiet time observed at the stationary sensor is

$$T_B(\text{observed}) = \frac{\frac{L_1}{U}\sum_{m=1}^{\infty} m(1-p)^m}{\sum_{m=1}^{\infty}(1-p)^m}.$$

Summing the series and using the definition of p above, this yields the following estimate for the expected time between observed bursts:

$$T_B(\text{observed}) = \frac{L_1}{U}\frac{1}{p} = \frac{L_1}{U}\frac{(T_{on} + \langle T_{off} \rangle)}{T_{on}}, \tag{11.5}$$

or, in terms of the eigenvalue and r.m.s. noise parameters of Equation (11.4):

$$T_B(\text{observed}) = \frac{L_1}{U}\left\{1 + \frac{1}{\lambda_u K_1}\left[K_0 + \ln\left(\frac{1}{\epsilon}\right)\right]\right\}. \tag{11.6}$$

Using the ratio $T_{\text{on}}/(T_{\text{on}} + \langle T_{\text{off}}\rangle) \approx 0.1$, typical of the computations involving pressure signals in [22,355], and the characteristic values (in wall units) of $U = 30$, $L_1 = 1000$, we obtain $T_B = 333$, close to the observed value 3×10^2, noted in Section 11.4. See Podvin *et al.* [281] for more details.

We observed in Section 9.2 (Equations (9.18) and (9.19)) that the distribution of times between heteroclinic excursions in a system with an attracting cycle perturbed by additive white noise, displays an exponential tail. There is considerable evidence for such tails in experimental distributions; Figure 11.14 shows data from four different boundary layers. However, no attempt has yet been made to identify the exponent with the eigenvalue of a linearly unstable mode, as our model would suggest.

These are not the only ways in which this model mimics the behavior of the turbulent boundary layer. Experimentally, the boundary layer is observed to be more stable in the presence of favorable pressure gradients (that is, pressure falling in the direction of the mean flow), while it is less stable in the presence of unfavorable pressure gradients (pressure rising in the direction of the mean flow). For favorable pressure gradients, turbulence production is partially suppressed, and drag reduced; if the gradient is sufficiently

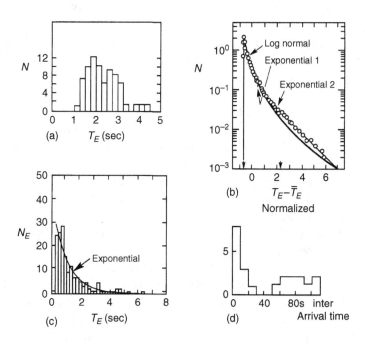

Figure 11.14 Probability distributions of inter-burst intervals in turbulent boundary layers: (a) from [189], (b) from [352], (c) from [53], (d) from [257]. (d) is from an atmospheric boundary layer, all others are laboratory flows. In (b) the events are zero crossings of a velocity component, in all others they are crossings of a preset $-u_1u_2$ threshold. See the references cited for details. Figure reproduced from Holmes and Stone [162].

favorable, turbulence production can be completely suppressed, and the layer becomes laminar again. On the other hand, when the gradient is unfavorable, turbulence production and drag are increased, and the layer rapidly becomes thicker. In her thesis, Stone ([354], Sections 5.2, 6.3 and Appendix III) showed that the attractivity of the heteroclinic cycles changes with the magnitude of the third order coefficients, and related the changing values of these coefficients to streamwise accelerations, showing that the heteroclinic cycles become more attractive when there is a positive streamwise acceleration, and less attractive when there is a negative streamwise acceleration. (One can see this in the simple case of two complex modes, by considering the variation of $\lambda_+ = \lambda_u$ with e_{ij} in Equation (7.63).) Of course, streamwise acceleration is often associated with a streamwise pressure gradient, which tends to accelerate or decelerate the mean flow. As we have seen, when the heteroclinic cycle is more attractive (lower λ_u), the mean time between bursts increases, and vice versa (see Equation (11.6)). Hence, a less attractive cycle, corresponding to reduced time between bursts, is associated with an unfavorable pressure gradient and increased drag, while a more attractive cycle, corresponding to increased time between bursts, is associated with a favorable pressure gradient and reduced drag. This correlates well with experimental evidence (see [195], p. 765, Figures 16(a, b), [181, 182] and [132], p. 375, Figure 118).

The connection between inter-burst time and drag reduction is well known experimentally. Lumley and Kubo [226] show that there is a correlation between the recurrence time for bursts T_B and the friction velocity, so that $T_B u_\tau^2 / \nu$ is constant (Section 10.8). The skin-friction drag, of course, is proportional to u_τ^2, so that T_B is inversely proportional to the drag: if the time between bursts is doubled, the drag is halved, and vice versa. It is the burst of high-frequency, high-wavenumber turbulence produced by the secondary instability during the burst, that is primarily responsible for transporting momentum normal to the wall in the boundary layer, and hence primarily responsible for the drag. Hence increasing or reducing the inter-burst time reduces or increases the drag.

The low-dimensional model displays yet more behavior connecting inter-burst time with drag, and with experimental observations of the boundary layer. It is known that drag can be reduced by the introduction of long-chain linear polymers in very small concentrations. These polymer chains are in a ball-like form in the sublayer close to the wall (due to the tumbling motion); because of this, they have essentially no effect on the properties of the fluid. However, they become expanded in the turbulent part of the flow away from the wall. Here their presence increases the effective viscosity very substantially, and they suppress the secondary instability associated with a burst. This of course reduces drag, because there is less high-frequency, high-wavenumber turbulence to transport the momentum. Drag reduction is associated with a thickening and general increase of the dimensions in the wall region, which remains structurally similar, but becomes larger in every respect. The thickening is a strictly geometrical effect: for a fixed value of friction velocity, if drag is reduced, the mean flow velocity must be increased; if the profile outside the wall region remains the same (which it must, since it is determined by fluid inertia), it will simply be translated parallel to itself to a higher value of velocity, requiring a thickening of the wall region to match the new value at the inner end of the outer region. One often refers to the value of this velocity difference as an effective slip. The low-dimensional model mimics this behavior in the following way: if the model is stretched in the wall region, so that all the dimensions are larger, then all the bifurcations are found to take place at higher values of the loss parameter α, but the form of the bifurcation diagrams remains structurally similar, see [24]. This suggests that solutions become less stable, requiring a higher value of the loss parameter to stabilize them. This is consistent with experimental observations of drag reduction, in which the presence of long chain linear polymers provides this additional stabilization, associated with increased dimensions in the wall region.

When modes with non-zero streamwise wavenumber are included in the model, there are additional sources of instability available for the steady streamwise rolls, and effects of the outer layer are perhaps seen to be less important in triggering bursts. See [25, 323] for discussions of the physical rôle of modes with streamwise variation.

Much is missing from these simple models. While we have included the effects on the local mean velocity profile due to increases in Reynolds stress during a burst, our drastic truncation cannot account for the increased rate of energy transport to higher (neglected) modes that also accompanies it. To address this, one could employ a variable loss parameter $\alpha(t)$, driven by its own dynamical equation, which would increase when more modes become active and relax back during quiescent periods. Dankowicz *et al.* [89] have made preliminary studies of such a model in the context of local representations of the Kuramoto–Sivashinsky equation (Section 8.4), in which variable viscosity corresponds to an inverse length scale.

We have already commented on the effect on time scales of including only a single structure in the localized physical domain of our model. In reality one has a population of quasi-independent structures, interacting as they go through their own evolutions and are swept downstream. Thus we must be cautious in identifying heteroclinic cycles directly with the burst-sweep phenomena observed at stationary sensors. Nonetheless, we believe that cycles are of rather general relevance to fluid flow problems.

The cycles we have found here are structurally stable and cannot therefore be dismissed as fragile phenomena that might vanish with small changes of parameters. In fact [22] noted (unpublished) work in which an entirely independent set of basis functions was derived from direct numerical simulations of channel flow due to Moin and Moser. When these were employed in the Galerkin projection of the wall layer equations (10.28), the coefficients of the resulting five mode model differed by up to 20%. Nonetheless, cycles were still observed for an open range of α values, albeit they now connected weakly oscillating (standing wave) solutions. The qualitative geometrical structure of the ODEs, of course, does not depend on the coefficient values, but is determined by the symmetries alone, as in the normal form discussions of Chapters 6 and 7 and Section 10.3. One can say that the symmetries of the physical flow predispose certain classes of behavior, while the detailed geometry and dimensions determine it more precisely.

Heteroclinic and homoclinic cycles occur rather commonly in dynamical systems with the translation, rotation, and reflection symmetries characteristic, not only of the boundary layer, but of other open turbulent flows such as mixing layers and jets. We therefore feel that they, and the intermittent fluid motions they imply, are important for the understanding of turbulence generation in such systems. They are also likely to be generally relevant in other, more constrained fluid systems in which relatively few modes interact. Cycles involving the same and other symmetry groups have been observed numerically (and even proven to occur) in Kolmogorov flow – the two-dimensional Navier–Stokes equations subject to a periodic body force [260, 261]. They have also been found in normal form equations for helical wave interactions in pipe flow [146, 232]. Similar behaviors are also seen in experiments on constrained flows, such as the Taylor–Couette apparatus with rotating endplates studied by Mullin ([252, 253, 287], also see [251] and [196]), although the rôle of symmetries in them is not fully understood. Nonetheless, in that work there is even clear experimental evidence of inter-event distributions with the exponential tails characteristic of noisy cycles, as discussed in Chapter 9. For a review of some of these applications, see [160].

The studies just referred to do not employ empirical eigenfunctions per se; classical Fourier mode projections are generally used. The symmetries of a (physical) system are, of course, more fundamental than any particular modal basis used for its description, although, as pointed out in Chapter 3, one should take care to choose a basis which respects the appropriate symmetries.

In the next and final part of this book, we return to our main theme and give a brief survey of a number of other applications of the POD and projection to obtain low-dimensional models of turbulent flows. We also draw some general conclusions and suggest directions for future study.

11.7 Appendix: coefficients

Here we list the coefficients of the five mode model of [22]. As noted in the text (Section 11.1), use of the complex conjugate reality condition to reduce from $2K_3$ to K_3 equations implies that each of the quadratic and cubic coefficients of the equation is a sum of two terms occurring in the "primitive" equations (10.54): $Q^j_{k\,k-k'}$, $Q^j_{k-k'\,k}$ and $C_{k\,k'}$, $C_{k\,-k'}$ respectively. In the following, as in [22], we include the multiplicative factor $\sqrt{L_1 L_3} = 333$. See text for additional discussion.

The linear terms are:

$k =$	1	2	3	4	5
$A^1_k =$	16.66	28.27	39.83	52.08	60.98
$A^2_k =$	−1.07	−1.75	−2.69	−3.95	−5.46

The quadratic terms of each (complex) equation in turn are:

	$k, l =$	$-1, 2$	$-2, 3$	$-3, 4$	$-4, 5$
a_1 :	$Q^1_{kl} =$	2.09	1.13	0.46	0.21
	$\frac{2}{3}l^2_> Q^2_{kl} =$	0.017	0.020	0.015	0.078

$k, l =$	1, 1	3, −1	4, −2	5, −3
$a_2:$ $\quad Q^1_{kl} =$	−1.30	2.41	1.26	0.56
$\frac{2}{3}l^2_> Q^2_{kl} =$	−0.024	0.018	0.012	0.023

$k, l =$	1, 2	4, −1	5, −2
$a_3:$ $\quad Q^1_{kl} =$	−4.49	3.13	1.77
$\frac{2}{3}l^2_> Q^2_{kl} =$	−0.04	0.009	0.014

$k, l =$	1, 3	2, 2	5, −1
$a_4:$ $\quad Q^1_{kl} =$	−5.81	−3.39	4.40
$\frac{2}{3}l^2_> Q^2_{kl} =$	−0.061	−0.110	−0.010

$k, l =$	1, 4	2, 3
$a_5:$ $\quad Q^1_{kl} =$	−6.44	−8.15
$\frac{2}{3}l^2_> Q^2_{kl} =$	−0.16	0.18

The cubic coefficients are given as a 5×5 matrix:

$$
[C_{kk'}] = \begin{bmatrix}
-3.08 & -3.76 & -2.46 & -1.30 & -0.68 \\
-4.80 & -6.20 & -4.23 & -2.34 & -1.27 \\
-6.26 & -8.44 & -5.97 & -3.43 & -1.91 \\
-7.37 & -10.36 & -7.62 & -4.60 & -2.67 \\
-7.86 & -11.39 & -8.62 & -5.43 & -3.26
\end{bmatrix} .
$$

PART FOUR

Other applications and related work

12

Some other fluid problems

The methods developed in this book can be applied rather generally to model the dynamics of coherent structures in spatially extended systems. They are gaining acceptance in many areas in addition to fluid mechanics, including mechanical vibrations, laser dynamics, non-linear optics, and chemical processes. They are even being applied to studies of neural activity in the human brain. Numerous studies of closed flow systems have been done using empirical eigenfunctions, some of which were discussed in Section 3.7. A considerable amount of work has also been done on model PDEs for weakly nonlinear waves, such as the Ginzburg–Landau and Kuramoto–Sivashinsky equations, but this work falls largely outside the scope of this book. We do not have the abilities (or space) to provide a survey of these multifarious applications, but we do wish to draw the reader's attention to some of the other recent work on open fluid flows.

We restrict ourselves to studies in which empirical eigenfunctions are used to construct low-dimensional models and some attempt is made to analyze their dynamical behavior. There is an enormous amount of work in which the POD is applied and its results assessed in a "static," averaged fashion. Some of this we have reviewed in Section 3.7. Yet even thus restricted, our survey cannot pretend to be complete: new applications to fluid flows are appearing at an increasing rate. We have selected five problems on which a reasonable amount of work has been done, the first of which (the jet) is a "strongly" turbulent flow. Of the next four flows, two involve transition, spatial growth, and evolution of coherent structures, and two have complex geometry. Our sixth example is closely related, and offers an alternative approach, to the boundary layer models discussed earlier, and in this second edition we have added a seventh example – a time-periodic flow. Both experimental and numerically generated databases are represented in these examples.

In general, the strategies applied in the studies outlined here parallel those employed for the boundary layer in Chapters 10 and 11, but each problem presents its own particular challenges. Here we attempt to put these different problems in context and, in the comments which follow each outline, suggest a systematic way of examining and evaluating low-dimensional models.

12.1 The circular jet

In Section 2.4 we discussed the plane mixing layer as a canonical example of an open flow which displays coherent structures. A related example, with more direct technological application, is the circular jet and the annular mixing layer that it induces. The work on low-dimensional models of the annular mixing layer that we outline here, was started at SUNY Buffalo by W. K. George, M. N. Glauser, and their co-workers and is described in [121–126, 141, 392].

The flow and large-scale phenomena of interest

The annular jet mixing layer is the region of mixing between a jet exiting from a nozzle and the ambient fluid. Figure 12.1 schematically depicts this region; also recall Figures 2.4 and 2.5. In a turbulent jet, the annular mixing region encloses the potential core, which is typically five jet diameters long. The bulk of the mixing is achieved by large, lobe-like structures, familiar to the reader from the flow visualizations of the plane mixing layer in Section 2.4. The goal of the work described in the present section is to study the mechanisms that govern the dynamics of these large-scale structures in the jet.

In the azimuthal direction the flow is homogeneous. Circular symmetry of the jet implies that two point correlation tensors should depend only on angle differences and not on the angles themselves. In the radial direction – across the layer – the flow is clearly inhomogeneous. Strictly speaking, in the streamwise direction it is also inhomogeneous. However, recall that the plane turbulent mixing layer described in Section 2.4 is *self-similar*. The jet partially shares this property: see Figure 12.1(a) and recall Figures 2.4 and 2.5. Adaptation of the POD to self-similar flows, perhaps by linking space and time via "traveling" coordinates as proposed in [127, 191], or via more general symmetry groups [317, 318], would be useful in studying such flows. See Section 7.5 for further information.

In the work described below, George, Glauser, and their co-workers treated the streamwise direction as if it were homogeneous, essentially assuming that the downstream growth of structures is gradual compared to the scales across the layer. They performed a Fourier decomposition in both the azimuthal and streamwise directions, much as did Herzog [155] in his derivation of the wall layer eigenfunctions from a pipe flow. Their assumption of streamwise homogeneity introduced an approximation whose effects have not been fully quantified. In this connection, see the discussions on the plane mixing layer of Section 2.4, especially that following Equations (2.13) and (2.14). In fact, Glauser *et al.* have also taken alternative approaches. In [126], they appealed to the Taylor hypothesis to develop a model for the streamwise (spatial) evolution of coherent structures.

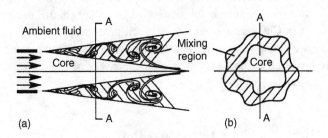

Figure 12.1 (a) Longitudinal, and (b) cross-stream sections of a circular jet, showing the mixing region and annular shear layer.

The experimental data were collected in several phases, the ones most relevant to our discussion being as follows. In the first, two rakes of seven wires each were used to measure the streamwise component of velocity. One rake was fixed at an arbitrarily chosen azimuthal location. The second rake was moved to different positions in the azimuthal direction. Because of reflection (antipodal) symmetry, it is sufficient to measure angle differences up to only 180 degrees. These measurements were made at 16 different azimuthal positions. A second phase used two rakes, each with four crosswires, to measure the radial and the streamwise velocity components simultaneously. In this phase, measurements were taken at 25 azimuthal locations, again covering 180 degrees. The exit diameter Reynolds number of the jet used in the experiments was 110 000.

The POD procedure and key results

The derivation of the empirical eigenfunctions from experimental jet data is described in [122]. The dominant family of eigenfunctions was found to contain over 50% of the energy on average, and only this single family of eigenfunctions was used in the modeling. One of the interesting findings of this work is the form of the dominant family of empirical eigenvalues as a function of azimuthal (m) and streamwise (k_1) wavenumber, which we reproduce in Figure 12.2. The structure of the plot is more complex than the analogous Figure 10.2 for the wall layer, but the authors observed that the peak in the k_1-direction for an azimuthal wavenumber $m = 0$ corresponds to the Strouhal frequency of the jet (as determined by the jet diameter and exit velocity), and that the peak in the azimuthal mode number m for small streamwise wavenumber k_1 is close to mode 5 [124]. The authors interpreted these two peaks as indicators of a dynamical mechanism by which energy is transferred between streamwise and azimuthal modes in a nonlinear fashion. In any case, the presence of a "ridge" connecting the peaks across the interior of the k_1, m wavenumber plane indicates that one must include a triangle of mixed azimuthal/streamwise modes in any reasonable low-dimensional model.

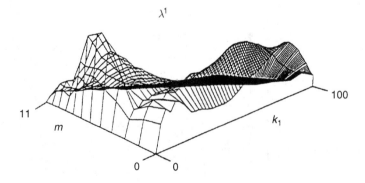

Figure 12.2 The spectrum of empirical eigenvalues for the dominant family in the jet study. From Glauser *et al.* [124].

The governing equations

The PDE and the boundary conditions used in the jet mixing layer study are described in [124, 141, 392]. The equations are the analog of those used in the boundary layer study of Chapter 10, but expressed in a cylindrical coordinate system appropriate to the jet geometry. In particular, the mean velocity is related to the Reynolds stress, although here there is no term corresponding to the second (pressure drop) component of Equation (10.10). The assumption of homogeneity in the streamwise direction was used, as in the wall layer model. A boundary condition on the pressure at the edge of the model domain was also imposed, similar to that in the boundary layer case. Thus the equations have linear, quadratic, and cubic terms as well as the "external" additive pressure term, just as does the boundary layer model developed in Section 10.1. Since the domain considered includes most of the mixing layer and the fluctuating velocities are practically zero at the boundaries, the pressure terms are significantly smaller than in the wall layer model. They were therefore neglected.

Choice of basis functions and resulting equations

Because of the double peak structure of the spectrum of the leading family of POD eigenvalues, and the authors' interpretation of these two peaks, a family of dominant eigenfunctions that includes both modes with non-trivial streamwise and radial variations was chosen, see Figure 12.3. A notable feature of this work is that the authors did not set the value of the lowest streamwise wavenumber to zero; rather they used the wavenumber that arose naturally from the length scale of the experimental measurements, so that there are no modes with $k_1 = 0$ included in the model. The number of (complex) modes included was 18, for a total of 36 real ODEs.

Dynamical behavior

The authors studied the behaviors of the resulting dynamical systems for different values of a bifurcation parameter that plays a rôle identical to the analogous α in the boundary layer

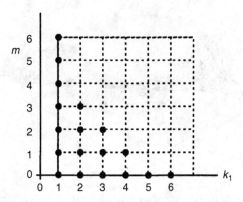

Figure 12.3 Streamwise and azimuthal wavenumbers for inclusion in a dynamical model. From Glauser *et al.* [124].

model of Chapter 10. For some values of this parameter they observed a cascade-like phenomenon in which there was an apparent interaction between streamwise and azimuthal modes, with a net transfer of energy from modes with $m = 0$, $k_1 > 1$ to modes with $m = 4, 5, 6$, $k_1 = 1$. Reconstructed velocity fields for this parameter range show pairs of approximately azimuthally symmetric ($m = 0$) vortices pairing and subsequently becoming unstable to the higher azimuthal modes. Vorticity field reconstructions also show the apparent creation of smaller-scale motions during intervals in which bursts occur in the modal coefficients.

The equations are invariant under infinitesimal azimuthal rotations and reflections and, as formulated, also invariant under streamwise translations. The projected equations therefore exhibit essentially the same symmetries as the boundary layer models of [22, 25, 323], and we can expect parameter ranges in which heteroclinic cycles occur. The time histories and phase portraits of modal coefficients presented in Figures 9–15 of [124] suggest that such intermittent, bursting solutions do indeed occur, but in the absence of detailed studies of the type described in Sections 11.1 and 11.2, this remains conjectural. In other parameter regions, solutions exhibit modulated traveling waves similar to those found at low loss values in the boundary layer model of Chapter 11 [22].

Comments

An interesting question arising from this study is that of universality and scaling of the dominant features of the jet. For example, one might conjecture that the energy dominance of the azimuthal wavenumbers 5 and 6 is a universal phenomenon. However, in the wall layer we saw that the typical roll spacing of 100 was in *wall units*. What is the analogous scaling for the jet? It seems unlikely that in a very high Reynolds number jet the same azimuthal wavenumbers dominate. Such a jet is closer to a plane shear layer and one might expect to see higher azimuthal modes. In a plane shear layer we can think of the analog of wall coordinates by normalizing with respect to the mean peak shear in the layer and the viscosity.

We remark that the same group – Glauser *et al.* – has also studied the plane mixing layer. In particular, they propose to derive time-dependent modal coefficients $a_k(t)$ directly from measurements taken with a limited number of probes. To provide adequate spatial resolution, they extend and fill in spatial data using stochastic estimation (Section 3.5). Their ultimate aim is to compare time histories of the experimental modal coefficients with those derived from solution of a low-dimensional model. One of the novelties of this work – reported in [374] – is that an attempt is being made to compare experimental and low-dimensional model data on the same basis. In the next sections we see examples of such direct comparisons between models and direct numerical simulations.

12.2 The transitional boundary layer

A second important canonical problem upon which much experimental, theoretical, and numerical effort has been expended is the laminar to turbulent flow transition in a flat plate boundary layer. A POD analysis and the corresponding construction of dynamical systems

for this system was recently undertaken by Fasel and Rempfer. The work is described
in [296, 297, 299–302].

The flow and large-scale phenomena of interest

The database used in this work derives from a direct numerical simulation (DNS) [305] of
an experiment carried out earlier by Kachanov *et al.* [180]. The simulation covered a rectangular box 8.77 mm high and 24.5 mm wide that extended from 300 to 500 mm from the
leading edge of the plate. The free-stream velocity was 9.09 m/s and the Reynolds number
based on momentum thickness was 283 at the inflow boundary, increasing to 573 at the outflow boundary. (In [301], a smaller subregion and concomitantly smaller Reynolds number
range was considered.) In addition, the simulation imposed a reflection symmetry with
respect to a central plane in the streamwise-normal (x, y)-plane. This corresponds to the
(approximate) symmetry induced experimentally by use of suction and blowing symmetric
with respect to the same plane [180]. The constraint of reflection symmetry has important consequences to which we shall return; in particular it breaks the spanwise translation
symmetry and prohibits occurrence of cross-stream traveling waves. The simulation used
finite differences in the (x, y)-planes (4000×120 grid points respectively) and a Fourier
representation with 16 modes in the spanwise direction.

The POD procedure and key results

The authors computed empirical eigenfunctions on several subdomains centered at different downstream locations, within the larger domain noted above. In this problem the flow
is inhomogeneous in all three directions. In the streamwise direction it is, of course, developing and it was this transition process that was studied. As we have noted, the natural
spanwise translation symmetry is removed by imposition of a reflectional constraint. Two
results are particularly worthy of mention. First, the authors showed that the POD is able
to extract a structure similar to a lambda vortex from the developing flow (see Section 2.5),
as indicated in Figure 12.4. Second, (see Figure 12.5) they compared the leading empirical
eigenfunctions at several streamwise locations, with the classical Orr–Sommerfeld eigenfunctions for a two-dimensional laminar layer. The latter represent the streamwise velocity
component of Tollmien–Schlichting waves of linear stability theory. At the upstream location, the empirical eigenfunction is nearly identical to the Orr–Sommerfeld eigenfunctions;
however, as locations further downstream are examined, the empirical eigenfunctions begin

(a)   (b)

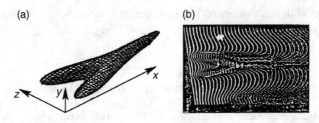

Figure 12.4 Comparison of the leading eigenfunction with an experimental visualization
of a lambda vortex. From Rempfer and Fasel [302].

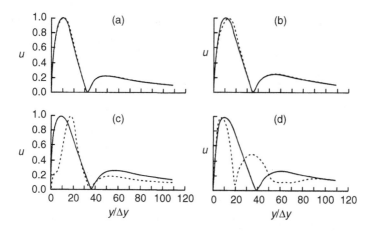

Figure 12.5 Comparison of streamwise velocity profiles of the leading pair of empirical eigenfunctions with Orr–Sommerfeld eigenfunctions: ——, O–S eigenfunction; - - - -, r.m.s. profile of empirical eigenfunction. Panels (a) through (d) represent successive downstream locations. From Rempfer and Fasel [302].

to differ significantly. The authors pointed out that this is to be expected, as the nearly linear initial instabilities evolve into their fully nonlinear forms.

An important point must be made here. Rempfer and Fasel found that the leading eigenfunctions occur approximately in pairs, with almost equal eigenvalues. As we saw in Section 3.3.3, if the layer were perfectly translation invariant in the streamwise direction, we would obtain Fourier modes. Each (complex) Fourier mode may be resolved into real sine and cosine components associated with a double eigenvalue, as described in Section 7.4. The empirical functions presented in Figure 2 of [302] are remarkably similar to such modes, indicating an approximate invariance in each subdomain, consistent with slowly developing structures. The authors therefore defined a "coherent structure" as the sum of the corresponding pair of eigenfunctions, multiplied by the appropriate modal coefficients. It is such pairwise combinations that appear in Figures 12.4 and 12.5.

The governing equations

The authors performed two types of dynamical analysis. In [296, 297] a straightforward projection of the Navier–Stokes equations was made on spaces spanned by 30 to 40 modes. At the wall, the no-slip condition is satisfied by the eigenfunctions, as in [22], while at the free-stream side, the exponential decay employed in the simulations of [305] is also appropriate. (Unlike the wall layer model of Chapter 10, the domain encompasses essentially the whole thickness of the developing layer.) Periodic boundary conditions, consistent with the symmetry constraint, were imposed in the spanwise direction. The problematic condition here arises at the inflow and outflow (streamwise) boundaries of each subdomain. Here the influences were simply neglected, essentially by appeal to the fact that the correct *combinations* of empirical eigenfunctions satisfy appropriate velocity boundary conditions on average, as remarked in Section 10.6. Energy losses to neglected modes were modeled by an effective viscosity which increases with wavenumber.

In [301] the authors derived energy equations for the various coherent structures and used the instantaneous evaluation of the coefficients of the eigenfunctions to study the characteristics of energy transfer between structures. Rather than integrating model equations for the coefficients (as in [296]), they obtained the modal coefficients by projection of the DNS solutions onto the appropriate empirical eigenfunctions. They studied both instantaneous and time-averaged energy flows in projections comprising 34 eigenfunctions and drew conclusions regarding both the energy budget which maintains the mixture of coherent structures in the fluctuating velocity field, and mechanisms involving higher-order structures in the characteristic spike events of transition.

Choice of basis functions and resulting equations

In [301, 302] the authors did not perform a dynamical systems analysis in the style described in the previous chapters. The criteria for choosing the basis functions seem to be based on numerical experimentation to yield good tracking and to capture the key contributions to the energy budget. In [296], in which low-dimensional systems were studied more closely, the form of the equations is, as can be expected, very close to the original Navier–Stokes equations, except for the viscosity that varies with wavenumber. In particular, it was not necessary to model the "quasilocal" mean velocity profile as in Chapter 10, [22]. The model ODEs are thus quadratic, like the original Navier–Stokes equations.

Dynamical behavior

The study of dynamical models of the transition layer, begun in [296, 297], was still in its early stages in 1996. In these two papers, the author demonstrated excellent short-time tracking of projections of the full (DNS) solution by the low-dimensional dynamical system, as in [296] (Figure 1(a), also shown in [301], Figure 2) and reproduced here as Figure 12.6. Rempfer and Fasel also display the increasing complexity of the time-dependent modal coefficients, as one moves to subdomains further downstream. Strongly periodic solutions are found for the model corresponding to the upstream subdomain, in which one is essentially seeing Tollmien–Schlichting waves. Further downstream, more complex,

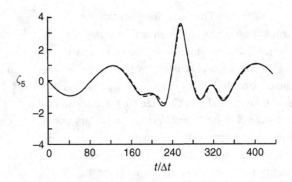

Figure 12.6 Comparison of modal amplitudes of experimental and low-dimensional model data for the boundary layer: - - - -, direct numerical simulation; ———, model. From Rempfer and Fasel [301].

quasiperiodic, and chaotic solutions are observed, although long-term integrations and the resulting attracting sets have not been studied in depth.

Little work has been done on the geometric structure of the models in phase space, but we note that in [297], Rempfer identifies an analogy between dynamical equations for the coherent structures and those of nonlinearly coupled multi-degree-of-freedom oscillators.

As mentioned earlier, the short-term dynamical tracking capabilities of the model derived in [296] appear to be very good. In this study there is no explicit bifurcation parameter, although the effective viscosity that varies with wavenumber is, in principle, a tunable parameter. However, no results are shown for different choices of this viscosity.

Comments

This study is "clean" in the sense that the property of the POD of rapid convergence in the energy norm was used to create a lower-dimensional DNS surrogate. No modeling of the mean flow or boundary conditions was necessary. In this light the results are pleasing, if unsurprising. The viscosity that varies with wavenumbers and increases towards the higher wavenumbers is an interesting construct, which might be of more general use. Such phenomena have also been suggested in the recursive renormalization group (RNG) approach to turbulence advocated by Zhou and Vahala, [395, 396]. There it is shown that the RNG approach yields a cubic term which provides a rapid drain of energy from wavenumbers closer to the cut-off wavenumber, much as would the variable viscosity of [296]. This rapid energy drain is corroborated by analysis of the energy equation for DNS data.

The work on the flat plate transitional layer outlined here differs in several respects from that of the turbulent boundary layer described in Chapters 10 and 11 (also see Section 12.5, below). In the latter, the assumption of spanwise and streamwise homogeneity leads to a Fourier decomposition in both these directions. Thus the experimentally observed *individual* coherent structures do not appear as single eigenfunctions, or even simple pairwise combinations, per se. Rather, the various Fourier components, weighted by the instantaneous dynamics of the low-dimensional models, produce the quasisteady rolls, streaks, and bursts displayed in Figures 11.2, 11.5, 11.7, and 11.13. In contrast, a pair of eigenfunctions associated with a near multiple empirical eigenvalue derived by [302] captures much of the lambda vortex structure surprisingly well, as indicated in Figure 12.4. This is due to the enforced reflection symmetry in the spanwise direction, inherited from the DNS database, leading to a subspace of reflection-symmetric eigenfunctions. We have already remarked on the approximate streamwise invariance which leads to pairs of eigenvalues.

Of course, the spanwise symmetry constraint is restrictive, since it rules out propagating waves in that direction and may not permit all the physically relevant instability modes to emerge. If such a constraint were imposed on the five mode model of [22], for example, it would be equivalent to restricting solutions to the real subspace in phase space and would therefore preclude the formation of heteroclinic cycles. (Recall that, of the two branch 2/4 equilibria restricted to that subspace, one is a sink; Sections 10.6, 11.1, and 11.2.)

12.3 A forced transitional mixing layer

Just as one may promote particular disturbances in a transitional boundary layer, other developing flows can be "locked" to particular disturbances. The following example, in which we outline the work of Rajaee *et al.* [292], concerns an acoustically excited plane mixing layer, which was also studied in [290,291]. The database derives from experimental hot wire measurements of streamwise and normal (cross layer) velocities and only two-dimensional representations were considered.

The flow and large-scale phenomena of interest

The mixing layer was excited via a spanwise slot in the floor of the test section, beneath which a loudspeaker was mounted. Two superposed sine waves were applied, at the unstable natural frequency and its first subharmonic frequency, separated by a (controllable) phase shift. In the work reported here, the phase difference was 270 degrees. The authors report that this phase difference suppresses the development of three-dimensional motions. They used a phase locking technique in which a single crosswire probe may be moved with a resolution of 0.0005 in. (The resolution actually used is 1 mm.) Two point correlations were constructed by appeal to the phase locking property. The Reynolds number based on the average of the free-stream velocities and the high-speed shear thickness 2 mm down from the splitter plate was approximately 360. Time traces encompassing 12 cycles of the subharmonic were measured, for a total of 540 composite snapshots. The velocity components on each two-dimensional section (with different phases with respect to the forcing) were measured on a grid of 251×121 points, representing a streamwise length from $x = 10$ to 35 cm and cross-stream from $y = -6$ to 6 cm, where the edge of the splitter plate was at $(x, y) = (0, 0)$.

The POD procedure and key results

In this case the streamwise extent is relatively large and the flow was treated as inhomogeneous in both streamwise and cross-stream directions. A two-dimensional POD analysis of the measured data was carried out using the method of snapshots [330], see Section 3.4.1. This enabled the reduction of the problem to an eigenvalue calculation for a matrix of size 540×540. The authors found that the first four modes capture 86% of the energy, the first eight modes contain 96% of the energy and modes nine to sixteen contribute only 1.3% of the energy, on average. The first two modes correspond to the subharmonic and the next two modes correspond to the natural frequency. Projection of the experimental data onto these pairs of eigenfunctions revealed almost perfect sinusoidal oscillations ([292], Figure 6). The average energies (eigenvalues) of the modes within each pair are approximately equal. As in the work of Rempfer and Fasel outlined above, this is a manifestation of the *approximate* streamwise homogeneity of the problem, also see [48].

The governing equations

The authors performed a Reynolds decomposition of the velocity field into steady and fluctuating components. They then wrote the equation for the fluctuating component, and modeled the time-averaged stress term in the equation for the fluctuations, appealing to the uncorrelatedness of the time-dependent expansion coefficients to express them as a sum of the empirical eigenvalues. The mean flow is not regarded as time dependent, hence it does not respond to the growth in the coherent structures. Moreover, the authors neglected viscosity as it is not dynamically important in this (non-turbulent) problem. As in the initial studies of [22] (Sections 11.1–11.3, above), the pressure term at the boundary of the domain of integration was set to zero.

Choice of basis functions and resulting equations

The authors described dynamical simulations for systems with eight and sixteen (real) modes. The behavior is described below. These systems appear to have no particular symmetry structure in phase space. The absence of a feedback term in the mean component makes the equations quadratic.

Dynamical behavior

Results from both the sixteen and eight mode systems were compared to experimental data, via projections of the latter onto individual eigenfunctions. Figure 12.7 shows a sample of the results. For both models, the first two modes display remarkably good tracking of the experimental time evolution, while higher modes are less accurate, although their general level remains correct for the sixteen mode problem. In the eight mode system, the fifth and sixth modes exhibit growing magnitudes significantly above the experimental data. The authors interpret this to indicate that an energy transfer mechanism, which would normally feed to the ninth and higher modes, is blocked in the eight mode truncation. (Presumably transfer could be modeled by an effective viscosity as in [22] and [296].) Both the eight and sixteen mode systems display modulation by a low frequency which is not observed in the data. The authors suggest that this may be due to numerical spatial differentiation errors, presumably leading to errors in the coefficients of the ODEs.

Comments

A particularly strong point of this study is that it includes a comparison of instantaneous modal coefficients with projections of experimental data. The dynamical behavior of these models is, however, considerably less complex than those in the transitional boundary layer of Section 12.2. Indeed, the dynamics of the experimental system are fairly straightforward. They primarily have the characteristics of advected vortices, although there is strong evidence of vortex pairing in some of the snapshots ([292], Figure 3). This is borne out in the structure of the first four eigenfunctions. The higher eigenfunctions do exhibit more complicated structures, but their significance is unclear. Nor is it clear if the measured data are a direct response to the bi-harmonic forcing of the layer by the loudspeaker, or to what extent intrinsic nonlinear fluid interactions are taking place.

The fact that the flow is essentially two-dimensional may explain why the dynamics of the systems, without the addition of effective viscosity, do not blow up: there is no three-dimensional vortex stretching of the type that can transfer energy into fluctuations from the mean flow. However, only relatively short integration times are presented, and long-term behavior is not considered. It would be interesting to perform an analysis of the energy budget of the flow, as described in Section 10.7 [43, 44].

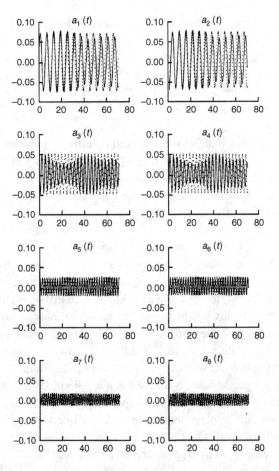

Figure 12.7 Comparison of modal amplitudes of experimental and low-dimensional model data for the forced mixing layer: - - - -, experiment; ———, model. From Rajaee *et al.* [292].

An interesting suggestion in this paper is the use of Reynolds stresses as determined from time averages of the dynamical systems, to close terms in the Reynolds-averaged equations. While the information required for producing low-dimensional models is clearly much more than that required for closing the Reynolds-averaged equations, the possibility of using dynamical systems to provide information for closure of the Reynolds equations is worth considering.

12.4 Flows in complex geometries

We have thus far thought of the POD and asssociated low-dimensional models as tools for the study of "canonical" flows, with a view to gaining general understanding of turbulence generation and instability mechanisms. One can also ask if the optimality of empirical bases can be advantageous in modeling of flows in complex geometries. Such flows are

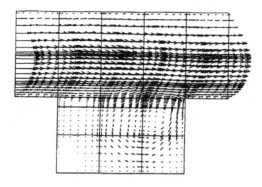

Figure 12.8 An instantaneous velocity field in the grooved channel. From Deane *et al.* [91].

of great technological importance, in rotating machinery, for example. The study outlined here, due to Deane and co-workers [91], appears to be the first to investigate this possibility.

The flows and large-scale phenomena of interest

The authors considered two geometries: flow in a grooved channel and the flow around and in the wake of a circular cylinder. Both databases were generated numerically, by spectral element simulations, and were two-dimensional. The flow in the wake of a cylinder will be familiar to many readers. It is dominated for a large range of Reynolds numbers by the celebrated Kármán street: pairs of vortices shed alternately from the upper and lower sections of the cylinder. The channel has a rectangular cross section with a rectangular cavity in its bottom wall. A typical velocity field for the grooved channel is shown in Figure 12.8. In the channel the boundary conditions are no-slip at the walls and periodic in the streamwise direction, so that the domain represents one cell in an array. Periodic boundary conditions are used in the cross-stream direction for the cylinder, with a prescribed inflow profile and a viscous "sponge" at the outflow boundary. For further details, see [91].

The grooved channel was simulated at a Reynolds number of $Re = 350$ based on 0.75 of the mean flow rate [91]. The onset of periodic oscillations occurs at $Re = 300$. In the cylinder wake simulation, the authors analyzed three flows at Reynolds numbers of 100, 150, and 200, based on cylinder diameter.

In both cases the flow was set up so that the system exhibited a limit cycle, and the temporal dynamics were, as in the example above, fairly simple. Here the concern is primarily with spatial complexity. On the face of it, the spatial form of trapped or shed vortices in problems of this type might seem highly Reynolds-number dependent, and an important question addressed in these studies concerns the range of validity of models that are based on data collected at a particular Reynolds number, and what may be done to extend that range.

The POD procedure and key results

Two-dimensional POD analysis was carried out, using the method of snapshots to reduce the magnitude of the calculation. The data ensembles used were of modest size and restarts

from different initial conditions were made, to provide an ensemble average without rely-
ing on ergodicity. In flows of this type, with simple temporal dynamics, this is clearly an
important consideration.

For the grooved channel flow the first two modes captured over 99% of the energy. For
the cylinder wake at all three Reynolds numbers, the first two modes captured over 97%
and the first four or five modes (depending on the Reynolds number) exceeded the 99%
mark. In spite of the apparent strong inhomogeneity in the streamwise direction, the leading
eigenfunctions for both flows again exhibited distinct pairwise patterns and the eigenvalues occurred in approximate pairs, as in the transition and mixing layers of Sections 12.2
and 12.3.

The governing equations

The authors performed a straightforward projection of the fluctuating component of the
Navier–Stokes equation, with a constant mean profile. They used a mixed velocity/vorticity
formulation. Due to the particular geometry and boundary conditions in both problems, and
the fact that the full computational domain was resolved, there was no contribution from
the pressure term in either problem.

In their studies of the performance of the dynamical systems for the cylinder wake over
a range of Reynolds numbers, the authors found that the constant-shape mean flow was
the greatest single factor that caused poor performance: if the correct mean profile were
used, even with the wrong eigenfunctions, acceptable results could still be obtained. Similar observations applied to the grooved channel model. They consequently introduced an
empirical relation that allowed for the mean velocity profile to vary with Reynolds number.

Choice of basis functions and resulting equations

The authors constructed dynamical systems with four modes for the grooved channel
flow and observed that increasing the number of modes to eight or sixteen gave comparable results. The dynamical systems were based on the eigenfunctions derived from the
$R_e = 350$ simulations. They also constructed dynamical systems with the same eigenfunctions but with a different mean velocity profile. For the cylinder wake, dynamical
systems containing at least six modes were necessary to obtain stable oscillations; when
only four modes were included, the oscillations appeared to grow without bound. Models
were derived for Reynolds numbers of 100, 150, and 200, with the mean velocity profiles from the Reynolds number of the decomposition. Models using eigenfunctions from
a Reynolds number of 100, with mean velocity profiles derived from higher Reynolds
numbers, were also constructed.

Dynamical behavior

The authors analyzed the short-term tracking capabilities of their models, as well as investigating long-term behavior. They compared projections of the attractors into various phase
planes as well as the bifurcation diagrams of the model ODEs with the full simulation.
They particularly emphasized the potential for extrapolation in Reynolds number range. In
general they found that a dynamical system displayed very good short-term tracking for
the Reynolds number from which it was derived. As we have already remarked, for the

grooved channel the authors found that modifying the mean velocity profile was essential to obtain good short-term tracking for extrapolated Reynolds numbers. This was also the case for the cylinder wake, although the results for that flow were more sensitive and it appears that the range of extrapolation is narrower. Similar statements may be made regarding the phase portraits of the attractors and the bifurcation diagrams. In particular, Figure 15 of [91] is a bifurcation diagram showing the dependence of the limit cycle amplitude on Reynolds number for the cylinder wake, in which model predictions are compared with the full simulations.

Comments

This study addressed an important aspect of low-dimensional models: their capability for use at Reynolds numbers different from those for which the eigenfunctions were constructed. An early study of a model problem, the complex Ginzburg–Landau equation [308, 339], addressed the same issue. It is important to emphasize that for the dynamical systems models constructed in [22] this is not a concern since those models are expressed in *wall variables*, see Section 10.1. Hence they should be universal. In particular, as described in Section 2.5, once scaled in wall variables, all near-wall flows exhibit similar behavior for similar pressure gradient characteristics. In general it is important to account for Reynolds number effects, but the choice of wall units in the boundary layer models obviates this, since in those units the Reynolds number is always equal to unity.

Regarding the issue of complex geometry per se, the work described above shows that empirical eigenfunctions can significantly reduce the amount of data needed to describe and model spatially complex flows, and that they enable one to identify at least the "pre-turbulent" bifurcations which occur in such flows. Ideas similar to these have also been used by Noack and Ekelmann to construct and analyze low-dimensional models of the cylinder wake problem [263–266].

Subsequent work on cylinder wakes by Noack and others [230, 262] pointed to the need for inclusion of additional basis functions in low-dimensional Galerkin projections, to correctly reproduce transient instability growth. This led to the definition of shift modes, which extend the span of the empirical eigenfunctions to include unstable steady states (fixed points in phase space) of the Navier–Stokes equations, as described in Section 4.4 of [262]. Shift modes are also closely related to mean field corrections due to Reynolds stresses, providing a more direct alternative to the feedback cycle described in Section 10.8 of [22], and they are especially important in producing appropriate modal energy flows during transients [362].

12.5 "Full channel" wall layer models

The main illustrative application of the present book, treated in Chapters 10 and 11, is to the wall region of a turbulent boundary layer. Our next example of related work, described in the present section, addresses the same problem. It is due to Zhou and Sirovich [394], and was motivated by a desire to improve the treatment of the boundary conditions at the interface between the model domain and the outer layer in the earlier work of Aubry *et al.* [22, 24] and others [43, 44].

As we saw in Sections 10.1 and 10.4, in projecting the governing equations in a subdomain close to the wall, one introduces a "free surface" on which a boundary condition for pressure alone is imposed, in contrast to the expected boundary condition for the velocity field: recall Equations (10.9) and (10.43). In general, an elliptic PDE in which an unknown vectorial boundary condition is replaced by a scalar will be ill-posed, and ODE projections of such a system may well be meaningless.

In the present context, we have already noted in Section 10.6 that, provided solutions are spatially smooth and enough basis functions are included, the properties of the POD guarantee that an appropriate scalar field will induce vectorial eigenfunction combinations at the boundary which are representative of typical vector fields. Rempfer and Fasel also rely on this in their derivation of models in different streamwise subdomains of the transition layer (Section 12.2). This may be seen as a "practical" justification for the procedure. Moreover, if the subdomain is part of a larger region in which the governing equations possess an inertial manifold [367] (which is essentially a tautology for data derived from numerical simulations of the Navier–Stokes equations), then the PDE with a suitable pressure boundary condition is likely to be well posed in the subdomain, provided that a non-singularity condition holds on the Stokes eigenfunctions in the subdomain. This argument is outlined in Section 13.4, below, although it does not completely resolve the mixed boundary conditions of interest here, and inertial manifolds have not been proven to exist for the Navier–Stokes equations in general three-dimensional domains.

In any case, the missing pressure term must still be supplied "externally," and the methods presented in [394] suggest an alternative approach to such subdomain problems, at the cost of tying in properties of the full domain and consequent loss of localization. Some of the issues raised by [394] were addressed in [46] and again in [340].

The flow and large-scale phenomena of interest

For their database the authors used a moderate Reynolds number flat-wall channel direct numerical simulation originally due to Kim *et al.* [190]. This has the advantage over the experimental database of Herzog [155], used in [22], in that it provides spatial data across the full channel and the resulting eigenfunctions satisfy the natural no-slip conditions on both channel walls. Fourier decompositions were used in streamwise and spanwise directions, with periodic boundary conditions, as in [155].

The POD procedure and key results

The authors performed two POD analyses: the first for the complete channel and the second for a subdomain in the wall region of the same streamwise and spanwise dimensions, but limited in the normal direction. Essentially, they repeated the POD procedure for the full channel, including a weighting function in the inner products which was set equal to one in the wall region and zero elsewhere. The resulting "wall eigenfunctions" are defined across the full channel and satisfy no-slip conditions at both walls, but are "concentrated" in the wall layer, see Figure 12.9.

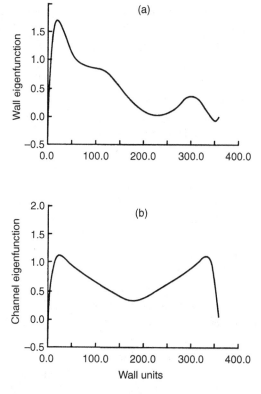

Figure 12.9 (a) A wall eigenfunction, and (b) a full channel eigenfunction. From Zhou and Sirovich [394].

The authors then constructed "channel eigenfunctions valid in the wall region" by approximating the optimal wall eigenfunctions by linear combinations of the channel eigenfunctions. For each (k_1, k_3) streamwise/spanwise wavenumber pair, a sum of 32 full channel eigenfunctions was used to approximate the corresponding wall eigenfunction. Taking into account all the Fourier modes included, this implies that about one thousand basis functions were actually used in forming the subspace for the low-dimensional model. We return to comment on this below.

The governing equations

The evolution equation for the coherent structures used in this study was the one proposed in [22] (Section 10.1, Equation (10.16)). In particular, the mean velocity feedback term proposed in [22] to account for the response of the spatially localized mean flow to the growth of turbulent fluctuations was adopted without change. However, the domain of the PDE was now the whole channel, and no boundary pressure term needed to be supplied.

Choice of basis functions and resulting equations

The authors constructed dynamical systems similar to those described in Chapter 10. In fact, since they also adopted the Heisenberg energy transfer model of [22], their ODEs are *identical* in structure to those derived in Section 10.4, lacking only the pressure and pseudo-pressure terms and with, of course, differing coefficients. (Zhou and Sirovich also normalized their modal coefficients in a different manner [394].)

They first studied a five-(complex-) dimensional model with no streamwise variations ($k_1 = 0$), as in [22], and then included streamwise variations ($k_1 = 1$ and 2) to derive and integrate models with 16 and 27 complex modes.

Dynamical behavior

As we have noted, from the point of view of symmetries and the geometrical structure of phase space, the systems of [394] are identical to those of [22, 25, 323]. Thus it is not surprising that generally similar behavior, including heteroclinic cycles in the five mode model, was found. Zhou and Sirovich speculated on the effects of noise (presumably, numerical errors) on the timing and phasing of these heteroclinic excursions [394]. These speculations were addressed in [46], in which it is shown how integrator step-size can interact delicately with the size of the unstable eigenvalue to produce spurious irregular bursting, when in fact there is an attracting cycle and inter-burst intervals should grow without bound. Also see Section 7.3 above and the discussion in Section 10 of [22].

Zhou and Sirovich then focused on models with one and two streamwise modes, similar to those of [25, 323]. In these, the $k_1 \neq 0$ modes, called "propagating modes" by the authors, were deemed to be of significant dynamical importance. A propagation frequency emerged clearly in power spectra computed from the model ODEs.

Here, as in the studies of Aubry *et al.* [22, 25, 323], no attempt was made to compare short-time tracking of the models to projections of the original turbulent field onto particular eigenfunctions. For strongly turbulent flows, and relatively low-dimensional models, it might be unclear how to interpret such data, but it seems worth carrying out such comparisons.

Comments

The work of [394] was originally motivated by the desire to correct apparent shortcomings and possible errors in that of [22]. However, in view of arguments developed in Section 13.4, below, the original, physically-motivated formulation seems to be on reasonably secure mathematical ground. Moreover, as pointed out in Chapter 10 and in [46], isolation of the wall region is an attractive feature of the models of [22] since it permits one to distinguish between mechanisms intrinsic to the wall region and those dependent wholly or partially on influences from the outer layer. An examination of Figure 12.9, above, reveals that the channel eigenfunctions valid in the wall region carry with their representation of the wall layer itself a substantial weighted average of the dynamical behavior in the main body of the channel. Thus, while an unknown pressure signal need not be supplied to close the model of [394], the ODE coefficients themselves include outer flow effects in a form which is difficult to disentangle and interpret.

The approximation of wall eigenfunctions by combinations of full channel eigenfunctions is open to two further physical objections. First, it is Reynolds number dependent. The fact that the whole channel is used at once introduces Reynolds number dependency into the eigenfunctions and hence to the numerical values of the ODE coefficients. The universal structure of the near-wall layer, in wall units, is ostensibly lost. Second, the approximation of wall eigenfunctions as linear combinations of channel eigenfunctions is singular in the limit of high Reynolds numbers. It is well known that the dominant wall eigenfunctions exhibit one or more turning points in the wall region (Figure 10.3). In contrast, the leading channel eigenfunctions start turning much closer to the center of the channel. In the limit of high Reynolds number, all effectively computable channel eigenfunctions will be very close to linear in the wall region. Combinations of such "almost linear" eigenfunctions will not be able to reproduce the necessary near-wall structure, including the streamwise rolls. Zhou and Sirovich's study is carried out at relatively low Reynolds number, and the wall region occupies an appreciable part of the channel (see Figure 12.9(a)), so this issue does not arise in their computations, but generally one would expect the truncated matrix representing the transformation relating full channel to wall eigenfunctions to become nearly singular at high Reynolds numbers. Even in their low Reynolds number case, as we have noted, the authors of [394] have to include about 1000 eigenfunctions to produce in the order of 10–50 (real) wall basis functions.

12.6 Flows in internal combustion engines

In reciprocating machinery, not only are fluid flows dominated by time-periodic structures, but the fluid domain itself is time-periodic. In [111] and [110] Fogleman *et al.* developed phase-invariant and phase-averaged POD methods to treat such flows, and demonstrated their use on experimental particle image velocimetry (PIV) data and CFD simulations of cylinder flows in internal combustion (IC) engines. Here a major interest is in describing and analyzing the large scale swirling or tumbling motions induced during the intake cycle, since these dominate the mixing of fuel and air, thereby influencing efficiency and emissions. In [247] PIV data from a rig equipped with an injector nozzle were analyzed using phase-averaged PODs.

Two three-dimensional DNS databases were used, computed for a 6:1 compression ratio cylinder with a single valve, placed either centrally or offset to one side, acting as both intake and exhaust [152, 153]. Figure 12.10 shows the geometry and parameters for the central-valve case. In [110, 111] only the intake and compression strokes (respectively 0–180 and 180−360 crank-angle degrees (CAD)) were analyzed. Velocity fields were stored at every 3 CAD, and computations run through 33 full 4-stroke cycles (720 CAD each), but without modeling ignition and firing. We briefly describe the phase-invariant POD method and then focus on the second of two phase-averaged POD methods and the low-dimensional model derived from it in [110]. For background and early history of IC engines and vortical motions in cylinders, see [225].

A phase-invariant POD for periodically-forced flows

Since the flow is approximately periodic, to compute empirical eigenfunctions for a given phase in the 4-stroke cycle one can select a crank angle and average over all realizations of the correlation tensor at that angle, thereby fixing a particular domain geometry. This was done for 120 velocity fields at 3 CAD intervals, using the method of snapshots (Section 3.4)

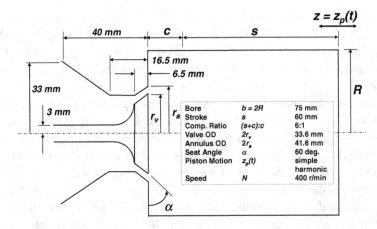

Figure 12.10 Cylinder geometry for the central-valve IC engine simulation; the piston face, at right, moves sinusoidally. Reproduced with permission from [110].

on 32 cycles of data over the intake and compression strokes (the first cycle was discarded to allow transients to decay). Averaging over the resulting sets of eigenvalues, one finds that the first modes capture on average only 12.7% and 10.6% of the kinetic energy for the central- and offset-valve cases respectively, and that ten to twenty modes are required to capture 75% [111, Figures 9 and 11].

During the first half of the intake stroke and the final third of compression the first mode contains substantially more energy (up to 30% for the central-valve case), showing evidence of a strong intake jet and tumble vortex breakup. The cyclic variation in the number of modes required to capture a given fraction of energy allows one to quantify the flow's complexity, and to assess the contribution of large coherent structures. However, in spite of their value in data analysis, phase-invariant modes are unsuitable for construction of low-dimensional models, which must be defined on subspaces spanned by a single set of basis functions. To obtain such a set the phase-dependent velocity fields can be averaged over an appropriate part of the engine cycle. A similar approach was used in [366].

Phase-averaged PODs

The first method presented in [110, 111] averages over the intake and compression strokes. The velocity fields are first mapped onto a fixed spatial domain. This was facilitated by use of a CFD grid that stretched linearly in the axial direction (z: Figure 12.10) while preserving the number of grid points: one simply rescales the axial velocity component W proportionally to the stretch:

$$W_{\text{fixed}}(\mathbf{x}) = \left(\frac{L_{\text{fixed}}}{L_{\text{original}}} \right) W_{\text{original}}(\mathbf{x}). \qquad (12.1)$$

To minimize the artificial effect of changing kinetic energy due to domain size, the mid-stroke grid was chosen as the reference phase L_{fixed}. To account for the considerable variation in phase-invariant modal energies revealed by the eigenvalue analysis summarized above, it remains to weight the contributions of the different phases appropriately.

Prior to forming the correlation tensor, each velocity snapshot was normalized by dividing by the square root of its total energy. Lacking such normalization, the first few modes could only represent structures which exist during the highest energy phases of the flow.

The resulting POD modes can equally approximate structures from all CAD phases, but they are still subject to the problem that, while divergence is preserved during stretching and normalization of the velocity snapshots, each snapshot did not start with the same divergence. The modes therefore do not have constant divergence and it is difficult to match the total divergence of each mode with its velocity boundary conditions (recall that there is inflow to the domain during the intake stroke).

Fogleman [110] therefore defined a second set of phase-averaged modes using only velocity fields during the compression stroke (180−360 CAD). This also simplifies the pressure term calculation in projection of the Navier–Stokes equations since divergence is almost spatially uniform during compression. The spatially-uniform component can be removed from the original snapshots by subtracting a divergence field with a single axial velocity component equal to the piston velocity at its face and decreasing linearly to zero at the top of the cylinder, yielding an almost-divergence-free velocity field whose wall-normal components vanish at all boundaries, including the piston face. Although the linear divergence mode, shown in Figure 12.11(a), is not a POD mode, it must be included in the basis for low-dimensional models since it is needed to reconstruct the original velocity fields and to describe energy flow from the piston to the other modes.

Cumulative eigenvalues for the two cases are shown in Figure 12.13 below, after subtraction of the linearly-divergent field. Over 60% of the divergence-free energy is now captured by the first 5 modes in the offset-valve case, and 35% in the central-valve case. Figures 12.11 and 12.12 show cross sections of the first four modes in the plane of maximum tumble for both cases. The second modes (Figures 12.11(b) and 12.12(a) actually contain the most energy. Complexity, as quantified by the number of zero-crossings of

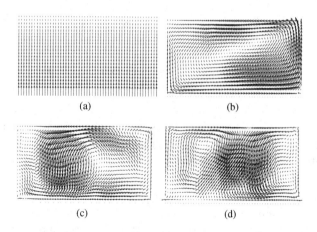

(a)   (b)

(c)   (d)

Figure 12.11 The first four POD modes shown in tumble cross section. (a) Linear divergence mode 1 for both central- and offset-valve cases, ((b)–(d)) modes 2–4 for central-valve case. Reproduced with permission from [110].

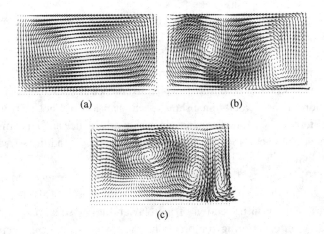

(a)   (b)

(c)

Figure 12.12 Modes 2 (a), 3 (b), and 4 (c) for offset-valve case shown in tumble cross section; note that they contain 1, 2, and 3 vortices respectively. Reproduced with permission from [110].

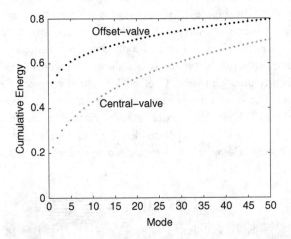

Figure 12.13 Cumulative eigenvalues of the compression stroke phase-invariant POD. Reproduced with permission from [110].

velocity components, increases with mode number. The only instance of non-zero velocity at a boundary is apparent in mode 1, in which axial velocity is uniform across the piston face. This simplifies the choice of velocity boundary condition in deriving low-dimensional models.

Projecting the rescaled velocity fields onto these modes and averaging over the 32 compression cycles produces the time histories of each modal coefficient shown in Figure 12.14(a) and (b). At bottom dead center (BDC, 0 degrees) most of the energy is initially contained in the tumble vortex (mode 2), with some energy in the higher-order modes, most likely representing turbulence remaining from the intake stroke. Mode 2 at first decays as compression proceeds, although it persists longer and even modestly strengthens in

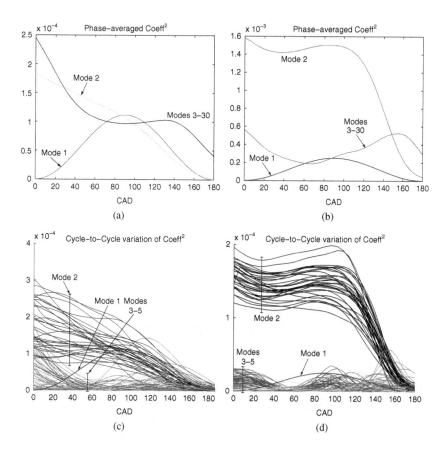

Figure 12.14 Squared coefficients of modes 1–3 averaged over 32 compression cycles, for the central-valve (a) and offset-valve (b) cases. Panels (c) and (d) show cycle-to-cycle variability of the squared coefficients of modes 1–5. Note that vertical scales differ. Reproduced with permission from [110].

the offset-valve case. As piston speeds decrease towards the end of compression, mode 1 becomes weaker and the tumble vortex collapses, passing its energy to higher order modes as it breaks into smaller structures.

Figures 12.14(c) and (d) show the cyclic variability of modal energies. By definition there is no variation in the linear piston-speed-dependent divergence field of mode 1, but substantial variability is evident in the strength of the tumble vortex at BDC in both cases, and in the higher modes activated near the end of compression, although variability decays sharply as the piston approaches top dead center. To further analyze the tumble breakdown process, times series of the averaged coefficients of modes 3–6 for the less-variable offset-valve case were compared with CAD values at which the aspect ratio of the domain passes integer values. It was found that at aspect ratio 2, mode 3, with 2 vortices (Figure 12.12(b)) peaks in energy; similarly modes 4, 5, and 6 peak near aspect ratios 3, 4, and 5, although the correspondence is not perfect, see Figure 12.15. This suggests that energy flow through higher order modes depends on the geometry and number of vortices present.

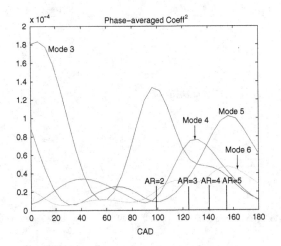

Figure 12.15 Squared coefficients of modes 3–6 compared to geometric aspect ratio (vertical bars marked AR). Reproduced with permission from [110].

A low-dimensional model

In [110] the second set of POD modes described above was used to construct a low-dimensional model of flow during the compression stroke. Projection of the Navier–Stokes equations proceeds much as in our previous examples except that the time-dependent domain geometry implies that the acceleration term becomes

$$\frac{\partial \mathbf{u}}{\partial t} \mapsto \frac{\partial \mathbf{u}}{\partial t} + \frac{\partial \mathbf{u}}{\partial z}\frac{\partial z}{\partial t}, \tag{12.2}$$

where z is the cylinder axis (Figure 12.10). The decreasing axial dimension must similarly be accounted for in computing spatial gradients. However, since the time dependence of mode 1 (Figure 12.11(a)) is specified, and satisfies the velocity boundary condition $\dot{z}_p(t)$ on the piston face, the pressure term still drops out because the remaining modes are divergence-free.

The projected ODEs take the form

$$\sum_{j=1}^{K} I_{jk}\dot{a}_j = \sum_{j=1}^{K} M_{ij}a_j + \sum_{j,k=1}^{K} T_{ijk}a_j a_k + \nu_T D_i a_i \, , \ i = 2, \ldots K, \tag{12.3}$$

with the prescribed modal coefficient $a_1(t) = f(t)$ acting as a forcing function. Here the matrix M_{ij} contains the additional acceleration terms that follow from (12.2):

$$M_{ij} = \int_{\Omega} \phi_i \frac{\partial \phi_j}{\partial z}\frac{\partial z}{\partial t}\, \mathbf{dx}. \tag{12.4}$$

In (12.3) I_{ij} is non-diagonal, since inner products between modes do not identically vanish, due to their interpolation on a fixed grid that differs from the CFD mesh, T_{ijk} are the usual conservative energy transport terms, and $\nu_T D_i$ represents viscous dissipation (possibly including a model for interactions with unresolved modes). Since the empirical eigenfunctions are 3-dimensional, there is only a single modal index.

Integration was started from initial conditions that match the averaged modal coefficients of Figures 12.14(a,b) and 12.15 at BDC, and using a 6-mode truncation with high effective viscosity ν_T the model qualitatively reproduced the compression-stroke time series of those figures. The model was then used to study the effects of changes in truncation value K, engine speed, and initial conditions such as tumble vortex strength. The methods of [301] were also used to describe modal energy transfer. See [110, Chapter 5] for details.

Comments

This model makes a significant step towards extending POD analyses and low-dimensional modeling to time-varying geometries, but it is lacking in several respects. Only the compression stroke is represented: to construct a full 4-cycle engine model one must approximate inlet and exhaust valve boundary conditions and deal with the resulting non-uniform divergence fields. This problem must be overcome if models are to be used to analyze the mixing properties of different fuel injection systems, for example, to build on the results of [247].

However, many time-dependent geometries do not involve such complications. For example, phase-dependent and phase-averaged POD modes will probably be useful in analysis and modeling of rhythmic bio-fluid motions and the resulting unsteady but approximately-periodic velocity fields (see [167] for an example of phase-averaged PIV in anguilliform swimming).

12.7 A miscellany of results: 1995–2011

It would be excessive to claim an *explosion* of interest in POD-based low-dimensional modeling of turbulence and fluid instabilities, but many papers on this topic have appeared since the first edition of this book in 1996. As noted in the preface to this second edition, some developments that seem of general interest have been incorporated at appropriate points throughout the preceding text, most notably in the new Chapter 5. Here, with apologies to those whose work we have not been able to include, we provide an incomplete list of other studies, partial in at least two senses of the word.

Shear flows in channels

Numerous papers on simple channel- and shear-flow geometries have appeared in which a picture emerges of turbulent or "pre-turbulent" attractors involving repeated visits to unstable periodic orbits and fixed points. Some of these studies employ direct numerical simulations (e.g., [100, 102, 103, 151, 185, 255, 327, 380, 381]) and some propose low-dimensional models (e.g., [100, 327, 377–379]). In others, stable manifolds of unstable periodic solutions are identified as possible boundaries of the basin of attraction for a stable laminar state that shrinks as Reynolds number increases [100, 326, 327].

In a trio of papers [346–348] motivated by the above work, Smith *et al.* developed and analyzed models for plane Couette flow in a minimal flow unit [175]. In addition to streamwise and spanwise translation symmetries (as in the boundary layer model of Chapters 10 and 11), they pay careful attention to the discrete symmetries of reflection about the domain's spanwise midplane and rotation about its spanwise axis (isomorphic to the group D_2 [134]), expanding the ensemble of fields used to construct the correlation tensor as proposed in Section 3.3.3 [23, 48, 330].

In [347] bifurcation diagrams are used to compare the behavior of low-dimensional models with numerical simulations [102], and in the largest such (31-dimensional) model each empirical eigenfunction is split into a pair of mutually orthogonal components to relax the constraint implicit in the POD process noted in Section 4.4, namely, that it can only reproduce mixtures of the input data. Here this procedure is necessary to restore stability to the laminar solution; a related method based on Squire's coordinates [351] is used in [179] to improve the representation of turbulent energy and Reynolds stress terms. Evidence of heteroclinic cycles similar to those of [22] emerges in [347, 348], and in a severely truncated 3-dimensional model [345].

While not using low-dimensional models per se, Gibson and his colleagues [120, 148, 149] employ dynamical systems ideas to visualize numerical solutions of plane Couette flow by projecting them into dynamically-invariant subspaces that are also representation-invariant. Such state space visualizations utilize symmetries of the system that allow them to reveal orbits leaving unstable equilibria and passing near saddle points en route to attractors. Traveling waves can also be shown. Orbits that pass through transient turbulence to relaminarize, or that display sustained quasiperiodic or turbulent behavior, can then be superimposed on the "skeleton" of heteroclinic connections among these unstable stationary and periodic states.

Overall, this work further supports the roll-streak-breakdown picture of shear-driven turbulence [151, 382], suggesting that unstable states and periodic orbits form a backbone for turbulence in a minimal flow unit [185], and even vindicating the early discovery of attracting heteroclinic cycles in [22].

12.8 Discussion

A number of common features emerge from the boundary layer modeling described in Chapters 10 and 11 and the work summarized in the present chapter. Low-dimensional models based on the dominant empirical eigenfunctions can reproduce the behavior of temporally simple systems, such as the forced mixing layer and low Reynolds number complex geometry flows, remarkably well. They display good quantitative short-time tracking capabilities, suggesting that they may be useful as models with sufficient predictive capacity for feedback control purposes, or for use in Kalman filters. Using phase averaging, the methodology can also be extended to time-periodic flows such as the internal combustion example discussed in Section 12.6.

For flows of greater (spatio-) temporal complexity, such as the turbulent boundary layer and jets, one should perhaps be more modest, seeking at first only *qualitative* understanding. However, as the discussion of Section 11.6 indicates, we believe that these models can provide important information on key mechanisms such as the burst-sweep cycle. The work on the transitional boundary layer of Section 12.2, an intermediate case in complexity between the "simple" flows of Sections 12.3 and 12.4 and fully developed turbulence, supports this view. The low-dimensional models enormously simplify and significantly restrict the full behavior, but such a simplification and the identification of key features which it permits are just what we set out to achieve.

In all cases, it seems important to include sufficiently many low-(average) energy modes to afford reasonable energy transfer and dissipation mechanisms, or to model these as suggested in Chapter 10. Modes which are inactive on average may well undergo brief bursts of high-energy action in which they play a crucial dynamical rôle. Indeed, this is essentially what the odd Fourier modes do during heteroclinic cycling in the five mode model of [22].

As we have indicated in Section 12.5, the approach of Zhou and Sirovich to the wall layer per se may be questioned on physical grounds. Nonetheless, their idea of deriving a set of basis functions which focus on a local subdomain within a spatially extended field, while satisfying boundary conditions on the larger domain, is of particular interest in fluid problems, where the pressure exerts a global constraint via incompressibility. In mathematical terms, one seeks functions whose support is concentrated in the subdomain, but which, to satisfy global constraints and boundary conditions, are expressed in terms of some convenient basis set for the *full* domain. From this viewpoint, one could equally well approximate wall eigenfunctions in terms of, say, the Chebyshev polynomials which were used in the numerical work of [190] on which [394] draws. There seems no obvious advantage in using functions optimized on a *different* domain. The development of wavelet bases [90, 104], which combine good localization in both physical space and wavenumber, may well be relevant here. Wavelets even provide a mathematical framework in which the notion of a "simple eddy" ([368], pp. 258–9) can be applied to the decomposition of turbulent flows. As we noted in Section 8.4, wavelet bases have been used for construction of local representations of spatio-temporal chaos in model problems [42, 101, 254].

Various mathematical issues remain to be resolved, especially concerning completeness and convergence of these "local" bases on the full domain. There is a difficulty here: one is really interested in performance on the subdomain, but since functions are defined on the full domain, one can investigate such questions only on that domain. Suitably weighted norms are necessary. In physical terms, as we have remarked, one ultimately seeks a better understanding of how such weighted inner products incorporate information from outside the subdomain into the local model ODEs.

The identification of burst-sweep dynamics with a heteroclinic attractor in a spatially localized model almost certainly simplifies these complex events, but we believe that the cycle identified by [22] does provide a skeleton to which muscles and flesh will ultimately be added by models which include more modes. The work of Aubry and Sanghi [25, 323] and, in part, that of Zhou and Sirovich [394] are steps in this direction. A more realistic picture yet, might include a "sea" of such models, loosely coupled in space, leading to a quasirandom space-time array of bursts and sweeps, as is actually observed. Such a model would also presumably provide better quantitative agreement with experimental observations of phenomena such as burst-sweep durations (see Section 11.6).

The suggestion implicit in this notion of a weakly coupled sea may be practically useful. It implies that we may be able to make local modifications to an open flow, based on local information, to achieve local results, without having the full flow field (system state) in hand. The evidence amassed above shows that local models of extended flows with coherent structure can indeed capture key features of individual structures, implying that such models may be useful in designing local control algorithms for turbulent flows. Such an effort for the boundary layer has already begun. In [78–81], Coller *et al.* address mathematical and control-theoretic issues relevant to stabilizing heteroclinic cycles in the presence of noise. In these papers, motivated by the idea that controlled streamwise vorticity can be added in the wall region, models of the type studied in Sections 11.1–11.4 are considered. Controlled vorticity, essentially expressed as a cross-stream velocity profile [81], adds rotational terms to the ODEs which permit one to steer incoming solutions closer to the unstable saddle points in the cycle and hence to delay bursting, see [274]. Such vorticity might be created by suction and blowing, or by micro-mechanical actuators, see [30].

In [38, 41, 47] control of heteroclinic cycles is considered for ODE models derived to include the effects of actuators which change the domain geometry. For the actuators considered, $O(2)$ symmetry is broken to reflection symmetry and it is shown how a fixed actuator location precludes control of events having a singular phase relation to it. In [47] Berkooz *et al.* demonstrate and emphasize the need for nonlinear estimation of dynamics in phase space from limited sensor information.

13

Review: prospects for rigor

As we near the end of our story, the reader will now appreciate that there are many steps in the process of reducing the Navier–Stokes equations to a low-dimensional model for the dynamics of coherent structures. Some of these involve purely mathematical issues, but most require an interplay among physical considerations, judgement, and mathematical tractability. While our development of a general strategy for constructing low-dimensional models has been based on theoretical developments such as the POD and dynamical systems methods, the general theory is still sketchy and, in specific applications, many details remain unresolved.

The mathematical techniques we have drawn on lie primarily in probability and dynamical systems theory. In this closing chapter we review some aspects of the reduction process and attempt to put them into context. Some prospects for rigor in the reduction process are also mentioned. This is by no means a comprehensive review or discussion of future work; instead, we have chosen to highlight a few applications of dynamical and probabilistic ideas to illustrate lines along which a general theory might be further developed.

We start by discussing some desirable properties for low-dimensional models, and criteria by which they might be judged. We then outline in Section 13.2 an a-priori short-term tracking estimate which describes, in a probabilistic context, how rapidly typical solutions of the model equations are expected to diverge from those of the full Navier–Stokes equations restricted to the model domain. Here and in the following section we view low-dimensional models as perturbations of the full evolution equations. Section 13.3 also addresses reproduction of statistics by low-dimensional models. In Section 13.4 we revisit a problem that arose in Chapter 10, and sketch how the notion of inertial manifolds could be used to show that models defined on "small" subdomains of the fluid flow can be well posed as PDEs, in spite of the difficulty of prescribing boundary conditions. Section 13.5 returns to general issues, reviewing the methods in broad terms and suggesting where low-dimensional models may have their greatest impact in turbulence studies.

13.1 The quality of models

In studying the dynamics of coherent structures we desire our model to exhibit one or more of the following properties:

345

Short-term tracking

By short-term tracking we mean that, given a realization of the fluid velocity field, and using the state of the coherent structures in that realization as an initial condition for the low-dimensional model, then the dynamics of the model should track the state of the coherent structures in the full fluid flow in the short term. The *error* or difference between the real and modeled flows can be measured at each point in time by any suitable phase space norm. (This generally implies spatial averages – L^2-norms – over the physical domain of the model: see "Localization in physical space" below.)

Errors will almost certainly grow with time. We wish to control them in terms of the number of modes included, the size of the domain, and other parameters of the model, and wish thus to be able to bound a priori the tracking error of the model as a function of time. We may be interested either in an "average" tracking error, i.e. the error growth over time when averaged over all likely initial conditions, or an instantaneous error for a given initial condition. Short-time errors may be bounded in terms of the difference between the model dynamical system and the projection of the actual vector field onto the phase space of the model. The closer the two vector fields, the smaller the short-time tracking error. We describe such a result in Section 13.2.

Long-term tracking and inertial manifolds

Long-term tracking implies that, given a realization of the fluid flow, an initial condition may be picked in the phase space of the model so that, as both the model solution and the full flow evolve, the difference between them *decreases* in time. (Ideally, one desires an exponential decrease.) It is a very stringent requirement: much more demanding than good short-term tracking, for it implies that the model in some way approximates (part of) the attractor of the full equations. This might seem a baseless hope, but for the fact that a large class of parabolic PDEs are now known to possess *inertial manifolds* [85, 114, 367]. These are smooth, compact, invariant, finite-dimensional manifolds in the (infinite-dimensional) phase spaces of such PDEs, which attract *all* solutions at an exponential rate. They can be seen as global versions of the center manifolds introduced in the context of local bifurcations in Chapter 6, but they contain locally unstable orbits and generally a set of stable modes as well.

Evidently, if such an inertial manifold exists, it must contain all the attractors of the PDE, as well as any unstable non-wandering sets and their unstable manifolds. Thus, solutions lying in it are capable of reproducing all the long-term behavior of the system. In this situation, a low-dimensional model attempts to approximate the vector field restricted to the inertial manifold. Such systems have been called *approximate inertial forms* [115, 369]. We briefly discuss inertial manifolds in the context of localization in Section 13.4, below, but this is an active field and a proper introduction would go far beyond the scope of this book. Proofs of the existence of inertial manifolds have been carried through for the two-dimensional Navier–Stokes equations in various special geometries (periodic boundary conditions, etc.).

If the attractors in question are strange, containing solutions that locally separate exponentially fast, as described in Section 6.5, then it is of course futile to ask that *individual* solutions of the model and the full system, started at the "same" initial condition, remain close. Almost all solutions of the full system itself, even if started nearby on the attractor, rapidly diverge. One can therefore only hope for results of shadowing type [55, 140, 144], in which one proves the existence of *some* model solution which tracks a given full solution. In such an analysis, the local separation of solutions is used to show that an initial condition can be selected for the model such that the solution based there follows the desired path, making the "right" turns at each stage. Much as in the doubling map example of Section 6.5, the existence of solutions with all possible symbol sequences implies that one can pick a sequence with any desired behavior.

As we remarked in Chapter 4, the range of wavenumbers active in fully developed turbulence is so broad that any inertial manifolds will be of enormous dimension. They are thus unlikely to be of *direct* help in deriving low-dimensional models, but as we indicate in Section 13.4, they can provide useful theoretical underpinnings.

Good statistics

By "good statistics" we mean that, starting from a given initial condition, the long-time dynamics of the model produce statistics that are close to those of the original system. We outline four approaches that can ensure that a dynamical model produces good statistics.

1. "Statistical structural stability" (SSS). This is a mathematical theory presented in [273]. SSS describes conditions under which perturbations of a system will retain statistics close to those of the original system. For our purposes, we view the model as a perturbation of the original system. It is hard to prove such results for the Navier–Stokes equations or, in fact, for any PDE. In [37] an attempt was made to connect these approaches with low-dimensional models, but much more work needs to be done.

2. "Time scales for convergence." One may reasonably expect that different statistics might require different elapsed times for acceptable convergence. In particular, the mean is likely to converge most rapidly, with higher order moments with "long tails" (i.e. low probability events that alter the statistics in a significant way) requiring longer averaging times, in view of the rare events that define them. It is hard to estimate a priori such times for convergence. In turbulence some of these issues were addressed in [221]. For general dynamical systems, early steps in this direction are described in [140].

3. "Matching the PDF of the resolved modes." Since we are interested primarily in time-averaged or "one point in time" statistics, such as means, moments, two point correlations, etc., it is sufficient to approximate the probability density function of the resolved modes, as this determines all such time-averaged statistics. (The formalism can be enhanced to deal with time-dependent flows.) To obtain good statistics, one can therefore appeal to a low-dimensional model whose density transport equation is identical to the marginal density transport equation. This

approach shares some of the philosophy behind PDF methods [284]. We discuss it further in Section 13.3.

4. "Long-term tracking." Provided that it is ergodic, a model with the long-term tracking property discussed above will obviously reproduce *all* statistics, since it essentially contains the global attractor of the original system. However, unless we are concerned only with weakly unstable, "preturbulent" flows, such models are unlikely to qualify as low-dimensional.

Turbulence models such as Reynolds stress closures suggest a further avenue for reproducing statistics. Reynolds stress models replace portions of phase space (typically the high wavenumber range of the spectrum) by moments of the fluctuating velocity field. As of 1966, this approach did not seem to have been examined from a dynamical systems viewpoint. In terms of probability measures on the attractor, which we introduce and use in the next two sections, this corresponds to averaging over the fibers above each point in the phase space of resolved velocity fields and replacing the full dynamics of the neglected modes by an algebraic or differential model which expresses moments of those modes in terms of the resolved modes.

Localization in physical space

Localization in physical space implies that the model describes the interaction of coherent structures in a *limited* region of physical space. This is particularly important if we seek a low-dimensional description, for as we have stressed repeatedly, the turbulent flows of interest in this book are "open" flows on large or notionally infinite domains. When the domain greatly exceeds the size of typical structures, one observes a "sea" of many interacting structures (see Figure 2.14, for instance), the description of which would require very many modes, even if each individual structure is economically described by just a few. We recall Figure 4.1, which illustrates how the number of POD basis functions needed to capture a given fraction of kinetic energy increases rapidly as one moves away from the wall in a turbulent boundary layer.

The desire for localization immediately raises the question of how to specify or model boundary conditions at the edges of the resolved region of physical space. We have already encountered this question in the context of the boundary layer study discussed in Chapter 10. As we saw there, the Galerkin projection method is most suited to cases in which the basis functions satisfy the natural boundary conditions of the problem, and one is therefore faced with modeling boundary conditions or otherwise attempting to replace missing spatial locations outside the model domain. We discuss this further in Section 13.4.

Resolution of coherent structures alone

In modeling the dynamics of coherent structures, it is important to resolve and separate them both from the mean flow, and from the smaller scales. Such triple decompositions have been proposed in turbulence studies in the past [168], and they have been used in derivation of optimal energy-bearing modes from first principles by energy growth arguments [283]. As of 1996, the models of the wall layer described in Chapters 10 and 11 and Section 12.5 and the jet studies of Glauser and co-workers (Section 12.1), had used dynamical models to account for the effect of coherent structures on the mean flow, but the only theory that attempted to substantiate a model for the interaction of coherent structures with the mean flow was that of [22] and [44], presented in Chapter 10 (Section 10.8). In that work, energy transfer to neglected modes was accounted for by simple Heisenberg type eddy viscosity models. More work on such models appeared in [323], and in this second edition we have referenced further, more recent studies that employ shift modes to model interactions (e.g. [207,218], and see Sections 4.4 and 10.8). However, additional work on both these issues is required.

13.2 A short-time tracking estimate

In this section we present an abstract (and rather technical) analysis of short-time tracking. We mention "pointwise" estimates that correspond to the divergence of model solutions from the projection of full solutions onto the phase space of the model for particular initial conditions, but concentrate on results relating to an average over all initial conditions likely to occur. In the latter, averaging is done with respect to the natural probability measure of the long-term states of the flow. The material in this section originally appeared in [37] and later in [89], where the short-term tracking question was also explored numerically.

We can consider the problem of isolating the Navier–Stokes equation on a subdomain in physical space in the following light. (Indeed, the separation of scales in wavenumber space and retention of the coherent structures alone can also be viewed in the same way.) Let Ω denote the full domain and $\Omega' \subset \Omega$ the subdomain, and consider the space $L^2(\Omega')$ of (incompressible) velocity fields defined on Ω'. This function space contains two subspaces: (1) the subspace of time-dependent boundary conditions arising from restriction of a given solution on Ω to Ω'; and (2) the subspace of velocity fields with boundary conditions imposed by the model, which represent the coherent structures alone. We shall denote the latter (finite-dimensional) subspace by R and velocity fields in it $u_<$, with $u_> \in R^\perp$ being the orthogonal complement: the "neglected modes." These include the correct boundary conditions as well as incoherent turbulent motions.

We now decompose $L^2(\Omega')$ as

$$L^2(\Omega') = \cup_{u_< \in R} u_< + R^\perp.$$

This in turn induces a decomposition of the invariant measure (or probability density) on the restriction of the full attractor in $L^2(\Omega')$ into a *measurable partition*; the associated measures being μ_R, the marginal on R, and $\mu_{u_<}(u_>)$, the conditional measure on the fibers "above" each point $u_< \in R$. (See the material on invariant measures in Sections 3.3 and 3.8 and, in greater detail, in [312].) Each such fiber is an identical copy of the space $R^\perp$, displaced by a coherent structure state $u_<$. (Here for simplicity we adopt a binary decomposition, instead of the triple decomposition that would be more appropriate.) Finally, we denote by P the projection into R.

In this context *short-time tracking* means the following: suppose the full system is started from an initial condition corresponding to the full flow field in Ω, restricted to

Ω': $u_0 \in L^2(\Omega')$. The model is started from the projection of u_0 on R: $u_{0<} = P(u_0)$. The decomposition $u = u_< + u_>$ splits u into the "coherent" component $u_<$ and its orthogonal complement, which has no coherent part. The orthogonal component is equal to the difference between the full state of the flow (restricted to Ω') and the coherent component of that particular state. The optimality property of the POD guarantees that this difference is, on average, as small as possible for any subspace with dimension equal to that of R.

Low-dimensional models typically exhibit divergence between the solution $v(t)$ of the model started at $u_{0<} \in R$ and the projection of the solution $u(t)$ of the full system started at u_0. We denote by $\epsilon(t)$ the discrepancy between $v(t)$ and the projection of the evolution of u_0 under the full system at time t. Working in the phase space $L^2(\Omega')$ implies that $\epsilon(t) = \|v(t) - P(u(t))\|$ is spatially averaged *over the subdomain* Ω', but in terms of the phase space, $\epsilon(t)$ represents a pointwise error. As we indicated above, especially when strange attractors and chaotic solutions are present, the divergence of individual solutions in phase space is too stringent a notion to be of much use. Consequently, we examine the growth of $\epsilon(t)$ as a function of time, *additionally averaged over all appropriate initial conditions*. This characterizes the short-time tracking capability of the model.

The need to study short-time tracking is motivated by the desire to control the dynamics of coherent structures by micromanipulators or other devices which move fluid around. For systems in which there may be a significant time lag between identification of the state of the system and the control action, one must forecast the state of the system at the (future) time when control will be effected. Thus the controller must be capable of short-term predictions.

Let $\dot{u} = F(u)$ be an evolution equation in $L^2(\Omega)$ and $\dot{v} = G(v)$ be an evolution equation in R – the low-dimensional model of $\dot{u} = F(u)$. We define the *short-time divergence constant* ϵ^* as follows:

$$\epsilon^* = \int_R \left[\int_{R^\perp} \|G(v) - PF(v + u_>)\| \, d\mu_v(u_>) \right] d\mu_R(v). \qquad (13.1)$$

Here ϵ^* is a measure of the average difference between the model vector field and the projection of the full vector field onto the subspace of the model.

We next define $\epsilon_s(t)$, the *average short-time tracking discrepancy at time* t:

$$\epsilon_s(t) = \int_R \left[\int_{R^\perp} \epsilon(t) \, d\mu_v(u_>) \right] d\mu_R(v)$$

$$= \int_R \left[\int_{R^\perp} \|v(t) - Pu(t)\|_{v(0)=v_0, u(0)=v_0+u_>} d\mu_v(u_>) \right] d\mu_R(v),$$

$$(13.2)$$

where $v(t)$ is the solution to the model $\dot{v} = G(v)$ and $u(t)$ is the solution to the full system $\dot{u} = F(u)$.

As we have noted, this latter quantity, averaged with respect to the natural measure on the attractor, seems the most suitable to gauge discrepancies between model and full solutions. Equipped with the short-time divergence constant, we can now prove an ensemble averaged generalized Gronwall inequality that controls $\epsilon_s(t)$ in terms of ϵ^*. In particular, we show that if L is the Lipschitz constant of G – the function defining the low-dimensional model – then we have

$$\epsilon_s(t) \leq \frac{\epsilon^*}{L} \exp(Lt). \tag{13.3}$$

Note that, while the linear term depends on averaged properties of the full solution restricted to Ω', the exponential growth factor depends only on the functions defining the model vector field.

To derive (13.3), we start with a pointwise Gronwall inequality for projections versus models. This provides control of errors for individual initial conditions. Let

$$\xi(v_0, u_>)(t) = \|v(t) - Pu(t)\|_{v(0)=v_0,u(0)=v_0+u_>}$$

$$= \left\| \int_0^t G(v(s)) - PF(u(s)) \, ds \right\| \tag{13.4}$$

be a formal expression for the difference between the model and the projected solution as a function of time. Observe that the right-hand side of (13.4) may be written

$$\left\| \int_0^t [G(v(s)) - G(Pu(s)) + G(Pu(s)) - PF(u(s))] \, ds \right\|, \tag{13.5}$$

which in turn is bounded above by

$$\int_0^t L\xi(s) \, ds + \int_0^t \|PF(u(s)) - G(Pu(s))\| \, ds. \tag{13.6}$$

The expression resulting from substitution of (13.6) in (13.4) is exactly in the form required to apply the Gronwall lemma (see [77], p. 37, [144], p. 169 and Section 4.4, Equations (4.80, 4.81), above). Applying the lemma, we obtain the pointwise bound at time t:

$$\xi(v_0, u_>)(t) \leq \int_0^t \|PF(u(s)) - G(Pu(s))\| \exp[L(t - s)] \, ds. \tag{13.7}$$

Integrating (13.7) with respect to the two components of the invariant measure, we therefore deduce that

$$\int_R \left[\int_{R^\perp} \xi(v_0, u_>)(t) d\mu_{v_0}(u_>) \right] d\mu_R(v_0)$$

$$\leq \int_R \left(\int_{R^\perp} \left\{ \int_0^t \|PF(u(s)) - G(Pu(s))\|_{u(0)=v_0+u_>} \right. \right.$$

$$\left. \left. \times e^{[L(t-s)]} \, ds \right\} d\mu_{v_0}(u_>) \right) d\mu_R(v_0). \tag{13.8}$$

Fubini's theorem allows us to exchange the order of the integrals, yielding

$$\int_0^t \int_R \left[\int_{R^\perp} \|PF(u(s)) - G(Pu(s))\|_{u(0)=v_0+u_>} d\mu_{v_0}(u_>) \right] d\mu_R(v_0)$$

$$\times \exp[L(t - s)] \, ds. \tag{13.9}$$

From the invariance of the measure μ under the solution semigroup corresponding to $S_t(u(0)) = u(t)$ (i.e. the fact that μ is a stationary measure), we deduce that the inner two integrals are independent of s (this is the non-trivial point of the proof). We can therefore use the definition of the short-time tracking constant to derive (13.3):

$$\epsilon_s(t) \le \int_0^t \epsilon^* \exp[L(t-s)]\,ds \le \frac{\epsilon^*}{L} \exp(Lt). \tag{13.10}$$

This result gives us some control over short-time tracking, which is what we set out to achieve. The term $\|PF(u(s)) - G(Pu(s))\|$ in (13.7) provides a clear picture of how the tracking error arises from the difference between the projection of the full field, $PF(u(s))$, and the low-dimensional evolution of the projected field $G(Pu(s))$.

13.3 Stability, simulations, and statistics

Although in the preceding section we have treated low-dimensional models as perturbations of a "full" system, in making such models of turbulent flows we are vastly simplifying the true dynamics. Indeed, the tracking estimate (13.3) suggests, in the worst case, exponential divergence, so we cannot expect solutions of models to follow true solutions for long. However, we can still hope that the former may reproduce some of the statistics of the latter. In the present section, following [39], we outline an approach to this question.

We start with a cautionary example – a one-dimensional map similar to the doubling map of Section 6.5 – which shows that even ostensibly small perturbations can display very different long-term behavior. Instead of doubling, we consider quadrupling: the map $x \mapsto f(x) = 4x \bmod 1$, defined on the unit interval $I = [0, 1]$. Modestly generalizing the analysis of Section 6.5, it is not hard to see that f is chaotic in the sense that almost all orbits realize non-periodic sequences in four symbols, corresponding to their visits to the four quarters of I. We next perturb f to the map f_p, defined by:

$$x \to f_p(x) = \begin{cases} 8x - 4\frac{1}{2} & x \in \left[\frac{5}{8}, \frac{247}{384}\right], \\ \frac{1}{2}x + \frac{1}{3} & x \in \left[\frac{247}{384}, \frac{265}{384}\right], \\ 8x - 4\frac{5}{6} & x \in \left[\frac{265}{384}, \frac{17}{24}\right], \\ 4x \bmod 1 & \text{otherwise}, \end{cases} \tag{13.11}$$

see Figure 13.1.

This map is identical to $f(x)$ outside the interval $\left[\frac{5}{8}, \frac{17}{24}\right]$, but the fixed point $x = \frac{2}{3}$, which is unstable for the original map, is stable for $f_p(x)$. Moreover, $x = \frac{2}{3}$ is a *global attractor* for the perturbed map, *i.e.* almost every point in $[0,1]$ (in the sense of Lebesgue measure) approaches $x = \frac{2}{3}$. To see this, observe that $x = \frac{2}{3} \pm \frac{1}{336}$ are two unstable fixed points. All points in the interval $\left[\frac{2}{3} - \frac{1}{336}, \frac{2}{3} + \frac{1}{336}\right]$ clearly converge to $x = \frac{2}{3}$. To see that the "remainder" of $[0, 1]$ converges to $x = \frac{2}{3}$, one has to construct the return map on the interval $\left[\frac{5}{8}, \frac{17}{24}\right]$, and undertake a simple (graphical) analysis, showing that the pre-images

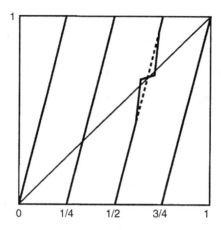

Figure 13.1 The map f_p of (13.11), with the original map f shown dashed.

of $\left[\frac{2}{3} - \frac{1}{336}, \frac{2}{3} + \frac{1}{336}\right]$ cover almost all of I. This is tangential to our discussion and is omitted.

The perturbation is localized in the interval $\left[\frac{5}{8}, \frac{17}{24}\right]$, the size of which is a parameter that may be made as small as we wish, while still keeping $x = \frac{2}{3}$ a stable fixed point. Orbit segments of f and f_p which remain outside this interval are *identical*: they suffer no short-term tracking errors. However, the long-term behaviors of most orbits of f and f_p differ utterly.

Almost all orbits of f are chaotic, but simulation of the perturbed mapping reveals trajectories with an initial chaotic transient, which eventually converge to $x = \frac{2}{3}$. The length of the transient depends on the initial condition and the size of the perturbation interval. During the transient state, finite time averages may reproduce some statistics of the original system quite well. But long-time simulations, extending well beyond the transient, yield poor reproductions of the statistics of the unperturbed system.

In what sense is the above perturbation small? It is small (or may be made as small as one wishes) in the L^2 sense (i.e., $\int_0^1 [f(x) - f_p(x)]^2\,dx$), and in the H^1 (or any H^m) sense (i.e. $\int_0^1 [f'(x) - f'_p(x)]^2\,dx$); it is even small in the L^∞ sense (i.e. $\sup |f(x) - f_p(x)|$). However, it is not small in the C^1 sense (i.e. $\sup |f'(x) - f'_p(x)|$): the slope of the map is locally changed by an order one amount.

Recall the concept of structural stability, introduced in Chapter 6. A mapping f is structurally stable if small perturbations (in the C^1 sense) are topologically equivalent to f (i.e., there exists a continuous one to one map that maps the trajectories of f to the trajectories of the perturbed map), see Section 6.3 above and Section 1.7 of [144]. The perturbation in the example above lies outside the range allowed by structural stability theory.

Allowable perturbations of structurally stable dynamical systems, one might hope, would not exhibit such discrepancies between short- and long-term behavior. Indeed, certain such systems possess a yet stronger property, namely *shadowing*, in which every trajectory of the perturbed dynamical system is closely approximated by *some* trajectory of the unperturbed system, see Section 5.3 of [144]. This may suggest that the statistics of the perturbed system should approximate those of the original system. However, structural stability and the concept of shadowing alone (see [55, 140, 144]), cannot explain the reproduction of statistics by simulations, for the following reasons:

- Many simulated systems (such as homogeneous turbulence) have symmetries which preclude them from being hyperbolic, as required for "simple" structural stability and shadowing (Chapter 7).
- The attractors for most systems of interest are thin (in the sense of dimension) in the function space. Numerical errors in the simulations may carry trajectories off the hyperbolic set, a scenario not covered by shadowing theory.
- Shadowing does not assure us that the shadowing trajectory is generic. The invariant measure on the ergodic component corresponding to the motion of most other points may be different from the measure generated by the orbit shadowing the one produced by the simulation. For example, shadowing may pick a periodic orbit as (part of) a shadowing orbit for a chaotic system. Indeed, from a theoretical point of view, when applying shadowing to computer simulations, one must recall that a computer is a finite state machine and *all* numerical trajectories are eventually periodic, see [273].

To appreciate the relevance of our simple example to fluid turbulence and simulations, we observe that in direct numerical simulations (i.e. simulations that attempt to resolve the full turbulence spectrum, cf. section 4.4 and [190]), one considers a computation well resolved if the dissipation range is resolved. The dissipation spectrum $|\nabla \mathbf{u}|^2$ is a time-averaged quantity that invokes an assumption of ergodicity, as is done often in turbulence. Resolving the dissipation range corresponds to an H^1 discrepancy being small. But in our example closeness in the H^1 sense does not guarantee good long-term statistics.

Another factor which affects the reproduction of statistics is the existence of unresolved modes. Neglect of such modes may be intentional, as in the low-dimensional models considered in this book, or in large eddy simulations [309]. In these cases models for the unresolved modes are generally included, as we have seen in Chapters 10 and 12. In other cases neglect may result from considering unresolved modes as "small" and physically insignificant, although as our example suggests, this may not be as innocent an assumption as it may appear.

We now outline the approach to this problem suggested in [39]. Consider a dynamical system with a bounded attracting set which supports an invariant measure (Section 6.5). The attracting set may be the union of several recurrent pieces, each separate attractor having its corresponding basin of attraction. The measure should correspond to time averages. Specifically, we suppose the system takes the form:

$$\left.\begin{array}{l} \dot{x} = f(x, y), \\ \dot{y} = g(x, y), \end{array}\right\} \tag{13.12}$$

where $x \in P$ is a finite-dimensional vector and $y \in Q$ may be high- or even infinite-dimensional. Here P is the subspace of the resolved modes and Q the complementary subspace of neglected modes, the "full" phase space being $P \oplus Q$. In the fluid-mechanical context, $P \oplus Q = L^2(\Omega)$.

Let μ be an invariant measure for the flow of (13.12), and δ denote its density in the sense of distributions (the existence of such a δ is assumed for the sake of simplicity of exposition; a technical problem here is that the underlying space $L^2(\Omega)$ is infinite-dimensional). Denote by ∇_x and ∇_y the gradients with respect to x and y respectively. Then it can be shown that

$$(\nabla_x \delta) \cdot f(x, y) + (\nabla_y \delta) \cdot g(x, y) = 0. \tag{13.13}$$

This is known as the Liouville equation, the stationary density transport equation, or stationary probability density function equation, see [284] and references therein. As in Section 13.2, we decompose the invariant measure μ in a manner corresponding to $P \oplus Q$. Denote by $\mu_<$ and $\delta_<$ the marginal measure and density on P, and $\mu_>(\cdot|x)$ the conditional measure on the fiber above x : $x + Q$. For the existence of such measurable partitions see [312]. Note that $\delta_<$ is now well defined, since P is finite-dimensional. It is possible to show that $\delta_<$ satisfies the equation

$$(\nabla_x \delta_<) \cdot \int_Q f(x, y) \, d\mu_>(y|x) = 0. \tag{13.14}$$

We now observe that the conditionally averaged dynamical system

$$\dot{x} = \int_Q f(x, y) \, d\mu_>(y|x) \overset{\text{def}}{=} \tilde{f}(x) \tag{13.15}$$

has the same invariant density equation (13.14) as that for the density of the marginal $\mu_<$. The boundary conditions for $\delta_<$ are zero at infinity and $\int_P \delta_<(x) \, dx = 1$. Thus (13.15) provides a model in P whose invariant density satisfies the same invariant density equation as $\delta_<$. (This was pointed out to us by Pope, see also [284]). Averaging methods in ODEs have a long history. In the simplest case of a system subject to periodic forcing fast in comparison with the autonomous time scale:

$$\dot{x} = \epsilon f(x, t), \qquad f(\cdot, t) = f(\cdot, t + T), \tag{13.16}$$

it can be shown that solutions of the averaged system

$$\dot{x} = \epsilon \frac{1}{T} \int_0^T f(x, t) \, dt \overset{\text{def}}{=} \epsilon \bar{f}(x) \tag{13.17}$$

remain $\mathcal{O}(\epsilon)$ close to those of (13.16) for times of order $(1/\epsilon)$. For certain solutions, the domain of validity may be extended to semi-infinite or infinite intervals, see Sections 4.1–4.4 of [144]. In (13.15) we have a much more complicated averaging process, but are asking only for the preservation of statistics, rather than approximation of individual solutions.

Equations (13.14) and (13.15) may lead one to hope that a simulation of (13.15) will produce the density $\delta_<$. However, satisfying Equation (13.14) does not, in itself, ensure correct reproduction of $\delta_<$. This is due to the possibility of multiple solutions to (13.14), even in cases where Equation (13.13) has a unique solution! See [39] for further discussion.

The reason for this unyielding behavior of the PDF equation is quite deep and has to do with the question of unique ergodicity of the original and modeled flows. Such questions are, a priori, difficult. The reproduction of statistics of the original model is intimately related to the reproduction of $\delta_<$. For turbulence (and many other applications as well) most statistics of interest are moments of a particular functional in phase space. In turn, most functionals, such as velocity or pressure at a point, are dominated by the resolved modes of the problem. Hence, reproducing statistics of interest is equivalent to reproducing moments of $\delta_<$, which is equivalent, through the characteristic functional, to the reproduction of $\delta_<$.

Although the above arguments suggest a bleak picture for reproducing statistics from simulations, experience is more encouraging. Certain statistics *are* reproduced quite well by turbulence simulations. This is noteworthy, and in itself tells us something about statistical robustness of the Navier–Stokes equations or the robustness of the statistics measured. Understanding this robustness may lead to significant progress in unraveling the mathematical nature of processes like turbulence, and may result in practical implications. Some progress in this direction was made in [37].

How might one approach computation of the averaged vector field $\tilde{f}(x)$ of (13.15) from a numerical simulation of the full system (13.12)? A priori, this requires knowledge of the family of conditional measures $\mu_>(\cdot|x)$, which does not seem practical for numerical implementation. However, closer examination of (13.15) reveals that all one really needs are *averages* with respect to the conditional measures: significantly less information than the measures themselves.

In [39], a computational tool was developed for approximating the averaged vector field $\tilde{f}(x)$ as a multi-variable polynomial, directly from time series of the full data. In particular, to probe the limits of the theory, an attempt was made to reproduce solutions of the Lorenz system [144, 217]. As described in Section 6.5 above, for certain parameter values, this three-dimensional ODE has a strange attractor that lies on a thin, multi-sheeted structure. Taking x to be two-dimensional and y one-dimensional, one can attempt to reproduce statistics of this essentially three-dimensional chaotic flow by a sequence of two-dimensional models, which at most can possess simple limit cycles. The results of [39] show that, even if the marginal field is well approximated, reproduction of the statistics of the resolved modes is not assured without additional assumptions. However, key geometric features, such as the characteristic pattern of the Lorenz attractor, were reproduced quite well. This study provides insight into the nature of the problem of reproducing statistics via lower-dimensional models, in highlighting the questions of uniqueness and sensitivity of the results to the estimate of $\tilde{f}(x)$.

13.4 Spatial localization

The low-dimensional models for the wall region of the boundary layer, described in Chapters 10 and 11, as well as other such models outlined in the previous chapter, are localized in physical space. There are several reasons for constructing such localized models. We have already remarked on the fact that large spatial domains, containing numerous, essentially independent coherent structures of possibly different forms and sizes, require relatively high-dimensional models, thus vitiating the advantages due to compact description of each individual structure via the POD. Two further reasons are:

- We are interested in basic physical mechanisms underlying the dynamical interactions between coherent structures. Physical intuition and quantitative data from spatial correlations indicate that these interactions are predominantly local in physical space. Moreover, there is a certain elegance in being able to describe the nonlinear phenomena of turbulence in open flows as a superposition of well-understood nonlinear building blocks. The work referenced and reported in [389, 390] provides examples, also see Section 8.4.

- For practical purposes, such as control, we must estimate the state, in phase space, of the fluid. In practice, we are likely to have only limited information from shear (or other) sensors mounted on the wall. It is unlikely that we will be able to estimate from this the state of the fluid in the entire domain, nor is it practical to expect that a processor at the heart of the control system will be able to store or compute in real time the vast amounts of information required to describe the state of the flow throughout the domain. Hence, the model must be restricted a priori to a smaller subdomain in physical space.

As a partial differential equation, the Navier–Stokes system is elliptic with respect to the spatial variables, due to the viscous Laplacian term, and parabolic in time. Ellipticity typically demands prescription of velocity components on all boundaries of the domain if the equation is to be well posed, although other boundary conditions can result in well posedness. The low-dimensional models of [22] specify only pressure as a boundary condition at the upper boundary of the (sub-) domain. This paradox was pointed out in [394], as we described in Section 12.5.

In the present section we outline an argument that partially resolves this concern by appealing to methods that assume the finite, albeit very large, dimensionality of viscous fluid flow. Specifically, if the Navier–Stokes equations on the full domain Ω possess an inertial manifold $\mathcal{M}$ (cf. Section 4.4 and see [367]), and if certain technical conditions hold on the subdomain $\Omega' \subset \Omega$ in question and on $\mathcal{M}$, then the Navier–Stokes equations viewed as a PDE on Ω', with only a scalar pressure field prescribed on the boundary $\partial\Omega'$, do indeed constitute a well-posed problem.

In Chapter 10 we saw that results from DNS suggest that, for the wall layer models, the pressure term at the outer boundary acts as a perturbation force. This, together with the arguments given below, suggests that, choosing a subdomain Ω' such that the pressure term on $\partial\Omega'$ is small, dynamical systems defined on Ω' via the Galerkin procedure of Chapters 4 and 10 are likely to be valid for tracking true solutions on time scales that increase with the decrease of the relative importance of the pressure perturbations. In particular, for flows driven by a pressure gradient, such as channel flow, this implies that the subdomain must advect with the flow in a manner that cancels the forcing by the mean pressure, which is the case for the models of Chapters 10 and 11 [22].

We now describe conditions which imply that *finite dimensional projections* of the Navier–Stokes equations defined on a subdomain, Ω', and with only pressure fields defined on $\partial\Omega'$, possess the same dynamics for lower order modes as the full system. This does not completely resolve the problem for the wall layer model, which has mixed velocity and pressure boundary conditions, but it does point towards a resolution. We indicate the necessary assumptions and the conditions under which they appear reasonable.

We start with the Navier-Stokes equations:

$$\left.\begin{aligned} \mathbf{u}_t + \mathbf{u} \cdot \nabla \mathbf{u} &= -\nabla p + \frac{1}{R_e}\nabla^2 \mathbf{u} + \mathbf{f}, \\ \nabla \cdot \mathbf{u} &= 0, \end{aligned}\right\} \qquad (13.18)$$

where the divergence-free velocity field $\mathbf{u}$ is defined on the full domain Ω with the appropriate boundary conditions (e.g. no-slip or suitable "forcing" $\mathbf{u} = \mathbf{g}(\mathbf{x})$ on $\partial\Omega$). We suppose that there is a modal decomposition of $\mathbf{u}$ in terms of an orthonormal basis of $L^2(\Omega)$:

$$\mathbf{u}(\mathbf{x}, t) = \sum_{i=1}^{\infty} a_i(t)\boldsymbol{\varphi}_i(\mathbf{x}), \qquad (13.19)$$

and that the evolution equation (13.18) has a smooth, invariant, inertial manifold $\mathcal{M}$ that may be parameterized globally as a graph over the finite-dimensional subspace spanned by the first N modal coefficients as follows:

$$\{a_i = \Phi_i(a_1, \ldots, a_N) | a_i = \Phi_i(a_1, \ldots, a_N), \; i = N+1, \ldots, \infty\}. \qquad (13.20)$$

Thus each higher modal coefficient is specified by a (nonlinear) function of the first N. The existence of $\mathcal{M}$ implies that, to describe the complete state of the fluid in the domain Ω, it is sufficient to specify only the values of these first N coordinates or modal amplitudes: $(a_1, \ldots, a_N)$.

Suppose $\Omega' \subset \Omega$ is the subdomain in physical space for which we wish to construct a low-dimensional model. We ask under what conditions do the dynamics of the system in Ω' determine the coordinates $(a_1, \ldots, a_N)$, which will in turn, via $\Phi_i(a_1, \ldots, a_N)$, determine the full fluid flow in Ω?

Let $\mathbf{u}$ and $\mathbf{v}$ be any two solutions of the Navier–Stokes equations (13.18) in the domain Ω, lying in $\mathcal{M}$, and with the corresponding pressure field solutions p_u and p_v, respectively. Suppose that on the boundary $\partial\Omega'$ of the subdomain Ω' we have:

$$p_u(\mathbf{x}, t) = p_v(\mathbf{x}, t) \quad \text{for every } \mathbf{x} \in \partial\Omega' \text{ and } t \geq 0. \qquad (13.21)$$

Moreover, suppose that $\mathbf{u}(\mathbf{x}, 0) = \mathbf{v}(\mathbf{x}, 0)$ for every $\mathbf{x} \in \Omega'$. Our question is equivalent to asking if, it is true under these conditions, that for all $t > 0$ and $\mathbf{x} \in \Omega'$ we also have $\mathbf{u}(\mathbf{x}, t) = \mathbf{v}(\mathbf{x}, t)$. In fact, as we shall see, equality in Ω' implies equality in Ω also.

Let us denote by $(\mathbf{f}, \mathbf{g})_{\Omega'} = \int_{\Omega'} \mathbf{f}(\mathbf{x}) \cdot \mathbf{g}(\mathbf{x})^* d\mathbf{x}$ the L^2 inner product of velocity fields in the subdomain Ω'. Note that, in contrast to their behavior in the domain Ω, the basis functions $\{\boldsymbol{\varphi}_i\}_{i=1}^{\infty}$ that were orthonormal in $L^2(\Omega)$ are generally not orthonormal in $L^2(\Omega')$. We perform a Galerkin projection of (13.18), in Ω', on the first N modes, to obtain evolution equations for the leading modal coefficients $(a_1, \ldots, a_N)$. Once these ODEs are solved for the time histories $(a_1(t), \ldots, a_N(t))$, we can then appeal to (13.20) to reconstruct the full velocity field:

$$\mathbf{u}(\mathbf{x}, t) = \sum_{j=1}^{N} a_j(t)\boldsymbol{\varphi}_j(\mathbf{x}) + \sum_{j=N+1}^{\infty} \Phi_j(a_1, \ldots, a_N)\boldsymbol{\varphi}_j(\mathbf{x}). \qquad (13.22)$$

Note that, due to the global nature of the graph Φ_j, (13.22) is implicitly invertible, so that there is a one to one correspondence between velocity fields $\mathbf{u}$ *defined on* $\mathcal{M}$ and N-tuples

$(a_1, \ldots, a_N)$ defining points in the phase space of the model. No such correspondence exists in general for arbitrary velocity fields.

To deduce the ODEs, we take the inner product of (13.18) in Ω' but using the basis functions φ_i defined on the full domain Ω. This is the analog of a windowed Fourier transform, an extreme case of a weighted inner product on Ω, concentrated entirely on Ω'. We carry out the following steps:

- substitute (13.19) into (13.18) and take the $L^2(\Omega')$ inner product with the functions $\varphi_i(\mathbf{x})$, $i = 1, \ldots, N$;
- substitute the expressions for a_i, $i > N$, as in (13.20);
- perform integration by parts of the pressure term as in Section 10.4, Equations (10.42) and (10.43).

The resulting ODEs take the form (using the Einstein notation):

$$
\left(\frac{d}{dt} \left[\sum_{j=1}^{N} a_j(t)\varphi_j + \sum_{j=N+1}^{\infty} \Phi_j(a_1, \ldots, a_N)\varphi_j(x) \right], \varphi_i \right)_{\Omega'}
$$
$$
+ \left(B(a_1, \ldots, a_N, \Phi_{N+1}, \ldots, a_1, \ldots, a_N, \Phi_{N+1}, \ldots), \varphi_i \right)_{\Omega'}
$$
$$
= - \int_{\partial\Omega'} p(\mathbf{n} \cdot \varphi_i) \, ds + \left(\frac{1}{R_e} A(a_1, \ldots, a_N, \Phi_{N+1}, \ldots), \varphi_i \right)_{\Omega'}.
$$

$$(13.23)$$

The terms coming from the bi-linear term (B), the pressure term, and the viscous term (A) are all well defined in terms of $(a_1, \ldots, a_N)$, the pressure field on the boundary $\partial\Omega'$, and the inertial manifold graph $\Phi_i = \Phi_i(a_1, \ldots, a_N)$. For existence and uniqueness of solutions to this ODE, the $N \times N$ matrix implicit in the time-derivative term must be invertible, so that the N-vector $(\dot{a}_1, \ldots, \dot{a}_N)$ can be uniquely deduced from $(a_1, \ldots, a_N)$. Using the chain rule to evaluate the time derivatives, the time-derivative of (13.23) becomes:

$$
\sum_{j=1}^{N} \dot{a}_j \left(\varphi_j, \varphi_i \right)_{\Omega'} + \sum_{j=N+1}^{\infty} \left[\sum_{k=1}^{N} \frac{\partial \Phi_j(a_1, \ldots, a_N)}{\partial a_k} \dot{a}_k \right] \left(\varphi_j, \varphi_i \right)_{\Omega'}. \qquad (13.24)
$$

Here we assume that $\mathcal{M}$ is sufficiently smooth for the derivatives to exist. The expression of (13.24) may in turn be rearranged and written as the following matrix product:

$$
\underbrace{\left[\left(\varphi_j + \sum_{l=N+1}^{\infty} \frac{\partial \Phi_l(a_1, \ldots, a_N)}{\partial a_j} \varphi_l, \varphi_i \right)_{\Omega'} \right]}_{i=1,\ldots,N \times j=1,\ldots,N} \begin{bmatrix} \dot{a}_1 \\ \vdots \\ \dot{a}_N \end{bmatrix}. \qquad (13.25)
$$

Hence, the well posedness of the evolution of the flow in the subdomain Ω' with the pressure alone prescribed on $\partial\Omega'$ reduces to invertibility of the matrix in (13.25), together with the one to one correspondence between the initial conditions in the full flow and in the N-dimensional phase space implicit in (13.22). Choosing $\mathbf{u}(\mathbf{x}, 0) = \mathbf{v}(\mathbf{x}, 0)$ both in $\mathcal{M}$

ensures that the corresponding phase space points $\mathbf{a}(\mathbf{u}(\mathbf{x}, 0))$ and $\mathbf{a}(\mathbf{v}(\mathbf{x}, 0))$ coincide, so that the desired uniqueness follows.

The matrix of (13.25) is certainly unlikely to be invertible for *every* choice of subdomain Ω', graph functions Φ_l, and dimension N. However, we observe that each term in this matrix contains two contributions. One comes from the inner product of the first N eigenfunctions in the subdomain Ω' and the other involves derivatives of the graph Φ_l defining $\mathcal{M}$. In the limit as $\Omega' \to \Omega$, the contribution from the eigenfunction inner product approaches the identity matrix, which is clearly invertible. In fact, for all (reasonable) subdomains Ω', $N < \infty$, and *analytic* basis functions $\{\varphi_i\}$, the $N \times N$ matrix $C_{i,j} = (\varphi_i, \varphi_j)_{\Omega'}$ will also be invertible, because the POD modes should be linearly independent on any subdomain. The partial derivative terms $\partial \Phi_l(a_1, \ldots, a_N)/\partial a_j$ should also be small for sufficiently large N, because all modes in the complement of the phase space spanned by $\{\varphi_i\}_1^N$ decay rapidly, implying that the smooth manifold $\mathcal{M}$ lies close to this linear subspace [367]. Hence it is likely that, provided that N is sufficiently large and Ω' is not too small or singular, the matrix of (13.25) will be invertible, thus establishing well-posedness of the dynamics in Ω'. However, for small domains and large N, the matrix is likely to have a poor condition number, and it may therefore be difficult to compute an accurate inverse.

In this argument the subspace dimension N is not directly related to the dimension of low-order dynamical systems to be derived by projection onto empirical eigenfunctions defined in Ω'; rather it appears as an intermediate mathematical construct to allow well-posedness of the problem. Of course, the fact that such a finite N can be defined relies on the existence of an inertial manifold $\mathcal{M}$, which remains an unsolved problem for the Navier–Stokes equations in arbitrary three-dimensional domains. Even assuming existence of $\mathcal{M}$, more work is required to make the above arguments mathematically rigorous.

These ideas are nonetheless relevant for the low-dimensional model program proposed in this book. With such an argument in hand, one could pursue a rigorous numerical study of which domains permit low-dimensional models and which do not. A first goal might be to produce low-dimensional models of the boundary layer without streamwise or spanwise periodic boundary conditions and which, with the specification of the pressure alone, can track the dynamics in the wall layer.

13.5 The utility of models

We have proposed an ambitious program in this book. In this closing section, it is therefore appropriate to speculate on the utility of low-dimensional models to the intellectual and practical problems of turbulence identified at the beginning of Chapter 1.

Our program is experimentally and computationally demanding. To derive empirical basis sets of reasonable dimension, one must either perform lengthy experiments to measure two point correlation tensors (using multiple probes or repeating for each separation distance), or make extensive DNS or LES computations to obtain suitable ensemble averages. Despite rapid advances in computational power and speed, it is still a major task to simulate sufficiently-long time histories of fully developed flows, even in simple geometries, to guarantee reliable statistics. In deriving experimental databases, measurement of three-dimensional velocity fields on sufficiently fine spatial scales remains challenging. One must acquire, store, and process enormous amounts of data. With analog to digital conversion and on-line processing, the task has eased somewhat, but it is still likely to be necessary to examine the raw measurements for sensor malfunctions, etc. In the case of Herzog's database [155], used for the boundary layer modeling of Chapter 10, 100 reels of nine-track magnetic tape were required to store pointwise velocity components.

The weakness of each method, experimental or numerical, is, in fact, a strength of the other. Experiments readily provide good temporal statistics, simply by extending each run for a sufficient duration, while DNS can provide excellent spatial information. Thus one might intelligently combine both methods to provide the averaged autocorrelation database necessary for the POD.

Nonetheless, collection and processing of data remain a substantial task, and one that must, in principle, be repeated whenever the flow domain geometry, boundary conditions, or even the Reynolds number is changed. In Chapters 3 and 12 (Sections 3.7 and 12.4), we have mentioned some studies in which empirical bases computed at one Reynolds number have been used to construct models for flows at different Reynolds numbers, but relatively little has been done in this area. Clearly, significant changes in geometry or boundary conditions demand completely new bases.

In this connection, we may mention once again the work of Poje and Lumley [283], who have adapted energy method stability theory to predict the approximate form of empirical eigenfunctions and their associated eigenvalue spectrum in the turbulent boundary layer. The modification of the energy method involves an anisotropic eddy viscosity, and requires that the mean velocity profile be responsive to the momentum transport by the eigenfunctions. Beginnings have been made on applying the same method to sheared thermal convection. Of course, the energy method stability theory identifies the *fastest growing* disturbance, whereas we are looking for the most energetic disturbance. The technique, however, does take account of distortion of the mean velocity profile by the growing disturbance, and the changes in stress induced in the background turbulence by the growing disturbance; with these modifications, at least in the boundary layer, the fastest growing disturbance has the same form as the most energetic one. This outcome makes reasonable physical sense, and if the method turns out to work as well in other flows as it does in the boundary layer, it will permit great savings relative to either experiment or to DNS or LES. The information input to the method can be obtained from a simple Reynolds-averaged Navier–Stokes calculation of the flow, which provides the necessary mean profiles and turbulent component energies and scales.

Once computed, empirical basis functions are typically stored as numerical tables, so that the inner products involved in Galerkin projections require numerical integration. Before this can be done, as Sections 10.1–10.4 indicate, extensive preprocessing analysis is generally necessary. While this can reveal and allow exploitation of symmetries inherent in the averaged data (Section 10.3), empirical eigenfunctions do not generally enjoy analytically convenient properties, such as those of Fourier modes, which permit fast Fourier transforms and other computational shortcuts. Thus use of empirical bases, in spite of their apparent dimensional economy, may not significantly speed up computational methods.

This may change if one expresses empirical functions in terms of analytical bases, possibly using wavelets or wavelet packets [90, 238].

We note that the work needed to prepare the Navier–Stokes equations for projection and to compute the various inner products can be automated to some degree [40]. This is more than a mere convenience: a sizeable part of the three-year project leading to [19] and, ultimately, papers such as [22, 25, 323], was consumed by the Galerkin projection process.

Even though POD bases afford a significant advantage in very low-dimensional truncations, this rapidly falls off as one goes to higher dimensions. For example, in a turbulent channel flow, Moin and Moser [244] find that the first family of eigenfunctions (summed over spanwise and streamwise Fourier modes) carries 23% of the kinetic energy on average, compared with only 4% for the leading Chebyshev polynomial. To represent 90%, however, one needs ten eigenfunctions or 12 Chebyshev polynomials. Increasing the resolution further, the figures remain close. They remark that "... for ... simulations of the type performed by [190], the small improvement in accuracy obtained by using the Karhunen–Loève expansion would not offset the increased computational cost ... ." Of course, for the POD to be advantageous at all, the flow must be inhomogeneous in at least one direction, and so will typically contain spatially localized coherent structures which dominate the energy spectrum.

It is therefore likely that the ideas introduced here will have their greatest impact, at least initially, in study of turbulence production in a few canonical, shear-dominated flows. It is surely no accident that the bulk of the work described in Chapters 10–12 involves such flows – boundary layers, jets, wakes, and mixing layers. Certainly, as the work of Rempfer and Fasel discussed in Section 12.2 indicates, they will be useful in transitional and pre-turbulent flows. We do not anticipate that it will be feasible or worthwhile to derive models of sufficiently high dimension via the POD to capture significant portions of the inertial range in fully developed flows. Nonetheless, we believe the work described in this book does reveal that, with suitable modeling of energy transfer to neglected modes, low-dimensional models have already demonstrated their ability to illuminate key aspects of the turbulence production process.

We see these methods as contributing both to the intellectual and practical challenges referred to above. Freeman Dyson [99] notes the ironic fact that some of the world's greatest physicists, Einstein and Oppenheimer in particular, spent fruitless years searching for unified theories – overarching laws and equations that model "everything," while ignoring solutions of particular problems. He remarks that " ... it often happens that the understanding of the mathematical nature of an equation is impossible without a detailed understanding of its solutions."

In turbulence theory, we have possessed a fine model for over one hundred and sixty years, the Navier–Stokes equations. There is little doubt that we have the correct laws, we just cannot fully read the language in which they are written. Anything that takes us closer to understanding solutions of these governing equations is surely worthwhile. We feel that their translation into low-dimensional dynamical systems as suggested here, highlighting the influence of spatial localization and the interplay between spatio-temporal effects and symmetries, is likely to lead to a deeper understanding.

Improved understanding of important issues such as the mechanisms of bursting and instability in the wall region is certainly intellectually satisfying; moreover it can also help us solve practical problems. There are many situations in which knowledge of the dynamical behavior of the coherent structures in a turbulent flow provides all the information that is needed, and resolution of the small-scale, incoherent part of the flow is unnecessary. At the same time, in many of these situations, more information is required than can be obtained from the simple statistics of a Reynolds-averaged Navier–Stokes calculation. The kinds of low-dimensional models we have discussed in this book are well suited to give this kind of information.

Practical examples are easy to find: vehicle panel vibration induced by an overlying turbulent boundary layer (generating noise in either air or water, both inside and outside the vehicle); equally, fluctuations in heat or mass transfer in a turbulent boundary layer. These latter ideas can be extended to flows in the blade passages of turbomachines and aircraft gas turbines in particular, in which a periodically driven, highly unsteady flow is dominated by coherent structures, and prediction of both noise and heat transfer is a matter of urgency. Prediction and control of noise radiated by subsonic and supersonic jets is another case in point: here most of the noise is also produced by coherent structures.

As already mentioned in Section 12.8 and above, in control of the turbulent boundary layer (and of many other flows), a low-dimensional model can serve as a short-term tracker/predictor for the flow, permitting one to plan actuator motions in time for their effects to counteract developing instabilities. The boundary layer models of [22] have already been used to suggest improved control schemes for drag reduction: cf. [38, 41, 47, 78, 80, 81]. Increasing demands for active flow control are likely to promote implementation of such models in real-time control schemes.

Prediction of fluctuations in index of refraction in the surface mixed layer of the atmosphere, which is also dominated by coherent structures, is another problem amenable to low-dimensional modeling – optical communication links are affected by the dynamics in this layer; the resultant fluctuations in index of refraction cause multipath interference. The list could go on and on. It is probably not too irresponsible to imagine in the near future low-dimensional models of many of these flows that could run in real-time on a workstation, generating pseudo-data, and permitting relatively simple analysis of many situations that now require at least DNS, or which remain entirely out of reach.

This book is a first essay on application of dynamical systems ideas to understanding of turbulence in open flows dominated by coherent structures. We hope that, in setting the stage and introducing the ideas, methods and flavor of the approach, we shall encourage development of these ideas into a comprehensive theory and a useful engineering tool.

References

[1] R. J. Adrian. Conditional eddies in isotropic turbulence. *Physics of Fluids*, **22**(11):2065–70, 1979.

[2] R. J. Adrian and P. Moin. Stochastic estimation of organized turbulent structure: Homogeneous shear flow. *J. Fluid Mech.*, **190**:531–59, 1988.

[3] R. J. Adrian, P. Moin, and R. D. Moser. Stochastic estimation of conditional eddies in turbulent channel flow. In *CTR, Proceedings of the 1987 Summer Program*, Stanford University, CA, 1987. Center for Turbulence Research.

[4] S. Ahuja and C. W. Rowley. Feedback control of unstable steady states of flow past a flat plate using reduced-order estimators. *J. Fluid Mech.*, **645**:447–478, Feb. 2010.

[5] V. R. Algazi and D. J. Sakrison. On the optimality of the Karhunen–Loève expansion. *IEEE Trans. Inform. Theory*, **15**:319–21, 1969.

[6] C. D. Andereck, S. S. Liu, and H. Swinney. Flow regimes in a circular Couette system with independently rotating cylinders. *J. Fluid Mech.*, **164**:155–83, 1986.

[7] C. A. Andrews, J. M. Davies, and G. R. Schwartz. Adaptive data compression. *Proc. IEEE*, **55**:267–77, 1967.

[8] A. A. Andronov, E. A. Vitt, and S. E. Khaikin. *Theory of Oscillators*. Pergamon Press, Oxford, UK, 1966. Reprinted 1987 by Dover Publications, New York.

[9] D. Armbruster, J. Guckenheimer, and P. Holmes. Heteroclinic cycles and modulated travelling waves in systems with O(2) symmetry. *Physica D*, **29**:257–82, 1988.

[10] D. Armbruster, J. Guckenheimer, and P. Holmes. Kuramoto–Sivashinsky dynamics on the center unstable manifold. *SIAM J. on Appl. Math.*, **49**:676–91, 1989.

[11] D. Armbruster, R. Heiland, and E. J. Kostelich. KLTOOL: a tool to analyze spatio-temporal complexity. *Chaos*, **4**(2):421–4, 1994.

[12] D. Armbruster, R. Heiland, E. J. Kostelich, and B. Nicolaenko. Phase-space analysis of bursting behavior in Kolmogorov flow. *Physica D*, **58**:392–401, 1992.

[13] D. Armbruster, B. Nicolaenko, N. Smaoui, and P. Chossat. Analysing bifurcations in the Kolmogorov flow equations. In P. Chossat, editor, *Dynamics, Bifurcations and Symmetries*, pages 11–33, Dordrecht, Kluwer, 1994.

[14] L. Arnold. *Stochastic Differential Equations*. John Wiley, New York, 1974.

[15] V. I. Arnold. *Ordinary Differential Equations*. MIT Press, Cambridge, MA, 1973.

[16] V. I. Arnold. *Mathematical Methods of Classical Mechanics*. Springer-Verlag, New York, 1978.

[17] V. I. Arnold. *Geometrical Methods in the Theory of Ordinary Differential Equations*. Springer-Verlag, New York, 1983.

[18] D. K. Arrowsmith and C. M. Place. *An Introduction to Dynamical Systems*. Cambridge University Press, Cambridge, UK, 1990.

[19] N. Aubry. *A Dynamical System/Coherent Structure Approach to the Fully Developed Turbulent Wall Layer.* PhD thesis, Cornell University, 1987.

[20] N. Aubry, R. Guyonnet, and R. Lima. Spatio-temporal analysis of complex signals: theory and applications. *J. Stat. Phys.*, **64**(3/4):683–739, 1991.

[21] N. Aubry, R. Guyonnet, and R. Lima. Spatio-temporal symmetries and bifurcations via bi-orthogonal decompositions. *J. Nonlinear Sci.*, **2**:183–215, 1992.

[22] N. Aubry, P. Holmes, J. L. Lumley, and E. Stone. The dynamics of coherent structures in the wall region of a turbulent boundary layer. *J. Fluid Mech.*, **192**:115–73, 1988.

[23] N. Aubry, W.-L. Lian, and E. S. Titi. Preserving symmetries in the proper orthogonal decomposition. *SIAM J. on Sci. Comput.*, **14**:483–505, 1993.

[24] N. Aubry, J. L. Lumley, and P. Holmes. The effect of modeled drag reduction in the wall region. *Theoret. Comput. Fluid Dynamics*, **1**:229–48, 1990.

[25] N. Aubry and S. Sanghi. Bifurcations and bursting of streaks in the turbulent wall layer. In M. Lesieur and O. Métais, editors, *Turbulence and Coherent Structures*, pages 227–51, Dordrecht, Kluwer, 1991.

[26] A. Back, J. Guckenheimer, M. Myers, F. J. Wicklin, and P. Worfolk. DsTool: Computer assisted exploration of dynamical systems. *AMS Notices, April 1992*, **39**(4):303–9, 1992.

[27] S. Bagheri, J. Hœpffner, P. J. Schmid, and D. S. Henningson. Input–output analysis and control design applied to a linear model of spatially developing flows. *Applied Mechanics Reviews*, **62**(2):020803, Mar. 2009.

[28] P. Bakewell and J. L. Lumley. Viscous sublayer and adjacent wall region in turbulent pipe flows. *Physics of Fluids*, **10**:1880–9, 1967.

[29] K. S. Ball, L. Sirovich, and L. R. Keefe. Dynamical eigenfunction decomposition of turbulent channel flow. *Int. J. for Num. Meth. in Fluids*, **12**:585–604, 1991.

[30] P. R. Bandyopadhyay, editor. *Application of Microfabrication to Fluid Mechanics 1994*, volume FED-Vol.197. Amer. Soc. Mech. Eng., New York, 1994.

[31] H. Bartlet, A. W. Lohmann, and B. Wirnitzer. Phase and amplitude recovery from bispectra. *Appl. Opt.*, **23**:3121–9, 1984.

[32] G. K. Batchelor. *The Theory of Homogeneous Turbulence.* Cambridge University Press, Cambridge, UK, 1956.

[33] G. K. Batchelor and I. Proudman. The large scale structure of homogeneous turbulence. *Phil. Trans. Roy. Soc.*, **A248**(949):369–405, 1956.

[34] M. Bergmann and L. Cordier. Optimal control of the cylinder wake in the laminar regime by Trust-Region methods and POD reduced order models. *J. Comp. Phys.*, **227**:7813–40, 2008.

[35] M. Bergmann, L. Cordier, and J.-P. Brancher. Optimal rotary control of the cylinder wake using proper orthogonal decomposition reduced order model. *Physics of Fluids*, **17**:097101, 2005.

[36] G. Berkooz. A numerical evaluation of the dynamical systems approach to wall layer turbulence. In *CTR, Proceedings of the 1990 Summer Program*, Stanford University, CA, 1990. Center for Turbulence Research.

[37] G. Berkooz. *Turbulence, Coherent Structures, and Low Dimensional Models.* PhD thesis, Cornell University, 1991.

[38] G. Berkooz. Controlling models of the turbulent wall layer by boundary deformation. Technical Report FDA-92-16, Cornell University, 1993.

[39] G. Berkooz. An observation on probability density equations, or, When do simulations reproduce statistics? *Nonlinearity*, **7**:313–28, 1994.

[40] G. Berkooz, L. P. Chew, R. Palmer, and R. Zippel. Generating spectral methods solvers for partial differential equations. In *IMACS International Conference on Computational Methods for Partial Differential Equations*, pages 471–6, New Brunswick, NJ, June 1992.

[41] G. Berkooz, T. Corke, M. N. Glauser, M. Psiaki, and M. Fisher. Design for control of flow instabilities: first principles and an application. In J. Paduano, editor, *Sensing, Actuation, and Control in Aeropropulsion*. SPIE, 1995.

[42] G. Berkooz, J. Elezgaray, and P. Holmes. Coherent structures in random media and wavelets. *Physica D*, **61**:47–58, 1992.

[43] G. Berkooz, P. Holmes, and J. L. Lumley. Intermittent dynamics in simple models of the wall layer. *J. Fluid Mech.*, **230**:75–95, 1991.

[44] G. Berkooz, P. Holmes, and J. L. Lumley. On the relation between low dimensional models and the dynamics of coherent structures in the turbulent wall layer. *Theoret. Comput. Fluid Dynamics*, **4**:255–69, 1993.

[45] G. Berkooz, P. Holmes, and J. L. Lumley. The proper orthogonal decomposition in the analysis of turbulent flows. *Ann. Rev. Fluid Mech.*, **25**:539–75, 1993.

[46] G. Berkooz, P. Holmes, J. L. Lumley, N. Aubry, and E. Stone. Observations regarding "Coherence and chaos in a model of turbulent boundary layer" by X. Zhou and L. Sirovich. *Physics of Fluids A*, **6**:1574–8, 1994.

[47] G. Berkooz, M. Psiaki, and M. Fisher. Estimation and control of models of the turbulent wall layer. In *Active Control of Noise and Vibration*, New York, 1995. ASME.

[48] G. Berkooz and E. S. Titi. Galerkin projections and the proper orthogonal decomposition for equivariant equations. *Phys. Lett. A*, **174**:94–102, 1993.

[49] L. P. Bernal, R. E. Breidenthal, G. L. Brown, J. H. Konrad, and A. Roshko. On the development of three dimensional small scales in turbulent mixing layers. In *Second Symposium on Turbulent Shear Flows*, pages 8.1–8.6, 1979.

[50] W.-J. Beyn and V. Thümmler. Freezing solutions of equivariant evolution equations. *SIAM Journal on Applied Dynamical Systems*, **3**(2):85–116, 2004.

[51] G. D. Birkhoff. *Dynamical Systems*. American Mathematical Society, Providence, RI, 1927.

[52] R. F. Blackwelder and R. E. Kaplan. On the wall structure of the turbulent boundary layer. *J. Fluid Mech.*, **76**:89–112, 1976.

[53] D. G. Bogard and W. G. Tiederman. Burst detection with single point velocity measurements. *J. Fluid Mech.*, **162**:389–413, 1986.

[54] J. P. Bonnet, D. R. Cole, J. Delville, M. N. Glauser, and L. Ukeiley. Stochastic estimation and proper orthogonal decomposition: complementary techniques for identifying structure. *Experiments in Fluids*, **17**:307–14, 1994.

[55] R. Bowen. *Equilibrium States and the Ergodic Theory of Anosov Diffeomorphisms*, volume 470 of *Springer Lecture Notes in Mathematics*. Springer-Verlag, New York, 1975.

[56] W. E. Boyce and R. C. DiPrima. *Elementary Differential Equations and Boundary Value Problems*. Wiley, New York, fifth edition, 1992.

[57] L. Breiman. *Probability*. Addison Wesley, Reading, MA, 1968.

[58] J. Bridges, H. S. Husain, and F. Hussain. Whither coherent structures? Comment 1. In J. L. Lumley, editor, *Whither Turbulence? Turbulence at the Crossroads*, pages 132–51. Springer-Verlag, New York, 1990.

[59] D. R. Brillinger and M. Rosenblatt. Asymptotic theory of estimates of kth order spectra. In B. Harris, editor, *Spectral Analysis of Time Series*. Wiley, New York, 1967.

[60] G. L. Brown and A. Roshko. The effect of density difference on the turbulent mixing layer. In *A.G.A.R.D. Conference on Turbulent Shear Flows, Conf. Proceedings No. 93*, pages 23.1–23.12. NATO Advisory Group for Aerospace Research and Development, 1971.

[61] G. L. Brown and A. Roshko. On density effects and large structure in turbulent mixing layers. *J. Fluid Mech.*, **64**:775–816, 1974.

[62] A. D. Bruno. *Local Methods in Nonlinear Differential Equations*. Springer-Verlag, New York, 1989.

[63] V. Brunsden, J. Cortell, and P. Holmes. Power spectra of chaotic oscillations of a buckled beam. *J. Sound and Vibration*, **130**:1–25, 1989.

[64] V. Brunsden and P. Holmes. Power spectra of strange attractors near homoclinic orbits. *Phys. Rev. Lett.*, **58**:1699–702, 1987.

[65] J. M. Burgers. A mathematical model illustrating the theory of turbulence. *Adv. Appl. Mech*, **1**:171–99, 1948.

[66] F. M. Busse and R. M. Clever. Nonstationary convection in a rotating system. In U. Muller, K. G. Roessner, and B. Schmidt, editors, *Recent Developments in Theoretical and Experimental Fluid Mechanics*, pages 376–85. Springer-Verlag, New York, 1979.

[67] F. M. Busse and K. E. Heikes. Convection in a rotating layer: a simple case of turbulence. *Science*, **208**:173–5, 1980.

[68] S. A. Campbell and P. Holmes. Bifurcation from O(2) symmetric heteroclinic cycles with three interacting modes. *Nonlinearity*, **4**:697–726, 1991.

[69] S. A. Campbell and P. Holmes. Heteroclinic cycles and modulated travelling waves in a system with D(4) symmetry. *Physica D*, **59**:52–78, 1992.

[70] C. Canuto, M. Y. Hussaini, A. Quarteroni, and T. A. Zang. *Spectral Methods in Fluid Dynamics*. Springer-Verlag, New York, 1988.

[71] J. Carr. *Applications of Center Manifold Theory*. Springer-Verlag, New York, 1981.

[72] D. H. Chambers, R. J. Adrian, P. Moin, D. S. Stewart, and H. J. Sung. Karhunen–Loève expansion of Burgers model of turbulence. *Phys. Fluids*, **31**:2573–82, 1988.

[73] J. Chomaz. Global instabilities in spatially developing flows: Non-normality and nonlinearity. *Annual Review of Fluid Mechanics*, **37**:357–92, 2005.

[74] J. M. Chomaz, P. Huerre, and L. G. Redekopp. Bifurcations to local and global modes in spatially developing flows. *Phys. Rev. Lett.*, **60**(1):25–8, 1988.

[75] P. Chossat and G. Iooss. Primary and secondary bifurcations in the Couette–Taylor problem. *Japan J. Appl. Math.*, **2**:37–68, 1985.

[76] P. Chossat and G. Iooss. *The Couette–Taylor Problem*. Springer-Verlag, New York, 1994.

[77] E. A. Coddington and N. Levinson. *Theory of Ordinary Differential Equations*. McGraw-Hill, New York, 1955.

[78] B. D. Coller and P. Holmes. Suppression of bursting. *Automatica*, **33**(1):1–11, 1997.

[79] B. D. Coller, P. Holmes, and J. L. Lumley. Control of bursting in boundary layer models. *Appl. Mech. Rev.*, **47**(6):part 2, S139–43, 1994.

[80] B. D. Coller, P. Holmes, and J. L. Lumley. Controlling noisy heteroclinic cycles. *Physica D*, **72**:135–60, 1994.

[81] B. D. Coller, P. Holmes, and J. L. Lumley. Interaction of adjacent bursts in the wall region. *Phys. Fluids*, **6**(2):954–61, 1994.

[82] P. Collet and J.-P. Eckmann. *Iterated Maps on the Interval as Dynamical Systems*. Birkhauser, Basel, 1980.

[83] P. Constantin and C. Foias. *Navier–Stokes equations*. Chicago University Press, Chicago, IL, 1988.

[84] P. Constantin, C. Foias, and R. Temam. *Attractors Representing Turbulent Flows*. Memoirs of the AMS 53, 314. American Mathematical Society, Providence, RI, 1985.

[85] P. Constantin, C. Foias, R. Temam, and B. Nicolaenko. *Integral Manifolds and Inertial Manifolds for Dissipative Partial Differential Equations*. Springer-Verlag, New York, 1989.

[86] I. P. Cornfeld, S. V. Fomin, and Y. G. Sinai. *Ergodic Theory*. Springer-Verlag, New York, 1982.

[87] S. Corrsin and A. L. Kistler. *The Free-stream Boundaries of Turbulent Flows*. NACA TN 3133. National Advisory Committee for Aeronautics, 1954.

[88] C. Cossu and J. M. Chomaz. Global measures of local convective instabilities. *Phys. Rev. Lett.*, **78**(23):4387–4390, June 1997.

[89] H. Dankowicz, P. Holmes, G. Berkooz, and J. Elezgaray. Local models of spatio-temporally complex fields. *Physica D*, **90**:387–407, 1996.

[90] I. Daubechies. *Ten Lectures on Wavelets*. CBMS–NSF Regional Conference Series in Applied Mathematics, 61. SIAM Publications, Philadelphia, PA, 1992.

[91] A. E. Deane, I. G. Kevrekidis, G. E. Karniadakis, and S. A. Orszag. Low-dimensional models for complex flows: Application to grooved channels and circular cylinders. *Physics of Fluids A*, **3**(10):2337–54, 1991.

[92] A. E. Deane and L. Sirovich. A computational study of Rayleigh–Bénard convection. Part I. Rayleigh number scaling. *J. Fluid Mech.*, **222**:231–50, 1991.

[93] J. W. Deardorff. A numerical study of three-dimensional turbulent channel flow at large Reynolds numbers. *J. Fluid Mech.*, **41**:453–80, 1970.

[94] R. G. Deissler. Is Navier–Stokes turbulence chaotic? *Physics of Fluids*, **29**(5):1453–7, 1986.

[95] Y. Demay, G. Iooss, and P. Laure. Wave patterns in small gap Couette–Taylor problem. *European J. Mech. B. Fluids*, **11**(5):621–34, 1992.

[96] C. R. Doering and J. D. Gibbon. *Applied Analysis of the Navier–Stokes Equations*. Cambridge University Press, Cambridge, UK, 1994.

[97] P. G. Drazin and W. H. Reid. *Hydrodynamic Stability*. Cambridge University Press, Cambridge, UK, 1981.

[98] G. E. Dullerud and F. Paganini. *A Course in Robust Control Theory: A Convex Approach*, volume 36 of *Texts in Applied Mathematics*. Springer-Verlag, 1999.

[99] F. Dyson. The scientist as rebel. *New York Review of Books*, **XLII**(9):31–3, May 25, 1995.

[100] B. Eckhardt and A. Mersmann. Transition to turbulence in a shear flow. *Phys. Rev. E*, **60**:509–17, 1999.

[101] J. Elezgaray, G. Berkooz, and P. Holmes. Wavelet analysis of the motion of coherent structures. In Y. Meyer and S. Roques, editors, *Progress in Wavelet Analysis and Applications*, pages 471–6, Gif-sur-Yvette, France, 1993. Editions Frontières.

[102] H. Faisst. The transition from the Taylor–Couette system to the plane Couette system. Diplomarbeit, Universität Marburg, 1999.

[103] H. Faisst and B. Eckhardt. Transition from the Couette–Taylor system to the plane Couette system. *Phys. Rev. E*, **61**:7227–30, 2000.

[104] M. Farge. Wavelet transforms and their applications to turbulence. *Ann. Rev. Fluid Mech.*, **24**:395–457, 1992.

[105] B. F. Farrell and P. J. Ioannou. Stochastic forcing of the linearized Navier–Stokes equations. *Physics of Fluids A*, **5**(11):2600–9, 1993.

[106] B. F. Farrell and P. J. Ioannou. A theory for the statistical equilibrium energy spectrum and heat flux produced by transient baroclinic waves. *J. Atmos. Sci.*, **51**(19):2685–98, 1994.

[107] B. F. Farrell and P. J. Ioannou. Variance maintained by stochastic forcing of non-normal dynamical systems associated with linearly stable shear flows. *Phys. Rev. Lett.*, **72**(8):1188–91, 1994.

[108] W. Feller. *An Introduction to Probability Theory and its Applications*. Wiley, New York, 1957.

[109] J. A. Ferre and F. Girlat. Pattern-recognition analysis of the velocity field in plane turbulent wakes. *J. Fluid Mech.*, **198**:27–64, 1989.

[110] M. Fogleman. *Low-Dimensional Models of Internal Combustion Engine Flows using the Proper Orthogonal Decomposition*. PhD thesis, Cornell University, 2005.

[111] M. Fogleman, J. Lumley, D. Rempfer, and D. Hawarth. Application of the proper orthogonal decomposition to datasets of internal combustion engine flows. *J. Turbulence*, **5**:023, 2005.

[112] C. Foias, O. Manley, and L. Sirovich. Empirical and Stokes eigenfunctions and the far dissipative turbulent spectrum. *Physics of Fluids A*, **2**:464–7, 1990.

[113] C. Foias, O. Manley, and R. Temam. Approximate inertial manifolds and effective viscosity in turbulent flows. Technical Report 9011, Dept. of Math., Indiana University, Bloomington, IN, 1990.

[114] C. Foias, G. Sell, and R. Temam. Inertial manifolds for nonlinear evolutionary equations. *J. Diff. Eqns*, **73**:199–224, 1989.

[115] C. Foias, G. R. Sell, and E. S. Titi. Exponential tracking and approximation of inertial manifolds for dissipative equations. *J. of Dynamics and Diff. Eqns*, 1:199–224, 1989.

[116] C. Foias and R. Temam. Gevrey class regularity for the solutions of the Navier–Stokes equations. *J. Functional Anal.*, 87:359–69, 1989.

[117] S. K. Friedlander and L. Topper, editors. *Turbulence: Classical Papers on Statistical Theory*. Interscience, New York, 1962.

[118] D. H. Gay and W. H. Ray. Identification and control of linear distributed parameter systems through the use of experimentally determined singular functions. In *Proc. IFAC Symp. Control of Distributed Parameter Systems, Los Angeles, CA, 30 June–2 July 1986*, pages 173–9, 1986.

[119] D. H. Gay and W. H. Ray. Application of singular value methods for identification and model based control of distributed parameter systems. In *Proc. IFAC Workshop on Model Based Process Control, Atlanta, GA, June 1988*, pages 95–102, 1988.

[120] J. F. Gibson, J. Halcrow, and P. Cvitanović. Equilibrium and travelling-wave solutions of plane Couette flow. *J. Fluid Mech.*, 638:243–66, 2009.

[121] M. N. Glauser and W. K. George. Orthogonal decomposition of the axisymmetric jet mixing layer including azimuthal dependence. In G. Comte-Bellot and J. Mathieu, editors, *Advances in Turbulence*, pages 357–66. Springer-Verlag, New York, 1987.

[122] M. N. Glauser and W. K. George. An orthogonal decomposition of the axisymmetric jet mixing layer utilizing cross-wire velocity measurements. In *Sixth Symposium on Turbulent Shear Flows, Toulouse, France*, pages 10.1.1–10.1.6, Toulouse, France, 1987. Ecole Nationale Superieure de l'Aeronautique et de l'Espace and ONERA Centre d'Etudes et de Recherches de Toulouse.

[123] M. N. Glauser, S. J. Leib, and W. K. George. Coherent structures in the axisymmetric turbulent jet mixing layer. In F. Durst, B. E. Launder, J. L. Lumley, F. W. Schmidt, and J. Whitelaw, editors, *Turbulent Shear Flows 5*, pages 134–45. Springer-Verlag, New York, 1987.

[124] M. N. Glauser and X. Zheng. A low-dimensional dynamical systems description of the axisymmetric jet mixing layer. Technical Report MAE-247, Clarkson University, 1991.

[125] M. N. Glauser, X. Zheng, and C. R. Doering. The dynamics of organized structures in the axisymmetric jet mixing layer. In M. Lesieur and O. Métais, editors, *Turbulence and Coherent Structures*, pages 253–65. Kluwer, Dordrecht, 1991.

[126] M. N. Glauser, X. Zheng, and W. K. George. The streamwise evolution of coherent structures in the axisymmetric jet mixing layer. In T. B. Gatski, S. Sarkar, and C. G. Speziale, editors, *Studies in Turbulence*, pages 207–22. Springer-Verlag, New York, 1992.

[127] S. Glavaški, J. E. Marsden, and R. M. Murray. Model reduction, centering, and the Karhunen–Loève expansion. In *Proc. IEEE Conf. Decision and Control*, volume 37, pages 2071–2076, 1998.

[128] J. Gleick. *Chaos: Making a New Science*. Viking Penguin, Inc., New York, 1987.

[129] P. Glendinning. *Stability, Instability and Chaos*. Cambridge University Press, Cambridge, UK, 1995.

[130] A. Glezer, A. J. Kadioglu, and A. J. Pearlstein. Development of an extended proper orthogonal decomposition and its application to a time periodically forced plane mixing layer. *Physics of Fluids A*, 1:1363–73, 1989.

[131] M. A. Gol'dshtik. *Structural turbulence*. Institute of Thermophysics, Novosibirsk, USSR, 1982.

[132] S. Goldstein. *Modern Developments in Fluid Dynamics*. Oxford University Press, Oxford, UK, 1952.

[133] M. Golubitsky and W. F. Langford. Pattern formation and bistability in flow between counterrotating cylinders. *Physica D*, 32:362–92, 1988.

[134] M. Golubitsky and D. G. Schaeffer. *Singularities and Groups in Bifurcation Theory, Volume I*. Springer-Verlag, New York, 1985.

[135] M. Golubitsky and I. Stewart. Symmetry and stability in Taylor–Couette flow. *SIAM J. on Math. Anal.*, **17**(2):249–88, 1986.

[136] M. Golubitsky, I. Stewart, and D. G. Schaeffer. *Singularities and Groups in Bifurcation Theory, Volume II*. Springer-Verlag, New York, 1988.

[137] M. Gorman and P. J. Widmann. Nonlinear dynamics of a convection loop: a quantitative comparison of experiments with theory. *Physica D*, **19**:255–67, 1986.

[138] D. Gottlieb and S. A. Orszag. *Numerical Analysis of Spectral Methods: Theory and Applications*. CBMS–NSF Regional Conference Series in Applied Mathematics, 26. SIAM Publications, Philadelphia, PA, 1977.

[139] W. R. Graham, J. Peraire, and K. Y. Tang. Optimal control of vortex shedding using low-order models. Part I: Open-loop model development. *Int. J. Numer. Meth. Engrng*, **44**:945–72, 1999.

[140] C. Grebogi, S. M. Hammel, J. A. Yorke, and T. Sauer. Shadowing of physical trajectories in chaotic dynamics: containment and refinement. *Phys. Rev. Lett.*, **65**(13):1527–30, 1990.

[141] F. F. Grinstein, M. N. Glauser, and W. K. George. A low-dimensional dynamical systems description of coherent structures in the axisymmetric jet mixing layer. In S. I. Green, editor, *Fluid Vortices*, pages 65–94. Kluwer, Dordrecht, 1995.

[142] J. Guckenheimer. A strange, strange attractor. In J. E. Marsden and M. McCracken, editors, *The Hopf Bifurcation and its Applications*, pages 368–81. Springer-Verlag, New York, 1976.

[143] J. Guckenheimer. The role of geometry in computational dynamics. In P. Chossat, editor, *Dynamics, Bifurcations and Symmetries*, pages 155–66. Kluwer, Dordrecht, 1994.

[144] J. Guckenheimer and P. Holmes. *Nonlinear Oscillations, Dynamical Systems and Bifurcations of Vector Fields*. Springer-Verlag, New York, 1983.

[145] J. Guckenheimer and P. Holmes. Structurally stable heteroclinic cycles. *Math. Proc. Cambridge Phil. Soc.*, **103**:189–92, 1988.

[146] J. Guckenheimer and A. Mahalov. Resonant triad interactions in symmetric systems. *Physica D*, **54**:267–310, 1992.

[147] J. Guckenheimer and R. F. Williams. Structural stability of Lorenz attractors. *Publ. Math. IHES*, **50**:59–72, 1979.

[148] J. Halcrow, J. F. Gibson, and P. Cvitanovič. Visualizing the geometry of state space in plane Couette flow. *J. Fluid Mech.*, **611**:107–30, 2008.

[149] J. Halcrow, J. F. Gibson, P. Cvitanovič, and D. Visawanth. Heteroclinic connections in plane Couette flow. *J. Fluid Mech.*, **621**:365–76, 2009.

[150] J. Hale. *Ordinary Differential Equations*. Wiley, New York, 1969.

[151] J. M. Hamilton, J. Kim, and F. Waleffe. Regeneration mechanisms of near-wall turbulence structures. *J. Fluid Mech.*, **287**:317–48, 1995.

[152] D. C. Hawarth. Large-eddy simulation of in-cylinder flows. *Oil and Gas Science and Technology*, **54**:175–85, 1999.

[153] D. C. Hawarth and K. Jensen. Large-eddy simulation on unstructured deforming meshes: Towards reciprocating IC engines. *Computers & Fluids*, **29**:023, 2000.

[154] D. Henry. *Geometric Theory of Semilinear Parabolic Equations*, volume 840 of *Springer Lecture Notes in Mathematics*. Springer-Verlag, New York, 1980.

[155] S. Herzog. *The Large Scale Structure in the Near Wall Region of a Turbulent Pipe Flow*. PhD thesis, Cornell University, 1986.

[156] M. W. Hirsch, S. Smale, and R. L. Devaney. *Differential Equations, Dynamical Systems and an Introduction to Chaos*. Academic Press/Elsevier, San Diego, CA, 2004.

[157] M. Högberg, T. R. Bewley, and D. S. Henningson. Linear feedback control and estimation of transition in plane channel flow. *J. Fluid Mech.*, **481**:149–75, 2003.

[158] P. Holmes. Can dynamical systems approach turbulence? In J. L. Lumley, editor, *Whither Turbulence? Turbulence at the Crossroads*, pages 195–249. Springer-Verlag, New York, 1990.

[159] P. Holmes. On Moffatt's paradox or, can empirical projections approach turbulence? In J. L. Lumley, editor, *Whither Turbulence? Turbulence at the Crossroads*, pages 306–9. Springer-Verlag, New York, 1990.

[160] P. Holmes. Symmetries, heteroclinic cycles and intermittency in fluid flow. In G. R. Sell, C. Foias, and R. Temam, editors, *Turbulence in Fluid Flows: a Dynamical Systems Approach*, pages 49–58. Springer-Verlag, New York, 1993.

[161] P. Holmes, G. Berkooz, and J. L. Lumley. Turbulence, dynamical systems and the unreasonable effectiveness of empirical eigenfunctions. In *Proceedings of the International Congress of Mathematicians, Kyoto, 1990*, pages 1607–17. Springer-Verlag, Tokyo, 1991.

[162] P. Holmes and E. Stone. Heteroclinic cycles, exponential tails and intermittency in turbulence production. In T. B. Gatski, S. Sarkar, and C. G. Speziale, editors, *Studies in Turbulence*, pages 179–89. Springer-Verlag, New York, 1992.

[163] E. Hopf. A mathematical example displaying the features of turbulence. *Comm. Pure Appl. Math.*, **1**:303–22, 1948.

[164] E. Hopf. On the application of functional calculus to the statistical theory of turbulence. In *Proc. Symp. Appl. Math.*, Providence, RI, 1957, American Math. Soc.

[165] S. Hoyas and J. Jiménez. Scaling of the velocity fluctuations in turbulent channels up to $Re_\tau = 2003$. *Physics of Fluids*, **18**: 011702, 2006.

[166] P. Huerre and P. A. Monkewitz. Local and global instabilities in spatially developing flows. *Annual Review of Fluid Mechanics*, **22**:473–537, 1990.

[167] M. Hultmark, M. Leftwich, and A. J. Smits. Flowfield measurements in the wake of a robotic lamprey. *Experiments in Fluids*, **43**:683–90, 2007.

[168] A. K. M. F. Hussain and W. C. Reynolds. The mechanisms of an organized wave in turbulent shear flow. *J. Fluid Mech.*, **41**:241–58, 1970.

[169] J. M. Hyman, B. Nicolaenko, and S. Zaleski. Order and complexity in the Kuramoto–Sivashinsky model of weakly turbulent interfaces. *Physica D*, **23**:265–92, 1986.

[170] M. Ilak. *Model Reduction and Feedback Control of Transitional Channel Flow*. PhD thesis, Princeton University, 2009.

[171] M. Ilak, S. Bagheri, L. Brandt, C. W. Rowley, and D. S. Henningson. Model reduction of the nonlinear complex Ginzburg–Landau equation. *SIAM Journal on Applied Dynamical Systems*, **9**(4):1284–1302, 2010.

[172] M. Ilak and C. W. Rowley. Modeling of transitional channel flow using balanced proper orthogonal decomposition. *Physics of Fluids*, **20**:034103, 2008.

[173] G. Iooss. Secondary bifurcations of Taylor vortices into wavy inflow or outflow boundaries. *J. Fluid Mech.*, **173**:273–88, 1986.

[174] G. Iooss and A. Mielke. Time-periodic Ginzburg–Landau equations for one dimensional patterns with large wave length. *Z. Angew. Math. Phys.*, **43**:125–38, 1992.

[175] J. Jiménez and P. Moin. The minimal flow unit in near-wall turbulence. *J. Fluid Mech.*, **225**:213–40, 1991.

[176] C. Jones and M. K. Proctor. Strong spatial resonance and travelling waves in Bénard convection. *Phys. Lett. A*, **121**:224–7, 1987.

[177] B. H. Jørgensen, J. N. Sørensen, and M. Brøns. Low-dimensional modeling of a driven cavity flow with two free parameters. *Theoret. Comput. Fluid Dynamics*, **16**:299–317, 2003.

[178] J.-N. Juang and R. S. Pappa. An eigensystem realization algorithm for modal parameter identification and model reduction. *Journal of Guidance, Control, and Dynamics*, **8**(5):620–7, 1985.

[179] V. Juttijudata, J. L. Lumley, and D. Rempfer. Proper orthogonal decomposition in Squire's coordinate system for dynamical models of channel turbulence. *J. Fluid Mech.*, **534**:195–225, 2005.

[180] Y. S. Kachanov, V. V. Kozlov, V. Y. Levchenko, and M. P. Ramazov. On the nature of K-breakdown of a laminar boundary layer. In V. Kozlov, editor, *Laminar–Turbulent Transition*, pages 61–73. Springer-Verlag, New York, 1985.

[181] B. A. Kader. Change in the thickness of an incompressible turbulent boundary layer in the presence of a longitudinal pressure gradient. *Izvestiya Akademii Nauk SSSR, Mekhanika Zhidkosti i Gaza*, 2:150–6, 1979. Translated in *Fluid Dynamics*, 14(2):283–9, 1979.

[182] B. A. Kader. Hydrodynamic structure of accelerated turbulent boundary layers. *Izvestiya Akademii Nauk SSSR, Mekhanika Zhidkosti i Gaza*, 3:29–37, 1983. Translated in *Fluid Dynamics*, 18(3):360–7, 1983.

[183] K. Karhunen. Zur Spektraltheorie stochastischer Prozesse. *Ann. Acad. Sci. Fennicae*, Ser. A1, 34, 1946.

[184] C. Kasnakoglu, A. Serrani, and M. O. Efe. Control input separation by actuation mode expansion for flow control problems. *Int. J. Control*, 81(9):1475–92, 2008.

[185] G. Kawahara and S. Kida. Periodic motion embedded in plane Couette turbulence: regeneration cycle and burst. *J. Fluid Mech.*, 449:291–300, 2001.

[186] L. Keefe, P. Moin, and J. Kim. The dimension of an attractor in turbulent Poiseuille flow. *Bull. Am. Phys. Soc.*, 32:2026, 1987.

[187] A. Kelley. The stable, center stable, center, center unstable and unstable manifolds. *J. Diff. Eqns*, 3:546–70, 1967.

[188] I. G. Kevrekidis, B. Nicolaenko, and C. Scovel. Back in the saddle again: a computer assisted study of the Kuramoto–Sivashinsky equation. *SIAM J. on Appl. Math.*, 50:760–90, 1990.

[189] H. T. Kim, S. J. Kline, and W. C. Reynolds. The production of turbulence near a smooth wall in a turbulent boundary layer. *J. Fluid Mech.*, 50:133–60, 1971.

[190] J. Kim, P. Moin, and R. J. Moser. Turbulence statistics in a fully developed channel flow at low Reynolds number. *J. Fluid Mech.*, 177:133–66, 1987.

[191] M. Kirby and D. Armbruster. Reconstructing phase space from PDE simulations. *Z. Angew. Math. Phys.*, 43:999–1022, 1992.

[192] M. Kirby, J. Boris, and L. Sirovich. An eigenfunction analysis of axisymmetric jet flow. *J. of Computational Physics*, 90(1):98–122, 1990.

[193] M. Kirby, J. Boris, and L. Sirovich. A proper orthogonal decomposition of a simulated supersonic shear layer. *International J. for Numerical Methods in Fluids*, 10:411–28, 1990.

[194] S. J. Kline. Observed structure features in turbulent and transitional boundary layers. In G. Sovran, editor, *Fluid Mechanics of Internal Flow*, pages 27–68, Amsterdam, 1967. Elsevier.

[195] S. J. Kline, W. C. Reynolds, F. A. Schraub, and P. W. Runstadler. The structure of turbulent boundary layers. *J. Fluid Mech.*, 30:741–73, 1967.

[196] J. J. Kobine and T. Mullin. Low dimensional bifurcation phenomena in Taylor–Couette flow with discrete azimuthal symmetry. *J. Fluid Mech.*, 275:379–405, 1994.

[197] D. D. Kosambi. Statistics in function space. *J. Indian Math. Soc.*, 7:76–88, 1943.

[198] M. Krupa. Robust heteroclinic cycles. Forschungsbericht 2, Inst. für Angewandte und Numerische Mathematik, TU Wein, Austria, 1994.

[199] M. Krupa and I. Melbourne. Asymptotic stability of heteroclinic cycles in systems with symmetry. *Ergodic Theory and Dynamical Systems*, 15:121–47, 1995.

[200] Y. Kuramoto. Diffusion-induced chaos in reaction systems. *Suppl. Prog. Theor. Phys.*, 64:346–67, 1978.

[201] O. A. Ladyzhenskaya. *The Mathematical Theory of Viscous Incompressible Flow*. Gordon and Breach, New York, 1969.

[202] S. Lall, J. E. Marsden, and S. Glavaški. A subspace approach to balanced truncation for model reduction of nonlinear control systems. *Int. J. Robust Nonlinear Control*, 12:519–35, 2002.

[203] H. Lamb. *Hydrodynamics*. Dover, New York, sixth edition, 1945.

[204] L. D. Landau and E. M. Lifschitz. *Fluid Mechanics.* Pergamon Press, Oxford, UK, second edition, 1987.

[205] O. Lanford. Appendix to Lecture VII: computer pictures of the Lorenz attractor. In A. Chorin, J. E. Marsden, and S. Smale, editors, *Turbulence Seminar, Berkeley, 1976/77*, volume 615 of *Springer Lecture Notes in Mathematics*, pages 113–16. Springer-Verlag, New York, 1977.

[206] P. Laure and Y. Demay. Symbolic computation and the equation on the centermanifold: application to the Taylor–Couette problem. *Comput. Fluids*, **16**:229–38, 1988.

[207] O. Lehmann, M. Luchtenburg, B. R. Noack, *et al.* Wake stabilization using POD Galerkin models with interpolated modes. In *44th IEEE Conference on Decision and Control and European Control Conference*, pages 500–5, Dec. 2005.

[208] S. Leibovich. The form and dynamics of Langmuir circulations. *Ann. Rev. Fluid Mech.*, **15**:391–427, 1983.

[209] S. Leibovich. Structural genesis in wall bounded turbulent flows. In T. B. Gatski, S. Sarkar, and C. G. Speziale, editors, *Studies in Turbulence*, pages 387–411. Springer-Verlag, New York, 1992.

[210] A. Leonard and A. Wray. A new numerical method for simulation of three dimensional flow in a pipe. In E. Krause, editor, *Proc. Int. Conf. on Numerical Methods in Fluid Dynamics*, volume 170 of *Lecture Notes in Physics*, pages 335–42. Springer-Verlag, New York, 1982.

[211] H. W. Liepmann. Aspects of the turbulence problem, Part II. *Z. Angew. Math. Phys.*, **3**:407–26, 1952.

[212] K. S. Lii, M. Rosenblatt, and C. Van Atta. Bispectral measurements in turbulence. *J. Fluid Mech.*, **77**:45–62, 1976.

[213] J. T. C. Liu. Contributions to the understanding of large-scale coherent structures in developing free turbulent shear flows. *Advances in Applied Mechanics*, **26**:183–309, 1988.

[214] Z.-C. Liu, R. J. Adrian, and T. J. Hanratty. Reynolds number similarity of orthogonal decomposition of the outer layer of turbulent wall flow. *Physics of Fluids*, **6**:2815–19, 1994.

[215] M. Loève. Functions aléatoire de second ordre. *Comptes Rendus Acad. Sci. Paris*, **220**, 1945.

[216] E. N. Lorenz. Empirical orthogonal functions and statistical weather prediction. In *Statistical Forecasting Project*, Cambridge, MA, 1956, MIT Press.

[217] E. N. Lorenz. Deterministic nonperiodic flow. *J. Atmos. Sci.*, **20**:130–41, 1963.

[218] D. M. Luchtenburg, B. Günther, B. R. Noack, R. King, and G. Tadmor. A generalized mean-field model of the natural and high-frequency actuated flow around a high-lift configuration. *J. Fluid Mech.*, **623**:339–65, 2009.

[219] D. M. Luchtenburg, M. Schlegel, B. R. Noack, *et al.* Turbulence control based on reduced-order models and nonlinear control design. In R. King, editor, *Active Flow Control II*, vol 108 *of Notes on Numerical Fluid Mechanics and Multidisciplinary Design*, pages 341–56 Springer-Verlag, Berlin, 2010.

[220] J. L. Lumley. The structure of inhomogeneous turbulence. In A. M. Yaglom and V. I. Tatarski, editors, *Atmospheric Turbulence and Wave Propagation*, pages 166–78. Nauka, Moscow, 1967.

[221] J. L. Lumley. *Stochastic Tools in Turbulence.* Academic Press, New York, 1971.

[222] J. L. Lumley. Two-phase and non-Newtonian flows. In P. Bradshaw, editor, *Turbulence. Topics in Applied Physics, Volume 12*, pages 290–324, Springer-Verlag, New York, 1978.

[223] J. L. Lumley. Coherent structures in turbulence. In R. E. Meyer, editor, *Transition and Turbulence*, New York, 1981, Academic Press. Mathematics Research Center Symposia and Advanced Seminar Series.

[224] J. L. Lumley, editor. *Whither Turbulence? Turbulence at the Crossroads*, volume 357 of *Lecture notes in Physics*. Springer-Verlag, New York, 1990.

[225] J. L. Lumley. Early work on fluid mechanics in the IC engine. *Ann. Rev. Fluid Mech.*, **33**:319–38, 2001.

[226] J. L. Lumley and I. Kubo. Turbulent drag reduction by polymer additives: a survey. In B. Gampert, editor, *The Influence of Polymer Additives on Velocity and Temperature Fields*, pages 3–21. Springer-Verlag, New York, 1985.

[227] J. L. Lumley and H. A. Panofsky. *The Structure of Atmospheric Turbulence*. Interscience, New York, 1964.

[228] J. L. Lumley and A. Poje. Low-dimensional models for flows with density fluctuations. *Physics of Fluids*, **9**(7):2023–31, 1997.

[229] A. Lundbladh, P. Schmidt, S. Berlin, and D. Henningson. Simulations of bypass transition for spatially evolving disturbances. In B. Cantwell, J. Jiménez, and S. Lekoudis, editors, *Application of Direct and Large Eddy Simulation to Transition and Turbulence*, pages 18.1–18.3. Fluid Dynamics Panel, NATO Advisory Group for Aerospace Research and Development, AGARD CP 551, 1994.

[230] X. Ma and G. E. Karniadakis. A low-dimensional model for simulating three-dimensional cylinder flow. *J. Fluid Mech.*, **458**:181–90, 2002.

[231] Z. Ma, S. Ahuja, and C. W. Rowley. Reduced order models for control of fluids using the eigensystem realization algorithm. *Theoret. Comput. Fluid Dynamics*, **25**(1):233–47, 2011.

[232] A. Mahalov and S. Leibovich. Multiple bifurcation of rotating pipe flow. *Theoret. Comput. Fluid Dynamics*, **3**:61–77, 1992.

[233] S. Malo. *Rigorous Computer Verification of Planar Vector Field Structure*. PhD thesis, Cornell University, 1994.

[234] R. Mane. *Ergodic Theory and Differentiable Dynamics*. Springer-Verlag, New York, 1987.

[235] B. Marasli, F. H. Champagne, and I. Wygnanski. Effect of traveling waves on the growth of a plane turbulent wake. *J. Fluid Mech.*, **235**:511–28, 1991.

[236] J. E. Marsden. *Lectures on Mechanics*, volume 174 of *London Mathematical Society Lecture Note Series*. Cambridge University Press, 1992.

[237] T. Matsuoka and T. L. Ulrych. Phase estimation using the bispectrum. *Proc. IEEE*, **72**:1403–22, 1984.

[238] Y. Meyer. *Wavelets: Algorithms and Applications*. SIAM Publications, Philadelphia, PA, 1993.

[239] K. Mischaikow and M. Mrozek. Chaos in the Lorenz equations: a computer assisted proof. *Bull. AMS (New Series)*, **32**(1):66–72, 1995.

[240] H. K. Moffatt. Fixed points of turbulent dynamical systems and suppression of nonlinearity. In J. L. Lumley, editor, *Whither Turbulence? Turbulence at the Crossroads*, pages 250–7. Springer-Verlag, New York, 1990.

[241] P. Moin. Probing turbulence via large eddy simulation. AIAA paper 84-0174, 1984.

[242] P. Moin. Similarity of organized structures in turbulent shear flows. In S. J. Kline and N. H. Afgan, editors, *Near-Wall Turbulence; 1988 Zoran Zaric Memorial Conference*. Hemisphere Publishing, Washington, DC, 1990.

[243] P. Moin, R. J. Adrian, and J. Kim. Stochastic estimation of organized structures in turbulent channel flow. In *Sixth Symposium on Turbulent Shear Flows*, Toulouse, France, 1987. Ecole Nationale Superieure de l'Aeronautique et de l'Espace and ONERA Centre d'Etudes et de Recherches de Toulouse.

[244] P. Moin and R. D. Moser. Characteristic-eddy decomposition of turbulence in a channel. *J. Fluid Mech.*, **200**:471–509, 1989.

[245] A. S. Monin and A. M. Yaglom. *Statistical Fluid Mechanics: Mechanics of Turbulence*, Volumes I and II. MIT Press, Cambridge, MA, 1971–1975.

[246] B. C. Moore. Principal component analysis in linear systems: Controllability, observability, and model reduction. *IEEE Transactions on Automatic Control*, **26**(1):17–32, Feb. 1981.

[247] J. Moreau, M. Fogleman, G. Charnay, and J. Boree. Phase invariant proper orthogonal decomposition for the study of a compressed vortex. *J. Thermal Sciences*, **14**(2):108–13, 2005.

[248] M. Morzyński, W. Stankiewicz, B. R. Noack, F. Thiele, and G. Tadmor. Generalized mean-field model for flow control using continuous mode interpolation. 3rd AIAA Flow Control Conference, AIAA paper 2006-3488, June 2006.

[249] M. Morzyński, W. Stankiewicz, B. R. Noack, F. Thiele, and G. Tadmor. Continuous mode interpolation for control-oriented models of fluid flow. In R. King, editor, *Active Flow Control*, vol 95 of *Notes on Numerical Fluid Mechanics and Multidisciplinary Design*, pages 260–78. Springer-Verlag, Berlin, 2007.

[250] R. D. Moser and M. M. Rogers. The three-dimensional evolution of a plane mixing layer: pairing and transition to turbulence. *J. Fluid Mech.*, **247**:275–320, 1993.

[251] T. Mullin. Disordered fluid motion in a small closed system. *Physica D*, **62**:192–201, 1993.

[252] T. Mullin and A. G. Darbyshire. Intermittency in a rotating annular flow. *Europhys. Lett.*, **9**(7):669–73, 1989.

[253] T. Mullin, S. J. Taverner, and K. A. Cliffe. An experimental and numerical study of a codimension-2 bifurcation in a rotating annulus. *Europhys. Lett.*, **8**(3):251–6, 1989.

[254] M. Myers, P. Holmes, J. Elezgaray, and G. Berkooz. Wavelet projections of the Kuramoto–Sivashinsky equation I: Heteroclinic cycles and modulated traveling waves for short systems. *Physica D*, **86**:396–427, 1995.

[255] M. Nagata. Three-dimensional finite-amplitude solutions in plane Couette flow: bifurcation from infinity. *J. Fluid Mech.*, **217**:519–27, 1990.

[256] R. Narasimha. The utility and drawbacks of traditional approaches. In J. L. Lumley, editor, *Whither Turbulence? Turbulence at the Crossroads*, pages 13–48. Springer-Verlag, New York, 1990.

[257] R. Narasimha and S. V. Kailas. Turbulent bursts in the atmosphere. *Atmospheric Environment*, **24A**(7):1635–45, 1990.

[258] S. E. Newhouse, D. Ruelle, and F. Takens. Occurrence of strange axiom A attractors near quasiperiodic flows on T^m, $m \geq 3$. *Comm. Math. Phys.*, **64**:35–40, 1978.

[259] B. Nicolaenko, B. Scheurer, and R. Temam. Some global dynamical properties of the Kuramoto–Sivashinsky equations: non-linear stability and attractors. *Physica D*, **16**:155–83, 1985.

[260] B. Nicolaenko and Z. She. Temporal intermittency and turbulence production in the Kolmogorov flow. In *Topological Dynamics of Turbulence*, pages 256–77. Cambridge University Press, Cambridge, UK, 1990.

[261] B. Nicolaenko and Z. She. Turbulent bursts, inertial sets and symmetry breaking homoclinic cycles in periodic Navier–Stokes flows. In G. R. Sell, C. Foias, and R. Temam, editors, *Turbulence in Fluid Flows: a Dynamical Systems Approach*, pages 123–36. Springer-Verlag, New York, 1993.

[262] B. R. Noack, K. Afanasiev, M. Morzyński, G. Tadmor, and F. Thiele. A hierarchy of low-dimensional models for the transient and post-transient cylinder wake. *J. Fluid Mech.*, **497**:335–63, 2003.

[263] B. R. Noack and H. Eckelmann. On chaos in wakes. *Physica D*, **56**:151–64, 1992.

[264] B. R. Noack and H. Eckelmann. A global stability analysis of the steady and periodic cylinder wake. *J. Fluid Mech.*, **270**:297–330, 1994.

[265] B. R. Noack and H. Eckelmann. A low dimensional Galerkin method for the three-dimensional flow around a circular cylinder. *Physics of Fluids*, **6**:124–43, 1994.

[266] B. R. Noack and H. Eckelmann. Theoretical investigation of the bifurcations and the turbulence attractor of the cylinder wake. *Z. Angew. Math. Mech.*, **74**:T396–T397, 1994.

[267] B. R. Noack, M. Morzyński, and G. Tadmor (editors). *Reduced-Order Modelling for Flow Control*. Springer-Verlag, Berlin, 2010.

[268] B. R. Noack, P. Papas, and P. A. Monkewitz. The need for a pressure-term representation in empirical Galerkin models of incompressible shear flows. *J. Fluid Mech.*, **523**:283–316, 2005.

[269] B. R. Noack, G. Tadmor, and M. Morzyński. Actuation models and dissipative control in empirical Galerkin models of fluid flows. In *American Control Conference,* Boston, MA, June 30-July 2, 2004, pages 1–6, 2004.

[270] A. Novick-Cohen. Interfacial instabilities in directional solidification of dilute binary alloys: the Kuramoto–Sivashinsky equation. *Physica D,* **26**:403–10, 1987.

[271] A. Novick-Cohen and G. I. Sivashinsky. On the solidification front of a dilute binary alloy: thermal diffusivity effects and breathing solutions. *Physica D,* **20**:237–58, 1986.

[272] A. M. Obukhov. Statistical description of continuous fields. *Trudy Geophys. Int. Aked. Nauk. SSSR,* **24**:3–42, 1954.

[273] D. Ornstein and B. Weiss. Statistical properties of chaotic systems. *Bull. of the AMS (New Series),* **24**(1):11–116, 1991.

[274] E. Ott, C. Grebogi, and J. A. Yorke. Controlling chaos. *Phys. Rev. Lett.,* **64**:1196–9, 1990.

[275] A. Papoulis. *Probability, Random Variables, and Stochastic Processes.* McGraw-Hill, New York, 1965.

[276] H. Park and L. Sirovich. Turbulent thermal convection in a finite domain, Part II. Numerical results. *Physics of Fluids A,* **2**(9):1659–68, 1990.

[277] M. Pastoor, B. R. Noack, R. King, and G. Tadmor. Spatiotemporal waveform observers and feedback in shear layer control. 44th AIAA Fluids Conference and Exhibit, AIAA paper 2006-1402, Jan. 2006.

[278] V. Perrier and C. Basdevant. Periodical wavelet analysis: A tool for inhomogeneous field investigation. theory and algorithms. *Recherche Aerospatiale,* **3**:54–67, 1989.

[279] N. Platt, L. Sirovich, and N. Fitzmaurice. An investigation of chaotic Kolmogorov flows. *Physics of Fluids A,* **3**(4):681–96, 1991.

[280] V. A. Pliss. A reduction principle in the theory of stability of motion. *Izv. Akad. Nauk. SSSR. Math. Ser.,* **28**:1297–324, 1964.

[281] B. Podvin, J. Gibson, G. Berkooz, and J. L. Lumley. Lagrangian and Eulerian views of the bursting period. *Physics of Fluids,* **9**(2):433–7, 1997.

[282] H. Poincaré. *Les Méthodes Nouvelles de la Mécanique Céleste, Tomes I–III.* Gauthier-Villars, Paris, 1892, 1893, 1899. Reprinted 1987 by Librarie Albert Blanchard, Paris.

[283] A. Poje and J. L. Lumley. A model for large scale structures in turbulent shear flows. *J. Fluid Mech.,* **285**:349–69, 1995.

[284] S. B. Pope. PDF methods for turbulent reactive flows. *Prog. Energy Combust. Sci.,* **11**:119–92, 1985.

[285] V. S. Pougachev. General theory of the correlations of random functions. *Izv. Akad. Nauk. SSSR. Math. Ser.,* **17**:401–2, 1953.

[286] R. W. Preisendorfer. *Principal Component Analysis in Meteorology and Oceanography.* Elsevier, Amsterdam, 1988.

[287] T. J. Price and T. Mullin. An experimental observation of a new type of intermittency. *Physica D,* **48**:29–52, 1991.

[288] M. K. Proctor and C. Jones. The interaction of two spatially resonant patterns in thermal convection 1: exact 1:2 resonance. *J. Fluid Mech.,* **188**:301–35, 1988.

[289] K. Promislow. Time analyticity and Gevrey regularity for solutions of a class of dissipative partial differential equations. *Nonlinear Analysis: Theory, Methods and Applications,* **16**(11):959–80, 1991.

[290] M. Rajaee and S. K. F. Karlsson. Shear flow coherent structures via Karhunen–Loève expansion. *Physics of Fluids A,* **2**:2249–51, 1990.

[291] M. Rajaee and S. K. F. Karlsson. On the Fourier space decomposition of free shear flow measurements and mode degeneration in the pairing process. *Physics of Fluids A,* **4**:321–39, 1992.

[292] M. Rajaee, S. K. F. Karlsson, and L. Sirovich. Low-dimensional description of free-shear-flow coherent structures and their dynamical behaviour. *J. Fluid Mech.,* **258**:1–29, 1994.

[293] R. H. Rand. *Computer Algebra in Applied Mathematics: An Introduction to MACSYMA*, volume 94 of *Research Notes in Mathematics*. Pitman, Boston, MA, 1984.

[294] R. H. Rand and D. Armbruster. *Perturbation Methods, Bifurcation Theory and Computer Algebra*. Springer-Verlag, New York, 1987.

[295] S. C. Reddy and D. S. Henningson. Energy growth in viscous channel flows. *J. Fluid Mech.*, **252**:209–38, 1993.

[296] D. Rempfer. Low dimensional models of a flat-plate boundary layer. In R. M. C. So, C. G. Speciale, and B. E. Launder, editors, *Near-Wall Turbulent Flows*, pages 63–72. Elsevier, Amsterdam, 1993.

[297] D. Rempfer. On the structure of dynamical systems describing the evolution of coherent structures in a convective boundary layer. *Physics of Fluids*, **6**(3):1402–4, 1994.

[298] D. Rempfer. On low-dimensional Galerkin models for fluid flow. *Theoret. Comput. Fluid Dynamics*, **14**:75–88, 2000.

[299] D. Rempfer and H. Fasel. Evolution of coherent structures during transition in a flat-plate boundary layer. In *Eighth Symposium on Turbulent Shear Flows*, volume 1, pages 18.3.1–18.3.6, 1991.

[300] D. Rempfer and H. Fasel. The dynamics of coherent structures in a flat-plate boundary layer. *Applied Scientific Research*, **51**:73–7, 1993.

[301] D. Rempfer and H. Fasel. Dynamics of three-dimensional coherent structures in a flat-plate boundary layer. *J. Fluid Mech.*, **275**:257–83, 1994.

[302] D. Rempfer and H. Fasel. Evolution of three-dimensional coherent structures in a flat-plate boundary layer. *J. Fluid Mech.*, **260**:351–75, 1994.

[303] S. O. Rice. Mathematical analysis of random noise. *Bell System Technical J.*, **23** and **24**:1–162, 1944. Reprinted in *Selected Papers on Noise and Stochastic Processes*, Wax, N., editor, Dover Publications Inc., New York, 1954.

[304] F. Riesz and B. S. Nagy. *Functional Analysis*. Ungar, New York, 1955.

[305] U. Rist and H. Fasel. Direct numerical simulation of controlled transition in a flat-plate boundary layer. *J. Fluid Mech.*, **298**:211–48, 1995.

[306] C. Robinson. Homoclinic bifurcation to a transitive attractor of Lorenz type. *Nonlinearity*, **2**:495–518, 1989.

[307] S. K. Robinson. Coherent motions in the turbulent boundary layer. *Ann. Rev. Fluid Mech.*, **23**:601–39, 1991.

[308] J. D. Rodriguez and L. Sirovich. Low-dimensional dynamics for the complex Ginzburg–Landau equation. *Physica D*, **43**:77–86, 1990.

[309] R. S. Rogallo and P. Moin. Numerical simulation of turbulent flows. *Ann. Rev. Fluid Mech.*, **16**:99–137, 1984.

[310] M. M. Rogers and P. Moin. The structure of the vorticity field in homogeneous turbulent flows. *J. Fluid Mech.*, **176**:33–66, 1987.

[311] M. M. Rogers and R. D. Moser. Spanwise scale selection in plane mixing layers. *J. Fluid Mech.*, **247**:321–37, 1993.

[312] V. A. Rokhlin. On the fundamental ideas of measure theory. *AMS Transl. (1)*, **10**:1–52, 1962.

[313] A. Rosenfeld and A. C. Kak. *Digital Picture Processing*. Academic Press, New York, 1982.

[314] C. W. Rowley. *Modeling, Simulation, and Control of Cavity Flow Oscillations*. PhD thesis, California Institute of Technology, 2002.

[315] C. W. Rowley. Model reduction for fluids using balanced proper orthogonal decomposition. *International Journal of Bifurcation and Chaos*, **15**(3):997–1013, 2005.

[316] C. W. Rowley, T. Colonius, and R. M. Murray. Model reduction for compressible flows using POD and Galerkin projection. *Physica D*, **189**(1–2):115–29, 2004.

[317] C. W. Rowley, I. G. Kevrekidis, J. E. Marsden, and K. Lust. Reduction and reconstruction for self-similar dynamical systems. *Nonlinearity*, **16**:1257–75, 2003.

[318] C. W. Rowley and J. E. Marsden. Reconstruction equations and the Karhunen–Loève expansion for systems with symmetry. *Physica D. Nonlinear Phenomena*, **142**:1–19, 2000.

[319] H. L. Royden. *Real Analysis*. Macmillan, London, 1963.

[320] D. Ruelle. *Chaotic Evolution and Strange Attractors*. Lezioni Lincee, Accademia Nazionale dei Lincei. Cambridge University Press, Cambridge, UK, 1989.

[321] D. Ruelle. *Chance and Chaos*. Princeton University Press, Princeton, NJ, 1991.

[322] D. Ruelle and F. Takens. On the nature of turbulence. *Comm. Math. Phys.*, **20**:167–92, 1970. Addendum, **23**, 343–4.

[323] S. Sanghi and N. Aubry. Mode interaction models for near-wall turbulence. *J. Fluid Mech.*, **247**:455–88, 1993.

[324] J. M. A. Scherpen. Balancing for nonlinear systems. *Systems and Control Letters*, **21**(2):143–53, 1993.

[325] P. J. Schmid and D. S. Henningson. *Stability and Transition in Shear Flows*. Springer-Verlag, New York, 2001.

[326] A. Schmiegel. *Transition to Turbulence in Linearly Stable Shear Flows*. PhD thesis, Universität Marburg, 1999.

[327] A. Schmiegel and B. Eckhardt. Fractal stability border in plane Couette flow. *Phys. Rev. Lett.*, **79**(26):5250–3, 1997.

[328] T.-H. Shih, J. L. Lumley, and J. Janica. Second order modeling of a variable density mixing layer. *J. Fluid Mech.*, **180**:93–116, 1987.

[329] S. G. Siegel, J. Seidel, C. Fagley, *et al.* Low dimensional modelling of a transient cylinder wake using double proper orthogonal decomposition. *J. Fluid Mech.*, **610**:1–42, 2008.

[330] L. Sirovich. Turbulence and the dynamics of coherent structures, Parts I–III. *Quarterly of Applied Math.*, **XLV**(3):561–82, 1987.

[331] L. Sirovich. Chaotic dynamics of coherent structures. *Physica D*, **37**:126–43, 1989.

[332] L. Sirovich, K. S. Ball, and R. A. Handler. Propagating structures in wall-bounded turbulent flows. *Theoret. Comput. Fluid Dynamics*, **2**:307–17, 1991.

[333] L. Sirovich, K. S. Ball, and L. R. Keefe. Plane waves and structures in turbulent channel flow. *Physics of Fluids A*, **2**(12):2217–26, 1990.

[334] L. Sirovich and A. E. Deane. A computational study of Rayleigh–Bénard convection. Part II. Dimension considerations. *J. Fluid Mech.*, **222**:251–65, 1991.

[335] L. Sirovich, M. Kirby, and M. Winter. An eigenfunction approach to large scale transitional structures in jet flow. *Physics of Fluids A*, **2**(2):127–36, 1990.

[336] L. Sirovich and B. W. Knight. The eigenfunction problem in higher dimensions: Asymptotic theory. *Proc. Nat. Acad. Sci.*, **82**:8275–8, 1985.

[337] L. Sirovich, M. Maxey, and H. Tarman. An eigenfunction analysis of turbulent thermal convection. In J.-C. André, J. Cousteux, F. Durst, *et al.*, editors, *Turbulent Shear Flows 6*, pages 68–77. Springer-Verlag, New York, 1989.

[338] L. Sirovich and H. Park. Turbulent thermal convection in a finite domain, Part I. Theory. *Physics of Fluids A*, **2**(9):1649–58, 1990.

[339] L. Sirovich and J. D. Rodriguez. Coherent structures and chaos: A model problem. *Phys. Lett. A*, **120**(5):211–14, 1987.

[340] L. Sirovich and X. Zhou. Reply to "observations regarding 'Coherence and chaos in a model of turbulent boundary layer' by X. Zhou and L. Sirovich". *Physics of Fluids A*, **6**:1579–82, 1994.

[341] G. I. Sivashinsky. Nonlinear analysis of hydrodynamic instability in laminar flames, Part I: Derivation of the basic equations. *Acta Astronautica*, **4**:1176–206, 1977.

[342] S. Skogestad and I. Postlethwaite. *Multivariable Feedback Control Analysis and Design*. John Wiley and Sons, 2nd edition, 2005.

[343] S. Smale. Differentiable dynamical systems. *Bull. AMS*, **73**:747–817, 1967.

[344] T. Smith and P. Holmes. Low dimensional models with varying parameters: A model problem and flow through a diffuser with variable angle. In J. L. Lumley, editor, *Fluid Mechanics and the Environment: Dynamical Approaches*, pages 315–36. Springer-Verlag, New York, 2001. Springer Lecture Notes in Physics 566.

[345] T. R. Smith, J. Moehlis, and P. Holmes. Dynamics of an 0:1:2 O(2)-equivarant system: Heteroclinic cycles and periodic orbits. *Physica D*, **211**:347–76, 2005.

[346] T. R. Smith, J. Moehlis, and P. Holmes. Low-dimensional modelling of turbulence using the proper orthogonal decomposition: A tutorial. *Nonlinear Dynamics*, **41**(1–3):275–307, 2005.

[347] T. R. Smith, J. Moehlis, and P. Holmes. Low-dimensional models for turbulent plane Couette flow in a minimal flow unit. *J. Fluid Mech.*, **538**:71–110, 2005.

[348] T. R. Smith, J. Moehlis, P. Holmes, and H. Faisst. Models for turbulent plane Couette flow using the proper orthogonal decomposition. *Physics of Fluids*, **14**(7):2493–507, 2002.

[349] W. H. Snyder and J. L. Lumley. Some measurements of particle velocity autocorrelation functions in a turbulent flow. *J. Fluid Mech.*, **48**:41–71, 1971.

[350] C. Sparrow. *The Lorenz Equations*. Springer-Verlag, New York, 1982.

[351] H. B. Squire. On the stability for three-dimensional disturbances of viscous fluid flow between parallel walls. *Proc. R. Soc. Lond. A*, **142**:621–8, 1933.

[352] K. R. Sreenivasan, R. Narashima, and A. Prabhu. Zero–crossings in turbulent signals. *J. Fluid Mech.*, **137**:251–72, 1983.

[353] M. M. Stanišić. *The Mathematical Theory of Turbulence*. Springer-Verlag, New York, 1987.

[354] E. Stone. *A Study of Low Dimensional Models for the Wall Region of a Turbulent Layer*. PhD thesis, Cornell University, 1989.

[355] E. Stone and P. Holmes. Noise induced intermittency in a model of a turbulent boundary layer. *Physica D*, **37**:20–32, 1989.

[356] E. Stone and P. Holmes. Random perturbations of heteroclinic cycles. *SIAM J. on Appl. Math.*, **50**(3):726–43, 1990.

[357] E. Stone and P. Holmes. Unstable fixed points, heteroclinic cycles and exponential tails in turbulence production. *Phys. Lett. A*, **155**:29–42, 1991.

[358] G. Strang. *Linear Algebra and Its Applications*. Academic Press, New York, 1980.

[359] D. Stretch, J. Kim, and R. Britter. A conceptual model for the structure of turbulent channel flow. In S. Robinson, editor, *Notes for Boundary Layer Structure Workshop*, Langley, VA, Aug. 1990. NASA.

[360] J. T. Stuart. On the non-linear mechanics of hydrodynamic stability. *J. Fluid Mech.*, **4**:1–21, 1958.

[361] H. L. Swinney and J. P. Gollub, editors. *Hydrodynamic Instabilities and the Transition to Turbulence*. Springer-Verlag, New York, second edition, 1985.

[362] G. Tadmor, O. Lehmann, B. R. Noack, and M. Morzyński. Mean field representation of the natural and actuated cylinder wake. *Physics of Fluids*, **22**(3):034102, 2010.

[363] R. Tagg. The Couette–Taylor problem. *Nonlinear Science Today*, **4**(3):1–25, 1994.

[364] R. Tagg, D. Hirst, and H. Swinney. Critical dynamics near the spiral-Taylor vortex transition. Unpublished report, University of Texas, Austin, 1988. See [363].

[365] F. Takens. Detecting strange attractors in turbulence. In D. A. Rand and L.-S. Young, editors, *Dynamical Systems and Turbulence, Warwick 1980*, volume 898 of *Springer Lecture Notes in Mathematics*, pages 366–81. Springer-Verlag, New York, 1981.

[366] J. A. Taylor and M. N. Glauzer. Towards practical flow sensing and control via POD and LSE based low-dimensional tools. *A.S.M.E. J. Fluids Engineering*, **126**:337–45, 2004.

[367] R. Temam. *Infinite-Dimensional Dynamical Systems in Mechanics and Physics*. Springer-Verlag, New York, 1988.

[368] H. Tennekes and J. L. Lumley. *A First Course in Turbulence*. MIT Press, Cambridge, MA, 1972.

[369] E. S. Titi. On approximate inertial manifolds to the Navier–Stokes equations. *J. Math. Anal. Appl.*, **149**:540–57, 1990.

[370] S. Toh. Statistical model with localized structures describing the spatio-temporal chaos of Kuramoto–Sivashinsky equation. *J. Phys. Soc. Jap.*, **56**(3):949–62, 1987.

[371] A. A. Townsend. *The Structure of Turbulent Shear Flow*. Cambridge University Press, Cambridge, UK, 1956.

[372] A. A. Townsend. Flow patterns of large eddies in a wake and in a boundary layer. *J. Fluid Mech.*, **95**:515–37, 1979.

[373] L. N. Trefethen and D. I. Bau. *Numerical Linear Algebra*. Society for Industrial and Applied Mathematics, Philadelphia, PA, 1997.

[374] L. Ukeiley. *Dynamics of Large Scale Structures in a Plane Turbulent Mixing Layer*. PhD thesis, Clarkson University, 1995.

[375] B. van der Pol. Forced oscillations in a circuit with nonlinear resistance (receptance with reactive diode). *London, Edinburgh and Dublin Phil. Mag.*, **3**:65–80, 1927.

[376] M. I. Vishik. *Asymptotic Behaviour of Solutions of Evolutionary Equations*. Lezioni Lincee, Accademia Nazionale dei Lincei. Cambridge University Press, Cambridge, UK, 1992.

[377] F. Waleffe. Hydrodynamic stability and turbulence: Beyond transients to a self-sustaining process. *Stud. Appl. Math.*, **95**:319–43, 1995.

[378] F. Waleffe. Transition in shear flows. Nonlinear normality versus non-normal linearity. *Physics of Fluids*, **7**(12):3060–6, 1995.

[379] F. Waleffe. On a self-sustaining process in shear flows. *Physics of Fluids*, **9**:883–900, 1997.

[380] F. Waleffe. Three-dimensional coherent states in plane shear flows. *Phys. Rev. Lett.*, **81**: 4140–3, 1998.

[381] F. Waleffe. Exact coherent structures in channel flow. *J. Fluid Mech.*, **435**:93–102, 2001.

[382] F. Waleffe, J. Kim, and J. M. Hamilton. On the origin of streaks in turbulent shear flows. In F. Durst, R. Friedrich, B. E. Launder, *et al.*, editors, *Turbulent Shear Flows 8*, pages 37–49. Springer-Verlag, New York, 1991.

[383] Y. Wang and H. H. Bau. Period doubling and chaos in a thermal convection loop with time periodic wall temperature variation. In G. Hetzroni, editor, *Proc. 9th International Heat Transfer Conf. Vol II*, pages 357–62, 1990.

[384] Y. Wang, J. Singer, and H. H. Bau. Controlling chaos in a thermal convection loop. *J. Fluid Mech.*, **237**:479–98, 1992.

[385] J. Weller, E. Lombardi, and A. Iollo. Robust model identification of actuated vortex wakes. *Physica D*, **238**:416–27, 2009.

[386] G. B. Whitham. *Linear and Nonlinear Waves*. Wiley, New York, 1974.

[387] P. J. Widmann, M. Gorman, and K. A. Robbins. Nonlinear dynamics of a convection loop II: chaos in laminar and turbulent flows. *Physica D*, **36**:157–66, 1989.

[388] M. Winter, T. J. Barber, R. M. Everson, and L. Sirovich. Eigenfunction analysis of turbulent mixing phenomena. *AIAA Journal*, **30**(7):1681–8, 1992.

[389] R. W. Wittenberg and P. Holmes. Scale and space localisation in the Kuramoto–Sivashinsky equation. *Chaos*, **9**(2):452–65, 1999.

[390] R. W. Wittenberg and P. Holmes. Spatially localized models of extended systems. *Nonlinear Dynamics*, **25**:111–32, 2001.

[391] M. Yokokawa, K. Itakura, A. Uno, T. Ishihara, and Y. Kaneda. 16.4 Tflops direct numerical simulation of turbulence by a Fourier spectral method on the Earth simulator. In *Proceedings of the ACM/IEEE Conference on Supercomputing*, 2002.

[392] X. Zheng and M. N. Glauser. A low dimensional description of the axisymmetric jet mixing layer. *ASME Computers in Engineering*, **2**:121–7, 1990.

[393] K. Zhou, G. Salomon, and E. Wu. Balanced realization and model reduction for unstable systems. *International Journal of Robust and Nonlinear Control*, **9**(3):183–98, 1999.

[394] X. Zhou and L. Sirovich. Coherence and chaos in a model of turbulent boundary layer. *Physics of Fluids A*, **4**:2855–74, 1992.

[395] Y. Zhou and G. Vahala. Local interaction in renormalization methods for Navier–Stokes turbulence. *Phys. Rev. A*, **46**:1136–9, 1992.

[396] Y. Zhou and G. Vahala. Reformulation of recursive renormalization group based subgrid modeling of turbulence. *Phys. Rev. E*, **47**:2503–19, 1993.

Index

Page numbers that are underlined denote the main definition of a term.

Printed in the United States
by Baker & Taylor Publisher Services